Drug Development
for Gene Therapy

Drug Development for Gene Therapy

Translational Biomarkers, Bioanalysis, and Companion Diagnostics

Edited by

Yanmei Lu
Sangamo Therapeutics
Richmond, California
USA

Boris Gorovits
Gorovits BioSolutions, LLC
Andover, Massachusetts
USA

WILEY

Published by John Wiley & Sons, Inc., Hoboken, New Jersey.
Published simultaneously in Canada.

Library of Congress Cataloging-in-Publication Data

Names: Lu, Yanmei, 1966- editor. | Gorovits, Boris, editor.
Title: Drug development for gene therapy : translational biomarkers, bioanalysis, and companion diagnostics / edited by Yanmei Lu, Boris Gorovits.
Description: Hoboken, New Jersey : John Wiley & Sons, Inc., [2024] | Includes bibliographical references and index.
Identifiers: LCCN 2023049872 (print) | LCCN 2023049873 (ebook) | ISBN 9781119852780 (cloth) | ISBN 9781119852797 (adobe pdf) | ISBN 9781119852803 (epub)
Subjects: MESH: Genetic Therapy–methods | Biomarkers, Pharmacological–analysis | Drug Development–methods
Classification: LCC RB155 (print) | LCC RB155 (ebook) | NLM QU 560 | DDC 616/.042–dc23/eng/20231214
LC record available at https://lccn.loc.gov/2023049872
LC ebook record available at https://lccn.loc.gov/2023049873

Cover Design: Wiley
Cover Image: © Jonathan Knowles/Getty Images

Contents

List of Contributors

Editors

Yanmei Lu
Biomarker and BioAnalytical Sciences
Sangamo Therapeutics
Richmond, California
USA

Boris Gorovits
Translational Sciences, Bioanalysis & Biomarkers
Gorovits BioSolutions, LLC
Andover, Massachusetts
USA

Authors

Paul Bartel
Companion Diagnostics
Myriad Genetics, Inc.
Salt Lake City, Utah
USA

Manuela Braun
Bayer AG
Berlin
Germany

George Buchlis
Department of Medicine
University of Pennsylvania
Philadelphia, Pennsylvania
USA

Liching Cao
Biomarker and Bioanalytical Sciences
Sangamo Therapeutics
Richmond, California
USA

Kennon Daniels
Precision for Medicine
Bethesda Metro Center
Maryland
USA

Maurus de la Rosa
Sangamo Therapeutics Allée de la Nertière
Valbonne
France

Robert Dodge
Department of BioMedical Research
Novartis
East Hanover, New Jersey
USA

Mica Elizalde
Regulatory Digital Health
Merck Sharp & Dohme LLC
Rahway, New Jersey
USA

Marina Falaleeva
Preclinical Department
Sangamo Therapeutics
Richmond, California
USA

Raffaele Fronza
ProtaGene CGT GmbH
Heidelberg
Germany

Irene Gil-Farina
ProtaGene CGT GmbH
Heidelberg
Germany

Jennifer Granger
PharmaDx
ARUP Laboratories
Salt Lake City, Utah
USA

Michael Havert
Gene Therapy Partners, LLC
Arlington, Virginia
USA

Vibha Jawa
Clinical Pharmacology,
Pharmacometrics, Disposition and
Bioanalysis (CPPDB)
Bristol Myers Squibb
Princeton, New Jersey
USA

Wibke Lembke
Celerion Switzerland AG
Fehraltorf
Switzerland

John Lin
Frontage Laboratories
Exton, Pennsylvania
USA

Hsing-Yin Liu
Molecular Biology, Johnson and Johnson
Innovative Medicine
Janssen Pharmaceuticals
Spring House, Pennsylvania
USA

Kathleen Meyer
Preclinical Department
Sangamo Therapeutics
Richmond, California
USA

John E. Murphy
Arbor Biotechnologies
Cambridge, Massachusetts
USA

Jane Owens
Rare Disease Research Unit
Pfizer Inc.
Cambridge, Massachusetts
USA

Karen L. Richards
Precision for Medicine
Bethesda Metro Center
Maryland
USA

Laura I. Salazar-Fontana
LAIZ Reg Science Consulting
Lausanne
Switzerland

Oscar Segurado
ASC Therapeutics
Milpitas, California
USA

Russell K. Soon Jr.
BioMarin Pharmaceutical, Inc.
Novato, California
USA

Kefeng Sun
Quantitative Clinical Pharmacology,
Data Sciences Institute
Takeda Development Center Americas
Cambridge, Massachusetts
USA

Magdalena Tary-Lehmann
CTL-Contract Laboratory
Cellular Technology Limited
Shaker Heights, Ohio
USA

Shengdar Q. Tsai
Department of Hematology
St Jude Children's Research Hospital
Memphis, Tennessee
USA

Venkata Vepachedu
Molecular Biology, Johnson and Johnson
Innovative Medicine
Janssen Pharmaceuticals
Spring House, Pennsylvania
USA

Christian Vettermann
BioMarin Pharmaceutical, Inc.
Novato, California
USA

Kai Wang
GlaxoSmithKline
Collegeville, Pennsylvania
USA

Laurence O. Whiteley
Pfizer Inc. Drug Safety Research and
Development
Cambridge, Massachusetts
USA

Bonnie Wu
Biologics Development Sciences, Janssen
Research and Development
LLC Pharmaceutical Companies of
Johnson & Johnson Innovative Medicine
Spring House, Pennsylvania
USA

Jing Yuan
Department of Toxicology
Kymera Therapeutics
Watertown, Massachusetts
USA

Preface

Having dedicated more than a couple of decades to the development of biomarkers and bioanalysis in the realm of biologics, including monoclonal antibodies and recombinant protein therapies, we embarked on a career change with the anticipation that our wealth of experience could readily translate into the field of gene therapy drug development. However, what we hadn't fully grasped at the outset was the considerable complexity and formidable challenges associated with translational biomarkers, bioanalysis, and companion diagnostics when deploying adeno-associated virus (AAV) as a vector to introduce transgenes, encompassing cDNAs and gene editing tools, into human subjects.

The successful advancement of a gene therapy drug necessitates the meticulous collection of pharmacokinetic and biomarker data to underpin efficacy and safety assessments, as well as the selection of suitable patients. The multifaceted nature of gene therapy, coupled with the vast troves of data involved, encompasses a wide spectrum of methods and technology platforms. This repertoire includes polymerase chain reaction (PCR)-based techniques, such as quantitative PCR and digital PCR, for scrutinizing viral biodistribution and shedding patterns, reverse transcription-PCR for analyzing transgene expression, enzyme activity assays, mass spectrometry, immunohistochemistry/in situ hybridization, and immunoassays for evaluating target engagement, substrate interactions, and distal pharmacodynamic biomarkers.

Moreover, ligation-mediated (LM)-PCR and linear amplification-mediated (LAM)-PCR are indispensable for the in-depth analysis of recombinant AAV integration, while next-generation sequencing (NGS) is employed to assess off-target gene editing activity. The assessment of humoral antibody response and cellular immune response to AAV capsid and transgene products requires the application of anti-drug antibody and neutralizing antibody assays, as well as ELISpot technology.

In addition, the evolving landscape of companion diagnostic development, particularly in relation to the anti-AAV antibody screening assay supporting clinical studies and drug approval, presents unique and rapidly evolving challenges. Furthermore, as clinical data continues to emerge from ongoing trials, the regulatory environment governing the evaluation of efficacy and safety in the gene therapy field is in a state of flux.

Over the past decade, the discovery and development of AAV gene therapy medicines have gained remarkable momentum. This surge in growth, marked by a proliferation of preclinical studies and clinical trials, has led to a shortage of qualified researchers in translational sciences. In this dynamic landscape, the adoption of best practices in biomarker and bioanalysis, combined with up-to-date knowledge of regulatory guidelines, is of paramount importance. Such information is invaluable for gene therapy developers, whether they are working in academia, industry, or government organizations, as it equips them with the timely insights required to navigate the constantly evolving challenges and opportunities in this dynamic field.

January 2024

Yanmei Lu
Sangamo Therapeutics

Boris Gorovits
Gorovits BioSolutions, LLC

Section I

Introduction

1

Introduction to AAV-based *in vivo* Gene Therapy

Oscar Segurado

ASC Therapeutics, Milpitas, CA, USA

1.1 Introduction

1.1.1 History of Gene Therapy

Watson and Crick first characterized the structure of DNA as a double helix in 1953 [1]. X-ray crystallography of DNA, performed by Franklin, confirmed this finding [2]. Knowing DNA's structure helped elucidate its functions, such as how it holds genetic information, can be copied, and gives rise to various proteins.

Although adeno-associated viruses (AAVs) were discovered in the 1960s [3], they would not be used as genetic vectors until the 1980s. The first attempt at genetic manipulation in humans is believed to be the work of Terheggen et al. in the 1970s. German scientists used the Shope papillomavirus in three children whose bodies were unable to produce arginase. Without arginase, arginine accumulates in the body, causing neurological and muscular defects. The virus, known to produce arginase, was injected intravenously (IV) in hopes that the genetic information from the virus could enter human cells, resulting in arginase production. Unfortunately, IV injections of the virus did not help any of the three sisters that had this rare disorder, and the youngest, who was given a larger dose as an infant, suffered a brief allergic reaction without any positive response to the treatment [4].

In the 1980s, retroviral gene therapy was in development [5–7], and the first recombinant AAV vectors were created [8]. Synthetic insulin was the first genetically engineered drug, reaching the market in 1982 [9]. Zinc fingers were discovered in 1985, later providing a method of targeted gene therapy through zinc

Drug Development for Gene Therapy: Translational Biomarkers, Bioanalysis, and Companion Diagnostics, First Edition. Edited by Yanmei Lu and Boris Gorovits.

finger nucleases (ZFNs) [10]. The hepatitis B vaccine was the first recombinant vaccine available in 1986 [11], and the discussion of the human genome project began two years later [12]. Also in 1988, the first genetically modified crop was grown in US fields [13].

In 1990, research began in the United States, studying human gene therapy [14]. Dolly, the sheep, was cloned in 1996 [15]. By the year 2000, around 400 gene therapies had been tested in clinical trials [16]. The first gene therapy was approved in China in 2003, using a replication-incompetent adenovirus vector for treating advanced head and neck cancer [17]. Modified lentiviral vectors began emerging in clinical trials around this time as well [18]. In 2007, human-induced pluripotent stem cells (iPSCs) were first isolated, and this method is now quite common, using genetic reprogramming to compare patient-derived cells to isogenic control cells [19]. The first gene therapy was approved in Europe in 2012 using an adenovirus [16]. In 2013, CRISPR/Cas9 was developed, where it was first used as a research tool [20]; it was not until 2018 that the first clinical trial in humans utilizing this technology completed their enrollment. Patients with refractory non-small-cell lung cancer were treated with CRISPR-edited T cells [21]. This timeline can be viewed in Figure 1.1.

In 2020, over 400 gene and genetically modified cell therapies were in development, and today (2022), there are over 1000 in recruitment or active studies (clinicaltrials.gov). Gene therapies may replace inadequate and complex therapies in the near future. For some diseases, it may be able to reduce the amount and, eventually, the cost of treatments a person needs. Thus, it is likely to benefit those with poor quality of life due to an untreatable condition or an intense therapy regimen the most.

1.1.2 AAV-based *in vivo* Gene Therapy: A Revolution in Medicine

Despite gene therapies being developed and tested in the United States since the 1990s, only 26 cell and gene therapies have been Federal Drug Administration (FDA)-approved until February 2023, seven of which are cord blood treatments (Table 1.1). Of the other 19 therapies, 14 are *ex vivo* cell therapies and five are *in vivo* gene therapy treatments. Genetic diseases, those driven by mutations in the human genome, are ideal targets for treatments using gene therapy modalities. Gene therapy can address diseases driven by well-defined genetic abnormalities where the biological function of the altered or missing gene is well understood. In many cases, these are rare diseases with unmet medical needs, often requiring complex medical regimens with limited options for effective treatments. However, in recent years, gene therapies have been investigated for the treatment of non-monogenic diseases, for example, cancers and degenerative diseases of the visual and nervous systems.

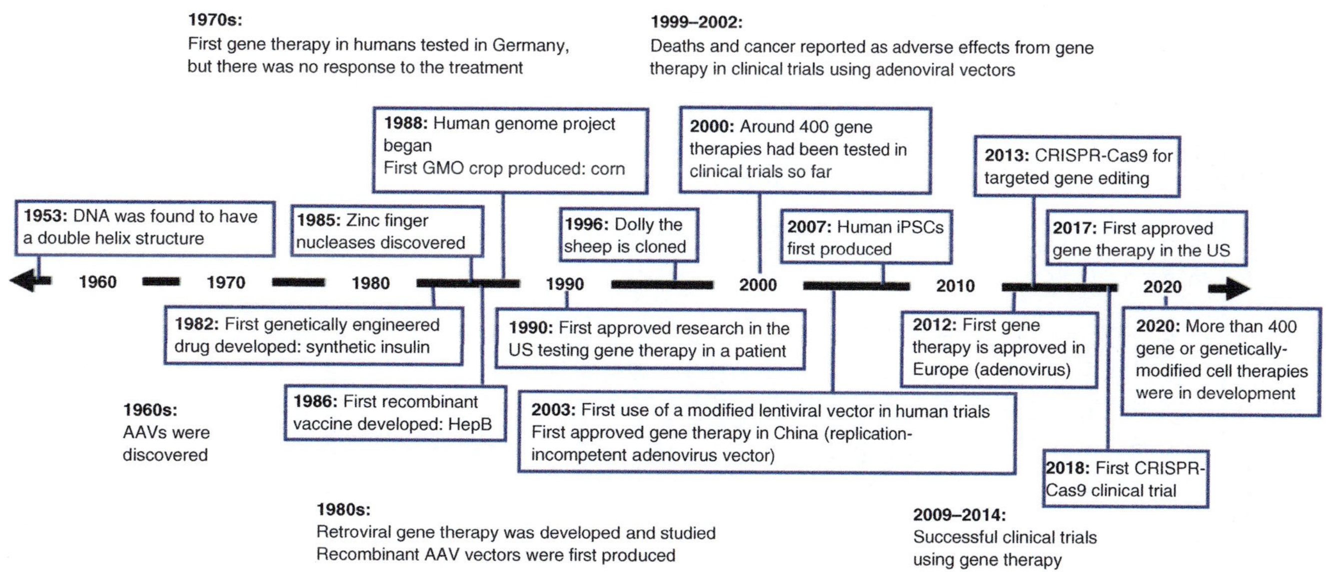

Figure 1.1 Timeline of scientific advances in gene therapy research [1].

Table 1.1 FDA-approved cellular and gene therapies.

Name	Indication	Type	Manufacturer
Abecma (idecabtagene vicleucel)	Adult relapse or refractory myeloma after ≥4 prior therapy lines, including immunomodulatory agent, proteasome inhibitor, anti-CD38 monoclonal antibody	*Ex vivo*; Lentivirus vector	Calgene Corporation, a Bristol-Myers Squibb Company
Adstiladrin	Adult high-risk Bacillus Celmette-Guerin-unresponsive non-muscle invasive bladder cancer with carcinoma in situ	Adenovirus vector	Ferring Pharmaceuticals A/S
HPC, Cord Blood; Allocord; Clevecord; Hemacord; HPC, Cord Blood – MD Anderson; HPC, Cord Blood – LifeSouth; HPC, Cord Blood – Bloodworks	Hematopoietic and immunologic reconstitution with disorders affecting the hematopoietic system that are inherited, acquired, or from myeloablative treatment	Hematopoietic progenitor cells	University of CO Cord Blood Bank; SSM Cardinal Glennon Children's Medical Center; Cleveland Cord Blood Center; Duke University School of Medicine; NY Blood Center; MD Anderson Cord Blood Bank; LifeSouth Community Blood Centers; Bloodworks
Breyanzi	Adult large B-cell lymphoma, including diffuse not otherwise specified high-grade primary mediastinal and follicular grade 3B	*Ex vivo*; Lentivirus vector-modified autologous CD4+ and CD8+ T cells	Juno Therapeutics, Inc., a Bristol-Myers Squibb Company
Carvykti (ciltacabtagene autoleucel)	Adult relapse or refractory multiple myeloma after ≥4 prior therapy lines, including proteasome inhibitor, immunomodulatory agent, anti-CD38 monoclonal antibody	*Ex vivo*; Lentivirus vector-modified autologous T cells	Janssen Biotech, Inc.

Name	Indication	Type	Manufacturer
Gintuit	Topical (non-submerged) application to surgically created vascular wound bed in adult mucogingival conditions	*Ex vivo*; Scaffold product of neonatal foreskin allogeneic fibroblasts & keratinocytes	Organogenesis, Inc.
Hemgenix	Adult hemophilia B (Factor IX deficiency)	AAV vector	CSL Behring LLC
Imlygic (talimogene laherparepvec)	Local treatment of unresectable, cutaneous, subcutaneous, and nodal lesions with melanoma recurrent after initial surgery	Modified HSV-1 isolate with oncolytic activity toward tumor cells (JS1)	BioVex, Inc., a subsidiary of Amgen, Inc.
Kymriah (tisagenlecleucel)	Adult relapsed or refractory follicular lymphoma after ≥2 lines of therapy	*Ex vivo*; Lentivirus vector-modified autologous T cells	Novartis Pharmaceuticals Corporation
Laviv (Azficel-T)	Improvement of adult moderate-to-severe nasolabial fold wrinkle appearance	*Ex vivo*; Autologous fibroblasts	Fibrocell Technologies
Luxturna	Biallelic RPE65 mutation-associated dystrophy	Recombinant AAV serotype 2 vector expressing RPE65	Spark Therapeutics, Inc.
Maci	Repair of adult single or multiple symptomatic, full-thickness cartilage defects of the knee	*Ex vivo*; Autologous knee cartilage chondrocytes in resorbable porcine type I/III collagen membrane	Vericel Corporation
Provenge (sipuleucel-T)	Asymptomatic or minimally symptomatic metastatic castrate-resistant (hormone refractory) prostate cancer	*Ex vivo*; Autologous cellular immunotherapy	Dendreon Corporation
Rethymic	Immune reconstitution in pediatric congenital athymia	*Ex vivo*; Allogeneic thymus from <9 months old heart surgery patients	Enzyvant Therapeutics GmbH

(Continued)

Table 1.1 (Continued)

Name	Indication	Type	Manufacturer
Skysona (elivaldogene autotemcel)	Slow progression of neurologic dysfunction in boys 4–17 years with early, active cerebral adrenoleukodystrophy	*Ex vivo*; Lenti-D lentivirus vector modified autologous CD34+ -enriched hematopoietic stem cells	Bluebird Bio, Inc.
Stratagraft	Adult thermal burns containing intact dermal elements, which surgical intervention is clinically indicated (deep partial-thickness burns)	*Ex vivo*; Allogeneic cultured keratinocytes and dermal fibroblasts in murine collagen type I scaffold	Stratatech Corporation
Tecartus (brexucabtagene autoleucel)	Adult relapsed or refractory mantle cell lymphoma and B-cell precursor acute lymphoblastic leukemia	*Ex vivo*; Gammaretrovirus vector modified antigen-specific autologous T cells	Kite Pharma, Inc.
Yescarta (axicabtagene ciloleucel)	Adult B-cell lymphoma refractory to first-line chemoimmunotherapy or relapse within 12 mo. Of first-line chemoimmunotherapy	*Ex vivo*; gammaretrovirus vector modified autologous T cells	Kite Pharma, Inc.
Zynteglo (betibeglogene autotemcel)	Adult and pediatric beta-thalassemia requiring red blood cell transfusions	*Ex vivo*; BB305 lentivirus vector modified autologous CD34+ -enriched hematopoietic stem cells	Bluebird Bio, Inc.
Zolgensma (onasemnogene abeparvovec-xioi)	Spinal Muscular Atrophy (Type I)	Recombinant AAV vector	Novartis Gene Therapies, Inc.

Information obtained on 1 February 2023 from https://clinicaltrials.gov/ct2/results?term=AAV+gene+therapy&Search=Apply&age_v=&gndr=&type=&rslt=.

Of the approved genetic therapies, five utilize lentivirus vectors, two use gammaretrovirus vectors, one uses an adenovirus vector, and three use an AAV vector. Two hundred and seven AAV gene therapy clinical trials can be found on the clinicaltrials.gov website, 111 of which are active trials, either in recruiting, enrolling, or collecting and analyzing data phases (Table 1.2). The majority of the AAV-based gene therapy trials are documented as Phase 1 or 2 trials (some are combined phase trials, i.e. Phase I/II). Sixty AAV-based gene therapy trials have been completed. Indications for completed AAV-based gene therapy trials include muscular disorders, neurodegenerative diseases, retinal diseases and other visual defects, lysosomal and glycogen storage disorders, blood coagulation disorders, cardiovascular diseases, amino acid metabolism disorder, and arthritis (Table 1.3). The active trials include the above diseases and other indications such as infections, adrenal diseases, developmental disorders, hearing loss, and cancer. Some early success has been seen in neurodegenerative diseases, such as Parkinson's and Alzheimer's disease; however, this is likely limited by the number of remaining neurons, many of which are lost prior to even diagnosis [1, 2].

Table 1.2 AAV gene therapy clinical trial status.

	# of trials
AAV Gene Therapy	207
Status	
Not yet recruiting	13
Recruiting	57
Enrolling by Invitation	8
Active, not recruiting	46
Terminated	17
Completed	60
Unknown	6
Phase	
Early Phase 1	3
Phase 1	138
Phase 2	109
Phase 3	20
N/A	7

Information obtained on 1st February 2023 from https://www.fda.gov/vaccines-blood-biologics/cellular-gene-therapy-products/approved-cellular-and-gene-therapy-products.

Table 1.3 Complete and active AAV gene therapy clinical trial indications.

Indications of clinical trials	Complete	Active
Muscular Disorders	12	8
Vision Defects	10	4
Neurodegenerative Diseases	10	13
Retinal Diseases	9	25
Lysosomal & Glycogen Storage Disorders	7	20
Blood Coagulation Disorders	5	20
Cardiovascular Diseases	2	4
Amino Acid Metabolism Disorders	1	5
Arthritis	1	1
Infections		3
Adrenal Diseases		3
Developmental Disorders		2
Digestive System Diseases	3	1
Hearing Loss		1
Cancer		1

Despite its promise and setbacks, gene therapy's potential across therapeutic areas remains enormous, offering the hope of "one and done" cures for serious diseases with significant unmet needs. AAV-based gene therapy has received the most attention in basic and clinical research recently, moving rapidly into the biotechnology industry, resulting in clinical trials designed to prove its efficiency, safety, cost-effectiveness, and range of use. Recombinant AAVs (rAAVs), engineered capsids with lower immunogenicity, the addition of synthetic promoters, and gene editing techniques are just some of the developments in AAV vector-based gene therapy. Although these developments have come to fruition relatively quickly, many challenges remain when developing and using gene therapy treatments. Several gene editing techniques have emerged over the past two decades, which have been crucial in researching and developing disease mechanisms, therapies, and more. Although there have been setbacks, the promise of gene therapies and the potential of direct gene editing remains. The technological advancements in labs, clinics, and manufacturing processes continue to make it more and more possible for long-lasting rare disease treatments. Thus, it is crucial to understand how AAV-based gene therapies are created and implemented, the benefits and risks of these treatments, as well as the identification of their most relevant indications.

1.1.3 The AAV Vector Structure

The AAV is part of the *Parvoviridae* family, being approximately 25 nm in diameter with its DNA contained within an icosahedral capsid, the protein shell. Single-stranded DNA of approximately 4.7 kb is composed of genes for three capsid proteins [22] four rep proteins, and an assembly-activating protein [23]. Palindromic inverted terminal repeats (ITRs) are located at each end [24], forming T-shaped hairpin structures. The *Rep* gene is required for DNA replication and packaging, and the *Cap* gene encodes the capsid assembly proteins. In studies of AAV2, it was found that the structural proteins, VP1, 2, and 3, exist in a 1 : 1 : 10 ratio, forming an icosahedral, symmetrical shape [25]. VP1 was essential for infection while VP2 was necessary for nuclear transfer of capsid proteins [26]. VP3 was found to allow the binding of the virus to cell receptors [27, 28].

The AAV was discovered in 1965 as a contaminant of an adenovirus preparation from rhesus monkey kidney cell cultures infected with simian adenovirus type 1 [3]. AAVs rely on helper viruses in order to replicate in mammalian cells, typically adenoviruses or herpes viruses.

When using AAV as a vector, the DNA of interest replaces the genomic DNA, the Rep and Cap genes, between the ITRs [29]. The exogenous DNA is referred to as the transgene expression cassette. When creating a recombinant AAV, the gene expression cassette needs to include a promoter, transgene, and a termination signal, which is specific to the goals of that particular gene therapy. Figure 1.2 displays the requirements for creating a rAAV vector. Tissue and cell-type specificity, packaging size limits, and necessary expression level all need careful consideration when creating the gene expression cassette.

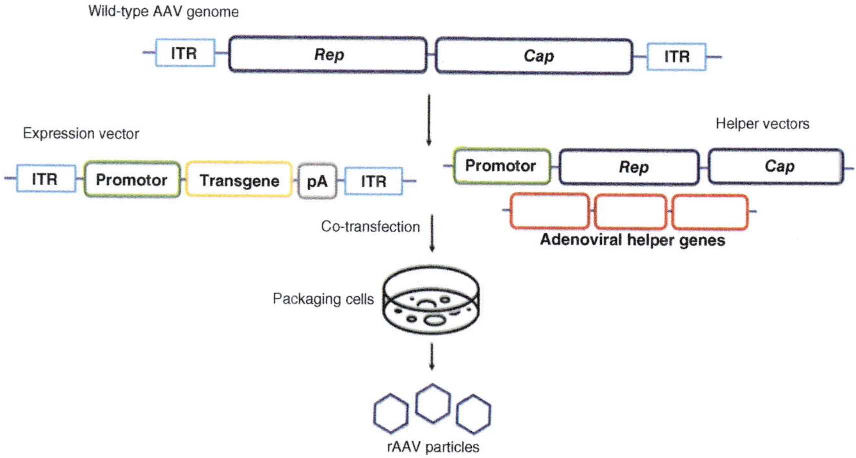

Figure 1.2 Creation of recombinant AAV particles.

However, the baculovirus-infected insect (*Sf9*) system is increasingly used with scale-up production. Specifically, the OneBac system has been developed which integrates *rep* and *cap* into one baculovirus, resulting in a decrease in the required number of baculovirus particles for efficient rAAV production [30].

1.1.4 Cell Entry and Transduction Pathway

Wild-type AAV viruses can infect dividing and nondividing cells. The majority of the AAV genomes, wild-type and recombinant, exist as circular episomes in tissues, though integration into the host cell chromosomes at AAV integration sites can happen in low frequency [31]. AAV will not replicate without Rep proteins, which regulate viral transcription. Rep proteins can be provided by a helper virus and remain latent until this occurs [32]. Once the helper virus is delivered, the Rep genes are expressed and replication can occur via the rolling hairpin mechanism, where a Rep protein binds the Rep binding element (RBE) within the hairpin, followed by duplication of the DNA strand and packaging. The cell entry and transduction process of an rAAV is shown in Figure 1.3. Cellular transduction of an AAV vector begins with viral capsid interaction with the target cell's receptors. The cell uses endocytosis (clathrin-coated pits) to internalize the virus, which

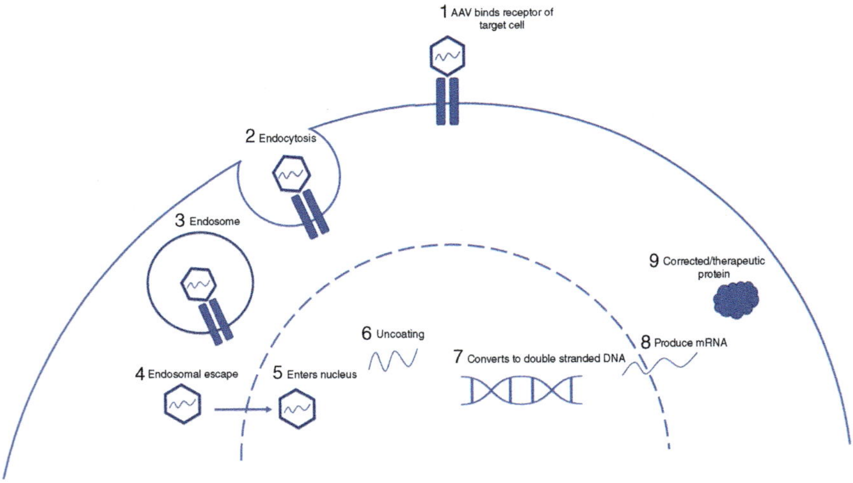

Figure 1.3 rAAV vector entry and transduction pathway. (1) AAV vectors bind to receptors of target cells, (2) initiating endocytosis via clathrin-coated pits. (3) The endosomal compartment contains the AAV, (5) which can escape to the nucleus through the nuclear pore. (6) Uncoating releases single-stranded DNA (7) for a second strand to be synthesized (de novo synthesis or base pair strand annealing). (8 and 9) This leads to the production of the corrected/therapeutic mRNA and subsequent protein to be produced.

then enters the endocytic/proteasomal compartment. These may be critical steps for transduction because injection of AAV2 directly into cells resulted in a reduced infection rate compared to exposure [33]. Interestingly, the efficiency of transduction is largely affected by acidity of the endosomal compartment [34]. Proteasome inhibitors also increase transduction efficiency [35].

The virus escapes the endosome and is imported through the nuclear pore complex into the nucleus, undergoes capsid uncoating, and the DNA is converted into double-stranded DNA via *de novo* synthesis or base pairing. The DNA can then be transcribed and expressed [36]. Particles that are not able to translocate to the nucleus are degraded by the proteasome and presented to cytotoxic T cells [37]. Thus, prior to genome integration, intracellular trafficking involves a number of events in which breakdown at any step can result in failure of gene delivery.

De novo synthesis of the complementary DNA strand and strand annealing result in the conversion of single-stranded DNA to double-stranded DNA. Strand annealing involves the base pairing of a coinfected separate AAV complementary single-strand genome. Due to the synthesis of the second DNA strand being considered the rate-limiting factor in transduction efficiency, second-generation AAV vectors with double-stranded DNA have been developed, called self-complementary AAV vectors, which have been shown to be safe and reliable for organ-specific transduction [38, 39].

1.2 Advantages and Disadvantages for AAV *in vivo*

1.2.1 Effectiveness and Advantages of AAV Vectors for *in vivo* Gene Therapy

AAV vector has been labeled as the safest and most effective vehicle for the delivery of *in vivo* gene therapies that produce long-term expression with a single injection. rAAV effectiveness is largely determined by the interaction of the capsid and the cell surface receptors as well as downstream events after internalization. Transduction efficiency can be altered due to the endosomal compartment pH and the use of proteasome inhibitors [34, 35]. Immune reactions to AAVs significantly impact a gene therapy's efficiency and will be discussed further in Section 1.5.3. Addressing Immunogenicity and Chapter 5.

There are many advantages to using AAV vectors for gene therapy. The human native AAV is not known to cause disease, has no pathogenicity [3], and has the ability to infect numerous mammalian cell types. AAVs efficiently deliver genetic material with low toxicity and immunogenicity, a good safety profile, and long-term effects [40, 41].

1.2.2 Challenges of AAV Vectors for *in vivo* Gene Therapy

Challenges, of course, still exist for gene therapy using AAV vectors. Large-scale manufacturing, particularly quality control and standardization, is difficult to maintain for gene therapies, resulting in high costs. Although AAVs provide efficient gene delivery, they are less immunogenic than adenoviruses. Before transduction, single-stranded DNA from AAVs must first be made into double-stranded DNA. AAVs have limited capacity, holding approximately 4.7 kilobases (kb) [22]. Implementing a split vector approach can bypass the limitation of small capacity; however, serogenicity and type of cell or tissue can markedly affect the expression of split vectors. Also, replacing *Rep* genes with exogenous DNA lowers the transduction frequency because, without them, ITRs have little enhancer and promotor activity. Self-complementary vector AAVs can be implemented to improve efficiency and onset of gene therapies.

There are potential sources of toxicity that must be considered in the development of AAV gene therapies, insertion mutagenesis, tumor induction, and immune response, dependent upon reactions to the capsid and DNA of the AAV. Toxicity has been reported for AAV clinical trials, causing hepatotoxicity, muscle toxicity, thrombotic microangiopathy, etc. [42]. Route of administration and number of viral particles used for gene therapy may contribute to toxicity and the immune response as well. In two studies testing a hemophilia B gene therapy, AAV gene transfer in the liver resulted in the development of capsid-specific CD8+ T cells and elimination of transduced hepatocytes. The reaction was halted by oral corticosteroids [40, 43], although in some cases, immune-modulating agents have been shown not able to prevent loss of transgene expression. Reactions to natural AAV capsids may predict the reaction and effectiveness of AAV-based gene therapies [44]. Transient B-cell depletion and inducing immune tolerance with rapamycin may be good strategies to reduce the host immune response to AAVs, though better strategies may be necessary to prevent a decrease or, in some cases, a complete loss in transgene expression.

Immunogenicity and malignancy are other challenges that must be addressed during development and preclinical studies to ensure patient safety. Manufacturing gene therapies can be challenging because of the quality control needed throughout and production costs. Scalability has been a major obstacle to gene therapy production. Due to the time it takes for clinical trials and regulatory procedures, drugs take ample time to reach the market, even for diseases without any current treatment options.

1.3 Technology Platforms of AAV-based *in vivo* Gene Therapy

Genetic editing is the act of revising, removing, or replacing DNA directly, which can be used to tailor recombinant AAVs for specific gene therapies. Gene therapy uses gene replacement, gene addition, genome editing, and gene regulation techniques. Typically, ZFN, transcription activator-like effector nucleases (TALENs), and clustered regularly interspaced short palindromic repeats (CRISPR) are used for various gene editing techniques. Each technology has specific uses and limitations, which are discussed in brief in this chapter. Further detail will be provided in Chapter 2.

1.3.1 cDNA Replacement

cDNA replacement is the method used in gene replacement therapies to express a functional copy of a defective gene episomally, directed by an exogenous promoter, which results in stable gene expression. Most programs use AAV to deliver vectors in gene replacement, although this method is limited to recessive or haploinsufficient disease targets. Artificial promoters allow for specificity but can compromise physiological gene expression.

1.3.2 Genome Editing

Breaks and DNA repair are used in genome editing to design specific, targeted vectors. Genome editing can be performed using various platforms, including ZFN, TALEN, CRISPR/Cas9, base and primer editing, and RNAi gene silencing. These technologies will be introduced next in this chapter with more detail explained in Chapter 2.

Many gene editing techniques utilize a DNA nuclease to cleave a specific site in the genome (Figure 1.4). ZFNs and TALENs have been used previously but are challenging and time-consuming to design or have large constructs with limited transduction.

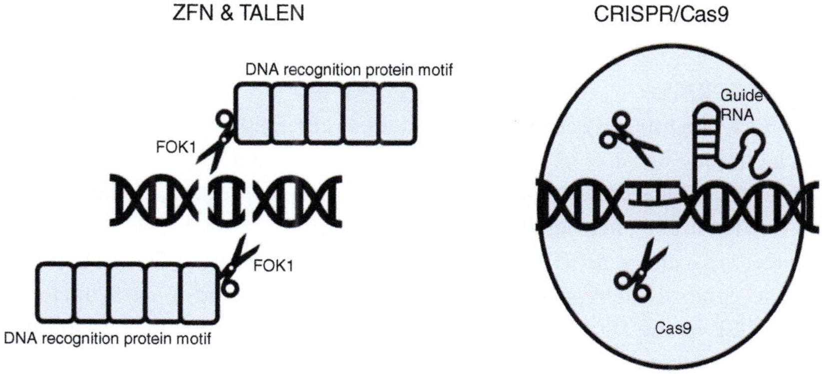

Figure 1.4 Gene editing techniques using double-stranded breaks. ZFN and TALEN use DNA recognition protein motifs and fused with a cleaving enzyme (FOK1). CRISPR/Cas9 uses a guide RNA that binds to a target sequence, then Cas9 binds the guide RNA to cleave both DNA strands.

Genome editing platforms can be combined with viral vectors to improve a gene therapy's efficacy and lower the risk of off-target effects.

1.3.2.1 ZFN

Targeted gene replacement was first developed using ZFNs. Both ZFN and TALEN technologies pair protein domains with nucleotides with a cleaving enzyme to create a double-stranded break (DSB). Approximately 30 amino acids make up one zinc-finger motif, each able to bind three nucleotides. ZFNs generally display lower specificity than TALENs [45, 46] and relatively similar specificity to improved CRISPR/Cas9 methods with certain guide RNAs (gRNAs) [47–49]. However, specificity can be difficult to compare and depends on the specific technological methods, including the gene editing target and nuclease architecture. It also produces less immune reaction compared to the newer CRISPR technology. In human studies using ZFN for gene editing was found to have a good safety profile with evidence of successful gene editing, although sustained expression was not detected [50].

1.3.2.2 TALENs

TALENs make use of non-specific DNA nucleases bound to a DNA-binding domain that can target a specific sequence. TALENs were more quick and efficient to use compared to earlier technologies [51]. The first use of TALEN technology in human treatment was in 2015, treating pediatric acute B lymphoblastic leukemia with TALEN-engineered CAR T cells [52]. TALEN can be advantageous for difficult-to-edit regions of DNA where it can be more efficient than CRISPR [53]. One disadvantage of the TALEN system is its incompatibility with AAV vectors. In general, the targeted use of TALEN *in vivo* is challenging because of its large size and repetitive characteristics. Using high-capacity adenovirus or non-viral delivery is best when implementing the TALEN system directly *in vivo* [54].

1.3.2.3 CRISPR/Cas9

CRISPR/Cas9 is a robust gene editing tool that has emerged as the preferred gene-editing technique because of its ease of use, low cost, and high efficacy. While ZFN and TALEN are protein-based DNA recognition techniques, CRISPR is an RNA-based technique. In brief, CRISPR/Cas9 binds to DNA fragments containing protospacer-adjacent motif (PAM) sites and creates DSBs in the DNA. A gRNA locates the gene of interest or location within the DNA and a Cas9 nuclease induces a DSB in the DNA. Then, DNA repair via non-homologous end joining (NHEJ) or homologous direct repair (HDR) is initiated [55].

As with most gene editing techniques, problems arise, particularly off-target effects and immunity. Off-target effects are an inherent issue in gene editing, being identified with use of ZFNs, TALENs, and CRISPR/Cas9 [56, 57]. Cas9 from bacteria, used on the human genome significantly increases the chance of off-target effects because the genome is much larger than that of bacteria [47]. Thus, many improvements in the CRISPR/Cas9 system have aimed at reducing off-target effects and providing stable genome expression. In 2007, cells treated with CRISPR were found to activate adaptive immunity, inspiring many improvements and modifications to the system [58].

Improving gRNA is another important area of gene therapy development because it has significant effects on sensitivity and specificity of the system. Truncated gRNA, or gRNA lengthened by two guanine nucleotides, has been found to reduce off-target effects [48, 49, 59]. Many tools have been developed to design gRNA and detect off-target effects *in vivo* [60]. Integrase defective lentiviral vectors-capture was one of the first assays created to monitor off-target effects *in vivo* following the use of ZFNs, and GUIDE-seq is the common tool used to identify off-target sites with CRISPR/Cas9 treatments [61–63]. The use of nickases to create a single-stranded rather than a double-strand break also greatly reduces off-target effects [49].

1.3.3 Base Editing and Prime Editing

As mentioned, base and prime editing do not necessitate any DSB, making it safer and more accurate than classical gene editing technologies. Base editing is a tool for permanently correcting base pair mismatches [64]. A more recently developed method called prime editing uses prime editing gRNA (pegRNA) and Cas-nickase, which can correct transition mutations, unlike base editing [65].

Base editors are classified as cytosine or adenine base editors, which need an inactive Cas or Cas nickase coupled with deaminase to make the edit and a gRNA to guide Cas to the targeted DNA binding site. Dual base-editor systems have also been created for a combinatorial editing approach [66]. Creating precise base edits of more than four transition mutations has not been feasible with base editing, leading the path to prime editing. An engineered reverse transcriptase bound to a Cas9 nickase and a prime editing gRNA are necessary for prime editing [65]. PegRNA not only holds the complementary sequence but also another sequence directing the exact sequence change. These methods will be very useful in editing large genes and the treatment of autosomal dominant diseases [67]. Because these technologies are quite new, much research is needed to determine their full safety and efficacy in a variety of uses.

1.3.4 RNAi Gene Silencing

Gene silencing can be performed with RNA interference (RNAi). RNAi can be utilized for gene regulation through the knockdown of a target gene, known as gene silencing. This occurs through a process where double-stranded RNA (dsRNA) is processed into short interfering RNAs (siRNAs), resulting in transcriptional or post-transcriptional gene silencing by degradation of complementary mRNA [68]. dsRNA of more than 30 base pairs can result in a significant antiviral response and apoptosis [69]. Probably the most promising area for RNAi is in antiviral strategy for infections [70].

1.3.5 Gene Addition

Gene therapy for more complex genetic diseases, infectious diseases, and cancer can be performed via gene addition methods to over-express an endogenous or synthetic gene. A herpes simplex virus (HSV) vector has been developed containing the granulocyte-macrophage colony-stimulating factor (GM-CSF) gene to help immune effector cells attack tumor cells [71]. Another example of gene addition has been done by transferring a CAR gene into T cells ex vivo for treating B-cell malignancies [72].

1.4 AAV Serotypes and Tissue Affinity

Twelve AAV serotypes have been identified, and over 100 variants have been found in human or nonhuman primate tissues; although, the best characterized is AAV2 [29]. Optimal serotypes are specific to tissue type, including AAVs 1, 2, 4, 5, 8, and 9 for the central nervous system (CNS), AAVs 1, 8, and 9 for the heart, AAV2 for the kidney, AAVs 7–9 for liver, AAVs 4–6 and 9 for the lung, AAV8 for the pancreas, AAVs 2, 5, and 8 for photoreceptor cells, AAVs 1, 2, 4, 5, and 8 for retinal pigment epithelium, and AAV1 and 6–9 for skeletal muscle. The varying serotypes with differing tissue specificity and infection rates, make AAV vectors a great candidate for gene therapy. AAV gene therapies commonly target the liver, CNS, and muscle, which can become biofactories for producing the required protein.

Other characteristics specific to an individual or a few AAV serotypes include their transduction efficiency, homology to other serotypes, and speed of tissue targeting. AAV1 and 6 share 99% homology, while AAV2 has close homology to most serotypes, not including AAV4, 5, 11, and 12. AAV 8 and 10 share 93% homology, and AAV11 and 12 have close homology to AAV4. AAV5 shares the least homology with other serotypes. Zincarelli et al. found that AAV2-5 have low transduction efficiency, and AAV3 and 4 are slow in targeting tissues [73], but this varies significantly depending on cell type [74].

1.4.1 The Liver as a Biofactory

The liver is targeted in gene therapies of hemophilia for the production of factor VIII and factor IX, missing coagulation proteins [75]. Hemophilia is an ideal target for gene therapy, especially as the genetics and pathophysiology of the disease are well understood. The goal of gene therapy is the long-term expression of the missing genes in the coagulation cascade at levels high enough to be therapeutic, a so-called functional cure. A growing body of clinical experience supports the use of liver-directed rAAV as a gene therapy transporter of the replacement gene for the treatment of hemophilia [75, 76].

1.4.2 The CNS as a Biofactory

Some neurodegenerative disease gene therapies have been successfully developed for spinal muscular atrophy (SMA) and are being developed for other CNS diseases, including Parkinson's and Alzheimer's disease and ALS. The blood–brain barrier (BBB) tends to be an obstacle for AAVs. Two AAVs are known to cross the BBB; IV injection of AAV9 and AAVrh.10 was shown to cross the BBB more effectively than AAV2 or AAV8 to transduce neurons and glial cells [77]. More research is critical to determine how the CNS can be targeted efficiently to use as a gene therapy biofactory for CNS diseases and disorders.

1.4.3 The Muscle as a Biofactory

Muscular dystrophies are another area of interest for gene therapy. AAV8 and AAV9 tend to be used for targeting muscle, although AAV1, 2, 5, 6, and 7 can also be utilized [78]. Muscle can be used as a biofactory, producing secretory factors for infectious diseases, diabetes, atherosclerosis, hemophilia, and cancer therapeutic agents. Muscle cells display the ability to secrete recombinant therapeutic proteins into the bloodstream after intramuscular AAV transduction, making it a useful tool for the treatment of distant organs [17].

1.5 Precision Medicine: Screening and Monitoring Biomarkers, Companion Diagnostics

Besides standard efficacy and safety assessments using clinical outcomes, usually primary endpoints in clinical trials, the incorporation of biomarkers can support the monitoring of patients on gene therapies. These biomarkers include translational, structural, functional, and tumorigenesis biomarkers, which can be used in diagnostics, safety, response, and monitoring. A more detailed discussion can be found in Chapter 3.

Translational biomarkers support the translation from data obtained in preclinical research to clinical trials. These assessments address the sensitivity, specificity, and toxicity of the AAV-based gene therapy and inform measures of therapeutic efficacy and prognosis. Structural biomarkers assess tissues and organs through imaging methodologies, such as ultrasound, MRI, and PET. Functional biomarkers can assess clinical endpoints related to organ functionality and prognosis, such as motor and cognitive clinical assessments in various CNSCNS and muscular degenerative diseases. Liver and kidney function biomarkers can be assessed in the blood to measure their function and monitor treatment responses of liver and kidney-targeted gene therapies. Preexisting antibodies to AAV can be used as biomarkers for the prediction of AAV gene therapy effectiveness and often need to be developed into companion diagnostics. AAV antibodies present may be a predictor for AAV gene therapy response and can indicate a need for adjuncts to prevent the rejection of transfected cells. Tumorigenesis biomarkers can assess the potential development of neoplasia. Tumor testing and profiling biomarkers can help identify the type of tumor and oncogenesis. Genetic testing can determine a person's risk of certain cancers or the type of cancer-causing tumor. Genetic testing, specifically of a tumor, can give further information about its growth and other characteristics.

1.5.1 Gene Therapy Clinical Trials: Spotlight on Hemophilia A

As a representative example of the clinical development processes for gene therapies, we will reference the trial design for patients with severe or moderately severe forms of hemophilia A. This has been extensively reviewed by Pipe et al., showing that, in contrast to several biopharmaceutical companies developing gene replacement therapies based on the B-domain deleted human factor VIII, ASC therapeutics is introducing a second-generation gene therapy based on a B-domain deleted human/porcine chimeric factor VIII [79]. This next-generation gene therapy has shown in preclinical studies the potential to decrease dose requirements, increase durability through improved factor VIII expression, and improve durability driven by a more efficient expression and reduced intracellular stress.

A first-in-human, open-label, dose-finding study is designed to assess the safety and preliminary efficacy of a single infusion of the AAV8-based chimeric transgene replacing the coagulation protein factor VIII in study participants with severe and moderately severe hemophilia A (FVIII activity ≤ 2 International Units/Deciliter).

In this study, the main safety monitoring parameters include:

1) Physical examination, including assessments of general appearance; head, eyes, ears, nose, and throat; the cardiovascular, dermatologic, lymphatic, respiratory, gastrointestinal, genitourinary, musculoskeletal, and neurologic systems. Height and weight will also be measured and recorded.
2) Vital signs, including oral temperature, pulse rate, respiratory rate, and blood pressure.
3) Viral shedding, assessing the evidence of potential viral transmission, will be tested on samples of blood, saliva, urine, stool, and semen.
4) Liver ultrasound.

The main efficacy monitoring assessments include:

1) FVIII activity, determined by validated assays, one-stage activated partial thromboplastin time (aPTT), and chromogenic FXa. FVIII levels should be taken at a trough or close to trough levels, meaning after a minimum of 72 hours has elapsed since the last infusion of FVIII protein concentrates.
2) Bleeding episodes and FVIII replacement therapy will be captured in the participant's diary. In addition, the number of bleeding episodes requiring treatment, and the number of bleedings that do not require treatment following the administration of ASC618 infusion, will be recorded throughout the study.

The main exploratory assessments include:

1) Hemophilia A quality of life (Haem-A-QoL) questionnaire was completed by participants during the study.
2) Blood samples will be collected for potential future research in the hemophilia A disease area to assess: (1) Liver structure, including Fibroscan and abdominal ultrasound; (2) Liver function, including liver enzymes, GRP78/BiP, and microRNA exploration analysis; (3) Liver tumorigenesis, including GALAD score (serum biomarker-based model that predicts the probability of having HCC*, incorporating gender, age, AFP-L3, AFP, and DCP), VirScan (comprehensively analyzing antiviral antibodies), and cell-free DNA assessment.

To complete these assessments, extensive monitoring, including on-site and remote visits, will be implemented. The levels of FVIII and liver function will be evaluated twice a week in the first 16 weeks, slowly decreasing as these measurements reach a plateau.

It is crucial during gene therapy to have measures of improvement, disease stage, or to prescreen for predicting its effectiveness and any need for adjunct therapy, for example, to suppress an immune reaction that may reject the AAV-infected cells (NCT04676048).

1.6 Predictions for Scientific and Medical Progress

Over the past decade, approximately 45% of gene therapy clinical trials have focused on *in vivo* as opposed to *ex vivo* gene therapy [76]. Many platforms are used for AAV-based gene therapy, including cDNA gene replacement, gene editing, genome regulation, and gene addition. The development and use of translational biomarkers are incredibly important for predictive and diagnostic measures during gene therapy clinical trials. A critical step in preclinical testing is a thorough analysis of off-target effects. As mentioned earlier, analysis software systems can aid in the detection and monitoring of off-target effects. Antibodies against the capsid, protein expression and activity levels, and viral shedding are just a few examples of measures used as predictive, diagnostic, and safety biomarkers. Immunogenicity is another important factor that should be monitored via biomarkers.

Biodistribution and shedding are typically analyzed through polymerase chain reaction (PCR), while transgene mRNA expression analysis can be performed using reverse transcription-PCR and in situ hybridization (ISH). Transgene protein concentration can be measured by numerous platforms – enzyme-linked immunosorbent assay (ELISA), electrochemiluminescence (ECL), immunoPCR, liquid chromatography–mass spectrometry (LC–MS), Western blot, and high-performance liquid chromatography (HPLC). Immune response measurements can be performed using ELISPOT, flow cytometry, electrochemiluminescence, and more. Genome integration must also be measured after treatment, which helps discern potential carcinogenesis. More safety, efficacy, and diagnostic biomarkers, companion diagnostics, and gene therapy treatments are sure to emerge continuously. Regulatory practices are essential in ensuring the study is worth the effort and the subjects' rights are protected. In addition, guidance surrounding the development of gene therapies, useful biomarkers, and companion diagnostics needs to be expanded.

1.6.1 Predictions for Challenges in the Field

Among the major factors preventing AAV-driven gene therapies from being more broadly adopted are challenges related to scalability, such as large-scale production and cost-effectiveness. Manufacturing gene therapies typically begins with plasmid development and production, then cell expansion, plasmid transfection, viral vector production, purification, and finally, formulation. Companies often use third parties to complete this process; however, competition for these services with other advanced therapies, such as mRNA vaccine development for COVID-19, may become a significant bottleneck. Thankfully, the development of the *Sf9* system has been integral in making large-scale manufacturing more possible by decreasing the amount of viral particles needed for efficient production [30].

This increased demand for third-party manufacturing services drives many companies to establish in-house manufacturing.

Several challenges are related to ensuring a balanced risk-benefit solution, in other words, achieving a safe and therapeutic gene therapy. Large doses may be necessary to reach a desired therapeutic effect, increasing the likelihood of side effects and costs. Thus, the lowest effective dose should be determined. Transgene expression needs to be specific to the target cell or tissue to minimize off-target effects and be long-lasting. We are likely to see more CRISPR use in the future of *in vivo* gene therapy, but for now, off-target effects are the main concern. The immune system needs to stay unresponsive to the viral vector in order to maximize efficacy, enable redosing, and prevent toxicity. Many strategies have been and are being developed to help AAV-based therapies evade the immune system, including codon optimization, CpG reduction, regulatory elements, integrative AAV vector approaches, lower dosing, and immunomodulation.

To overcome these challenges, it is vital to identify translational biomarkers to measure a particular therapy's durability, immunogenicity, malignancy, and toxicity. Recent adverse events and deaths have led to clinical holds on some gene therapies. Thus, the FDA has closely monitored gene therapy safety in the United States and the European Medicines Agency (EMA) in the EU.

Translational biomarkers, either primary, secondary, or exploratory, i.e. surrogate endpoints include efficacy biomarkers, defining the clinical activity of the gene replacement therapy (i.e. durability can be measured through the length of effectiveness observed by normal blood levels, staying in remission, or physical symptom abatement); safety biomarkers to assess dose-limiting toxicity; tumorigenic biomarkers to assess malignancy risk of organs targeted as biofactories for gene therapy; immunogenicity biomarkers, which can be used as inclusion/exclusion criteria, especially anti-AAV preexisting antibodies produced by previous exposure to an AAV. These antibodies can significantly impact the effectiveness and eligibility for a particular treatment.

Recent focus has been placed on developing safe strategies to block anti-AAV antibody effects. Among the alternatives to overcome these challenges are isolated liver perfusion, which can be effective but may induce inflammation; reduction of IgG levels, a method undergoing preclinical trials of autoimmunity; plasmapheresis, a potential approach that may require several therapeutic rounds for effective depletion of anti-AAV neutralizing antibodies; and capsid modifications as a method to prevent neutralization.

1.6.2 Addressing Durability

AAVs have the potential for long-lasting effects, which is one of the main goals of gene therapies. Gene therapy seeks to provide a onetime treatment for patients.

This is incredibly important for those who currently have inadequate treatment options that may have rigorous regimens or substantial side effects and who have no available treatments on the market. The immune response considerably influences durability and efficiency, as existing antibodies may activate immune cells that prevent viral transduction.

1.6.3 Addressing Immunogenicity

The percentage of people possessing preexisting anti-AAV antibodies is quite significant. It has been estimated that anti-AAV1 and anti-AAV2 antibodies can be found in 70% of the population [80]. Anti-AAV6, 8, and 9 antibodies have a lower incidence of approximately 30–50%, while the overall incidence of all serotypes is estimated at 30–70%. These rates are impacted by geographical location and increase with age [80]. A person's level of anti-AAV antibodies, typically measured by ligand binding assay, can be used as a biomarker for inclusion/exclusion criteria. Measurements can be performed using cell-based transduction inhibition assay for neutralizing activity as well. Thankfully, several strategies have been discussed and suggested for addressing immunogenic responses to gene therapies.

There is potential for an immune response to the capsid or the transgene itself posttreatment. Capsid modifications have been suggested as a potential method to prevent neutralization by antibodies [81]. Immunosuppressive drugs show promise for pretreatment in seropositive patients, but more testing is necessary. Strong promotors are vital to lowering the dose necessary for a therapeutic effect and, thus, reduce the potential for adverse effects [82]. Further detail of immunogenicity will be discussed in Chapter 5.

1.6.4 Addressing Malignancy

AAV-based gene therapies have shown less malignancy than the use of other vectors; however, recent research has shown more risks than were thought to exist prior. The development of sensitive tumorigenic markers is essential for understanding the risk of malignancy during development and clinical trials. As research advances and continual developments are made, gene therapy will become more targeted, effective, and safe.

1.7 Predictions for Market Adoption

With the rise in chronic diseases such as cancers, neurodegenerative diseases, genetic disorders, and rare diseases, the market for gene therapy has grown greatly. The most commonly targeted tissues of *in vivo* gene therapy clinical trials over the past decade have been the CNS, heart, liver, eye, bone marrow, and tumors [83]. Pharmaceutical and biotech companies have been the biggest contributors to the gene therapy market. Some large competitors in the gene therapy market include Novartis, Sangamo Therapeutics, and Biogen to name a few, although many partnerships seem to be transpiring between these companies to produce *in vivo* therapies [84]. The Q2 report of 2022 released by the American Society of Gene & Cell Therapy (ASGCT) and PharmaIntelligence reports that 2024 gene therapies were under development and testing in the preclinical, clinical trial, or preregistration phases. Several market assessment reports have estimated that the market for gene therapy will grow at a 20–30% annual rate throughout 2030.

1.7.1 Patients and Patient Advocacy Groups

Several gene therapy patient advocacy groups help advance research, educate people about gene therapy, and help with treatment costs. ASGCT is a professional organization made up of scientists, physicians, and patient advocates with the goal of enhancing knowledge, education, and awareness about gene and cell therapy. Gene and cell therapy clinical trials and landscape reports can be found on their website. Askbio is a similar site with a collaborative program called AskFirst, where patients and their families are "asked" how they can help. Askbio also has a list of advocacy partners that help with education, research, and events to benefit patients. To name a few, CureDuchenne, Speak Foundation, Angelman Syndrome Foundation, Defeat MSA Alliance, and Nationwide Children's Hospital.

1.7.2 Physicians, Clinical Guidelines, Regulatory Agencies

Regulatory approval for gene therapy involves several regulatory boards that vary by country. The EU and United States have different oversight in that the FDA oversees all US clinical trials and follows a drug's lifecycle from its conception. In contrast, the EMA is a reviewing body over many European nations. Each state has its own regulatory body that follows the clinical trial. New drug applications in the United States go through the FDA, and in the EU, they can either go through a state's regulatory committee or the EMA. The time it takes drugs to be approved is a long process, and the FDA's reviewing process has been thought to be much longer; however, a study published in 2011 found that the process through the EMA is longer. When looking at approval for cancer drugs, they found that new drugs were reaching patients in the United States faster than in Europe due to a faster review process by the FDA [85].

Before clinical trials begin in the United States, an investigational new drug application (IND) must be submitted to the FDA and approved. The Institutional

Review Board (IRB) reviews and monitors the clinical research to ensure compliance throughout a trial. The trial protocol, consent form, HIPPA authorization, etc., must be presented to the IRB for review and approval. Study modifications and amendments, adverse events, and annual reviews are required by the IRB to be submitted and approved promptly to stay compliant and complete the trial. More information will be presented in Chapter 18.

1.7.3 Payers

Most gene therapies are likely to be covered by insurance in the future, but currently, most policies have yet to cover them. Advocacy groups have been tremendously important for many who needed gene therapy but could not afford the high costs out of pocket. Experimental treatments are typically free options but put patients at unknown risks. Clinicaltrials.gov can provide a list of AAV *in vivo* gene therapies being tested in active clinical trials.

1.8 Final Thoughts

Inherited monogenic rare diseases are ideal targets for gene therapies based on the success of gene replacement therapies in adults and children; however, gene therapies are quickly emerging for various non-monogenic diseases as well. Preclinical studies have been using various gene editing platforms to correct a defective gene or insert a functional gene in animal models and/or clinical studies. Sustained expression of gene therapies in humans necessitates lengthy, thorough studies to measure long-term efficacy and off-target activity to provide a positive safety profile. Improved methods in genome editing have dramatically reduced off-target and deleterious effects successfully in preclinical models, needing translation to clinical programs. AAV vectors are currently being utilized by numerous clinical trials to treat numerous, various diseases, and gene editing will pave the way for achieving successful results with high specificity, efficacy, and safety.

1.8.1 Can We Afford *in vivo* Gene Therapies?

Large-scale manufacturing, availability of financial resources, and limited access to experts are causing the price of gene therapies to skyrocket. The current cost of Glybera is approximately 1.2 million and Luxturna, released in 2017, costs approximately $425,000 per eye. For many researchers, the development cost throughout the preclinical to clinical phase is not feasible. The need for toxicology, safety, and dosing studies and IND approval creates more hurdles with approval and cost.

Inherited monogenic rare disease treatment is costly and continual, with enormous financial burdens. Advancing therapies for restoring defective protein levels have a tremendous impact on people suffering from rare and currently untreatable diseases. Hopefully, with the development of advanced technologies, the creation of cheaper, safe, and successful methods will lower the costs for both production and patient treatment. This is promising as we see the development of novel tools for measuring, monitoring, and preventing adverse effects and the success of gene therapies.

1.8.2 Can *in vivo* Gene Editing Replace Gene Therapy?

While incredible progress in gene editing has been made in the past 15 years, there are many areas that still need substantial improvement. Key elements required for transitioning from gene therapy to gene editing include overcoming significant challenges. New therapies need to concentrate on overcoming transient expression, reducing off-target activity, and lowering the risk of genotoxicity following long-term exposure. Efforts should be made to minimize the need for repeated injections of gene therapies, which are more likely to trigger an immune response, lowering the efficacy while reducing its safety.

Delivery of gene-editing techniques has been achieved with both viral and non-viral delivery systems. Viral delivery includes AAV vectors while alternative vehicles like lipid nanoparticles can easily pass through cell membranes and have seen increased use in targeting the liver without activation of the immune system, allowing for repeated administration. Additionally, magnetic nanoparticles have successfully delivered gene-editing systems with little off-target effects and no toxicity [86, 87]. A nuclease-free system can be paired with CRISPR/*Sa*Cas9, which has shown significant improvements in targeting effectiveness and stable gene modification in mice [88]. Studies utilizing these newer technologies are still in their infancy for *in vivo* applications.

Unfortunately, the major barrier to both gene therapy and gene editing is the immune response. However, patients can be given prophylactic steroid treatments, prior, to reduce the immune response [89]. Another major concern is long-term stability. While AAV vectors are technically capable of cargo delivery to mitotic and postmitotic cells, in practice, the delivery is typically episomal, resulting in diminished effects as cells divide. This is particularly relevant in cases involving pediatric patients, where attempts to cure monogenic liver diseases have been met with concerns about hepatocyte proliferation [90–92]. As discussed previously, repeated injections of AAV vectors lead to the production of anti-AAV neutralizing antibodies, reducing efficacy and safety of subsequent treatments or therapies [40, 93–95]. In patients who have undergone AAV therapy and developed anti-AAV antibodies, utilizing non-viral delivery of CRISPR/Cas9 would be optimal as it would not be affected by anti-AAV antibodies. In summary, *in vivo* gene therapies have limitations that can be overcome with *in vivo* gene editing, but significant developments and research are needed before the safety and efficacy of treatments are shown and optimized.

References

1 Watson, J.D. and Crick, F.H. (1953). Molecular structure of nucleic acids; a structure for deoxyribose nucleic acid. *Nature* 171 (4356): 737–738.
2 Franklin, R.E. and Gosling, R.G. (1953). Evidence for 2-chain helix in crystalline structure of sodium deoxyribonucleate. *Nature* 172 (4369): 156–157.
3 Atchison, R.W., Casto, B.C., and Hammon, W.M. (1965). Adenovirus-associated defective virus particles. *Science* 149 (3685): 754–756.
4 Terheggen, H.G., Lowenthal, A., Lavinha, F. et al. (1975). Unsuccessful trial of gene replacement in arginase deficiency. *Z. Kinderheilkd.* 119 (1): 1–3.
5 Tabin, C.J., Hoffmann, J.W., Goff, S.P. et al. (1982). Adaptation of a retrovirus as a eucaryotic vector transmitting the herpes simplex virus thymidine kinase gene. *Mol. Cell. Biol.* 2 (4): 426–436.
6 Robert-Guroff, M., Markham, P.D., Popovic, M. et al. (1985). Isolation, characterization, and biological effects of the first human retroviruses: the human T-lymphotropic retrovirus family. *Curr. Top. Microbiol. Immunol.* 115: 7–31.
7 Bodine, D.M., Karlsson, S., Papayannopoulou, T. et al. (1989). Introduction and expression of human beta globin genes into primitive murine hematopoietic progenitor cells by retrovirus mediated gene transfer. *Prog. Clin. Biol. Res.* 319: 589–599. discussion 600.
8 Hermonat, P.L. and Muzyczka, N. (1984). Use of adeno-associated virus as a mammalian DNA cloning vector: transduction of neomycin resistance into mammalian tissue culture cells. *Proc. Natl. Acad. Sci. U. S. A.* 81 (20): 6466–6470.
9 Johnson, I.S. (1983). Human insulin from recombinant DNA technology. *Science* 219 (4585): 632–637.
10 Miller, J., McLachlan, A.D., and Klug, A. (1985). Repetitive zinc-binding domains in the protein transcription factor IIIA from Xenopus oocytes. *EMBO J.* 4 (6): 1609–1614.
11 Beardsley, T. (1986). Genetic engineering. Hepatitis vaccine wins approval. *Nature* 322 (6078): 396.
12 Palca, J. (1988). James Watson to head NIH human genome project. *Nature* 335 (6187): 193.
13 Rhodes, C.A., Pierce, D.A., Mettler, I.J. et al. (1988). Genetically transformed maize plants from protoplasts. *Science* 240 (4849): 204–207.

14 Anderson, W.F. (1990). September 14, 1990: the beginning. *Hum. Gene Ther.* 1 (4): 371–372.

15 Campbell, K.H. et al. (1996). Sheep cloned by nuclear transfer from a cultured cell line. *Nature* 380 (6569): 64–66.

16 Hanna, E. et al. (2017). Gene therapies development: slow progress and promising prospect. *J. Mark. Access Health Policy* 5 (1): 1265293.

17 Wang, D., Wang, K., and Cai, Y. (2020). An overview of development in gene therapeutics in China. *Gene Ther.* 27 (7–8): 338–348.

18 MacGregor, R.R. and Clinical protocol. (2001). A phase 1 open-label clinical trial of the safety and tolerability of single escalating doses of autologous CD4 T cells transduced with VRX496 in HIV-positive subjects. *Hum. Gene Ther.* 12 (16): 2028–2029.

19 Takahashi, K., Okita, K., Nakagawa, M. et al. (2007). Induction of pluripotent stem cells from fibroblast cultures. *Nat. Protoc.* 2 (12): 3081–3089.

20 Cong, L., Ran, F.A., Cox, D. et al. (2013). Multiplex genome engineering using CRISPR/Cas systems. *Science* 339 (6121): 819–823.

21 Lu, Y., Xue, J., Deng, T. et al. (2020). Safety and feasibility of CRISPR-edited T cells in patients with refractory non-small-cell lung cancer. *Nat. Med.* 26 (5): 732–740.

22 Srivastava, A., Lusby, E.W., and Berns, K.I. (1983). Nucleotide sequence and organization of the adeno-associated virus 2 genome. *J. Virol.* 45 (2): 555–564.

23 Sonntag, F., Schmidt, K., and Kleinschmidt, J.A. (2010). A viral assembly factor promotes AAV2 capsid formation in the nucleolus. *Proc. Natl. Acad. Sci. U. S. A.* 107 (22): 10220–10225.

24 Lusby, E., Fife, K.H., and Berns, K.I. (1980). Nucleotide sequence of the inverted terminal repetition in adeno-associated virus DNA. *J. Virol.* 34 (2): 402–409.

25 Xie, Q., Bu, W., Bhatia, S. et al. (2002). The atomic structure of adeno-associated virus (AAV-2), a vector for human gene therapy. *Proc. Natl. Acad. Sci. U. S. A.* 99 (16): 10405–10410.

26 Hoque, M., Ishizu, K., Matsumoto, A. et al. (1999). Nuclear transport of the major capsid protein is essential for adeno-associated virus capsid formation. *J. Virol.* 73 (9): 7912–7915.

27 Wu, P., Xiao, W., Conlon, T. et al. (2000). Mutational analysis of the adeno-associated virus type 2 (AAV2) capsid gene and construction of AAV2 vectors with altered tropism. *J. Virol.* 74 (18): 8635–8647.

28 Hermonat, P.L., Labow, M.A., Wright, R. et al. (1984). Genetics of adeno-associated virus: isolation and preliminary characterization of adeno-associated virus type 2 mutants. *J. Virol.* 51 (2): 329–339.

29 Wu, Z., Asokan, A., and Samulski, R.J. (2006). Adeno-associated virus serotypes: vector toolkit for human gene therapy. *Mol. Ther.* 14 (3): 316–327.

30 Mietzsch, M., Grasse, S., Zurawski, C. et al. (2014). OneBac: platform for scalable and high-titer production of adeno-associated virus serotype 1-12 vectors for gene therapy. *Hum. Gene Ther.* 25 (3): 212–222.

31 Schnepp, B.C., Jensen, R.L., Chen, C. et al. (2005). Characterization of adeno-associated virus genomes isolated from human tissues. *J. Virol.* 79 (23): 14793–14803.

32 Chadeuf, G., Favre, D., Tessier, J. et al. (2000). Efficient recombinant adeno-associated virus production by a stable rep-cap HeLa cell line correlates with adenovirus-induced amplification of the integrated rep-cap genome. *J. Gene Med.* 2 (4): 260–268.

33 Sonntag, F., Bleker, S., Leuchs, B. et al. (2006). Adeno-associated virus type 2 capsids with externalized VP1/VP2 trafficking domains are generated prior to passage through the cytoplasm and are maintained until uncoating occurs in the nucleus. *J. Virol.* 80 (22): 11040–11054.

34 Douar, A.M., Poulard, K., Stockholm, D. et al. (2001). Intracellular trafficking of adeno-associated virus vectors: routing to the late endosomal compartment and proteasome degradation. *J. Virol.* 75 (4): 1824–1833.

35 Yan, Z., Zak, R., Zhang, Y. et al. (2004). Distinct classes of proteasome-modulating agents cooperatively augment recombinant adeno-associated virus type 2 and type 5-mediated transduction from the apical surfaces of human airway epithelia. *J. Virol.* 78 (6): 2863–2874.

36 Nicolson, S.C. and Samulski, R.J. (2014). Recombinant adeno-associated virus utilizes host cell nuclear import machinery to enter the nucleus. *J. Virol.* 88 (8): 4132–4144.

37 Li, C., He, Y., Nicolson, S. et al. (2013). Adeno-associated virus capsid antigen presentation is dependent on endosomal escape. *J. Clin. Invest.* 123 (3): 1390–1401.

38 Wang, Z., Ma, H., Li, J. et al. (2003). Rapid and highly efficient transduction by double-stranded adeno-associated virus vectors in vitro and in vivo. *Gene Ther.* 10 (26): 2105–2111.

39 Natkunarajah, M., Trittibach, P., McIntosh, J. et al. (2008). Assessment of ocular transduction using single-stranded and self-complementary recombinant adeno-associated virus serotype 2/8. *Gene Ther.* 15 (6): 463–467.

40 Nathwani, A.C., Tuddenham, E.G.D., Rangarajan, S. et al. (2011). Adenovirus-associated virus vector-mediated gene transfer in hemophilia B. *N. Engl. J. Med.* 365 (25): 2357–2365.

41 Kessler, P.D., Podsakoff, G.M., Chen, X. et al. (1996). Gene delivery to skeletal muscle results in sustained expression and systemic delivery of a therapeutic protein. *Proc. Natl. Acad. Sci. U. S. A.* 93 (24): 14082–14087.

42 Mullard, A. (2021). Gene therapy community grapples with toxicity issues, as pipeline matures. *Nat. Rev. Drug Discov.* 20 (11): 804–805.

43 Nathwani, A.C., Reiss, U.M., Tuddenham, E.G.D. et al. (2014). Long-term safety and efficacy of factor IX gene therapy in hemophilia B. *N. Engl. J. Med.* 371 (21): 1994–2004.

44 Calcedo, R., Morizono, H., Wang, L. et al. (2011). Adeno-associated virus antibody profiles in newborns, children, and adolescents. *Clin. Vaccine Immunol.* 18 (9): 1586–1588.

45 Mussolino, C., Alzubi, J., Fine, E.J. et al. (2014). TALENs facilitate targeted genome editing in human cells with high specificity and low cytotoxicity. *Nucleic Acids Res.* 42 (10): 6762–6773.

46 Guilinger, J.P., Pattanayak, V., Reyon, D. et al. (2014). Broad specificity profiling of TALENs results in engineered nucleases with improved DNA-cleavage specificity. *Nat. Methods* 11 (4): 429–435.

47 Pattanayak, V., Lin, S., Guilinger, J.P. et al. (2013). High-throughput profiling of off-target DNA cleavage reveals RNA-programmed Cas9 nuclease specificity. *Nat. Biotechnol.* 31 (9): 839–843.

48 Fu, Y., Sander, J.D., Reyon, D. et al. (2014). Improving CRISPR-Cas nuclease specificity using truncated guide RNAs. *Nat. Biotechnol.* 32 (3): 279–284.

49 Ran, F.A., Hsu, P.D., Lin, C. et al. (2013). Double nicking by RNA-guided CRISPR Cas9 for enhanced genome editing specificity. *Cell* 154 (6): 1380–1389.

50 Harmatz, P., Prada, C.E., Burton, B.K. et al. (2022). First-in-human in vivo genome editing via AAV-zinc-finger nucleases for mucopolysaccharidosis I/II and hemophilia B. *Mol. Ther.* 30 (12): 3587–3600.

51 Osborn, M.J., Starker, C.G., McElroy, A.N. et al. (2013). TALEN-based gene correction for epidermolysis bullosa. *Mol. Ther.* 21 (6): 1151–1159.

52 Qasim, W., Zhan, H., Samarasinghe, S. et al. (2017). Molecular remission of infant B-ALL after infusion of universal TALEN gene-edited CAR T cells. *Sci. Transl. Med.* 9 (374): eaaj2013.

53 Jain, S., Shukla, S., Yang, C. et al. (2021). TALEN outperforms Cas9 in editing heterochromatin target sites. *Nat. Commun.* 12 (1): 606.

54 Wang, L., Li, F., Dang, L. et al. (2016). In vivo delivery systems for therapeutic genome editing. *Int. J. Mol. Sci.* 17 (5): 626.

55 Barrangou, R. and Doudna, J.A. (2016). Applications of CRISPR technologies in research and beyond. *Nat. Biotechnol.* 34 (9): 933–941.

56 Hagedorn, P.H., Pontoppidan, M., Bisgaard, T.S. et al. (2018). Identifying and avoiding off-target effects of RNase H-dependent antisense oligonucleotides in mice. *Nucleic Acids Res.* 46 (11): 5366–5380.

57 Jackson, A.L. and Linsley, P.S. (2010). Recognizing and avoiding siRNA off-target effects for target identification and therapeutic application. *Nat. Rev. Drug Discov.* 9 (1): 57–67.

58 Barrangou, R., Fremaux, C., Deveau, H. et al. (2007). CRISPR provides acquired resistance against viruses in prokaryotes. *Science* 315 (5819): 1709–1712.

59 Cho, S.W., Kim, S., Kim, Y. et al. (2014). Analysis of off-target effects of CRISPR/Cas-derived RNA-guided endonucleases and nickases. *Genome Res.* 24 (1): 132–141.

60 Bae, S., Park, J., and Kim, J.S. (2014). Cas-OFFinder: a fast and versatile algorithm that searches for potential off-target sites of Cas9 RNA-guided endonucleases. *Bioinformatics* 30 (10): 1473–1475.

61 Park, P.J. (2009). ChIP-seq: advantages and challenges of a maturing technology. *Nat. Rev. Genet.* 10 (10): 669–680.

62 Tsai, S.Q., Zheng, Z., Nguyen, N.T. et al. (2015). GUIDE-seq enables genome-wide profiling of off-target cleavage by CRISPR-Cas nucleases. *Nat. Biotechnol.* 33 (2): 187–197.

63 Gabriel, R., Lombardo, A., Arens, A. et al. (2011). An unbiased genome-wide analysis of zinc-finger nuclease specificity. *Nat. Biotechnol.* 29 (9): 816–823.

64 Gaudelli, N.M., Komor, A.C., Rees, H.A. et al. (2017). Programmable base editing of A*T to G*C in genomic DNA without DNA cleavage. *Nature* 551 (7681): 464–471.

65 Anzalone, A.V., Randolph, P.B., Davis, J.R. et al. (2019). Search-and-replace genome editing without double-strand breaks or donor DNA. *Nature* 576 (7785): 149–157.

66 Zhang, X., Zhu, B., Chen, L. et al. (2020). Dual base editor catalyzes both cytosine and adenine base conversions in human cells. *Nat. Biotechnol.* 38 (7): 856–860.

67 Kantor, A., McClements, M.E., and MacLaren, R.E. (2020). CRISPR-Cas9 DNA base-editing and prime-editing. *Int. J. Mol. Sci.* 21 (17).

68 Sontheimer, E.J. (2005). Assembly and function of RNA silencing complexes. *Nat. Rev. Mol. Cell Biol.* 6 (2): 127–138.

69 Gil, J. and Esteban, M. (2000). Induction of apoptosis by the dsRNA-dependent protein kinase (PKR): mechanism of action. *Apoptosis* 5 (2): 107–114.

70 Kelleher, A.D., Cortez-Jugo, C., Cavalieri, F. et al. (2020). RNAi therapeutics: an antiviral strategy for human infections. *Curr. Opin. Pharmacol.* 54: 121–129.

71 Goins, W.F., Huang, S., Cohen, J.B. et al. (2014). Engineering HSV-1 vectors for gene therapy. *Methods Mol. Biol.* 1144: 63–79.

72 Kochenderfer, J.N. et al. (2013). Donor-derived CD19-targeted T cells cause regression of malignancy persisting after allogeneic hematopoietic stem cell transplantation. *Blood* 122 (25): 4129–4139.

73 Zincarelli, C., Soltys, S., Rengo, G. et al. (2008). Analysis of AAV serotypes 1-9 mediated gene expression and tropism in mice after systemic injection. *Mol. Ther.* **16** (6): 1073–1080.

74 Ellis, B.L., Hirsch, M.L., Barker, J.C. et al. (2013). A survey of ex vivo/in vitro transduction efficiency of mammalian primary cells and cell lines with nine natural adeno-associated virus (AAV1-9) and one engineered adeno-associated virus serotype. *Virol. J.* 10: 74.

75 Markusic, D.M. and Herzog, R.W. (2012). Liver-directed adeno-associated viral gene therapy for hemophilia. *J. Genet. Syndr. Gene Ther.* 1: 1–9.

76 Flotte, T.R. (2021). Liver targeting with rAAV7: balancing tropism with immune profiles. *Gene Ther.* 28 (3-4): 115–116.

77 Klein, R.L., Dayton, R.D., Tatom, J.B. et al. (2008). Tau expression levels from various adeno-associated virus vector serotypes produce graded neurodegenerative disease states. *Eur. J. Neurosci.* 27 (7): 1615–1625.

78 Wang, D., Zhong, L., Nahid, M.A. et al. (2014). The potential of adeno-associated viral vectors for gene delivery to muscle tissue. *Expert Opin. Drug Deliv.* 11 (3): 345–364.

79 Pipe, S.W., Gonen-Yaacovi, G., and Segurado, O.G. (2022). Hemophilia A gene therapy: current and next-generation approaches. *Expert. Opin. Biol. Ther.* 22 (9): 1099–1115.

80 Vandamme, C., Adjali, O., and Mingozzi, F. (2017). Unraveling the complex story of immune responses to AAV vectors trial after trial. *Hum. Gene Ther.* 28 (11): 1061–1074.

81 Bartel, M., Schaffer, D., and Buning, H. (2011). Enhancing the clinical potential of AAV vectors by capsid engineering to evade pre-existing immunity. *Front. Microbiol.* 2: 204.

82 Naso, M.F., Tomkowicz, B., Perry, W.L. et al. (2017). Adeno-Associated Virus (AAV) as a Vector for Gene Therapy. *BioDrugs* 31 (4): 317–334.

83 Arabi, F., Mansouri, V., and Ahmadbeigi, N. (2022). Gene therapy clinical trials, where do we go? An overview. *Biomed. Pharmacother.* 153: 113324.

84 Lubroth, P., Colasante, G., and Lignani, G. (2021). In vivo genome editing therapeutic approaches for neurological disorders: where are we in the translational pipeline? *Front. Neurosci.* 15: 632522.

85 Roberts, S.A., Allen, J.D., and Sigal, E.V. (2011). Despite criticism of the FDA review process, new cancer drugs reach patients sooner in the United States than in Europe. *Health Aff. (Millwood)* 30 (7): 1375–1381.

86 Rohiwal, S.S., Dvorakova, N., Klima, J. et al. (2020). Polyethylenimine based magnetic nanoparticles mediated non-viral CRISPR/Cas9 system for genome editing. *Sci. Rep.* 10 (1): 4619.

87 Sizikov, A.A., Kharlamova, M.V., Nikitin, M.P. et al. (2021). Nonviral locally injected magnetic vectors for in vivo gene delivery: a review of studies on magnetofection. *Nanomaterials (Basel)* 11 (5): 1078.

88 Ran, F.A., Cong, L., Yan, W.X. et al. (2015). In vivo genome editing using Staphylococcus aureus Cas9. *Nature* 520 (7546): 186–191.

89 Chu, W.S. and Ng, J. (2021). Immunomodulation in administration of rAAV: preclinical and clinical adjuvant pharmacotherapies. *Front. Immunol.* 12: 658038.

90 Cunningham, S.C., Spinoulas, A., Carpenter, K.H. et al. (2009). AAV2/8-mediated correction of OTC deficiency is robust in adult but not neonatal Spf(ash) mice. *Mol. Ther.* 17 (8): 1340–1346.

91 Bortolussi, G., Zentillin, L., Vaníkova, J. et al. (2014). Life-long correction of hyperbilirubinemia with a neonatal liver-specific AAV-mediated gene transfer in a lethal mouse model of Crigler-Najjar syndrome. *Hum. Gene Ther.* 25 (9): 844–855.

92 Wang, L., Wang, H., Bell, P. et al. (2012). Hepatic gene transfer in neonatal mice by adeno-associated virus serotype 8 vector. *Hum. Gene Ther.* 23 (5): 533–539.

93 Calcedo, R. and Wilson, J.M. (2016). AAV natural infection induces broad cross-neutralizing antibody responses to multiple AAV serotypes in chimpanzees. *Hum. Gene. Ther. Clin. Dev.* 27 (2): 79–82.

94 Chirmule, N., Xiao, W., Truneh, A. et al. (2000). Humoral immunity to adeno-associated virus type 2 vectors following administration to murine and nonhuman primate muscle. *J. Virol.* 74 (5): 2420–2425.

95 Nathwani, A.C., Gray, J.T., McIntosh, J. et al. (2007). Safe and efficient transduction of the liver after peripheral vein infusion of self-complementary AAV vector results in stable therapeutic expression of human FIX in nonhuman primates. *Blood* 109 (4): 1414–1421.

2

Recent Development in *in vivo* Clinical Gene Therapy Platforms

John E. Murphy[1] *and Jane Owens*[2]

[1] *Research and Development, Arbor Biotechnologies, Cambridge, MA, USA*
[2] *Rare Disease Research Unit, Pfizer, Cambridge, MA, USA*

2.1 Introduction

Over the past decade significant evidence has accumulated for the safety and efficacy of adeno-associated virus (AAV) vectors providing complementary DNA (cDNA) gene replacement in both clinical trials and in clinical practice. This evidence is reviewed in this chapter, with emphasis on detailed case studies of more clinically advanced therapies. Recently, a number of trials have used AAV vectors to deliver gene editing modalities to patients. Background on these studies and initial clinical data are reviewed.

2.1.1 rAAV-cDNA Replacement Therapies

This section provides an update on rAAV-cDNA (or gene) replacement therapies that have been approved or are in clinical development at the time of writing. For those that were approved prior to 2022, brief case studies are provided in Section 2.1.1. For those not yet approved, a list of the candidate therapies is provided in Section 2.2, which includes tables documenting the cDNA (transgene) being delivered, the AAV capsid being used and the target tissue, in addition to the clinical indication for which the therapy is being developed and its stage of development.

Drug Development for Gene Therapy: Translational Biomarkers, Bioanalysis, and Companion Diagnostics, First Edition. Edited by Yanmei Lu and Boris Gorovits.

2.1.1.1 Introduction: Approved rAAV-cDNA Replacement Therapies

Six recombinant AAV (rAAV)-based cDNA replacement therapies have been approved for marketing. The first was Glybera (alipogene tiparvovec) which was approved by the European Medicines Agency (EMA) in October 2012. Glybera is an AAV1 vector that delivers cDNA encoding a functional lipoprotein lipase (LPL) transgene to skeletal muscle for patients with LPL deficiency due to gene mutations [1]. In December 2017, Luxturna (voretigene neparvovec) was the first rAAV-based gene therapy approved by the US Food and Drug Administration (FDA). It was also approved by the EMA in November 2018. Luxturna is an AAV2 vector that delivers cDNA encoding a functional retinal pigment epithelium 65 (RPE65) transgene to the eye for the treatment of Leber congenital amaurosis and Retinitis Pigmentosa in patients with confirmed biallelic RPE65-mutations [2]. In May 2019, Zolgensma (onasemnogene abeparvovec) was the second rAAV-based gene therapy approved by the FDA. It received conditional approval from the EMA in May 2020. Zolgensma is an AAV9 vector that delivers a cDNA encoding the human survival motor neuron (SMN) protein to treat pediatric patients less than 2 years of age with spinal muscular atrophy (SMA) with bi-allelic mutations in the survival motor neuron 1 (SMN1) gene [3]. A discussion in the form of a case study is provided below for each of these approved gene therapies.

In 2022, there were three more rAAV-based gene therapies approved in the EU and United Kingdom. Upstaza (eladocagene exuparvovec), developed by PTC Therapeutics, was approved for the treatment of Aromatic L-amino acid decarboxylase (AADC) deficiency [4]. Roctavian (valoctocogene roxaparvovec), developed by BioMarin, was approved for the treatment of hemophilia A [5] and hemgenix (etranacogene dezaparvovec), developed by uniQuire/CSL Behring) FDA Approves HEMGENIX for hemophilia B (cslbehring.com). There are two candidate AAV gene therapies at the "Pre-Registration" stage (ASCGT report 2022 asgct-pharma-intelligence-q1-2022-report.aspx), and therefore the promise of more gene therapy approvals in 2023:

- Delandistrogene moxeparvovec (AAVrh74-micro-dystrophin for DMD (Sarepta/Roche). Delandistrogene moxeparvovec was approved by the FDA in June 2023;
- Lenadogene nolparvovec, Lumevoq (AAV2-ND4 for Leber Hereditary Optic Neuropathy LHON), (Genethon, GenSight Biologics).

2.1.1.2 Glybera (alipogene tiparvovec), Marketed by uniQure

Although Glybera was famously recognized as the first rAAV cDNA replacement gene therapy to be approved, it was later associated with headlines stating that it was the most expensive drug in the world, and withdrawn from the market because it was a commercial failure.

Glybera was indicated for a very rare subset of patients with a lipoprotein lipase deficiency (LPLD). The disease is called Familial chylomicronemia syndrome

(FCS), which is a rare autosomal recessive disorder. Loss of function mutations in the LPL gene results in hyper-chylomicronemia, potentially life-threatening increases in the level of large fat-carrying particles (chylomicrons) in the blood after eating. The severity of FCS varies depending upon the degree of chylomicronemia, which fluctuates depending upon the amount of fat in an individual's diet. The main symptoms are abdominal pain, pancreatitis, eruptive xanthomas, and hepatosplenomegaly [6]. uniQure developed Glybera, or alipogene tiparvovec (AAV1-LPLS447X) gene therapy, to prevent complications and decrease the clinical morbidity of LPLD. The vector used to deliver the therapeutic gene was a non-replicating and non-integrating vector that comprises a protein shell derived from AAV serotype 1 (AAV1), the cytomegalovirus (CMV) promoter, a woodchuck hepatitis virus posttranscriptional regulatory element, and AAV2-derived inverted terminal repeats. The transgene in Glybera comprised a naturally occurring variant of the LPL gene that has higher enzyme activity than the endogenous version of the gene that encodes the enzyme. Glybera consisted of an engineered copy of the human LPL gene packaged with a constitutive promoter in a nonreplicating AAV1 vector, which has a known tropism for skeletal muscle cells. In this case, the skeletal muscle cells were targeted for the production and secretion of the LPL. The route of administration for Glybera was intra-muscular, given as a single onetime series of intramuscular injections in the leg muscles (single dose of 1×10^{12} genome copies/kg administered or multiple injections of 1.5×10^{12} genome copies) [7].

Glybera was tested in three interventional clinical studies conducted in the Netherlands and in Canada, where there are higher numbers of FCS patients. In all studies, the therapy proved to be well tolerated and no safety concerns were observed. Data from the clinical trials indicated that fat concentrations in blood were reduced after therapy in nearly all patients between 3 and 12 weeks after injection of Glybera. A single-dose administration of Glybera resulted in the long-term presence and biological activity of the protein in the injected muscle. Importantly, a single administration of Glybera resulted in a long-term, clinically important reduction in the occurrence and severity of acute pancreatitis episodes, the most debilitating complication of FCS(8) [8]. A case note review study showed a significant reduction in the frequency of hospital presentations for pancreatitis and severe abdominal pain related to the pancreas [9]. Glybera became the world's first million-dollar drug [10]. However, 5 years after it became the first gene therapy to win approval in Europe, uniQure did not apply for a renewal of the 5-year marketing authorization in 2017. Glybera's usage was extremely limited with no evidence for future increasing patient demand. "Glybera, the most expensive drug in the world, is to be withdrawn after commercial flop" – (pharmaphorum.com). With no pharmacologic treatment and management options that include an extremely restricted, very low-fat diet, along with avoidance of certain medications and alcohol, affected patients were left with no promise of affordable therapy. However, there is some hope on the horizon for FCS patients. With the higher prevalence of FGS patients in Quebec, which has been described due to a founder effect (when a small, isolated population of settlers, founders, expands over several generations leading to a high prevalence of a particular genetic trait), researchers in Quebec have been working to develop a more affordable gene therapy. Researchers at the National Research Council (NRC) in Canada are currently developing an affordable version of Glybera, using new viral vectors, in partnership with Virica Biotech who will provide enhanced manufacturing capabilities for producing this new AAV-LPL by deploying its custom Viral Sensitizer (VSE™) formulation to optimize a robust and scalable AAV vector manufacturing process – accelerating the next-generation gene therapy for FCS. Virica Biotech Receive Funding to Advance Gene Therapy | Phacilitate.

Virica Biotech – Virica Biotech announces collaboration with the Government of Canada in the press.

2.1.1.3 Luxturna (voretigene neparvovec-rzyl), Marketed by Spark Therapeutics

Luxturna claims the title of the first FDA-approved *in vivo* gene therapy for a genetic disease (December 2017) and also the first of its kind being approved in both the United States and Europe (November 2018). Also, the Phase 3 trial with Luxturna was the first randomized, controlled study completed in gene therapy for a genetic disease [11]. Subsequent headlines have primarily been focused on its cost and long-term effectiveness.

Luxturna is an AAV vector-based gene therapy indicated for the treatment of children and adults with Leber congenital amaurosis 2 (LCA2) or retinitis pigmentosa (RP) who have a confirmed biallelic RPE65 mutation. The RPE65 gene encodes the RPE65 enzyme which is produced in the retinal pigment epithelial (RPE) cells and converts all-trans-retinol to 11-cis-retinol, which subsequently forms the chomophore,11-cis-retinal during the visual cycle. The visual cycle is critical in phototransduction (the biological conversion of a photon of light into an electrical signal in the retina). Mutations in the RPE65 gene lead to reduced or absent levels of RPE65 isomerohydrolase activity, blocking the visual cycle and resulting in impairment of vision Package Insert – LUXTURNA (fda.gov). LCA2 and RP are degenerative diseases that progresses to near-total blindness. Spark Therapeutics developed Luxturna to improve light sensitivity, visual field, and navigational ability in LCA2 and RP patients with RPE65mutation–associated inherited retinal disease. Luxturna was designed to deliver a normal copy of the complementary DNA encoding the human RPE 65 kDa protein (hRPE65) to cells of the retina in these patients.

The vector used to deliver the therapeutic gene is a nonreplicating and non-integrating vector that comprises a protein shell derived from AAV serotype 2 (AAV2). The RPE65 cDNA has a modified Kozak sequence engineered at the translational start site and is under control of a hybrid chicken β-actin promoter with a cytomegalovirus enhancer. The recommended dose of Luxturna for each eye is 1.5×10^{11} vector genomes (vg), administered by subretinal injection. Each eye is injected on separate days within a close interval, but no fewer than 6 days apart Package Insert – LUXTURNA (fda.gov).

Luxturna's safety and efficacy was established in two Phase 1 clinical trials and an open-label, randomized phase 3 trial [11–13], in which a statistically significant and clinically meaningful difference was observed in the primary endpoint, mean bilateral multi-luminance mobility testing (MLMT) score change, between the intervention group and control group at 1 year. In addition, a statistically significant difference in two key secondary endpoints was observed, including full-field light sensitivity threshold testing averaged over both eyes and the mobility test score change for the first injected eye. A third secondary endpoint, change in visual acuity averaged over both eyes, was not statistically significant between the groups.

Spark developed the novel assay of mobility, the MLMT, for the phase 1 trial of Luxturna because traditional mobility metrics do not address the effects of illumination on speed and accuracy of navigation in a standardize and quantitative manner. MLMT integrates aspects of visual acuity, visual field, and light sensitivity into a quantifiable measure and comprises aspects of laboratory-based and real-world approaches. By increasing the complexity of the walking environment, and the effects of a range of different light levels on independent navigation, in a setting that maintains the controls more typical of a laboratory experiment, the MLMT was validated [14].

The durability of Luxturna was evaluated in Phase 3 patients at 3–4 years of follow-up. Improvements in ambulatory navigation, light sensitivity, and VF were observed. Overall, improvements were maintained up to 3–4 years, with ongoing observation [2]. A post-marketing trial called PERCEIVE was launched in 2019 in which patients dosed with Luxturna will be followed for up to 5 years to assess long-term safety and effectiveness in a real-world setting [15]. The primary objective is to collect adverse events. The secondary objectives include assessment of pregnancy outcomes and visual function over time. Overall, the safety and effectiveness of Luxturna observed in the PERCEIVE study, with up to 2 years of data are consistent with the findings of Luxturna clinical trials. Chorioretinal atrophy has been identified as a new adverse drug reaction, which so far has not been associated with loss of visual function. These events and overall longer-term safety will be further characterized in the ongoing study.

Luxturna garnered the headline "The Most Expensive U.S. Medicine Now Has an Official Sticker Price of $850,000." At $425,000 per eye with added surgery costs, questions arose about the cost-effectiveness of this therapy. The manufacturer (Spark Therapeutics) justified it by Luxturna's ability to restore vision in a small number of people with an IRD after a onetime injection in each eye and ongoing follow-up assessments suggesting a long-term benefit. Furthermore, to address reimbursement concerns, it was reported that Spark committed to offer rebates to health insurers if patients failed to meet specified vision thresholds 30–90 days and 30 months following Luxturna treatment [16, 17]. Five years on, there have been no additional retinal gene therapies approved despite numerous development programs since Luxturna was approved (see Table 2.1). Some of the trials have reported safety events while others have reported efficacy hurdles associated with what's considered a high bar for approval by the FDA [20].

2.1.1.4 Zolgensma (onasemnogene abeparvovec), Marketed by Novartis

Zolgensma claims the title of the first FDA-approved *in vivo* gene therapy for a genetic disease, which is administered systemically. It was also the first AAV vector administered systemically at 1.1E14 vg/kg, which was the highest dose of AAV at that time. Zolgensma is now approved in more than 40 countries and more than 2000 patients have been treated with Zolgensma globally across clinical trials, managed access programs, and in the commercial setting.

Zolgensma was developed to treat SMA which affects motor neurons. SMA is a rare, genetic neuromuscular disease and a leading genetic cause of infant death [21]. SMA is actually a group of inherited neuromuscular disorders characterized by loss of lower motor neurons in the spinal cord. The loss of lower motor neurons leads to progressive muscle weakness, muscle wasting and low muscle tone (hypotonia) which affects muscle functions responsible for breathing, swallowing and basic movement [22, 23].

SMA is caused by deletion or mutation of the *SMN1* gene, which encodes a protein known as SMN. This protein plays an important role in the functioning and maintenance of motor neurons. The *SMN2* gene is a paralog of *SMN1* and also encodes the SMN protein, which can compensate for the loss of the *SMN1* gene. However, most SMN protein produced by the *SMN2* gene is not functional, which means that the *SMN2* gene can only partially compensate for the loss of the *SMN1* gene. For this reason, an individual with SMA who has more copies of the *SMN2* gene will produce more functional SMN protein and may be better able to compensate for the loss of the *SMN1* gene, leading to less severe disease. Generally, more copies of *SMN2* are associated with milder SMA disease [23].

SMA1 onset typically occurs before 6 months of age and is characterized by progressive weakness, including loss of respiratory and swallowing functions. Infants with SMA1 do not achieve major development milestones and experience progressive respiratory failure and frequent pulmonary infections [24]. Untreated SMA, in infants with two copies of *SMN2* usually results in premature death or the need for permanent ventilation by 2 years of age. Loss of motor neurons cannot be reversed, so it is imperative to diagnose SMA and begin treatment,

Table 2.1 Clinical trials with AAV-mediated gene replacement therapy or inherited retinal dystrophies.

Indication	cDNA	Capsid	Sponsor	Clinical Phase	NCT
Achromatopsia	CNGA3 (Cyclic Nucleotide Gated Channel Subunit Alpha 3)	AAV2tYF	AGTC	Phase ½	NCT02935517
		AAV2/8	Meira GTx		NCT03758404
		AAV8	STZ eyetrial		NCT02610582
	CNGB3 (Cyclic Nucleotide Gated Channel Subunit Beta 3)	AAV2tYF	AGTC	Phase ½	NCT02599922
		AAV2/8	Meira GTx		NCT03278873
Choroideremia	CHM (REP1)	AAV2	NightstaRx/ Biogen	Phase 3	NCT03584165
	CHM (REP1)	AAV2	Spark	Phase ½	NCT02341807
			STZ eyetrial,		NCT02671539
			University of Oxford		
		4D-R100	4D Molecular Therapeutics,		NCT04483440
Leber congenital amaurosis 1 (LCA1)	GUCY2D	AAV5	Atsena Therapeutics	Phase ½	NCT03920007
Leber congenital amaurosis 2 (LCA2)	RPE65	AAV2	Meira GTx	Phase 1/2 completed; long-term follow-up	NCT02946879
Leber Hereditary Optic Neuropathy (LHON)	ND4 (NADH-ubiquinone oxidoreductase chain 4 protein)	AAV2	GenSight (Genethon)	Phase 3 REFLECT	NCT03293524
	ND1	AAV2	Byron Lam/NEI	Phase 1	NCT02161380

(Continued)

Table 2.1 (Continued)

Indication	cDNA	Capsid	Sponsor	Clinical Phase	NCT
	ND4	AAV2	Neurophth Therapeutics	Phase 3	NCT04912843
Retinitis Pigmentosa (RP)	PDE65B (rod cGMP phosphodiesterase6 β subunit)	AAV2/5	Coave Therapeutics	Phase ½	NCT03328130
	PDE65A (rod cGMP phosphodiesterase 6 α subunit)	AAV2	STZ eyetrial	Phase ½	NCT04611503
	RLBP1 (retinaldehyde binding protein 1) (CEPK850)	AAV8	Novartis	Phase ½	NCT03374657
X-linked RP	RPGR (retinitis pigmentosa GTPase regulator)	AAV5	Meira Gtx	Phase 3	NCT04671433
		AAV2tYF	AGTC	Phase 2/3	NCT04850118
		AAV8	NightstarRx/ Biogen	Phase 1/2	NCT03116113
		4D-R100	4D Molecular Therapeutics	Phase ½	NCT04517149
Retinoschisis (X-linked)	RS1 (retinoschisin 1)	AAV2tYF	AGTC	Phase ½	NCT02416622
		AAV8	National Eye Institute		NCT02317887

AAV2tYF: AAV2 with mutations to 3 surface tyrosine residues [18]; **4D-R100**: novel AAV called R100 developed by 4D Molecular Therapeutics [19].

including proactive supportive care, as early as possible to halt irreversible motor neuron loss and disease progression. This observation has led to the addition of SMA to the Recommended Uniform Screening Panel (RUSP) for newborn screening programs [25].

Zolgensma (Onasemnogene abeparvovec) is a recombinant self-complementary AAV9 containing a human *SMN* transgene under the control of a chicken β-actin promoter and cytomegalovirus enhancer, which together ensure rapid and sustained transcription of *SMN* messenger RNA. The AAV9 capsid and promoter/enhancer ensures body wide biodistribution, including into the brain and spinal cord via the blood-brain barrier. Zolgensma is marketed by Novartis Pharmaceuticals and was designed to directly address the genetic cause of the disease by replacing the function of the missing or non-working *SMN1* gene to halt disease progression through sustained SMN protein expression with a single, onetime IV infusion. Zolgensma is indicated for the treatment of pediatric patients less than 2 years of age with all types of SMA (with bi-allelic mutations in the SMN1 gene), Package Insert – ZOLGENSMA (fda.gov).

The efficacy and safety of Zolgensma in patients with SMA type 1 was assessed in a 2-year, open-label, phase 1 study (START). Data demonstrated that a onetime infusion of dose 6.7×10^{13} vg/kg of body weight in 3 patients or a therapeutic dose 1.1×10^{14} vg/kg in 12 patients resulted in longer survival than observed in historical controls. Two of three patients in the low-dose cohort and all patients in the therapeutic-dose cohort were alive without the need for permanent ventilation at 24 months of age [3]. Patients in the therapeutic-dose cohort also had significantly improved motor function and motor milestone achievements compared with historical cohorts. Eleven of twelve patients sat unassisted for ≥5 seconds, and two children pulled to a stand, stood, and walked independently. Regulatory approval was based on the data from this study. A long-term follow-up study is in progress but interim results were recently published in which a favorable safety profile was observed for up to 6.2 years after dosing, with sustained and durable efficacy (all patients were alive and were without the need for permanent ventilation [26]. This long-term follow-up study will monitor patients for 15 years (Long-Term Follow-up Study for Patients From AVXS-101-CL-101 – Full Text View – ClinicalTrials.gov).

Subsequently, Phase 3 (STR1VE) studies in different countries were initiated. The STR1VE-US study which enrolled 22 symptomatic SMA patients, provided further evidence for the safety and efficacy of Zolgensma [27]. There were two patient deaths reported during these Phase 3 studies. However, the deaths (due to respiratory complications) were deemed unrelated to the intervention [28]. Analysis of postmortem tissue samples from these patients enabled an assessment of the vector biodistribution. This was the first study to demonstrate in humans the widespread biodistribution of vector genomes and transgene expression throughout the CNS and peripheral organs after intravenous delivery of

AAV9-mediated gene therapy to treat a neurodegenerative disease. Zolgensma achieved an effective distribution, transduction and expression throughout the CNS and restored SMN expression [28].

Novartis is also developing an intrathecally (IT) administered version of Zolgensma called OAV-101 IT which uses a lower dose of the AAV vector than currently approved for IV administration. This has the potential to broaden access to older patients with SMA. A Phase 1 study (called STRONG) with OAV-101 IT led to significant increases in HFMSE (Hammersmith Functional Motor Scale-Expanded) scores and a clinically meaningful response in older patients ≥2 years and <5 years old with SMA Type 2. AveXis presents AVXS-101 IT data demonstrating remarkable increases in HFMSE scores and a consistent clinically meaningful response in older patients with SMA Type 2 | Novartis. In October 2019, the FDA partially suspended this trial because of findings in animals that showed dorsal root ganglia (DRG) mononuclear cell inflammation, which was sometimes accompanied by neuronal cell body degeneration or loss. The hold was lifted 2 years later after review of data from Novartis' comprehensive nonclinical toxicology study in nonhuman primates (NHP) that addressed all issues identified, including questions of DRG injury following IT administration. Following this announcement, Novartis started a Phase 3 trial (called STEER) in patients with SMA Type 2 aged between 2 and 18 years (see Table 2.2), which will build upon the Phase 1/2 STRONG study results.

With a $2.1 million price tag, Zolgensma is often described as one of the most expensive drugs in the world. However, Novartis has justified this based on its dramatic effects in stopping SMA progression and helping babies reach milestones such as sitting up and chewing that they otherwise would not reach. The high cost has been debated on multiple points, including exacerbating inequalities in access to medicines for patients living in different countries, together with long-term effectiveness of the gene therapy [30]. Zolgensma has been in the news more recently when two children with SMA died after receiving Zolgensma. The deaths resulted from acute liver failure, 5–6 weeks posttreatment and about one to 10 days after doctors began to taper corticosteroids. Systemic corticosteroids are administered before and after the gene therapy infusion to dampen the immune response to the AAV9 capsid, which would otherwise trigger liver enzyme elevations. Indeed, the label for Zolgensma includes a black box warning that the drug "can increase liver enzyme levels and cause acute serious liver injury or acute liver failure." Nevertheless, Novartis acknowledged these are the first fatal cases reported. Novartis Confirms Deaths of Two Patients Treated with Gene Therapy Zolgensma (genengnews.com).

Table 2.2 Clinical trials with AAV-mediated gene replacement therapy for brain/CNS disease.

Indication	scDNA	Capsid	Sponsor	Clinical Phase	NCT
Spinal Muscular Atrophy (SMA)	SMN1	AAV9	Novartis	Phase 3 (IV in pre-symptomatic SMA) Phase 3b (IV in symptomatic SMA) Phase 1 (IT) Phase 3 (IT in type 2 SMA)	NCT03505099 NCT0485187 NCT03381729 NCT05089656
Giant Axonal Neuropathy	GAN	AAV9	Taysha Gene Therapies	Phase 1	NCT02362438
Canavan disease	ASPA	AAV9 AAV-Olig-001	Aspa Myrtelle	Phase ½ Phase ½	NCT04998396 NCT04833907
Ceroid	CLN3	AAV9	Amicus	Phase ½	NCT03770572
Lipofuscinoses	CLN6	AAV9	Amicus	Phase ½	NCT02725580
CLN	CLN7	AAV9	University of Texas Southwestern Medical Center	Phase ½	NCT04737460
	CLN5	AAV9	Neurogene		NCT05228145
FTD with GRN mutations (FTD-GRN)	GRN	AAV9	Prevail	Phase ½	NCT04408625
Parkinson's disease	GBA1	AAV9	Prevail	Phase ½	NCT04127578
Huntington's disease	HTT	AAV5	UniQure	Phase ½	NCT04120493/NCT05243017

AAV-Olig001: novel AAV comprising a chimeric mixture of AAV1, 2, 6, 8, and 9 [29].

2.1.2 Introduction: rAAV-cDNA (gene) Therapy Candidates in Clinical Development

There are many rAAV-based cDNA replacement therapies that have reached the clinical development stage, which are not yet approved. These candidate AAV-mediated gene therapies have been developed to target a variety of diseases that affect the eye, brain, heart, liver, and skeletal muscle. These target tissues have many cells that need to receive the therapeutic gene for the therapy to be effective and therefore need to receive appropriate doses of the gene therapy product (measured in vector genomes, vg). Gene therapies that are administered to a local area of a smaller organ such as the eye, require a smaller dose compared to gene therapies for skeletal muscle diseases, which require systemic administration for delivery to all muscles and therefore require much larger doses (Figure 2.1).

In the next section, candidate gene therapies are grouped according to the tissue type/system they target, which includes the eye, brain/central nervous system, heart, liver, and skeletal muscle. It is noteworthy that although many AAV-mediated gene therapies target the liver, the target indications are associated with

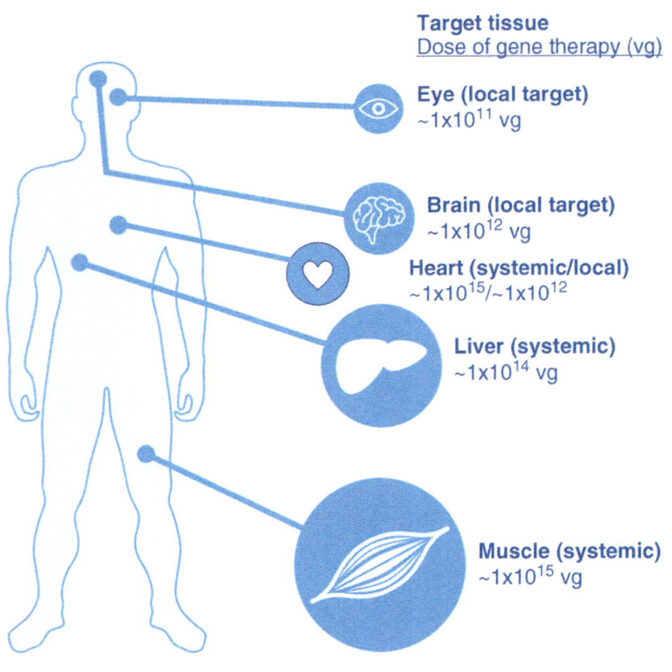

Figure 2.1 Target organs for AAV-mediated gene therapies in clinical development. Target tissue size and accessibility impacts dose. Source: Modified from Tretiakova 2019, Scientific American [31].

pathology in non-liver tissues. In these cases, the liver is used as the organ that produces the protein encoded by the transgene.

2.1.2.1 AAV-Gene Replacement Clinical Trials for the Eye

The approval of Luxturna for LCA2 and RP patients with a biallelic RPE65 mutation has resulted in a large number of additional gene replacement therapy clinical trials for IRD patients. The diseases targeted are Achromatopsia, Choroideremia, Leber congenital amaurosis types 1 and 2 (LCA1, LCA2), Leber Hereditary Optic Neuropathy (LHON), RP, and X-linked RP.

2.1.2.2 Clinical Trials for Heart Disease

There are several AAV-mediated gene therapy trials targeting the heart. However, the only trials in which a cDNA replacement approach is used are shown in Table 2.3. The other clinical trials use AAV to deliver transgenes which express proteins that operate as bypass mechanisms. For example, delivery of SERCA2a (sarco/endoplasmic reticulum Calcium-ATPase) with an AAV1 vector was developed to treat heart failure. SERCA2a regulates cardiac contraction and relaxation by its role in controlling the level of Ca2+ in the cytosol and SR in the cardiomyocyte [32]. It was observed in the failing heart in animal models and in humans that there is a reduction in expression of SERCA2a at both the mRNA and protein level [33]. This led to nonclinical studies assessing the effectiveness of increased levels of SERCA2a in ameliorating the heart failure phenotype and then to the human clinical trials (CUPID 1 and 2) with a single intracoronary infusion of AAV1-SERCA in HFrEF patients. Even though the CUPID trials failed to demonstrate an improvement in outcomes in HFrEF patients, another trial is underway

Table 2.3 Clinical trials with AAV-mediated gene replacement therapy for heart disease.

Indication	cDNA	Capsid	Sponsor	Clinical Phase	NCT
Danon Disease	LAMP2B (lysosome-associated membrane protein 2B)	AAV9	Rocket Pharmaceuticals	Phase 1	NCT03882437
Friedreich's Ataxia	FXN (Frataxin)	AAVrh10 AAVrh10	Lexeo Therapeutics Weill Medical College of Cornell University	Phase 1/2 Phase 1	NCT05445323 NCT05302271

with the same vector and route of administration, but with a higher dose (CUPID 3), Modulation of SERCA2a of Intra-myocytic Calcium Trafficking in Heart Failure With Reduced Ejection Fraction – Full Text View – ClinicalTrials.gov.

2.1.2.3 Clinical Trials for Hematologic and Metabolic Disease (Targeting the Liver)

A large number of trials are ongoing in which liver is the target organ for AAV delivery (Table 2.4). In certain instances, such as hemophilia A and B and Gaucher disease, the liver is used to produce and secrete an enzyme that functions outside of the liver. In other instances, PKU, Wilson's disease, AAV delivery replaces a defective metabolic enzyme which functions within the liver.

2.1.2.4 Clinical Trials for Skeletal Muscle

Treating skeletal muscle disease that presents as a muscular dystrophy/myopathy (Table 2.5).

2.1.3 Introduction: rAAV-as a Vehicle for in vivo Gene Editing

One of the key limitations of canonical AAV delivery of a cDNA transgene is that the transgene exists in an episomal form that may not persist over time, particularly when delivered to very young patients with actively growing organs [37]. Additionally, many genetic diseases are caused by dominant mutations that cannot be corrected simply by ectopically expressing a wild-type cDNA transgene. To overcome these limitations, numerous gene editing approaches have been attempted, with strong proof of principle in animal models, and in a few cases initiation of clinical trials [38]. For the discussion in this section, gene editing refers to any method designed to alter the chromosomal DNA *in vivo* in a targeted fashion. These include passive methods that rely solely on cell-based DNA repair machinery, such as homologous recombination methods, and methods actively catalyzed by site-directed nucleases, such as zinc finger nucleases (ZFN) and CRISPR-Cas systems.

2.1.3.1 Non-nuclease Mediated Methods

One method by which chromosomal DNA can be altered to correct a gene or insert a novel sequence is through homologous recombination, an approach that is relatively inefficient and requires cells to be in a mitotic state [39]. AAV is used as a DNA donor to carry a sequence with the desired chromosomal modification flanked by homology arms, which target this sequence to a precise chromosomal location. Recombination between the donor arms and the homologous endogenous sequence leads to replacement of the endogenous sequence with that templated by the AAV vector [40, 41]. While this process can take place in the absence of exogenous nucleases, it can be enhanced by nucleases that create double-strand breaks at the site of homology, as homology-directed repair (HDR) is one pathway used by cells to repair double-stranded breaks [42]. Homologous recombination

Table 2.4 Clinical trials with AAV-mediated gene replacement therapy hematologic and metabolic disease.

Indication	cDNA	Capsid	Sponsor	Clinical Phase	NCT
Hemophilia B	Factor IX	AAV5 AAV-Spark100 AAVS3 AAV8	uniQure/CSL Behring Spark/ Pfizer Freeline Takeda/Shire	Phase 3 Phase 3 Phase 1/2 Phase 1/2	NCT03569891 NCT03307980 NCT05164471 NCT01687608
Hemophilia A	Factor VIII	AAV5 AAV2/6 AAV-LK03 AAVhu37 AAV-Spark200 AAV2/8 AAV8	BioMarin Pfizer/ Sangamo Roche/ Spark Bayer/ Ultragenix Spark UCL-St Jude Takeda/Shire	Phase 3 Phase 3 Phase 3 Phase 1/2 Phase 1/2 Phase 1 Phase ½	NCT03370913 NCT04370054 NCT03003533 NCT03588299 NCT03734588 NCT03001830 NCT03370172
Gaucher disease	GBA1	AAV9 AAVS3	PrevailFreeline	Phase 1/2 Phase 1/2	NCT0532494/NCT04411654 NCT05324943
Fabry disease	Alpha-GLA	AAV2/6 AAVS3 4D-C102	Sangamo Freeline 4D Molecular Therapeutics	Phase 1/2 Phase 1/2 Phase 1/2	NCT04046224 NCT04040049 NCT04519749
PKU	PAH	AAV5	BioMarin	Phase 1/2	NCT04480567
Wilson disease	ATP7B	AAV3B AAV9	Vivet/Pfizer Ultragenyx	Phase 1/2 Phase 1/2	NCT04537377 NCT04884815
Crigler-Najjar syndrome	UGT1A1	AAV8	Genethon	Phase 1/2	NCT03466463

(Continued)

Table 2.4 (Continued)

Indication	cDNA	Capsid	Sponsor	Clinical Phase	NCT
Pompe Disease	GAA (acid alpha-glucosidase)	AAV8	Audentes (Astellas)	Phase 1/2	NCT04174105
	GAA	AAV8	Asklepios Biopharmaceutical	Phase 1/2	NCT03533673
	GAA	AAV9	GeneCradle Therapeutics	Phase 1/2	NCT05567627
	GAA	AAVrh74	Spark Therapeutics (Roche)	Phase 1/2	NCT04093349
Glycogen storage disease 1A	Glucose-6-phosphatase	AAV8	Ultragenyx	Phase 1/2	NCT05139316
MPS I MPS II MPS IIIA	IDUA IDS SGSH	AAV9 AAV9 AAV9 AAVrh10	Regenxbio Regenxbio	Phase 1/2 Phase 2/3	NCT03580083 NCT03566043 NCT04571970 NCT02716246 NCT03612869
MPS VI	ARSB	AAV9 AAV2/8	Ultragenyx (Abeona) Lysogene Esteve Fondazione Telethon	Phase 1/2 Phase 1/2 Phase 1/2	EudraCT2015-000359-26 NCT03173521
Ornithine Transcarbamylase (OTC) Deficiency	OTC	AAV8	Ultragenyx	Phase 3	NCT05345171
Familial Hypercholesterolemia	LDLR	AAV8	Regenxbio	Phase 1/2	NCT02651675

AAV-SPARK100: engineered capsid derived from AAVrh74 [34]; **AAVS3**: engineered capsid derived from AAV8 and AAV3B [35]; AAV-LK03: AAVhu37 AAV-SPARK200: **4D-C102**: novel AAV variant identified through 'directed evolution' in non-human primates [36].

Table 2.5 Clinical trials with AAV-mediated gene replacement therapy for skeletal muscle disease.

Indication	cDNA	Capsid	Sponsor	Clinical Phase	NCT
Duchenne muscular dystrophy (DMD)	Mini/micro-dystrophin	AAV9 AAVrh74 AAV9 AAV9	Pfizer Sarepta Therapeutics Solid Biosciences Genethon	Phase 3 Phase 3 Phase 1/2 Phase 1/2/3	NCT03362502 NCT03375164 NCT03368742 EU Clinical Trials Register. Clinical trial 2020-002093-27
Limb Girdle Muscular Dystrophy 2E (LGMD2E)	β-sarcoglycan	AAVrh74	Sarepta Therapeutics	Phase ½	NCT03652259
Limb Girdle Muscular Dystrophy 2I (LGMD2I)	FKRP (Fukutin-related protein)	AAV9	Atamyo Therapeutics	Phase ½	NCT05224505
X-Linked Myotubular Myopathy (XL-MTM)	MTM1 (Myotubularin)	AAV8	Audentes/Astellas	Phase ½	NCT03199469

can be used to repair a gene at the endogenous location in the genome or to insert the desired therapeutic gene at a safe harbor locus, such as albumin.

One application of passive homologous recombination is the GeneRide technology developed by LogicBio, which is currently in clinical trials for pediatric patients with severe methylmalonic acidemia (MMA) [43]. MMA is an inborn error of metabolism that is caused by mutations in the methylmalonyl-CoA mutase gene (MMUT). MMA is characterized by life-threatening metabolic crises and multiorgan disease which can ultimately be lethal [44]. While addition of a wild-type MMUT cDNA could overcome the recessive defect, the need to treat very young patients with actively growing livers would likely lead to dilution of the episomal transgene overtime as has been seen with attempts to treat a neonatal mouse model [37, 44]. Furthermore, preclinical studies have shown treatment of very young mice with an AAV construct with an active liver-specific promoter can lead to hepatocellular carcinoma [45].

To overcome these limitations, investigators used a promoterless integrating cassette delivered by AAV. Specifically, the cassette contains homology arms for the albumin locus, a P2A ribosomal skipping sequence and a codon-optimized MMUT sequence. The construct was packaged into either AAV8 or AAVDJ serotypes and used at various doses to treat mice within 1 day of birth. Homologous recombination directs integration of the construct into the albumin locus adjacent to the albumin stop codon. Following integration, a transcript is made under control of the albumin promotor, which is then translated into 2 proteins, an albumin-2A fusion protein and a functional MMUT protein [43].

As mentioned above, homologous recombination is a relatively infrequent event but in this case, the treatment outcome was helped by both the young age of the mice and a selective advantage of growth for corrected hepatocytes. Treated MMUT knockout mice followed over time show a significant increase in corrected protein product between 2 months posttreatment and 13–15 months posttreatment. This increase occurred against the background of significant loss of AAV genome copy number over this time. The level of integrated genomes (>10%) increased over time in MMUT-treated mice, while it remained below 1% in treated wild-type mice, consistent with a selective advantage in mutant mice. The albumin 2A fusion formed a secreted biomarker of integration that also showed a dose-dependent increase over time. Examination of liver by reflectance in situ hybridization (RISH) also showed an increase in clusters of MMUT-positive cells over time. Finally, this editing outcome led to an increase survival and decrease in the MMA metabolite in treated mice [43].

These results provide the rationale for a clinical trial NCT04581785, which is currently in Phase 1 enrolling patients 6 months to 12 years of age. Following delivery of the albumin-2A-MMUT construct encapsulated with LK03 serotype rAAV, circulating albumin-2A fusion protein was observed in the initial treated patients, a biomarker of successful editing [46]. The positive data allowed extension into younger children where evidence of thrombotic microangiopathy (TMA) was seen resulting in a clinical hold which has since been lifted following inclusion of additional mitigations in protocol.

Homology Medicines is also using a form of non-nuclease mediated homologous recombination in their HMI-103 product for treatment of Phenylketonuria (PKU). Rather than integration into the albumin locus, the AAV construct is designed to be integrated into the native locus of the causal PAH gene through homologous recombination, replacing the mutant PAH sequence with a functional allele. However, rather than being driven by the endogenous promoter, the gene expression is driven by a liver-specific promoter, which is contained in the AAV cassette. This design allows functional expression of the PAH gene from vector copies that remain episomal, which are likely to be significant given the relatively low frequency of homologous recombination. A novel capsid, HSC15 is used to deliver the cassette. A Phase 1 trial NCT05222178 was initiated in 2022.

2.1.3.2 Nuclease-mediated Homology Directed Repair

Early clinical studies using active nucleases to target transgenes to the albumin locus have been performed by Sangamo to potentially provide treatments for

MPS1, MPS2, and hemophilia B [47]. This approach is similar to the GeneRide approach in that the therapeutic gene is targeted for integration into the albumin locus. The key difference is that an active nuclease is used to enhance the frequency of homologous recombination (by homology-directed repair) [48]. Through protein engineering, zinc finger domains can be designed to specifically bind to 3 nucleotide sequences, thus fusing 6 engineered zinc fingers can provide specificity for 18 nucleotides. The zinc finger is then fused to a FOK1 endonuclease. Binding of the zinc finger-Fok1 fusion results in DNA nicking, however delivery of a pair of zinc finger-FOK1 nuclease fusions, one binding to a top strand and one binding to an immediately adjacent bottom strand results in a highly specific double-stranded DNA break targeting the albumin locus in this case. An AAV construct with arms homologous to albumin flanking a cDNA encoding the therapeutic gene product is delivered and will integrate at the site of the break with a high frequency. Successful integration leads to production of a fusion mRNA expressing the albumin and the therapeutic transgene driven by the strong albumin promoter in the liver.

The feasibility of this approach was tested in preclinical disease models with encouraging results. AAV8 vectors expressing the human FIX cassette with the albumin homology arms and separate vectors expressing the ZFN pair specific for the albumin gene, were delivered to mice, and high levels (60% of normal) of human FIX were secreted to the plasma. This high level occurred despite the fact that the level of albumin-FIX mRNA fusion was only 0.5%. The design was ultimately optimized to express each ZFN on a separate vector which led to a three vector product with 2 ZFNs and a DNA donor. This configuration of three components was introduced into the clinic for treatment of severe hemophilia B (NCT02695160). A similar study was performed in a mouse model of Hurlers Syndrome (MPS1) in which the IDUA gene, which is defective in MPS1, is designed to be inserted into the albumin locus by ZFN-mediated HDR. Delivery of each component at $1.2–1.5\times10^{11}$ vg/mouse led to sustained transgene expression, reduction of the pathogenic glycosyl amino glycans and amelioration of neuropathology [49]. Similar promising results were seen when the corrective iduronate-2-sulfatase (I2S) gene was delivered along with albumin targeting zinc finger nucleases to an MPS II mouse model [50].

Despite the promising preclinical data, clinical trials using ZFN targeting in MPS 1, MPS II, and HemB were unsuccessful [47]. The studies were conducted in parallel, allowing dose escalation information from one study to inform dose selection for the other studies. The studies enrolled 9 MPS II patients at doses of 5×10^{12} vg/kg ($N = 2$), 1×10^{13} vg/kg ($N = 2$), 5×10^{13} vg/kg ($N = 5$), 3 MPS I patients at doses of 1×10^{13} vg/kg ($N = 1$), 5×10^{13} vg/kg ($N = 2$) and 1 HemB patient at a dose of 5×10^{13} vg/kg. In contrast to the mouse studies in which the AAV8 serotype was used, in these studies, the AAV6 serotype was used. The treatment was well tolerated with no serious treatment-related adverse events. While some evidence of editing, as determined by the presence of albumin-transgene fusion was seen, the expression of transgene was variable and generally transient. 2 MPSII and 1 MPS I patient treated with the 1×10^{13} dose showed the albumin-transgene fusion mRNA. There was also evidence of I2S expression in one patient treated with the highest dose in the MPS II study; however, this was transient and reduction may have coincided with transaminitis. None of the patients who had previously been treated with enzyme replacement therapy were able to stop therapy without a rise in pathogenic glycosylaminogycans. Similarly, the HemB patient was not able to cease FIX replacement. The reasons for lack of translation between the preclinical and clinical studies is unclear. The need to deliver three vectors to a cell for successful transgene insertion and expression may pose a barrier to successful dose translation between species.

2.1.4 Nuclease-mediated Gene Disruption following AAV Delivery

While there is extensive pre-clinical experience using AAV to deliver programable nucleases such as CRISPR-Cas9 to disrupt gene function for research studies [38], the clinical experience is limited. The most advanced clinical trial is being conducted for treatment of Lebers Congenital Amaurosis (LCA), a retinal degenerative disease, caused by a mutation in the CEP290 gene which leads to aberrant splicing of the mRNA and aberrant protein production. The therapeutic is known as EDIT-101 and consists of a single vector in which the Staphylococcus aureus Cas9 (SaCas9) nuclease gene is driven by a photoreceptor-specific promoter, GRK1 and two guide RNAs (gRNA) utilize expression U6 promoters to drive expression [51]. The guides bind to intronic sequences which flank the causative IVS26 mutation and direct Cas9 cleavage to these sites. This construct is encapsidated with an AAV5 capsid. The EDIT-101 vector was used to treat multiple explanted retinal samples from a patient and the gene-edited outcome was determined by a sequencing method known as Uditas. Several editing outcomes were observed including productive edits such as excision of the region containing the cryptic splicing site and inversion of the region surrounding that site, which will result in the cryptic splice no longer being recognized. Unproductive edits included small insertions or deletions (indels) at the guide target site which do not impact the aberrant splicing event. Interestingly, another common nonproductive splicing event was insertion of AAV sequences at the guide target site. This occurs at a frequency of about 5–10% of total edits and has been seen in other systems indicative of a tendency of AAV to integrate at double-strand breaks in DNA. Overall, in the explants, the editing rate was $41.7 + 15.9\%$ with the productive rate (excision and inversion) being about one-third of this at 16.6%. While the editing frequency at the target site correlates with efficacy, a second key concern is off-target editing creating mutations elsewhere in the genome. The researchers used a variety of techniques including in silico prediction, digenome seq on purified genomic DNA and GuideSeq to identify potential off-targets and then used directed sequencing to assess the frequency of off-target events. Directed sequencing revealed that the off-target events occurred at a frequency below the limits of detection.

Given the exquisite specificity of CRISPR_Cas9 for sequences with full homology to guide RNA spacer regions, studies in preclinical species can be difficult. To overcome this, a transgenic mouse model in which CEP290 IVS26 was knocked into the mouse genome was used. Levels of productive editing following subretinal injection were measured with a minimal target of 10% productive editing (based on natural history). Based on this a therapeutic range of 3×10^{11} to 3×10^{12} vg was established. While mice can be engineered to express the human sequence, the anatomy of the eye is different than in humans and thus nonhuman primate studies were required to validate dosing route. The NHP studies verified that the GRK-1 promoter restricted expression to photoreceptor cells (expression of nuclease within the target tissue is an important safety requirement) and also predicted that therapeutic levels of productive editing could be obtained with doses between 7×10^{11} and 3×10^{12} vg/mL.

This robust preclinical package has led to a Phase 1/2 clinical trial known as the Brilliance Trial (NCT03872479). This is a dose escalation study of EDIT-101 in which the product is injected sub-retinally initially in cohorts of adult patients and once safety is established in juvenile patients. The adult portion of the dose escalation has been completed with no dose-limiting toxicities or serious adverse events.

A clinical trial was recently initiated to use AAV-mediated delivery of SaCas9 and guides specific for sequences in the HIV genome for the treatment of HIV (NCT03872479). While anti-retroviral therapy (ART) is effective at controlling viremia, it does not impact integrated proviruses and thus removal of therapy results in renewed viremia. Gene editing holds the potential to remove the underlying proviral sequences. Preclinical data to support this approach comes from NHP infected with SIV [52]. Following infection animals were put on ART and viral load was measured. Approximately 5 months after infection an AAV9 vector expressing saCas9 and guides targeting the SIV long terminal repeat or the SIV gag gene was administered iv at 10^{13} gc/kg and necropsies were performed at 3 weeks posttreatment. The AAV9 vector was widely distributed with evidence of vector in spleen, lymph node, and in CD4 T cells. Evidence of editing, as determined by deletions between the guide-targeted sequences in the 5′ or 3′ LTR and gag was seen in multiple tissues. Comparison of lymph node posttreatment with a biopsy sample pretreatment showed editing at variable levels, including 95% editing in one sample. ART was not discontinued posttreatment so the impact of this treatment on viremia is not clear.

2.1.5 Challenges and Opportunities with AAV as a Delivery Vehicle for Nuclease-Mediated Gene Editing

It is important to acknowledge the challenges unique to AAV-mediated delivery of targeted nucleases. The first challenge is that components required for *in vivo* gene editing often exceed the capacity of packaging for AAV. A minimal configuration includes a nuclease, with control sequences such as a promoter and polyadenylation sequence and a guide RNA often expressed from a separate Pol III promoter. For the canonical SpCas9, the nuclease coding gene alone is 4.2 kb leaving minimal room for control regions and separate guide cassettes. This has led to a search for smaller nucleases through both genomic discovery and protein engineering and numerous small nucleases which are compatible with AAV have been identified [53, 54]. However, new precision editing modalities, such as base editing and prime editing, which require nuclease-effector protein fusions, will require further identification of smaller nucleases to be reliably packaged by AAV [55].

The persistence of AAV, which is one of the key advantages for replacement therapies, is a potential issue for AAV delivery of gene editing nucleases for two reasons. First, the nucleases are of nonhuman origin and thus persistent expression may generate an immune response. A second concern is the potential accumulation of rare off-target events over time. As mentioned above, the safety of guides used in clinical trials is highly vetted but the ability to detect off-targets is limited to the number of cells analyzed and the length of exposure to treatment. The consequences of persistence in a large number of cells are unknown.

Finally, AAV integration at the sites of double-stranded breaks has been described in multiple preclinical studies. In the case of EDIT-101 [51], this accounted for approximately 10% of all edits but was seen as a positive outcome as the integrated AAV could disrupt the aberrant splicing event underlying the pathology of LCA.

References

1 Haddley, K. (2013). Alipogene tiparvovec for the treatment of lipoprotein lipase deficiency. *Drugs Today* 49: 161–170.

2 Maguire, A.M., Russell, S., Chung, D.C. et al. (2021). Durability of voretigene neparvovec for biallelic RPE65-mediated inherited retinal disease: phase 3 results at 3 and 4 years. *Ophthalmology* 128 (10): 1460–1468.

3 Mendell, J.R., Al-Zaidy, S., Shell, R. et al. (2017). Single-dose gene-replacement therapy for spinal muscular atrophy. *N. Engl. J. Med.* 377 (18): 1713–1722.

4 Keam, S. (2022). Eladocagene exuparvovec: first approval. *Drugs* 82 (13): 1427–1432.

5 Blair, H.A. (2022). Valoctocogene roxaparvovec: first approval. *Drugs* 82 (14): 1505–1510.

6 Scott, L.J. (2015). Alipogene tiparvovec: a review of its use in adults with familial lipoprotein lipase deficiency. *Drugs* 75 (2): 175–182.

7 Bryant, L.M., Christopher, D.M., Giles, A.R. et al. (2013). Lessons learned from the clinical development and market authorization of Glybera. *Hum. Gene Ther. Clin. Dev.* 24 (2): 55–64.

8 Libby, A.E. (2014). An update on gene therapy for the treatment of lipoprotein lipase deficiency. *Orphan Drugs: Res. Rev.* 2014 (1): 47–54.

9 Gaudet, D., Stroes, E., Méthot, J. et al. (2016). A long-term (up to 6 years) retrospective analysis of gene therapy with alipogene tiparvovec and its effect on lipoprotein lipase deficiency (LPLD)-induced pancreatitis. *Hum. Gene Ther.* 27: 916–925.

10 Evans, C.H. (2019). The vicissitudes of gene therapy. *Bone Joint Res.* 8 (10): 469–471.

11 Russell, S. (2017). Efficacy and safety of voretigene neparvovec (AAV2-hRPE65v2) in patients with RPE65-mediated inherited retinal dystrophy: a randomised, controlled, open-label, phase 3 trial. *Lancet* 390: 849–860.

12 Maguire, A.M., High, K.A., Auricchio, A. et al. (2009). Age-dependent effects of RPE65 gene therapy for Leber's congenital amaurosis: a phase 1 dose-escalation trial. *Lancet* 374 (9701): 1597–1605.

13 Bennett, J., Wellman, J., Marshall, K.A. et al. (2016). Safety and durability of effect of contralateral-eye administration of AAV2 gene therapy in patients with childhood-onset blindness caused by RPE65 mutations: a follow-on phase 1 trial. *Lancet* 388 (10045): 661–672.

14 Chung, D.C., McCague, S., Yu, Z.F. et al. (2018). Novel mobility test to assess functional vision in patients with inherited retinal dystrophies. *Clin. Exp. Ophthalmol.* 46 (3): 247–259.

15 Fischer, M.D., Maier, R., Suhner, A. et al. (2022). PERCEIVE study report: Real-world safety and effectiveness of voretigene neparvovec. *Invest. Ophthalmol. Vis. Sci.* 63 (7).

16 Feuerstein, A.G. and Garde, D. (2018). *Spark Prices its Gene Therapy as Most Expensive U.S. medicine – but with Plans to Ease Cost Concerns*. STAT: Daily News Coverage of the Biopharma, Health Tech, and Life Sciences Industries.

17 Darrow, J.J. (2019). Luxturna: FDA documents reveal the value of a costly gene therapy. *Drug Discov. Today* 24 (4): 949–954.

18 Markusic, D.M., Herzog, R.W., Aslanidi, G.V. et al. (2010). High-efficiency transduction and correction of murine hemophilia B using AAV2 vectors devoid of multiple surface-exposed tyrosines. *Mol. Ther.* 18 (12): 2048–2056.

19 Kotterman, M., Beliakoff, G., Croze, R. et al. (2021). *Directed Evolution of AAV Targeting Primate Retina by Intravitreal Injection Identifies R100, a Variant Demonstrating Robust Gene Delivery and Therapeutic Efficacy in Non-Human Primates*. bioRxiv 2021.06.24.449775.

20 Kay, D.M. (2022). Experiences with gene therapy my perspective on referring patients for treatment with Luxturna. *Ophthalmol. Manag.* 26: 43–45.

21 Finkel, R.S., McDermott, M.P., Kaufmann, P. et al. (2014). Observational study of spinal muscular atrophy type I and implications for clinical trials. *Neurology* 83 (9): 810–817.

22 Sugarman, E.A., Nagan, N., Zhu, H. et al. (2012). Pan-ethnic carrier screening and prenatal diagnosis for spinal muscular atrophy: clinical laboratory analysis of >72,400 specimens. *Eur. J. Hum. Genet.* 20 (1): 27–32.

23 Lorson, C.L., Rindt, H., and Shababi, M. (2010). Spinal muscular atrophy: mechanisms and therapeutic strategies. *Hum. Mol. Genet.* 19 (R1): R111–R118.

24 D'Amico, A., Mercuri, E., Tiziano, F.D., and Bertini, E. (2011). Spinal muscular atrophy. *Orphanet J. Rare Dis.* 6: 71.

25 Kay, D.M., Stevens, C.F., Parker, A. et al. (2020). Implementation of population-based newborn screening reveals low incidence of spinal muscular atrophy. *Genet. Med.* 22 (8): 1296–1302.

26 Mendell, J.R., Al-Zaidy, S.A., Lehman, K.J. et al. (2021). Five-year extension results of the phase 1 START trial of onasemnogene abeparvovec in spinal muscular atrophy. *JAMA Neurol.* 78 (7): 834–841.

27 Day, J.W., Finkel, R.S., Chiriboga, C.A. et al. (2021). Onasemnogene abeparvovec gene therapy for symptomatic infantile-onset spinal muscular atrophy in patients with two copies of SMN2 (STR1VE): an open-label, single-arm, multicentre, phase 3 trial. *Lancet Neurol.* 20 (4): 284–293.

28 Thomsen, G., Burghes, A.H.M., Hsieh, C. et al. (2021). Biodistribution of onasemnogene abeparvovec DNA, mRNA and SMN protein in human tissue. *Nat. Med.* 27 (10): 1701–1711.

29 Powell, S.K., Khan, N., Parker, C.L. et al. (2016). Characterization of a novel adeno-associated viral vector with preferential oligodendrocyte tropism. *Gene Ther.* 23 (11): 807–814.

30 Nuijten, M. (2022). Pricing Zolgensma – the world's most expensive drug. *J. Mark. Access Health Policy* 10 (1): 2022353.

31 Tretiakova, A. (2019). Realizing the promise of gene therapy through Collaboration and partnering: Pfizer's view. *Sci. Am.* https://api.semanticscholar.org/CorpusID:221709734.

32 Greenberg, B., Yaroshinsky, A., Zsebo, K.M. et al. (2014). Design of a phase 2b trial of intracoronary administration of AAV1/SERCA2a in patients with advanced heart failure: the CUPID 2 trial (calcium up-regulation by percutaneous administration of gene therapy in cardiac disease phase 2b). *JACC Heart Fail.* 2 (1): 84–92.

33 Fragoso-Medina, J. and Zarain-Herzberg, A. (2014). SERCA2a: its role in the development of heart failure and as a potential therapeutic target. *Res. Rep. Clin. Cardiol.* 5: 43–54.

34 Ito, M., Takino, N., Nomura, T. et al. (2021). Engineered adeno-associated virus 3 vector with reduced reactivity to serum antibodies. *Sci. Rep.* 11 (1): 9322.

35 Chowdary, P. and Nathwani, A. (2022). Phase 1-2 trial of AAVS3 gene therapy in patients with hemophilia B reply. *N. Engl. J. Med.* 387 (14): 1336–1337.

36 Au, H.K.E., Isalan, M., and Mielcarek, M. (2021). Gene Therapy advances: a meta-analysis of AAV usage in clinical settings. *Front. Med. (Lausanne)* 8: 809118.

37 Cunningham, S.C., Dane, A.P., Spinoulas, A., and Alexander, I.E. (2008). Gene delivery to the Juvenile mouse liver using AAV2/8 vectors. *Mol. Ther.* 16 (6): 1081–1088.

38 Wang, D., Zhang, F., and Gao, G. (2020). CRISPR-based therapeutic genome editing: strategies and in vivo delivery by AAV vectors. *Cell* 181 (1): 136–150.

39 Cox, D.B., Platt, R.J., and Zhang, F. (2015). Therapeutic genome editing: prospects and challenges. *Nat. Med.* 21 (2): 121–131.

40 Russell, D.W. and Hirata, R.K. (1998). Human gene targeting by viral vectors. *Nat. Genet.* 18 (4): 325–330.

41 Gaj, T., Epstein, B.E., and Schaffer, D.V. (2016). Genome engineering using adeno-associated virus: basic and clinical research applications. *Mol. Ther.* 24 (3): 458–464.

42 Gaj, T., Staahl, B.T., Rodrigues, G.M.C. et al. (2017). Targeted gene knock-in by homology-directed genome editing using Cas9 ribonucleoprotein and AAV donor delivery. *Nucleic Acids Res.* 45 (11): e98.

43 Chandler, R.J., Venturoni, L.E., Liao, J. et al. (2021). Promoterless, nuclease-free genome editing confers a growth advantage for corrected hepatocytes in mice with methylmalonic acidemia. *Hepatology* 73 (6): 2223–2237.

44 Chandler, R.J. and Venditti, C.P. (2019). Gene therapy for methylmalonic acidemia: past, present, and future. *Hum. Gene Ther.* 30 (10): 1236–1244.

45 Chandler, R.J., LaFave, M.C., Varshney, G.K. et al. (2015). Vector design influences hepatic genotoxicity after adeno-associated virus gene therapy. *J. Clin. Invest.* 125 (2): 870–880.

46 LogicBio Therapeutics Announces Early Clinical Trial Results Demonstrating First-Ever In Vivo Genome Editing in Children.pdf.

47 Harmatz, P., Prada, C.E., Burton, B.K. et al. (2022). First-in-human in vivo genome editing via AAV-zinc-finger nucleases for mucopolysaccharidosis I/II and hemophilia B. *Mol. Ther.* 30: 3587–3600.

48 Sharma, R., Anguela, X.M., Doyon, Y. et al. (2015). In vivo genome editing of the albumin locus as a platform for protein replacement therapy. *Blood* 126 (15): 1777–1784.

49 Ou, L., DeKelver, R.C., Rohde, M. et al. (2019). ZFN-mediated in vivo genome editing corrects Murine Hurler syndrome. *Mol. Ther.* 27 (1): 178–187.

50 Laoharawee, K., DeKelver, R.C., Podetz-Pedersen, K.M. et al. (2018). Dose-dependent prevention of metabolic and neurologic disease in murine MPS II by ZFN-mediated in vivo genome editing. *Mol. Ther.* 26 (4): 1127–1136.

51 Maeder, M.L., Stefanidakis, M., Wilson, C.J. et al. (2019). Development of a gene-editing approach to restore vision loss in Leber congenital amaurosis type 10. *Nat. Med.* 25 (2): 229–233.

52 Mancuso, P., Chen, C., Kaminski, R. et al. (2020). CRISPR based editing of SIV proviral DNA in ART treated non-human primates. *Nat. Commun.* 11 (1): 6065.

53 McGaw, C., Garrity, A.J., Munoz, G.Z. et al. (2022). Engineered Cas12i2 is a versatile high-efficiency platform for therapeutic genome editing. *Nat. Commun.* 13 (1): 2833.

54 Kim, D.Y., Lee, J.M., Moon, S.B. et al. (2022). Efficient CRISPR editing with a hypercompact Cas12f1 and engineered guide RNAs delivered by adeno-associated virus. *Nat. Biotechnol.* 40 (1): 94–102.

55 Anzalone, A.V., Koblan, L.W., and Liu, D.R. (2020). Genome editing with CRISPR-Cas nucleases, base editors, transposases and prime editors. *Nat. Biotechnol.* 38 (7): 824–844.

Section II

Translational Biomarkers for Gene Therapy

3

Biomarker and Bioanalytical Readouts for the Development of AAV Gene Therapy

Yanmei Lu[1] and Wibke Lembke[2]

[1] *Biomarker and BioAnalytical Sciences, Sangamo Therapeutics, Richmond, CA, USA*
[2] *Bioanalytical Services, Celerion Switzerland AG, Allmendstrasse 32, 8320 Fehraltorf, Switzerland*

3.1 Introduction

3.1.1 AAV-Mediated *in vivo* Gene Therapy

In vivo gene therapy (GTx) has been proven to be a viable therapeutic approach and holds enormous potential to treat patients with rare diseases with monogenic mutations as well as complex acquired disorders. Recombinant adeno-associated virus (rAAV) has emerged as the most frequently used vehicle for *in vivo* GTx compared with other viral delivery systems such as adenoviral or lentiviral vectors. Based on 136 unique clinical trials posted in the US National Library of Medicine database (ClinicalTrials.gov) before 26 April 2021, a majority of the clinical applications fall into the therapeutic areas of ocular diseases, lysosomal storage disorders (LSD), blood disorders, neuromuscular disorders, and central nervous disorders [1]. Most of these are inherited rare diseases.

The natural occurring AAVs are small non-enveloped icosahedral virion of ~26 nm in diameter that contain a single-stranded DNA genome of 4.7 kilobase long. Flanked between two inverted terminal repeats (ITRs) are the two major open reading frames *Rep* (Replication) and *Cap* (Capsid). *Rep* encodes proteins required for viral replication and packaging, whereas the *Cap* gene produces structural proteins that encapsulate viral genome and direct the capsid's binding to receptors on host cells [1]. Various AAV serotypes that differ in capsid protein sequences have been isolated from a number of species, such as humans and

Drug Development for Gene Therapy: Translational Biomarkers, Bioanalysis, and Companion Diagnostics, First Edition. Edited by Yanmei Lu and Boris Gorovits.

nonhuman primates (NHP). These diverse serotypes have broad but different tissue tropism profiles determined by their binding to distinct set of receptors on host cell surfaces. As a result, AAVs can transduce and deliver the DNA payload to a wide range of mammalian cells and tissues, such as liver, muscle, retina, or brain.

With AAV vector, the coding sequence between the two ITRs is replaced by an engineered DNA expression cassette to introduce a therapeutic gene of interest (also known as transgene) to target tissues in humans. ITRs are responsible for transgene packaging and are the only viral sequences required in rAAV vectors. The technology platforms for clinical applications can be divided into four different approaches: gene replacement, gene addition, gene editing, and gene regulation. Corresponding to the four classes, the transgene may (1) encode a functional gene to replace the patient's defective gene; (2) overexpress an endogenous gene or synthetic gene; (3) encode gene editing components like zinc finger nucleus (ZFN), transcription activator-like effector nuclease (TALEN) or clustered regularly interspaced short palindromic repeats (CRISPR) associated nuclease 9 (Cas9) to modify the host genome; and (4) encode gene regulation agents like zinc finger protein-transcriptional factor (ZFP-TF), microRNA or small hairpin RNA to suppress or activate a gene [2].

Unlike most other viruses, AAV is naturally replication-deficient unless in the presence of helper virus, such as adenovirus or herpesvirus. The rAAV vectors lacks *Rep* and *Cap* genes and are incapable of replication and packaging in humans even when coinfected with helper viruses. The viral DNA persists in the nucleus mostly as an episomal form outside of the host cell's chromosome. The circular episomes are stable and provide long-term gene expression after a single administration. The rAAV genome has a low frequency of integration into the host genome, thus limiting risk of insertional mutagenesis and oncogenicity [3, 4]. In addition, AAVs are relatively less immunogenic compared with other viral vectors. These advantageous properties lead to a more favorable risk-benefit profile for rAAVs. Glybera (alipogene tiparvovec) was the first licensed product in Europe in 2012 for familial lipoprotein lipase deficiency. In the United States, the approval of three AAV GTx products by the Food and Drug Administration (FDA) has further advanced the field and boosted tremendous interest in this therapeutic approach: (1) Luxtuna® (voretigene neparvovec-rzyl) was approved in 2017 for treating patients greater than 12 months of age with inherited retinal disease due to biallelic *RPE65* mutations via targeted delivery to subretina of the eye; (2) Zolgensma® (onasemnogene abeparvovec-xioli) was approved in 2019 for treating patients less than 2 years of age with spinal muscular atrophy (SMA) caused by biallelic mutations in the *survival motor neuron 1 (SMN1)* gene via systemic intraventricular (IV) infusion; and (3) Hemgenix® (etranacogene dezaparvovec) was approved in 2022 for the treatment of hemophilia B in adults with congenital Factor IX deficiency.

For more in-depth background information, see Chapters 1 and 2.

3.1.2 Biomarker Category and Utility

The assessment of biomarkers became an undisputable tool in the development of therapeutics and play a key role in the treatment paradigm change toward precision medicine. A biomarker is not a clinical endpoint that measures how an individual feels, functions, or survives, but rather a defined characteristic that is measured as an indicator of normal biological processes, pathogenic processes, or responses to an exposure or intervention, including therapeutic interventions [5]. The type of biomarkers may include molecular, histologic, radiographic, or physiologic characteristics. Biomarker-driven drug development might have advantages over a conventional approach, as they have the potential to predict drug efficacy more quickly than conventional clinical endpoints as well as to accelerate drug development in certain disease areas. The concept of biomarkers is not new. The ancient Egyptians used pregnancy tests, a thousand years later Hippocrates noted the relationship between various visible manifestations of diseases, and finally in the 1950s the term "biological marker" was introduced followed by widespread use of "biomarkers" as of the 1980s. In 1998, the first biomarker-guided drug trastuzumab was approved by the FDA. Finally, in 2015, the FDA–National Institute of Health (NIH) Joint Leadership Council identified the harmonization of terms used in translational science and medical product development as a priority need, with a focus on terms related to study endpoints and biomarkers. Working together with the goals of improving communication, aligning expectations, and improving scientific understanding, the two agencies developed the BEST (Biomarkers, EndpointS, and other Tools) Resource [5]. The first phase of BEST comprises a glossary that clarifies important definitions and describes some of the hierarchical relationships, connections, and dependencies among the terms it contains to harmonize and clarify terms of translational science in medical drug development. The glossary is considered a living document.

According to the BEST glossary, there are seven biomarker categories, namely diagnostic, monitoring, predictive, prognostic, PD/response, safety, and susceptibility/risk biomarkers. Examples are given for the context of use for each of the biomarker categories in Figure 3.1.

Pharmacokinetic (PK) (*pharmakon* "drug" and *kinetikos* "moving, putting in motion") classically describes the absorption, distribution, metabolism, and excretion of a drug over time after administration. For the development of biologics, there is a clear distinction between PK and biomarker readouts. In the case of AAV-based GTx, it is a contentious topic if a particular readout or analyte belongs to PK or the biomarker category. One reason being that the active drug will be produced *in vivo* by the patients' cells, after administration of a vehicle system (AAV capsid) containing all relevant information (DNA). Generally, the assessment of biodistribution and shedding is seen as the equivalent of PK.

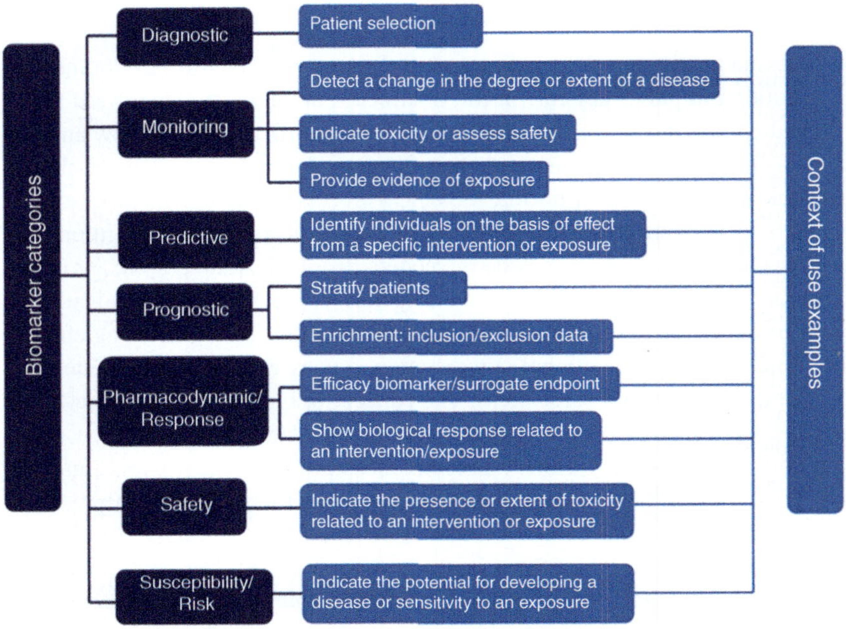

Figure 3.1 Biomarker categories according to BEST and examples of context of use. Source: https://www.fda.gov/drugs/biomarker-qualification-program/context-use. Public domain.

This chapter provides a high-level overview of the following topics: viral biodistribution and shedding, pharmacodynamic biomarkers of transgene expression and changes of substrate and downstream pathway markers, assessment of genotoxicity caused by rAAV integration and off-target gene editing, biomarkers for immune-mediated toxicity, safety biomarkers for nonimmune organ-specific toxicity, predictive and diagnostic biomarkers for study enrollment and patient stratification.

3.2 Pharmacokinetic (PK) and Pharmacodynamic (PD) Biomarkers

3.2.1 Viral Biodistribution and Shedding

Though different AAV serotypes have high transduction efficiency toward different set of cell types and tissue, the tropism is not completely restrictive and nontarget tissues may be transduced. Unlike conventional PK studies for biologics and small molecule drugs, GTx performs nonclinical biodistribution studies to evaluate drug exposure and safety [6]. Biodistribution studies in animals, standalone or more often combined with good laboratory practice (GLP)-compliant toxicity study, study the spread of virus within the body, including target and nontarget tissues from the site of administration. Assessments include the kinetics of vector distribution, persistence, and clearance profile at peak and one (or more) later time points.

Based on regulatory guidance (US Food and Drug Administration [FDA] guidance), the minimum panel of tissue collection should include liver, gonads, brain, heart, spleen, kidney, lung, blood, and injection site. Additional tissues may need to be collected depends on route of administration (ROA), vector type, and mechanism of action [7]. For example, intravenous administration and cerebrospinal fluid (CSF) delivery of AAV vectors has been shown to cause dorsal root ganglia (DRG) toxicity in multiple animal species [8–10] and therefore, DRG may need to be included in biodistribution study for the relevant ROAs. International Council for Harmonisation guideline S12 recommended a broader panel that includes adrenal gland and spinal cord collection [11]. In long-term follow-up studies in humans, surrogate samples can be used to test for vector sequence. Chapter 4 of this book has detailed descriptions of the design and conduct of biodistribution studies and PK and PD evaluations.

The wildly used bioanalytical methods to measure vector genome copy number from tissue DNA are quantitative polymerase chain reaction (qPCR) (also referred to as real-time PCR) and more recently, digital PCR (dPCR). Both methods can achieve FDA recommended lower limit of quantitation of 50 copies of vector genome per 1 μg of genomic DNA [12]. dPCR has superior precision than qPCR but narrower dynamic range and lower throughput.

A shedding study examines the dissemination of viral vectors through secretion and/or excreta (urine, feces, semen, saliva, etc.) from the patient to assess the risk of horizontal and environmental transmission. Nonclinical shedding data provide information on the likelihood of transmission to untreated human individuals (e.g. caregivers) and inform the design of shedding studies in clinical trial. Additionally, vector copy numbers in blood circulation must be determined as part of the PK analysis. The primary assay for shedding analysis is to measure vector genome copy number via qPCR or dPCR which is specific, sensitive, reproducible, high throughput with rapid turnaround. These methods, however, cannot differentiate intact vs. noninfectious or degraded vector sequences. An infectivity assay using *in vitro* culture of samples with a permissive cell line followed by qPCR endpoint analysis can accurately assess the nature of the shed material. For replication-competent virus, it is important to characterize infectivity as secondary analysis. For replication-incompetent virus like rAAV, the infectivity assay results are informational. The limitations of this type of assay are poor sensitivity and precision. It is plausible not to perform this test if qPCR/dPCR results are below the limit of detection of the infectivity assay [13]. For additional analytical-related details, see Chapter 7.

3.2.2 Transgene mRNA Expression

The basic elements of a transgene expression cassette typically include a promoter/enhancer at the 5' end, gene of interest, and polyadenylation sequence at the 3' end. Systemic IV delivery results in broad tissue distribution. The majority of the clinical trials using an IV delivery route for indications such as blood disorders (e.g. hemophilia A and B) employed tissue-specific promoters to minimize transgene expression outside of target tissues, which may help with a better safety profile. Most of the central nervous system disorders have used more targeted local delivery (e.g. intrathecal and intracerebral) incorporating a ubiquitous promoter [1]. After single-stranded AAV (ssAAV) genome is taken by a cell, the vector undergoes uncoating in the nucleus followed by *de novo* synthesis of complementary DNA strand to form double-stranded DNA before mRNA can be expressed. The engineered self-complementary AAV (scAAV) can bypass the host-cell DNA synthesis and immediately reanneal, thereby can produce transgene product much earlier than ssAAV. The kinetics of vector distribution and transgene expression coupled with other safety endpoints such as clinical pathology and histopathology can help determine whether vector presence correlates with adverse findings. Additionally, characterization of mRNA expression can be used as an efficacy endpoint when vector-encoded protein cannot be analytically distinguished from host native protein. High level of transgene mRNA expression may induce stress to the cell machinery [14] and therefore, in addition to vector genome copy numbers, transgene mRNA expression in different tissues is also a critical readout for biodistribution and/or GLP-toxicology studies when evaluating drug exposure and nonclinical safety.

The most frequently used method for mRNA expression is reverse transcription (RT)-qPCR. The mRNA is first transcribed into single-stranded complementary DNA (cDNA) by reverse transcriptase. The cDNA is subsequently used as the template for DNA polymerase in qPCR amplification or most recently dPCR. In situations to define proportion and/or uniformity of transduction and gene expression in target cell populations, image analysis of *in situ* hybridization signals of tissue sections can be a powerful technique to provide spatial information [6]. These and additional technologies for mRNA expression analysis can be found in Chapter 8.

3.2.3 Transgene and Target Protein Activity and Concentration

Gene therapies address the underlying cause of genetic disorders and is often considered a disease modifying therapy. For gene replacement and gene addition applications, the transgene encodes the therapeutic protein directly. For gene editing platforms, the transgene codes for molecule(s) to modify the host genome to correct mutation(s) in a target gene using technologies such as base editing and prime editing or insert a functional gene by creating a double-stranded break in the target DNA using nuclease of choice. Transgene can also encode proteins (transcription activator or repressor that are capable of binding to specific DNA sequences) or RNAs (shRNA or microRNA) to modulate levels of target protein. All approaches result in production or reduction of a target protein to exert a therapeutic effect. After being taken up by cells via receptor-mediated endocytosis, rAAV enters the nucleus and releases the DNA by uncoating the capsid. The single-stranded DNA genome undergoes second-strand synthesis and forms stable, circular, and double-stranded DNA episome, which provides transgene expression over long term [15].

To achieve the desired level of target proteins, much effort has been invested into early design of vector constructs. The first step is to choose either naturally occurring capsid types with high tropism toward target tissue or engineered novel capsids with improved transduction efficiency and selectivity for target tissue(s). Additionally, engineering DNA expression cassette by incorporating a strong promoter (ubiquitous or tissue-specific) and/or RNA stabilization motif (such as mutated Woodchuck Hepatitis Virus Posttranscriptional Regulatory Element) is a second approach to reach high level of transgene expression.

Transgene or target protein expression is an essential readout in preclinical and clinical pharmacological assessment because the expression can correct the molecular defects of the diseases. Proof-of-concept studies in animal disease models are important to demonstrate that a certain level of gene or protein expression can positively impact disease. Duration of the expression is another important characteristic of a successful treatment. Factor IX expression in human muscle tissue has been shown to last up to 10 years post a single administration [16]. However, loss of transgene expression of factor VIII and IX have been reported in humans [17, 18]. Multiple factors can affect the durability of transgene expression. The host cell-intrinsic innate immune system can detect the capsid and DNA genomes rich in hypomethylated 5'-cytosine-phosphate-guanine-3' (CpG) dinucleotides through pattern recognition receptors such as Toll-like receptors (TLRs) 2 and 9, respectively. The TLRs activate the myeloid differentiation factor 88 (MyD88) pathway leading to proinflammatory type 1 interferon response and augmented adaptive immune response to capsid and/or the transgene product, which may result in destruction of transduced cells and loss of transgene expression [18]. Detectable capsid-specific cytotoxic T lymphocytes in the circulation have been linked to liver enzyme elevation in several clinical trials. Some of the subjects have lost transgene expression likely due to CD8+ T-cell killing of transduced cells that display capsid antigen. There are also non-immunological factors that may affect the durability of transgene expression. Terminal differentiated (neurons) or long-lived cells (muscle myocytes and hepatocytes) can maintain circular episomes for long term, whereas rapid dividing cells may dilute the episomes through rounds of replication [19, 20]. Lastly, cellular stress from transgene overexpression may trigger apoptotic signaling leading to death of transduced cells [15].

Most inherited metabolic disorders such as LSDs are caused by "loss-of-function" genetic mutations in genes that encode enzymes. For GTx that corrects the enzyme deficiency, enzyme activity measurement is preferred over protein quantitation as this is a direct readout of the biological activity of the enzyme that indicates efficacy. For blood disorders such as hemophilia, coagulation activity of the transgene product is also preferred over protein levels. These activity readouts can serve as a surrogate endpoint for efficacy. For transgene and target gene products that are proteins with no activity, the amount of protein in biofluid or tissue biopsies can be measured by ligand binding assay or other technologies, e.g. immuno-affinity mass spectrometry, immunohistochemistry, and Western blotting or similar techniques [21]. The activity and protein expression of transgene or target gene products, together with PD biomarker readouts are critical parameters for PK/PD modeling and clinical dose selection.

GTx transgene products are expressed endogenously from the target cells. This is different from small molecules and biologics which exert therapeutic effects directly after administration without the need of further *in vivo* processing. Enzyme activity determination and protein quantitation for endogenous analytes pose bioanalytical challenges. Chapter 9 describes solutions to overcome various unique challenges.

3.2.4 Substrate and Other Distal PD Biomarkers

Genetic mutations of inherited metabolic disorders often affect enzymes in metabolic pathways. They are typically grouped by the affected substrates such as amino acids, carbohydrates, lipids, purines, and pyrimidines. A large class of inherited metabolic disorders are LSDs caused by deficiency in lysosomal enzymes of acidic hydrolases that are capable of digesting macromolecules such as glycoproteins and lipids. The absence of the functional enzyme typically results in toxic substrate buildup and/or loss of essential metabolites that cause organ damage [22]. Some of the other inherited metabolic disorders are phenylketonuria where deficiency of phenylalanine hydroxylase results in high levels of phenylalanine in the blood; glycogen storage diseases with mutant enzymes that affect glycogen metabolism, and peroxisomal disorders with defective enzymes resided in peroxisomes. *In vivo* production of a functional protein through GTx would potentially clear the substrate buildup and restore metabolites.

Substrates, metabolites, and other molecules in the downstream pathway are primary disease biomarkers that reflect disease pathogenesis, making them ideal

PD biomarkers to assess treatment response. Mass spectrometry is a powerful technology to detect the change of these molecules before and after treatment. Histology and imaging techniques to quantify substrate inclusions in tissues have been utilized as surrogate endpoints for efficacy in the clinic. For genome regulation platform, mRNA detection of downstream target RNA expression is an important target engagement biomarker for preclinical studies. See Chapter 10 for a comprehensive review of different methods for measuring PD biomarkers.

3.3 Safety and Monitoring Biomarkers and Readouts

Like other therapeutic modalities such as biologics and small molecule drug, safety assessment of *in vivo* GTx products is required at both preclinical and clinical development stages. The nonclinical safety evaluation uses small and/or large animal species to establish proof-of-concept and safety profile of the biologically active dose levels and route of administration [23]. The nonclinical safety profile, together with preclinical pharmacology data will inform clinical dose selection and safety monitoring. Clinical data collected over the last 20 years from more than 3000 patients treated with rAAV GTx indicating the therapeutic modality is generally well-tolerated and efficacious [24]. As a wealth of preclinical and clinical data becoming available in recent years, the field is experiencing emerging knowledge and gaining mechanistic understanding of toxicities related to rAAV-based GTx.

The treatment-emergent safety findings with AAV GTx in patents have been associated with hepatotoxicity and thrombotic microangiopathy (TMA), and brain magnetic resonance imaging (MRI) findings of uncertain significance with some treatment-emergent serious adverse events (TESAEs) resulting in deaths of study subjects. Other toxicities have primarily been reported in animal studies following AAV vector administration, including DRG and peripheral nerve toxicities, primarily seen in nonhuman primates [25]. Another potential risk of AAV vectors is oncogenicity due to integration and insertional mutagenesis. For in vivo genome editing therapies, engineered nuclease-induced off-target editing needs to be assessed.

The contributing factors to these toxicities are quite complex and may involve variables of vector design (capsid serotype and DNA construct), dose level, dose regimen, ROA, product quality, and intrinsic patient, and disease-specific characteristics. For example, rAAV serotypes and promoters have broad tissue tropism and high expression which could affect safety profiles. Higher vector dose levels have been shown to increase risk of toxicity with greater incidences and more serious adverse events (SAEs) [26].

Empty AAV capsids in the drug product are considered an impurity as they do not carry the genomic payload that exerts a therapeutic effect. There has been much debate on the effect of empty capsids on clinical outcome. Vector titers are determined using PCR-based methods of quantifying copy numbers of DNA genome packaged inside the capsid. Empty capsids are thus not accounted for in dose calculation yet add to increased total AAV exposure in patients. AAV8 empty capsids are fully capable of triggering strong B and T cell immune responses in humans which pose a safety concern [27]. It has been shown that AAV8 empty capsids inhibited liver transduction in mice [28]. Conversely, it was proposed that empty capsids may act as decoys for preexisting antibody to AAV to ameliorate transduction inhibition [29]. Overall, low empty capsid content is desirable to increase the safety margin by maximizing the DNA delivered per mass of capsid protein. Finally, a patient's preexisting immune status to wild-type AAV, genetic background and disease severity are additional host-related factors that may acerbate drug toxicity. For instance, certain lysosomal or metabolic diseases have increased autoimmunity risk [30, 31].

3.3.1 Assessment of genotoxicity

Evaluating a therapeutics' effect on the integrity of the genetic material is an important part of safety assessment. With regards to GTx, the main components assessed to understand genotoxicity risk include: (i) AAV vector-mediated insertional mutagenesis risk; (ii) risk of germline transmission of AAV vector, and (iii) for gene therapies that utilize engineered nuclease to edit the genome, an assessment of off-target genome editing is necessary. Genome alteration events need to be accurately detected and risks adequately assessed in preclinical and/or clinical development phases of GTx.

3.3.1.1 AAV Integration/Insertional Mutagenesis Risk

AAV genomes mostly persist in host cell nucleus as episomal circular forms [32]. Wild-type AAVs can integrate, in low frequency, into host genomic DNA, which is mediated by the Rep protein [33]. Recombinant AAVs that lack the *Rep/Cap* genes are even less efficient in integration. The integration sites of rAAV are largely random though may favor open chromatin regions near actively transcribed genes [34]. Vector design (serotype and promoter) and product quality (impurity) may affect the integration outcomes but factors that influence the risk of oncogenicity remain to be further investigated.

To date, oncogenesis of hepatocellular carcinoma (HCC) associated with rAAV insertional mutagenesis has only been reported in mice [35, 36]. HCC occurs as a background tumor in most laboratory mouse strains and mice have enhanced sensitivity to HCC relative to humans, which argues the translatability to human. In hemophilia A dogs treated with rAAV-FVIII and monitored for up to 10 years, 2 out of 9 dogs showed evidence of clonal expansion of hepatocytes with integrated rAAV 4 years post vector administration [19]. Though integrations were near oncogenes or growth control genes, no histological evidence of HCC or other lesions was observed. No clinical oncogenesis has been attributed to AAV integration in patients. Long-term follow-up monitoring and detection of rAAV integration should be risk-based. More detailed information on studies, methodologies, data analysis, and oncogenesis risk assessment associated with rAAV integration can be found in Chapter 13.

3.3.1.2 AAV Germline Transmission Risk

Nonclinical safety assessment of AAV germline transmission risk is generally supported by studies in the literature, a Sponsor's AAV biodistribution studies in animals evaluating gonad tissue for vector copy number, and on a case-by-case basis if additional developmental and reproductive toxicology studies are needed.

3.3.1.3 Off-Target Gene Editing

The field of using engineered nucleases to correct genetic disease is rapidly evolving and expanding. These genome editing components such as ZFN, TALEN, and CRISPR can be designed to efficiently create double-stranded breaks at the target DNA sites to accomplish the intended genome modification by triggering cell's intrinsic DNA repair mechanisms. Due to imperfect specificity of the designer nucleases, off-target cleavage sites have been detected at unintended genomic sites though in lower frequency. These off-target effects cause short insertions and deletions (indels) and/or large genomic rearrangements, which may lead to oncogenesis. Therefore, clinical applications of genome editing tools require thorough understanding of off-target activities and risks in both preclinical and clinical studies.

Much progress has been made in defining a workflow that incorporates multiple approaches to detect and quantify off-target sites. Various *in silico* tools based on computer algorithms have been developed and used to predict and/or rank off-target sites by scanning the reference human genome. It is important to note that genetic heterogeneity of individual genome variation also needs to be analyzed. *In vitro* cell-based and biochemical methodologies such as GUIDE-seq, oligonucleotide capture, CIRCLE-seq and SITE-seq are available to perform unbiased, genome-wide off-target cleavage site identification. The computational and experimental approaches can be combined to nominate potential off-target editing sites [37]. After careful ranking of these sites based on frequency, location, oncogenic risk, etc., top sites are further validated in cell type of interest and/or *in vivo* samples using more targeted techniques such as amplicon sequencing. In addition, large genomic alterations of chromosomal deletions, inversions and translocations also need to be evaluated. The standardized and conventional karyotyping methodologies can provide valuable analysis on chromosomal abnormalities at the DNA structure level. In recent years, next-generation sequencing techniques have been developed to identify large gene modifications with greater resolution, sensitivity, and specificity [38]. Comprehensive review can be found in Chapter 14.

3.3.2 Biomarkers for Immune-Mediated Toxicity

Host immune response against rAAV GTx represents a major hurdle for human applications. Natural infections of wild-type AAVs are generally benign and have not been reported to cause any disease or illness [39], therefore, are classified as nonpathogenic. GTx, however, is distinct in introducing non-replicative recombinant virus in a single high-titer bolus dose via an unnatural ROA. Compared with biologics such as monoclonal antibody therapy, rAAVs illicit more complex immune responses due to the multi-component nature of the therapeutic modality. rAAV GTx introduces multiple foreign antigens to humans (aside from impurities carried over from the production and purification process): the protein capsid, the DNA genome, and the transgene-expressed product (protein or RNA). These components are capable of inducing strong innate and adaptive (acquired) immune responses that can affect safety and efficacy. rAAV treatment induces a high level of anti-AAV neutralizing antibodies in animals and humans, which presents a significant limitation for vector re-administration. For this reason, it is more critical to establish a minimum effective dose to benefit patients than other therapeutic modalities. Before rAAV treatment, a significant proportion of human populations have been naturally infected with wild-type AAVs and acquired adaptive immunity, including humoral B cell-dependent antibody response and T-cell-mediated cellular response. The preexisting anti-AAV neutralizing antibody (NAb) response is more prevalent (as high as 74% prevalence for AAV2 [40]) than capsid-specific CD8+ T-cell response and the two do not necessarily correlate. Both humoral and cellular immunities to wild-type AAVs cross-react with rAAV capsids, consequently, this may prevent efficient cell transduction and/or lead to elimination of transduced cells [41], and possibly liver toxicity and TMA. Additionally, TLR-MyD88 sensing of high CpG content in vector DNA genome induces type 1 interferon secretion and subsequent cellular response to capsids, which may lead to hepatotoxicity and loss of transgene expression. Cellular and humoral responses to transgene expressed product could further complicate the outcome or efficiency of GTx.

3.3.2.1 Hepatotoxicity

Dose-dependent liver toxicity following systemic administration of rAAV has been the most common adverse event in clinical trials [42]. This was first reported in a patient receiving AAV2-ssFIX at 2E12 vector genome per kilogram (vg/kg) dose in hemophilia B trial [41]. The clinical presentation was transient serum transaminase (e.g. alanine aminotransferase [ALT] and aspartate aminotransferase [AST]) elevations between 4 and 6 weeks post vector infusion. The peak increase in liver enzymes coincided with decline and eventual loss of factor IX activity. This effect was associated with the appearance of AAV2 capsid-specific CD8+ T-cell response [43]. This cytotoxic cellular response can be detected by ELISpot where incubation of PBMCs with capsid peptides stimulates IFN-γ secretion. The underline mechanism could be that rAAV antigen peptides activate resting memory T cells originated from previous natural infection with wild-type AAV. There is also recent evidence that this capsid-specific CD8+ T-cell response reflects primary immune responses following vector administration [44]. The cytotoxic T-cell response may destroy transduced hepatocytes and result in loss of transgene expression, thereby limiting long-term benefits of the therapy. This type of adverse event can be managed by prophylactic or reactive intervention of glucocorticoid (prednisolone) or other immune-modulating agents such as mycophenolate mofetil or tacrolimus. Hepatotoxicity is a known safety risk associated with Zolgensma in the treatment of children with spinal muscular atrophy based on data collected during clinical trials and in post-marketing setting [45]. Elevated aminotransferases as well as acute liver failure warning have been added to Zolgensma label. No loss of SMN transgene expression has been reported possibly owing to the target cells of spinal motor neurons are protected from the immune response.

However, some clinical trial (NCT01687608) subjects who have received a scAAV8 vector expressing the hyperactive hFIX-R338L variant lost transgene expression even with immune-modulating agents [18]. The hepatotoxicity, capsid immunity, and loss of transgene expression was possibly attributed to elevated CpG content in the transgene (mostly introduced during codon optimization) stimulated TLR innate immune response which in turn potentiated adaptive response against the capsids. It has also been reported that some transaminase elevations in hemophilia trials had no evidence of capsid-specific cellular immune response and no loss of transgene expression [46]. Lastly, death due to cholestatic liver failure was reported in pediatric patients with X-linked myotubular myopathy at AAV doses greater than 1E14 vg/kg [47].

Though rAAV-induced hepatotoxicity presents a complex clinical picture, ALT and AST have been commonly used as a biomarker for reactive steroid treatment intervention. These liver enzymes are not specific to liver injury because skeletal and cardiac muscle injury often trigger ALT and AST elevation, making them unsuitable for detecting the onset of rAAV-induced liver injury in patients with underlying muscle impairment such as Duchenne muscular dystrophy (DMD). In addition, when the liver enzyme increase is detected, the irreversible elimination of transduced liver cells might have already happened. Sensitive and specific biomarkers that can predict or detect early-onset liver injury are desirable. Serum glutamate dehydrogenase (GLDH) has been established as one of the promising biomarkers out of novel biomarker identification effort for drug-induced liver injury. It would be worth testing whether GLDH is an early and specific biomarker to detect rAAV hepatotoxicity.

3.3.2.2 Thrombotic Microangiopathy

SAEs of hemolytic anemia, thrombocytopenia, and acute renal injury that are consistent with TMA have been observed for several clinical trials involving high doses of systemically delivered rAAVs. Two independent DMD trials (NCT03368742 and NCT03362502) of rAAV9 expressing a micro-dystrophin with a muscle-specific promoter reported several cases of TMA within two weeks of vector administration at doses greater than 5E13 vg/kg level. Evidence of complement activation has been presented by the sponsors and SAEs were resolved with a C5 complement inhibitor, Ecrulizumab, and other treatments, including glucocorticoids. Zolgensma also revealed a potential role for complement in rAAV host immune responses that induce TMA [48]. Novartis Safety Data as of 31 July 2021 reported 9 cases of TMA among over 1400 patients with one death due to sepsis after TMA recovery. The risk of TMA has been updated in Zolgensma product information. Additional trials that have reported TMA include a liver-targeted rAAV capsid LK03 to treat Severe Methylmalonic Acidemia at 5E13 vg/kg (NCT04581785) and a novel capsid 4D-C102 to treat Fabry disease at 1E13 vg/kg via IV infusion (NCT04519749).

Complement is a surveillance system of innate immunity that protects the host from pathogens. Classical, alternative, and lectin pathways are three distinct but overlapping proteolytic activation cascades that lead to complement activation. All three pathways converge on C3 convertase, which leads to the formation of membrane attack complex to cause cell lysis. The rapid antibody response against the capsid detected in patients at time of TMA and high complement activation suggest that anti-AAV and rAAV antigen-antibody complex activated classical pathway through C1q binding to Fc portion of the antibody complex. Alternative pathway may also be involved as preclinical studies showed that AAV capsids physically interacted with many components of complement pathway and alternative pathway activation has been linked to thrombocytopenia in NHP [49].

There have been no known preventative measures for TMA. Early detection is primarily based on clinical symptoms and frequent laboratory evaluations that monitor platelet count, renal function, and hemolytic anemia during the first two weeks. Alteration of serum complement components that are indicative of complement activation are useful biomarkers to confirm the involvement of complement in TMA. These include consumption of complement factors (C3, C4, Factor B, Factor H, and Factor I) and increase of split products (Bb, C5b-9) in serum. Complement activation may prime myeloid cells for cytokines secretion and further amplification of downstream immunity. It would be helpful to test if pro-inflammatory cytokine increase is detectable at time of TMA. Additional adaptive immunity monitoring may aid in the investigation of mechanisms behind AAV-induced TMA toxicity.

3.3.2.3 Muscle Toxicity

Serious adverse events involving muscle have been reported by four investigational rAAV gene therapies for the treatment of DMD (NCT05096221, NCT03368742, NCT04281485, and 2020-002093-27). These therapies use different AAV serotypes (AAVrh7.4, AAV9, AAV8) to deliver various versions of a micro-dystrophin transgene driven by different muscle-specific promoters (MHCK7, CK8, MSP, and Spc5.12) through IV administration. Five participants across studies experienced muscle-related toxicity between 3 and 7 weeks in dose range of 1E13-2E14 vg/kg. SAEs commonly exhibited extremity and bulbar muscle weakness. Three individual patients experienced severe respiratory muscle compromise and increased cardiac troponin-I levels. SAEs were resolved following immunosuppressive and other supportive treatments.

All patients affected by the SAEs share strikingly similar genotypes with genomic deletions in the N-terminal part of the dystrophin gene. The deleted sequences are present in the transgene, rendering the patients being cross-reactive immunological material (CRIM)-negative [50]. T-cell-mediated immune response to transgene protein epitopes corresponding to the N-terminal deleted region was detected by positive ELISpot test in all participants with these skeletal muscle and cardiac SAEs. Anti-dystrophin antibodies from 1 patient also mapped to N-terminal deleted epitope [51]. Sponsors are starting to exclude patients with these at-risk genotypes from clinical trials. This is the first report that associated anti-transgene immune response to adverse events in humans, highlighting the importance of measuring both humoral and cellular response to transgene in DMD patients receiving rAAV GTx.

3.3.2.4 Immunogenicity Assessment for rAAV Gene Therapy

Immunogenicity to AAV-based gene therapies is an important assessment to evaluate efficacy and safety of the drug. Observed adverse events during clinical development have been implicated with host immune responses against AAV gene therapies, resulting in comprehensive evaluation of immunogenicity during nonclinical and clinical development. Consequently, health authorities do see immunogenicity testing as a focus area in the nonclinical and clinical setting. A comprehensive summary of regulatory considerations for rAAV GTx is summarized in Yang et al. [52].

Immunogenicity of AAV gene therapies is complex due to the nature of the therapy bearing three components: the capsid, the DNA content as well as the *in vivo* generated active drug, paired with risk factors associated with manufacturing, treatment-related risks like dose and ROA [52] and the high prevalence of preexisting immunity in humans [40]. Besides the immunogenicity risks listed above, patient-related risk factors, including age and disease status, need to be considered to complete the picture of all potential immunogenicity risks and are captured in a single repository document, the immunogenicity risk assessment.

For more detailed information on AAVs immune responses in patients, mitigation strategies and bioanalytical assessments please refer to Chapter 5. Immune response to transgene-expressed proteins, including genome editing reagents and methods of detection can be found in Chapter 12.

3.3.3 Safety Biomarkers for Nonimmune Organ-Specific Toxicity

3.3.3.1 Dorsal Root Ganglia Toxicity

Dorsal root ganglia are bilateral structures that emerge from the dorsal root of the spinal nerves. DRGs are located in the peripheral nervous system outside of the blood-brain barrier and yet surrounded by a thin layer of CSF and have direct communication with the CSF compartment. DRG neurons relay somatosensory information, including those for pain and temperature from the peripheral nerves to the central nervous system of spinal cord and brain. In addition to primary sensory neuron cell bodies, DRGs also contain satellite glial cells that form a layer around neuronal cell bodies, small blood vessels constituted of endothelial cells and smooth muscle cells, and immune cells of macrophages and lymphocytes. Capillaries in DRG allow blood molecules to enter the DRG and interact with neuronal and nonneuronal cells.

Both systemic IV administration and CSF delivery can result in efficient transduction of DRGs. DRG pathology findings of mononuclear cell infiltration and inflammation, neuronal degeneration, and/or satellite glial cell proliferation have been reported for 33 NHP studies involving different capsid serotype, promoters, and transgenes [8]. The DRG pathology peaked at approximately 1 month and extended to 5 months followed by decrease in severity after 6 months post dose. While viral purification methods and sex had no influence, ROA and dose levels are significant contributors to the incidence and severity of the toxicity. CSF local delivery (via intra-cisterna magna or intrathecal injects) and greater than 1E13 vg/animal had greater impact than IV administration and lower dose. Similar DRG toxicity findings have been observed in piglet, mouse, and rat studies following rAAV administration [49].

Several lines of evidence suggest that underlying mechanism of the toxicity is cellular stress due to transgene mRNA and/or protein overexpression [14]. Initial experiment of co-administration of immunosuppressives did not mitigate DRG toxicity. Whether the immune response to rAAV vector or transgene product may also play a role remains to be investigated. The clinical relevance of DRG safety findings in preclinical animal studies is currently unclear.

Circulating neurofilament light chain (NFL) in serum/plasma has been identified as a noninvasive safety biomarker for AAV-induced DRG toxicity in NHP [53]. NFL is a versatile fluid biomarker across diverse neurological diseases. Neurofilaments are abundant scaffolding proteins to support neuronal cell structure. Neurofilament light chain is one of the subunits and primarily expressed in axon of a neuron. Damage to neurons due to injury, disease, or other toxicity release NFL into both CSF and blood compartments. The Quanterix Simoa NFL assay has received Breakthrough Device designation by the FDA as a prognostic aid in relapsing-remitting multiple sclerosis. Additional clinical validation of NFL as a prognostic biomarker has been reported for Huntington's [54], Alzheimer's [55], and spinocerebellar ataxia [56, 57]. NFL is a differential diagnostic biomarker in Parkinson's disease that can discriminate between Parkinson and atypical parkinsonian disorders [58]. In addition, NFL is also being used as a surrogate endpoint for efficacy in amyotrophic lateral sclerosis program [59].

3.3.3.2 Other Target Organ Toxicity Biomarkers

Toxicity biomarkers are biomarkers that are capable of detecting liver, kidney, bone marrow, and other target organ injuries. The essential characteristics of these type of biomarker(s) are excellent specificity to tissue damage of the indicated target organ, high sensitivity to low dose toxicity, and early detection or prediction that allows timely intervention and treatment decisions. Additional ideal features include easy accessibility, noninvasive sample acquisition, and reasonably low cost. Consortium efforts involving industry and health authorities for biomarker qualification of organ toxicity have taken a translational approach. Nonclinical safety biomarkers are identified by correlating with histopathology toxicity findings in animal species commonly used in safety assessment. For example, cardiac troponins T (cTnT) and I (cTnI) are safety biomarkers to indicate cardiotoxicity in animals and may be used to help estimate nontoxic human dose. The utility of these biomarkers in clinical studies is validated by comparison to the current standard biomarker performance and/or functional readouts. Nephrotoxicity and cardiotoxicity biomarkers have been fully qualified for use in the clinic to monitor drug-induced injury to kidney and heart, respectively [60]. GTx drug developers can incorporate these biomarker endpoints based on context of use into rAAV clinical trials to address patient safety and for decision-making. New safety biomarkers for liver, skeletal muscle, and vascular injury are in the process of being qualified with the regulatory agencies as drug development tools to support clinical trials [60]. These biomarkers, though not yet qualified as drug development tools, can be used as exploratory endpoints in GTx trials for information gathering, or for decision-making after discussion and gaining the agency's acceptance.

3.4 Predictive and Diagnostic Biomarkers for Study Enrollment and Patient Stratification

A predictive biomarker is used to identify individuals who are more likely than similar individuals without the biomarker to experience a favorable or unfavorable effect from exposure to a medicinal product. The utility of predictive biomarkers is not limited to clinical trial settings, where they can be used to either select patients for participation or to stratify patients into biomarker positive and negative groups, as they can also assist in informing patient care decision. Predictive biomarkers are for example protein levels and DNA mutations. They are often chosen initially based on the mechanism of action of the drug and understanding pathophysiology. Usually, predictive biomarkers are evaluated in clinical trials comparing the outcome of individuals with and without the biomarker. There are cases where there is sufficient evidence to use a certain biomarker only in populations enriched for the putative predictive biomarker.

Prognostic biomarkers on the other hand are used to identify the likelihood of a clinical event, disease recurrence, or progression in patients who have the disease or medical condition of interest. They are often used as eligibility criteria in clinical trials to identify patients who are more likely to have clinical events or disease progression. Thus, they are widely used as enrichment factors in drug development. Distinguishing predictive and diagnostic biomarkers can be difficult with some being both, predictive and diagnostic.

In general, all the concepts listed above apply to the clinical development of gene therapies and are very much driven by the program specifics. Especially the patient selection based on the predictive biomarkers of protein expression and/or gene mutation is of importance, given that to date AAV gene therapies are given as a single treatment with a "point of no return" once administered to a patient, reflected in the clinical development starting with an effective and safe dose to be administered into the patients predicted from an extensive preclinical program.

3.4.1 Preexisting Anti-Capsid Antibody

The prevalence of preexisting anti-AAV capsid antibodies in humans is generally high and can reach up to 80% of the individuals, depending on a variety of factors such as age, geographic location/ethnicity, and serotype [40]. To date, in most of the clinical trials testing an AAV-based gene therapeutic product, patients bearing preexisting antibodies above a certain threshold (titer) are excluded, to ensure an efficacious therapy and/or to avoid any possible adverse events. Such findings due to preexisting anti-capsid antibodies, namely efficacy loss and/or adverse events have been reported in nonclinical and clinical trials [9, 41]. As our current understanding of the effect of preexisting anti-capsid antibodies is still limited, the clinical relevance of excluding patients needs to be balanced with the eligibility for patients to receive beneficial treatments, especially under life-threatening conditions. The capsid serotype (natural occurring or engineered), the seroprevalence of the target population, the titer level in the individual subject, the dose, and the ROA as well as preclinical data will have an influence on the specific risk for each AAV-based GTx and guide the need of setting a titer as an exclusion criterion. In case patients with a certain level of preexisting anti-capsid antibodies need to be excluded based on efficacy and/or safety concerns, sponsors might consider the development of a companion diagnostic (CDx). Current available guidance documents from health authorities do not mandate the exclusion of patients in clinical trials based on a preexisting anti-capsid antibody titer, but if the sponsor does so, the consideration of developing a CDx to detect such antibodies is strongly recommended [61]. This leads to some flexibility on the sponsor side to carefully assess the benefit-risk profile of the program but may require an early dialogue with the agencies about preexisting anti-capsid antibody exclusion criteria and the need to develop a CDx.

3.4.1.1 Companion Diagnostic

A companion diagnostic is a medical device, usually an *in vitro* diagnostic device, that provides information that is essential for the safe and effective use of a corresponding therapeutic product. The use of a companion diagnostic with a therapeutic product is typically stipulated in the instructions for use in the labeling of both the diagnostic device and the corresponding therapeutic product, including the labeling of any generic equivalents of the therapeutic product [5]. Therefore, CDx is a test for a predictive biomarker that allows to select for patients that will benefit from the treatment and therefore, the information and recommendation of use of a therapeutic that requires a CDx is particularly important for healthcare professionals to identify the correct patients. CDx is classed by the FDA as Class III medical devices because the test results equate directly to the administration of the specific therapeutic product. For more information on CDx for GTx, see Chapter 15, Chapter 16, and Chapter 17.

3.4.2 Preexisting Anti-Transgene Protein Antibody

In nonclinical and clinical studies, humoral immune responses against the transgene protein have been reported [62]. There is even the possibility of preexisting immunity against the transgene protein, especially in patients that have been, prior administering an AAV GTx, treated with protein replacement therapies. Most hemophilia rAAV trials exclude patients with neutralizing antibodies to the factor replacement therapies including inhibitors to FVIII or FIX protein [62], while some lysosomal storage disease and hemophilia trials (NCT04046224, NCT04684940,

NCT03734588, etc.) enroll patients with preexisting antibodies to enzyme replacement therapy to understand the impact of preexisiting anti-transgene protein antibodies. Preexisting antibody and cellular response to Cas9 protein have been detected in healthy human adults [63]. The impact of these preexisting anti-transgene antibodies on the efficacy and safety of GTx remains to be investigated.

3.5 Summary

In summary, rAAV-based GTx has been shown to be effective and well-tolerated for treating patients with genetic diseases. The technology approaches for clinical applications include gene replacement, gene addition, gene editing, and gene regulation. The therapeutic technologies, preclinical/ clinical development, and regulatory landscapes are rapidly evolving, and the field is expanding with significant growth in the number of investigational new drug (IND) applications each year. GTx drug development requires complex and extensive collection of PK and biomarker readouts to support safety and efficacy evaluation as well as patient selection. Deeper understanding of biomarker, bioanalysis, and CDx development helps to drive the success and advancement of GTx.

References

1. Au, H.K.E., Isalan, M., and Mielcarek, M. (2021). Gene therapy advances: a meta-analysis of AAV usage in clinical settings. *Front. Med. (Lausanne)* 8: 809118.
2. Deverman, B.E., Ravina, B.M., Bankiewicz, K.S. et al. (2018). Gene therapy for neurological disorders: progress and prospects. *Nat. Rev. Drug Discov.* 17: 641–659.
3. Vandamme, C., Adjali, O., and Mingozzi, F. (2017). Unraveling the complex story of immune responses to AAV vectors trial after trial. *Hum. Gene Ther.* 28: 1061–1074.
4. Zaiss, A.-K., Liu, Q., Bowen, G.P. et al. (2002). Differential activation of innate immune responses by adenovirus and adeno-associated virus vectors. *J. Virol.* 76: 4580–4590.
5. Califf, R.M. (2018). Biomarker definitions and their applications. *Exp. Biol. Med.* 243: 213–221.
6. Hutt, J.A., Assaf, B.T., Bolon, B. et al. (2022). Scientific and regulatory policy committee points to consider: nonclinical research and development of in vivo gene therapy products, emphasizing adeno-associated virus vectors. *Toxicol. Pathol.* 50: 118–146.
7. Gene Therapy Clinical Trials – Observing Subjects for Delayed Adverse Events - Guidance for Industry [Internet]. (2023). [cited 14 January 2023], Available from: https://downloads.regulations.gov/FDA-2015-D-3399-0005/content.pdf.

8 Hordeaux, J., Buza, E.L., Dyer, C. et al. (2020). Adeno-associated virus-induced dorsal root ganglion pathology. *Hum. Gene Ther.* 31: 808–818.

9 Hinderer, C., Katz, N., Buza, E.L. et al. (2018). Severe toxicity in nonhuman primates and piglets following high-dose intravenous administration of an adeno-associated virus vector expressing human SMN. *Hum. Gene Ther.* 29: 285–298.

10 Hordeaux, J., Hinderer, C., Goode, T. et al. (2018). Toxicology study of intra-cisterna magna adeno-associated virus 9 expressing iduronate-2-sulfatase in rhesus macaques. *Mol. Ther. Methods Clin. Dev.* 10: 68–78.

11 Nonclinical biodistribution considerations for gene therapy products S12 – ICH guideline [Internet]. (2023). [cited 14 January 2023], Available from: https://www.ema.europa.eu/en/documents/regulatory-procedural-guideline/ich-guideline-s12-nonclinical-biodistribution-considerations-gene-therapy-products-step-2b_en.pdf.

12 Preclinical assessment of investigational cellular and gene therapy products – FDA guidance for industry [Internet]. (2023). [cited 14 January 2023], Available from: https://www.fda.gov/media/87564/download.

13 ICH Considerations General Principles to Address Virus and Vector Shedding [Internet]. (2023). [cited 14 January 2023], Available from: https://www.ema.europa.eu/documents/scientific-guideline/international-conference-harmonisation-technical-requirements-registration-pharmaceuticals-human-use_en-10.pdf.

14 Hordeaux, J., Buza, E.L., Jeffrey, B. et al. (2020). MicroRNA-mediated inhibition of transgene expression reduces dorsal root ganglion toxicity by AAV vectors in primates. *Sci. Transl. Med.* 12: eaba9188.

15 Muhuri, M., Levy, D.I., Schulz, M. et al. (2022). Durability of transgene expression after rAAV gene therapy. *Mol. Ther.* 30: 1364–1380.

16 Buchlis, G., Podsakoff, G.M., Radu, A. et al. (2012). Factor IX expression in skeletal muscle of a severe hemophilia B patient 10 years after AAV-mediated gene transfer. *Blood* 119: 3038–3041.

17 Pasi, K.J., Rangarajan, S., Mitchell, N. et al. (2020). Multiyear follow-up of AAV5-hFVIII-SQ gene therapy for hemophilia A. *N. Engl. J. Med.* 382: 29–40.

18 Konkle, B.A., Walsh, C.E., Escobar, M.A. et al. (2021). BAX 335 hemophilia B gene therapy clinical trial results: potential impact of CpG sequences on gene expression. *Blood* 137: 763–774.

19 Nguyen, G.N., Everett, J.K., Kafle, S. et al. (2021). A long-term study of AAV gene therapy in dogs with hemophilia A identifies clonal expansions of transduced liver cells. *Nat. Biotechnol.* 39: 47–55.

20 Nathwani, A.C., Rosales, C., McIntosh, J. et al. (2011). Long-term safety and efficacy following systemic administration of a self-complementary AAV vector encoding human FIX pseudotyped with serotype 5 and 8 capsid proteins. *Mol. Ther.* 19: 876–885.

21 Hathout, Y., Brown, K.J., Nagaraju, K., and Hoffman, E.P. (2019). Biomarkers for muscle disease gene therapy. *Muscle Gene Ther.* 3: 239–252.

22 Fernández-Pereira, C., Millán-Tejado, B.S., Gallardo-Gómez, M. et al. (2021). Therapeutic approaches in lysosomal storage diseases. *Biomolecules* 11: 1175.

23 Cavagnaro, J.A. (2002). Preclinical safety evaluation of biotechnology-derived pharmaceuticals. *Nat. Rev. Drug Discov.* 1: 469–475.

24 Kuzmin, D.A., Shutova, M.V., Johnston, N.R. et al. (2021). The clinical landscape for AAV gene therapies. *Nat. Rev. Drug Discov.* 20: 173–174.

25 Toxicity Risks of Adeno-associated Virus (AAV) Vectors for Gene Therapy (GT) – 2021 FDA Cellular, Tissue and Gene Therapies Advisory Committee [Internet]. (2023). [cited 14 January 2023], Available from: https://www.fda.gov/media/151599/download.

26 Samaranch, L., Sebastian, W.S., Kells, A.P. et al. (2014). AAV9-mediated expression of a non-self protein in nonhuman primate central nervous system triggers widespread neuroinflammation driven by antigen-presenting cell transduction. *Mol. Ther.* 22: 329–337.

27 AAV empty capsid immunogenicity and immune modulation in healthy volunteers [Internet]. 2023 [cited 24 February 2023], Available from: https://asgct.org/advocacy/fda-immune-responses-aav-vectors/watch-day-1?_zs=UXKHb&_zl=AcDQ3

28 Gao, K., Li, M., Zhong, L. et al. (2014). Empty virions in AAV8 vector preparations reduce transduction efficiency and may cause total viral particle dose-limiting side-effects. *Mol. Ther. Methods Clin. Dev.* 1: 20139.

29 Mingozzi, F., Anguela, X.M., Pavani, G. et al. (2013). Overcoming preexisting humoral immunity to AAV using capsid decoys. *Sci. Transl. Med.* 5: 194ra92.

30 Melis, D., Carbone, F., Minopoli, G. et al. (2017). Cutting edge: increased autoimmunity risk in glycogen storage disease Type 1b Is associated with a reduced engagement of glycolysis in T cells and an impaired regulatory T cell function. *J. Immunol.* 198: 3803–3808.

31 Shoenfeld, Y., Gallant, L.A., Shaklai, M. et al. (1982). Gaucher's disease: a disease with chronic stimulation of the immune system. *Arch. Pathol. Lab. Med.* 106: 388–391.

32 Nakai, H., Yant, S.R., Storm, T.A. et al. (2001). Extrachromosomal recombinant adeno-associated virus vector genomes are primarily responsible for stable liver transduction in vivo. *J. Virol.* 75: 6969–6976.

33 Im, D.S. and Muzyczka, N. (1990). The AAV origin binding protein Rep68 is an ATP-dependent site-specific endonuclease with DNA helicase activity. *Cell* 61: 447–457.

34 Kaeppel, C., Beattie, S.G., Fronza, R. et al. (2013). A largely random AAV integration profile after LPLD gene therapy. *Nat. Med.* 19: 889–891.

35 Walia, J.S., Altaleb, N., Bello, A. et al. (2015). Long-term correction of Sandhoff disease following intravenous delivery of rAAV9 to mouse neonates. *Mol. Ther.* 23: 414–422.

36 Chandler, R.J., LaFave, M.C., Varshney, G.K. et al. (2015). Vector design influences hepatic genotoxicity after adeno-associated virus gene therapy. *J. Clin. Invest.* 125: 870–880.

37 Bao, X.R., Pan, Y., Lee, C.M. et al. (2021). Tools for experimental and computational analyses of off-target editing by programmable nucleases. *Nat. Protoc.* 16: 10–26.

38 Sasu, B.J., Opiteck, G.J., Gopalakrishnan, S. et al. (2022). Detection of chromosomal alteration after infusion of gene-edited allogeneic CAR T cells. *Mol. Ther.* 31: 676–685.

39 Smith, R.H. (2008). Adeno-associated virus integration: virus versus vector. *Gene Ther.* 15: 817–822.

40 Kruzik, A., Fetahagic, D., Hartlieb, B. et al. (2019). Prevalence of anti-adeno-associated virus immune responses in international cohorts of healthy donors. *Mol. Ther. Methods Clin. Dev.* 14: 126–133.

41 Manno, C.S., Pierce, G.F., Arruda, V.R. et al. (2006). Successful transduction of liver in hemophilia by AAV-factor IX and limitations imposed by the host immune response. *Nat. Med.* 12: 342–347.

42 Ronzitti, G., Gross, D.-A., and Mingozzi, F. (2020). Human immune responses to adeno-associated virus (AAV) vectors. *Front. Immunol.* 11: 670.

43 Mingozzi, F. and High, K.A. (2013). Immune responses to AAV vectors: overcoming barriers to successful gene therapy. *Blood* 122: 23–36.

44 Xiang, Z., Kurupati, R.K., Li, Y. et al. (2020). The effect of CpG sequences on capsid-specific CD8(+) T cell responses to AAV vector gene transfer. *Mol. Ther.* 28: 771–783.

45 Chand, D., Mohr, F., McMillan, H. et al. (2020). Hepatotoxicity following administration of onasemnogene abeparvovec (AVXS-101) for the treatment of spinal muscular atrophy. *J. Hepatol. Suppl.* 74: 560–566.

46 Rangarajan, S., Walsh, L., Lester, W. et al. (2017). AAV5-factor VIII gene transfer in severe hemophilia A. *N. Engl. J. Med.* 377: 2519–2530.

47 Philippidis, A. (2020). After third death, Audentes' AT132 remains on clinical hold. *Hum. Gene Ther.* 31: 908–910.

48 Chand, D.H., Zaidman, C., Arya, K. et al. (2021). Thrombotic microangiopathy following onasemnogene abeparvovec for spinal muscular atrophy: a case series. *J. Pediatr.* 231: 265–268.

49 Palazzi, X., Pardo, I.D., Sirivelu, M.P. et al. (2022). Biodistribution and tolerability of AAV-PHP.B-CBh-SMN1 in Wistar Han Rats and cynomolgus macaques reveal different toxicologic profiles. *Hum. Gene Ther.* 33: 175–187.

50 Elangkovan, N. and Dickson, G. (2021). Gene therapy for duchenne muscular dystrophy. *J. Neuromuscul. Dis.* 8: S303–S316.

51 Bonnemann, C.G, Belluscio, B.A., Braun, S., et al., Collaborative Analysis by Clinical Trial Sponsors and Academic Experts of Anti-transgene SAEs in Studies of Gene Therapy for DMD [Internet]. 2022 ASGCT Annual Meeting. ASGCT Annual Meeting; 2023. [cited 14 January 2023], Available from: https://annualmeeting.asgct.org/global/am22/late-breaking-abstract-publication.aspx.

52 Yang, T.-Y., Braun, M., Lembke, W. et al. (2022). Immunogenicity assessment of AAV-based gene therapies: an IQ consortium industry white paper. *Mol. Ther. Methods Clin. Dev.* 26: 471–494.

53 Fader, K.A., Pardo, I.D., Kovi, R.C. et al. (2022). Circulating neurofilament light chain as a promising biomarker of AAV-induced dorsal root ganglia toxicity in nonclinical toxicology species. *Mol. Ther. Methods Clin. Dev.* 25: 264–277.

54 Byrne, L.M., Rodrigues, F.B., Johnson, E.B. et al. (2018). Evaluation of mutant huntingtin and neurofilament proteins as potential markers in Huntington's disease. *Sci. Transl. Med.* 10: eaat7108.

55 Preische, O., Schultz, S.A., Apel, A. et al. (2019). Serum neurofilament dynamics predicts neurodegeneration and clinical progression in presymptomatic Alzheimer's disease. *Nat. Med.* 25: 277–283.

56 Wilke, C., Haas, E., Reetz, K. et al. (2020). Neurofilaments in spinocerebellar ataxia type 3: blood biomarkers at the preataxic and ataxic stage in humans and mice. *EMBO Mol. Med.* 12: e11803.

57 Gendron, T.F., Heckman, M.G., White, L.J. et al. (2022). Comprehensive cross-sectional and longitudinal analyses of plasma neurofilament light across FTD spectrum disorders. *Cell Rep. Med.* 3: 100607.

58 Hansson, O., Janelidze, S., Hall, S. et al. (2017). Blood-based NfL: a biomarker for differential diagnosis of parkinsonian disorder. *Neurology* 88: 930–937.

59 Dreger, M., Steinbach, R., Otto, M. et al. (2022). Cerebrospinal fluid biomarkers of disease activity and progression in amyotrophic lateral sclerosis. *J. Neurol. Neurosurg. Psychiatry* 93: 422–435.

60 Sauer, J.-M. and Porter, A.C. (2021). Qualification of translational safety biomarkers. *Exp. Biol. Med.* 246: 2391–2398.

61 Human gene therapy for rare diseases – FDA guidance for industry [Internet]. (2023). [cited 14 January 2023], Available from: https://www.fda.gov/media/113807/download

62 Herzog, R.W. (2019). Complexity of immune responses to AAV transgene products – example of factor IX. *Cell. Immunol.* 342: 103658.

63 Charlesworth, C.T., Deshpande, P.S., Dever, D.P. et al. (2019). Identification of preexisting adaptive immunity to Cas9 proteins in humans. *Nat. Med.* 25: 249–254.

4

Nonclinical and Clinical Study Considerations for Biodistribution, Shedding, and Pharmacokinetics/Pharmacodynamics

Manuela Braun[1] *and Kefeng Sun*[2]

[1] *Preclinical Development, Project Management, Bayer AG, Berlin, Germany*
[2] *Quantitative Clinical Pharmacology, Data Sciences Institute, Takeda Development Center Americas, Cambridge, MA, USA*

4.1 Biodistribution and Viral Shedding

4.1.1 Introduction to Biodistribution and Viral Shedding

Biodistribution (BD) describes the distribution, persistence, and clearance of a gene therapeutic (GT) product within the body, whereas viral shedding is the release of a GT product outside the body.

BD data are required for the interpretation of nonclinical pharmacology and toxicology studies which are conducted to support early-phase clinical trials. Therefore, it is important to evaluate the BD profile of a GT product following *in vivo* administration during nonclinical development.

Shedding data are required to evaluate the secretion/excretion profile of a GT product and its potential dissemination to third parties and to the environment. Shedding studies are performed based on the outcome of the environmental risk assessment. Nonclinical shedding data can contribute to the design of clinical shedding studies.

For more in-depth background information, see Chapter 1: Introduction to AAV-based *in vivo* Gene Therapy and Chapter 2: Recent Development in *in vivo* Clinical Gene Therapy Platforms.

Drug Development for Gene Therapy: Translational Biomarkers, Bioanalysis, and Companion Diagnostics, First Edition. Edited by Yanmei Lu and Boris Gorovits.

4.1.1.1 Definition and Terminology for Biodistribution and Shedding

"Biodistribution is the *in vivo* distribution, persistence, and clearance of a GT product at the site of administration and in target and nontarget tissues, including biofluids (e.g. blood, cerebrospinal fluid, vitreous fluid)" [1]. Shedding is the release of a GT product outside the body via excreta (feces), secreta (urine, saliva, nasopharyngeal fluids, etc.), or through the skin (pustules, sores, wounds) [1].

4.1.1.2 Global Regulatory Guidance on Conducting Biodistribution and Shedding Studies

The guidance documents listed in Table 4.1 provide background on BD and shedding and give advice on the timing of the studies and the study design. For example, they provide considerations for the selection of relevant animal species, target and nontarget tissues, and sampling time points. They describe the relevant analytes (vector DNA, transgene expression products), analytical technologies, and relevant method parameters.

Table 4.1 Summary of guidelines, concept papers, and authority considerations on cell and gene therapy medicinal products addressing biodistribution and shedding.

Document title	Adopting Committee	Status (Date)
Guideline on the nonclinical studies required before first clinical use of gene therapy medicinal products [2]	EMA	Active (November 2008)
Reflection paper on quality, nonclinical, and clinical issues related to the development of recombinant adeno-associated viral vector [3]	EMA	Public (June 2010)
Guideline on the quality, nonclinical, and clinical aspects of gene therapy medicinal products [4]	EMA	Active (March 2018)
Guideline on quality, nonclinical, and clinical requirements for investigational advanced therapy medicinal products in clinical trials [5]	EMA	DRAFT (January 2019)
Guideline on the risk-based approach according to annex I, part IV of Directive 2001/83/EC applied to advanced therapy medicinal products [6]	EMA	Active (February 2013)
Preclinical assessment of investigational cellular and gene therapy products [7]	FDA	Active (November 2013)
Long-term follow-up after administration of human gene therapy products [8]	FDA	Active (January 2020)
Design and analysis of shedding studies for virus or bacteria-based gene therapy and oncolytic products [9]	FDA	Active (August 2015)

Document title	Adopting Committee	Status (Date)
Considerations on general principles to address virus and vector shedding [10]	ICH	Public (June 2009)
Concept Paper M6: Guideline on virus and gene therapy vector shedding and transmission [11]	ICH	Public (September 2009)
Guideline S12 on nonclinical biodistribution considerations for gene therapy products [1]	ICH	Active (September 2023)
Reflection paper on expectations for biodistribution (BD) assessment for gene therapy (GT) products [12]	IPRP	Final (April 2018)

EMA, European Medicines Agency; FDA, Food and Drug Administration; ICH, International Council for Harmonisation of Technical Requirements for Pharmaceuticals for Human Use; IPRP, International Pharmaceutical Regulators Programme.

The products addressed in these guidance documents are complex and diverse and the whole field is still evolving. Therefore, some content of the documents might not reflect the most recent view and, therefore, an early interaction with relevant authorities to align on the required studies and the appropriate design is recommended.

4.1.2 Nonclinical Biodistribution and Shedding Studies for AAV Vectors

BD data are required for the interpretation of the nonclinical study findings and help evaluating the benefit-risk profile of a GT product before administration in humans [12, 1]. BD data are used to understand the relationship of e.g. efficacy or safety findings to the exposure of target and nontarget tissues with the genetic material (vector DNA) and/or the expression product(s) (RNA, protein) of a GT product [13]. Therefore, BD studies often are integrated into nonclinical pharmacology or toxicology studies. Nonclinical characterization of the *in vivo* BD profile of a GT product can inform dose levels, dosing procedure, monitoring plan, and assessment of long-term follow-up in a first-in-human trial [1, 5, 8]. Stand-alone BD studies are not required but could support early product development (e.g. BD of a modified capsid) and inform sampling schedules in pivotal nonclinical studies. The known tropisms (preferred target tissues) of AAV serotypes are listed in Table 4.2.

Nonclinical shedding data may be required before the start of first-in-human trials. Currently, EMA guidance requests shedding assessment in nonclinical studies for AAV-based GT when no shedding information is available for a GT

Table 4.2 Natural tissue tropism of select AAV serotypes [14].

Serotype	Origin	Natural tropism
AAV1	Nonhuman primate	Muscle, CNS, heart, liver, lungs
AAV2	Human	Heart, CNS, liver, lungs, retina
AAV3	Nonhuman primate	Liver
AAV4	Nonhuman primate	Retina, lungs, kidney
AAV5	Human	Retina, CNS, liver
AAV6	Human	Heart, liver, muscle, retina
AAV7	Nonhuman primate	Liver
AAV8	Nonhuman primate	Muscle, heart, CNS, liver
AAV9	Human	Heart, CNS, liver
AAV10	Nonhuman primate	Muscle, myoblast tissue
AAV11	Nonhuman primate	Muscle, myoblast tissue
AAV12	Nonhuman primate	Salivary glands, muscle

product (e.g. no previous exposure, new ROA, modified tropism), while the FDA guidance does not as AAV vectors are generally non-replicating without a helper virus. Nevertheless, when sufficient data on the shedding behavior of the AAV vector used as component in the AAV GT product is available in literature or from nonclinical or clinical data of other GT products using the same AAV vector, nonclinical shedding evaluation may be omitted [4, 5, 9]. Shedding analysis may be integrated into nonclinical pharmacology or toxicology studies. Conduction of separate shedding studies is not necessary. Nonclinical shedding evaluation may help to select the sample types and to define sampling frequency and duration in a first-in-human trial [9].

4.1.2.1 Design, Execution, and Reporting

BD studies should be performed in an animal species or disease model that is biologically relevant for the GT product. The animal species should be permissive/susceptible to infection with the viral vector. The BD pattern of the vector DNA as well as of the transgene expression products (RNA, protein) in this species/disease model should be expected to mimic the BD pattern expected in the intended to treat patient population. Therefore, species differences regarding tissue tropism and viral transduction of target and nontarget tissues but also regarding regulatory elements like the promoter should be considered [12, 1, 4, 8]. The physiology and anatomy of the selected species should be comparable to humans and the use of the same delivery system/procedure as planned for the clinical study should be feasible, which may not always be the case in small animals like mice. The selected species should reflect the patient population regarding disease-specific physiological conditions (e.g. blood-ocular barrier could be impaired in patients), sex, and age and is biologically responsive to the expressed therapeutic transgene [1]. It also needs to be considered that, preexisting immunity as well as immune responses mounted to the administered GT product and its expression products may affect the BD profile of the vector and the transgene expression product(s) and may be associated with adverse events (e.g. liver enzyme increase) [15]. Nonhuman primates may be the best choice to assess immune-related effects on BD or to evaluate new regulatory elements, whereas disease models in most cases will be only available in mice. *In vitro* studies comparing transduction efficiency and transgene expression in human and animal target and nontarget cells can support species selection. This is of special importance when new AAV capsids and regulatory elements like promoters are used as components of the GT. In such cases, small stand-alone BD studies using qualitative or semiquantitative readouts like bioluminescence imaging may be used to confirm the *in vitro* data before pivotal nonclinical studies with larger numbers of animals are performed [1, 13]. Nonclinical or clinical studies performed with the same AAV capsid and/or promoter in combination with another transgene as well as literature data may also help to select a biologically relevant species.

An appropriate number of animals should be evaluated at each BD sampling time point. ICH S12 guideline (note 2) recommends the use of a minimum of 5 rodents or 3 non-rodents per sex/group/time point [1, 15]. The total number of animals can be achieved combining the BD data of several studies in case they were conducted using the same material, dose, and route of administration (ROA) [1].

Evaluation of preexisting immunity prior to inclusion in nonclinical studies should be considered to support data interpretation, especially in nonhuman primates and other non-rodent species [1]. Animals with preexisting immunity should be randomized to the study groups in nonclinical studies. It needs to be considered that exclusion of animals from nonclinical studies based on preexisting immunity might impact patient inclusion in clinical trials. Collection of samples for analysis of treatment-induced immune responses may also be considered to support data interpretation, if necessary [1].

The test article, administered in BD studies as well as the delivery device should be representative of the intended clinical device and clinical material considering important product characteristics, the manufacturing process, and final clinical formulation as changes in the manufacturing process or in the delivery procedure can affect product quality and thus the BD profile of a GT product [12, 1, 8]. For example, product impurities like empty capsids or non-functionally filled capsids can interfere with the uptake of the functional GT product and reduce exposure of target and nontarget tissues to the GT vector and its expression products. Other contaminations like host cell or helper virus-derived proteins, miss-folded proteins, aggregates, CpG-rich DNA impurities might induce additional immune reactions to the administered GT product [16].

The administered dose as well as the ROA can affect the BD profile of a GT product, including the cell types that are transduced and the immune response induced against the GT vector and its transgene expression product(s). For example, a local intrathecal administration of an AAV9 vector having a tissue tropism for CNS, lung, liver, heart, and muscle will result in a different BD profile (e.g. increased number of vector genomes in CNS) when compared to intravenous administration (e.g. majority of the vector DNA in peripheral tissues) [17–19] and even differences in the administered dose could result in different BD patterns. Therefore, BD in nonclinical studies should be assessed for the expected maximum dose level in toxicity studies, which should equate to or exceed the highest anticipated clinical dose level [1].

A vehicle control group, usually the formulation buffer, should also be included in the nonclinical studies. As the capsid is a component of the AAV GT, which can induce an immune response potentially associated with adverse observations, an empty capsid vehicle control is not warranted.

The sample collection time points should sufficiently characterize the BD profile of the GT product [1]. The collected samples should cover the maximum exposure of the GT product to target and nontarget tissues/biofluids and should sufficiently describe the steady-state period to estimate persistence. For example, in a study with a duration of 4 weeks, samples might be collected at least one and two weeks after administration of the GT product and at the end of the study. In a 13 weeks study, the second sample might be taken between week four and eight instead of week two. Early *in vivo* BD studies conducted with the same AAV vector using a reporter gene like luciferase instead of the transgene of interest in combination with bioluminescence imaging can support the selection of appropriate BD sampling time points after GT product administration [1]. To minimize potential cross-contamination, BD sampling should follow a prespecified process, e.g. starting with animals of the vehicle control group followed by the nontarget tissues of dosed animals expected to have the lowest exposure before collecting the tissues targeted by the vector according to its tissue tropism. It is important to document the order of sample collection [1] as this may support data interpretation in case unexpected BD patterns are observed.

The following core panel of tissues/biofluids should be collected: blood, injection site(s), kidney, liver, heart, lung, spleen, brain, spinal cord (cervical, thoracic, and lumbar), gonads, and adrenal gland [1]. Based on additional considerations, including GT product properties like vector type/tropism, expression product, ROA as well as disease pathophysiology, existing nonclinical data, and the intended clinical population the core panel may be expanded or even reduced. For example, additional tissues/biofluids may include vitreous fluid, eyes, optic nerve, bone marrow, dorsal root ganglia, cerebrospinal fluid, peripheral nerves, or draining lymph nodes [1]. In cases where systemic exposure is not expected (e.g. subretinal administration) collection of a reduced panel of tissues/biofluids might be sufficient. Usually, the vector DNA is measured and, depending on the GT product, the transgene expression product(s) (RNA, protein) are analyzed in BD samples containing the vector DNA [4, 15].

Animals and humans usually mount an immune response against the viral vector of the GT product and the probability that the human protein expressed from the transgene also induces an immune response is quite high in nonhuman species. Therefore, sampling of plasma/serum samples for immunogenicity analysis correlating with the BD sampling time points is recommended. In case of unexpected efficacy and/or safety findings, the immunogenicity data can support data interpretation [1, 5].

Shedding of a GT product depends on the vector's tissue tropism, the administered dose, and the ROA. Immune responses mounted against a GT product may increase the clearance of the vector resulting in reduced duration of vector shedding. Factors affecting the immune reaction to a GT product are the patient immune status, product design (e.g. CpG content, AAV capsid) as well as product impurities [9]. Recombinants of the replication-incompetent AAV vector and the replication-competent helper virus generated during manufacturing may result in a replication-competent GT product with a prolonged shedding period. Such product-related impurities should be avoided or eliminated.

Nonclinical shedding evaluation should be performed in an animal species or model that is biologically relevant for the tested GT product [9, 10]. This means that the shedding pattern of the vector DNA should mimic the shedding expected in humans. If the selected animals cannot mount an immune response to the GT product, the observed shedding pattern reflects the maximum duration of vector shedding. The highest dose and same ROA and regimen as planned in the clinical studies should be used and the administered test article as well as the delivery device should be representative of the intended clinical material and device [9, 10]. Nevertheless, the nonclinical shedding profile might not directly correlate with human shedding, reflecting different cellular and tissue sequestration, and will therefore not replace shedding evaluation in clinical studies.

The sampling frequency usually follows practical considerations and depends on each type of secreta and excreta [9, 10]. Sampling should start directly after dosing and may be more frequent in the first weeks (e.g. sampling on day 1, 3, 7,

10, 14 and then weekly, later monthly depending on study duration and shed material in the samples). Duration of the shedding assessment depends on factors like the natural course of infection of the parental virus, persistence of the vector in target and nontarget tissues, and potential immune responses to the GT product as well as of the duration of the nonclinical pharmacology or toxicology study. The shedding observation period can be shortened for a particular sample type if multiple consecutive negative samples (e.g. >3 consecutive values below LOD) are observed [9, 10].

Sample types to be analyzed for shedding are selected based on the characteristics of the vector, the ROA, and the animal species used in the study and may include urine, feces, saliva, and semen, but also buccal swabs, nasal swabs, bronchial lavage, tears, and nasopharyngeal fluids as appropriate [20]. Some of the sample types might not be accessible in every species.

For replication-incompetent viral vectors like AAV vectors, vector shedding is sufficiently assessed detecting the vector DNA (incl. the transgene sequence). An additional infectivity assay, as requested for replication-competent viruses to detect the intact virus, is not required [9, 10].

BD and shedding assessment can be integrated in Good Laboratory Practice (GLP) compliant toxicology studies and/or in pharmacology studies that might not be conducted under GLP. Even when BD/shedding assessment is integrated in a GLP toxicology study, BD/shedding samples can be analyzed without GLP compliance. However, good quality, integrity, and reliability of the generated data must be ensured [1, 5].

Currently, the vector genome DNA and/or the transgene mRNA of a GT product is measured in tissues/biofluids and secreta/excreta using nucleic acid amplification methods like quantitative PCR (qPCR) and digital droplet PCR (ddPCR) [15]. It is important to assess and document method performance parameters like sensitivity (LOQ, LOD), specificity, accuracy, reproducibility, and spike recovery during method development. Matrix interferences, expected in secreta/excreta (e.g. proteases, nucleases, ions, salts, bacterial DNA) need to be considered and addressed during method development and characterization [12, 4, 8–10]. The BD/shedding data usually are given relative to the genomic DNA as vector copy number/μg genomic DNA or in case of some biofluids like CSF or excreta like urine as vector copy number/volume (μL or mL) [1, 4, 8]. In situ hybridization (ISH) may be used to assess the morphology-based distribution of the DNA/mRNA to the different cell types within an organ/tissue.

The expressed transgene protein can be quantitatively or qualitatively detected using enzyme-linked immunosorbent assay (ELISA), LC/MS, flow cytometry, Western blot, or various *in vivo* and *ex vivo* imaging techniques. Immunohistochemistry (IHC) may be used for morphology-based BD analysis [12].

Immune responses can be analyzed using ligand binding assays to determine total antibodies or cell-based assays for determination of transduction inhibiting or neutralizing antibodies [21–24].

The preparation of a separate BD report is of advantage especially in cases where the BD data of two or more studies are combined to generate the full data set. Such a BD report can then be attached to the submission-relevant pharmacology and/or toxicology study report(s). The shedding data as well as considerations on risk of transmission to third parties should be provided with the environmental risk assessment [4].

4.1.2.2 Examples

In a GT development program, a liver targeting AAV2/8 vector was designed to express the human arylsulfatase B (ARSB) under the control of the liver-specific thyroxine-binding globulin (TBG) promoter to replace the missing enzyme in patients with Mucopolysaccharidosis Type VI [25]. In this program, a very comprehensive BD (vector DNA, mRNA, protein) and vector shedding evaluation was performed.

After intravenous administration of 2×10^{12} vg copies/kg AAV2/8.TBG.hARSB (corresponding to the expected highest clinical dose) to mice, BD was assessed on day 15 and 180 in the following tissues: adrenal gland, aorta, bone marrow, cervical spinal cord, duodenum, brain, forestomach, gallbladder, glandular stomach, thyroid with parathyroid, heart, inguinal lymph node, kidney, liver, left ovary, left testis, left tail epididymis, lung, mammary gland, esophagus, pancreas, pituitary gland, rectum, salivary gland, skin, spleen, sternum, seminal vesicles, and uterus. Additionally, blood, urine, and stool were collected for shedding analysis. Blood samples were collected before and on day 2, 9, 15, 23, 37, and 60 (optional) after administration of the test article, whereas urine and stool were collected before and on day 2, 4, 11, 14, 22, 37 (optional), and 60 (optional) after administration.

Vector DNA was analyzed using a quantitative PCR method with a LOQ of 50 vg copies/μg gDNA and a LOD of 15 vg copies/μg gDNA for tissues and stool. For plasma and urine, LOQ and LOD were 30 vg copies/well and 9 vg copies/well, respectively. hASRB RNA expression was analyzed by a quantitative qRT-PCR method with an LOQ of 625 RNA copies/μg total RNA and an LOD of 188 RNA copies/μg total RNA. The enzymatic activity of the transgene protein was measured using a fluorescence-labeled substrate directly in homogenized tissue.

The observed vector DNA BD pattern showed the typical liver tropism as expected for AAV8 with vector DNA levels of around 10^6 vg copies/μg gDNA on day 15 and between 10^5 and 10^6 vg copies/μg gDNA on day 180 after test article administration. Vector DNA levels detected in the gallbladder and adrenal gland were about 1–2 log scales lower compared to the liver. In all other tissues, the vector DNA level were below 10^4 vg copies/μg gDNA on day 15 and 180 after dosing. In line with other AAV BD studies in different species, low vector DNA levels above the LOD of the PCR method were observed in the gonads of the mice on day 15 and 180. The liver vector DNA level was quite stable until day 180 after dosing, whereas a general decline from day 15 to day 180 was observed in most of the other tissues. In line with other nonclinical studies, a statistically higher vector DNA level was observed in male liver when compared to females on day 15 and 180 after test article administration.

The highest hARSB mRNA expression with $>10^6$ RNA copies/μg total RNA was observed in liver and gallbladder with a slight decrease of expression from day 15 to day 180 after dosing. Interestingly, a quite high ARSB expression was also observed in adrenal gland, stomach, spinal cord, and intestine on day 15 although the liver (hepatocyte) specific TBG promoter was used mainly due to expression of TBG in these tissues. Transgene expression in these tissues mainly disappeared until day 180 in alignment with the decline of the vector DNA.

The enzymatic activity of the transgene protein was measured in liver, kidney, and spleen tissue. The highest activity was observed in the liver. hARSB activity was 8- and 40-fold lower in the spleen and kidney, respectively.

Vector shedding was observed until day 11 and day 14 after dosing in urine and stool, respectively. Plasma was cleared on day 23.

As vector DNA was detected in the mice gonads, a further evaluation was performed to investigate the risk of germline transmission after administration of the same test article dose to rabbits. In this study, sperm was collected to determine vector shedding in rabbit semen taking the duration of spermatogenesis (42–48 days in rabbits) into account. Samples for shedding analysis were taken on day 4, 8, 16, 31, 61, 91, 121, and 150 after dosing. In this study, vector DNA shedding in sperm disappeared after day 8 demonstrating that it is transient in rabbits. In other studies, AAV8 vector shedding in semen was observed up to 13 weeks after dosing in rabbits [26]. Shedding was shown to be dose-dependent and may depend on several factors like frequency of semen collection.

In another study in NHPs, an scAAV2/8 vector expressing human FIX under a liver-specific promoter, the vector DNA has cleared from plasma, urine, saliva, and stool by day 10 after administration of 2×10^{12} vg copies/kg [27].

4.1.3 Clinical Biodistribution and Shedding Studies for AAV Vectors

Clinical BD data can help understanding the exposure-efficacy relationship in humans and verify the predictions made based on nonclinical study data.

Shedding data are required in clinical trials to evaluate the potential risk associated with transmission of a GT product to third parties (e.g. study nurses, physicians, and patient's family members) and the potential risk to the environment [4, 9, 10].

4.1.3.1 General Considerations in Viral Shedding Studies in the Clinical Setting

The same factors as already discussed in the nonclinical section will affect vector shedding in the clinical studies. The design of clinical shedding studies therefore depends on the available information about the biological properties of the parental virus or vector (including tissue tropism, replication-competency), the administered dose and ROA, and the patient population. The immune status of the patient population and possible immunosuppressive treatments need consideration as this might impact vector shedding.

Shedding usually is assessed in initial clinical studies like first-in-human studies and can be omitted in later clinical trials provided that sufficient shedding data were collected in good quality in the early clinical studies and a consistent shedding pattern was observed in the patients [10]. If the dose, treatment regimen or anything else potentially affecting shedding is modified in later clinical studies additional shedding analysis may be required. Therefore, FDA suggests performing shedding analysis for non-replication competent vectors like AAVs in phase II instead of phase I studies [9].

Nonclinical data, data from relevant clinical studies (e.g. GT product with same AAV capsid, ROA, dose), and the characteristics of the virus/vector can inform the sampling duration and frequency and can guide the decision on which samples to collect in clinical studies. Sampling will probably be more frequent in the first days or weeks after administration and may be less frequent at later time points (months or years). A potential sampling scheme for a systemically administered AAV GT product could be as follows: weekly up to 16 weeks, then monthly up to one year and then every 3 months [28]. Sample collection and analysis usually is stopped for a particular sample type when multiple consecutive samples are tested negative (>3 samples below the LOD of the method) [9].

The same technologies and method parameters as described for sample analysis in the nonclinical section do apply to the clinical setting.

When shedding is observed the potential risk for the environment and third parties (e.g. study personnel, family members) associated with transmission of the vector must be evaluated and provided with the environmental risk assessment. The risk assessment may consider the properties of the GT product, e.g. natural route of infection, amount of shed material, duration of shedding, pathogenicity of the virus, replication competence, latency, and characteristics of the transgene [4]. Although AAV-based GT products may be associated with a lower risk for third parties as the virus is replication-deficient and non-pathogenic, precautions (e.g. contraceptive measures in case the GT product is shed through the seminal fluid) may be warranted [4, 28].

4.1.3.2 Biodistribution Characterization in Human: Necessity and Concerns

In nonclinical studies, efficacy and safety findings can be correlated to the exposure with the vector DNA and/or the transgene product(s). In clinical studies, such a correlation is often not possible as the target tissues/organs as well as most nontarget tissues/organs can only be accessed via biopsies. In most cases, such biopsies are not recommended from ethical perspectives. Therefore, the BD data evaluation in clinical studies often is limited to analysis of blood samples (serum/plasma, cellular fraction).

4.1.3.3 Examples

Viral shedding has been studied in the majority of AAV clinical trials so far [28–31], as requested by the respective guidelines [4, 5, 9, 10].

After infusion of 6×10^{12} vg copies/kg AAV2/8.TBG.hARSB, AAV vector shedding was detected in median for 13.8 weeks (13.1–23.8) in serum, 4.5 weeks (3.0–6.8) in saliva, 1.3 weeks (0.6–23.0) in urine, and 13.0 weeks (11.2–13.3) in stool above the LOD [31].

After infusion of 2×10^{12} vg copies/kg, corresponding to the dose administered to mice, described in Section 4. 1.2.2, AAV vector shedding above the LOD was detected for up to 15 weeks in serum compared to 23 days in mouse plasma, and for up to 6 or 7 weeks in urine or stool compared to 11 and 14 days in the corresponding mouse matrix [25, 31]. Overall, the shedding time windows observed in humans were not the same as in mice.

Vector DNA was analyzed using a quantitative PCR method with a LOQ of 50 vg copies/µg gDNA and a LOD of 15 vg copies/µg gDNA for tissues and stool. For serum, saliva, and urine, LOQ and LOD were 30 vg copies/well and 9 vg copies/well, respectively [25, 31].

After intravenous administration of 2×10^{12} vg/kg AAV2-FIX in a hemophilia B trial, rAAV2-FIX vector was detectable in urine up to 4 weeks and in semen up to 16 weeks [29]. In another trial, the vector has cleared from urine, semen, stool, and saliva 6 weeks after intravenous administration of 2×10^{12} vg/kg scAAV2/8-LP1-hFIXco [30]. Depending on the AAV serotype, the ROA and the administered dose, the shedding pattern as well as the time to clearance may vary.

Analysis of BD usually is limited to analysis of the vector DNA in plasma/serum and blood cells [28, 30]. In clinical trials in patients with Duchenne muscular dystrophy, biopsies of the muscle were taken to analyze at least the distribution of the transgene protein to the target tissue, but no information on vector copy numbers in muscle tissue was shared [32].

4.1.4 Gaps and Challenges on Biodistribution and Shedding Characterization

Analysis of BD is not possible in most clinical studies for ethical reasons and successful exposure to the target tissue/organ in humans can only be confirmed indirectly via expression of the transgene expression product(s) and/or efficacy readouts from the patients.

Especially when modified components like a redesigned tissue-specific promoter and/or an engineered capsid are used in a GT product, successful transgene expression observed in nonclinical species may not always translate into humans due to potential species differences in capsid tropism and/or regulation of transcription. Therefore, viral transduction efficiency and transgene expression usually is compared *in vitro* using cell lines from nonclinical species and human cell lines before nonclinical and clinical studies are performed. However, cell lines are not always predictive for the cells embedded in its tissue/organ *in vivo*. In cases where nonclinical transgene expression does not translate into clinic, complexity of the GT product will make root cause analysis difficult especially when more than one engineered component is used in that GT product. For example, when an engineered AAV capsid is used in combination with a redesigned tissue-specific promoter, both elements of the GT product contribute to the BD and transgene expression profile and each of them could be responsible for a lack of treatment response in humans. When BD of the vector DNA as well as the expression of the transgene cannot be assessed in such a case, it is challenging to identify the exact reason for the missing treatment efficacy.

Shedding samples may be collected over a timeframe of more than one year, especially when higher doses of a GT product were administered. As the patients may not always be at the clinical site when samples must be taken, appropriate sample storage and transport to the clinical site must be ensured to generate reliable shedding data. Such aspects need to be considered early during bioanalytical method development and should be reflected in the analyte stability assessment in the respective matrices. Matrices like stool heavily vary between patients and within an individual patient depending on dietary habits and medication intake. For example, the water content in the stool might be higher after a vegetable meal compared to a meat meal and the foreign DNA content might be heavily reduced after a treatment with antibiotics. Therefore, some authorities may ask to provide the shedding data for stool based on both, copies/μg DNA and copies/mg stool or copies/ml stool. Shedding results in urine also vary depending on the drinking habits of a patient and on the sampling time point, e.g. early in the morning vs. afternoon. Medication intake may also affect urine secretion and thus the shedding result. Therefore, varying results in such matrices within or between patients should not be overinterpreted.

4.2 Pharmacokinetic/Pharmacodynamic (PK/PD) Modeling and Clinical Dose Selection of Gene Therapy

4.2.1 Overview on PK/PD and Dose Selection Strategies for Gene Therapy

The foundation of PK/PD for recombinant AAV is its biological pathway after *in vivo* administration. AAV BD, transfection, and transgene expression processes that are relevant to downstream pathways of pharmacodynamics and potential adverse events are depicted in Figure 4.1. Clinical dose selection for recombinant AAV should be based on both safety and pharmacology/efficacy, and consider all applicable processes in Figure 4.1, especially those that are particularly relevant for safety and efficacy.

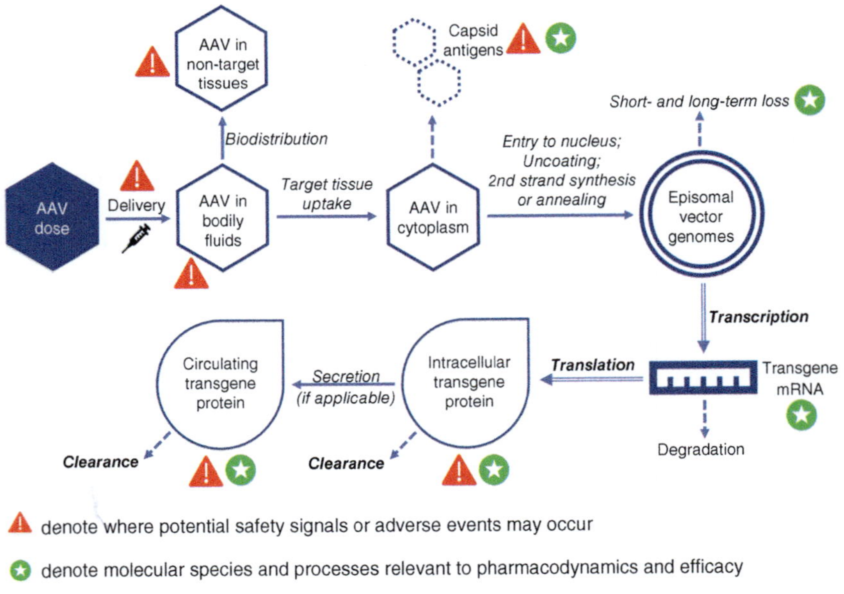

Figure 4.1 Biodistribution and transgene expression pathways of a recombinant AAV relevant to Its safety and pharmacodynamics/efficacy.

4.2.1.1 AAV Dosing Regimen – Safety Relationship and Safety-based Clinical Dose Projection

Cumulative evidence points to a strong relationship between the AAV dose (capsid and vector genome) and instances and frequency of adverse events. In addition, safety risks can also result from the delivery process of AAV, when surgery and other invasive techniques are required to deliver the recombinant AAV into the target organ/tissue. The mechanisms and examples of known safety risks related to AAV dosing (Figure 4.1) are listed in Table 4.3.

A no-adverse event level (NOAEL) dose is typically determined by nonclinical safety/toxicology studies that adhere to the Good Laboratory Practice (GLP) with the final gene therapy candidate. The NOAEL dose serves as the upper boundary of the dose range when scaling from nonclinical to human. However, due to

Table 4.3 Mechanisms of potential adverse events during and after recombinant AAV dosing.

AAV-related process	Adverse event	Relevant AAV dosing regimen	Example
Injection/ infusion into solid tissue(s)	Localized inflammation	Direct injection/ infusion	AMT-130 (uniQure) [33]
Activation of humoral immune system by AAV in blood	Thrombotic microangiopathy (TMA)	Intravenous, $\geq 1 \times 10^{14}$ cp/kg	Onasemnogene abeparvovec [34] Fordadistrogene movaparvovec [35]
AAV transfection of the liver	Inflammatory responses (increased aminotransferase levels, cytokine release); potential liver failure in severe cases	Any AAV dose; Severe cases at $\geq 5 \times 10^{13}$ cp/kg intravenous	AT-132 [36, 37] Onasemnogene abeparvovec [38]
Anti-capsid T-cell response against the host cell	Inflammatory responses (increased aminotransferase levels, cytokine release); potential cytokine storm in severe cases	Any AAV dose; Severe cases at $\geq 1 \times 10^{14}$ cp/kg intravenous	Onasemnogene abeparvovec [39]
Anti-transgene product T-cell response against the host cell			AAV2.5-CMV-minidystrophin [40]
Very high levels of circulating transgene product	Excessive pharmacology from the circulating transgene product	Depending on AAV transduction efficiency	Giroctocogene fitelparvovec [41]

cp, capsid protein.

inherent differences in lifespan, physiology, transduction efficiency by AAV, and inflammatory/immune responses to AAV, no nonclinical species is capable of full resemblance of these findings in human [42]. Therefore, the maximum tolerated dose (MTD) for any recombinant AAV should consider both its nonclinical NOAEL dose as well as clinical doses of other recombinant AAVs with the same capsid and ROA (Table 4.3).

4.2.1.2 AAV Dose – Pharmacodynamics/Efficacy Relationship and Projection of Pharmacologically-Active Dose (PAD)

The transgene product encoded by the vector genome is the primary source of pharmacodynamic responses from AAV gene therapy. PK/PD analysis and modeling of the transgene product should, therefore, include not only the kinetics of the transgene product, but also the downstream events such as biomarkers, organ functions, and clinical assessments (or equivalent nonclinical endpoints) (Table 4.4). To enable PD/efficacy-based dose projections, the endpoints in the nonclinical pharmacology model should closely mirror those in the patient population.

In addition to being reasonably safe and well tolerated, all dosing regimens for any recombinant AAV throughout its course of clinical development should also be likely to offer therapeutic benefits to trial participants. This is due to the single-dose nature of all current recombinant AAV modalities: if an AAV gene therapy fails to demonstrate sufficient efficacy to the administered subject, there is no well-tested re-dosing strategy for a subsequent augmentative or rescue administration with the same AAV. A positive benefit/risk assessment is of heightened importance if the delivery of the recombinant AAV requires complex surgical procedures, and/or if the trial population is largely pediatric [57–59]. Therefore, PAD-based dose projection should be diligently performed prior to first-in-human (FIH) dosing, preferably with multiple methods and considering more than one endpoint if available (Table 4.4).

In summary, a holistic and multi-pronged approach to clinical dose selection of Phase I or Phase I/II trials based on both safety and efficacy data is recommended (Figure 4.2). The theoretical margin of safety (MOS) in human for a particular investigational AAV modality should be well justified by observation in nonclinical species as well as with known safety events associated with high-dose AAV (Table 4.3).

4.2.2 Dose Scaling Approaches: Allometric and Activity-Based Methods

The general formulae for AAV dose scaling from animal to human are shown in Eqs. (1) and (2).

$$\text{Dose}_{\text{Human}} = \text{Dose}_{\text{Animal}} \times (\text{Morphological factor}) \times (\text{Transduction factor}) \quad (4.1)$$

Table 4.4 Select examples of recombinant AAV gene therapy by transgene product, biomarker, and efficacy endpoints for use in human efficacious dose projection.

AAV gene therapy program	Transgene, and type of the transgene product	Pharmacological and efficacy endpoint(s) in:	
		Animal model	**Human**
AMT-130 (uniQure)	miHTT (microRNA)	mHTT protein level (HD mouse and minipig) in brain and CSF [43]	mHTT level in CSF Neurofilament light level in CSF & serum HD-related clinical assessments [44, 45]
Eladocagene exuparvovec	AADC, intracellular protein	AADC distribution and enzyme activity in NHP brain[a]; [46]	Dopamin uptake in the putamen Motor function tests [46]
Onasemnogene abeparvovec	SMN1, intracellular protein	Survival in SMN1 knockout mouse model [47]	Survival at 14 months old Independent sitting [47]
Etranacogene dezaparvovec	hFIX (R338L), secreted (extracellular) protein	Plasma hFIX activity in NHP [48]	Plasma hFIX activity Annualized bleeding rate [49]
Valoctocogene roxaparvovec	hBDD-FVIII, secreted (extracellular) protein	Plasma hFVIII activity in mouse and NHP [50, 51]	Plasma hFVIII activity Annualized bleeding rate [28, 52, 53]
Giroctocogene fitelparvovec	hBDD-FVIII, secreted (extracellular) protein	Plasma hFVIII activity in mouse and NHP [54]	Plasma hFVIII activity Annualized bleeding rate [41]
Isaralgagene civaparvovec	hα-Gal, secreted (extracellular) protein	Plasma hα-Gal activity in mouse and NHP Substrate reduction in plasma and affected organs [55]	Plasma hα-Gal activity Substrate reduction in plasma [56]

AADC, aromatic L-amino acid decarboxylase; CSF, cerebrospinal fluid; hα-Gal, human α-galactosidase; HD, Huntington's Disease; hFIX, human factor IX; hFVIII, human factor VIII; mHTT, mutant Huntingtin protein; NHP, nonhuman primates; SMA, spinal muscular atrophy; VG, vector genome.

[a] Pharmacology-related NHP model was for Parkinson's disease; eladocagene exuparvovec was approved for use in AADC deficiency.

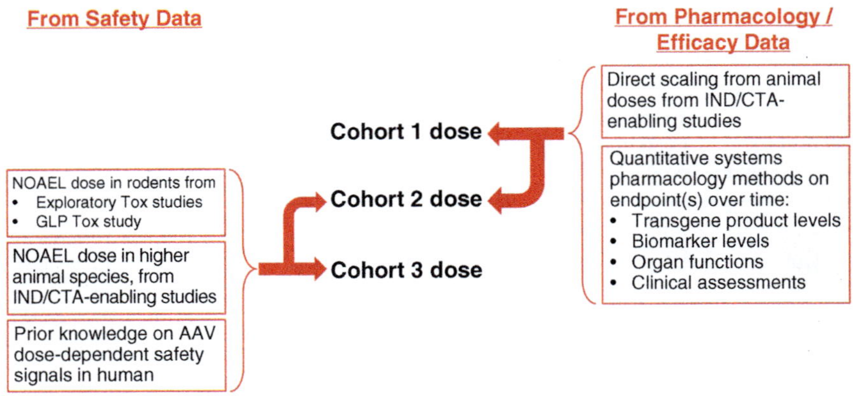

Figure 4.2 General strategy for selection of recombinant AAV doses in Phase I or Phase I/II (including first-in-human) of a clinical trial. An example of a three-cohort Phase I/II clinical study design is depicted. CTA, Clinical Trial Application; GLP, Good Laboratory Practice; IND, Investigational New Drug; NOAEL, No-Adverse Event Level.

Where

$$\text{Morphological factor} = \frac{(\text{Body or organ metric})_{\text{Human}}}{(\text{Body or organ metric})_{\text{Animal}}} \tag{4.2}$$

In Eq. (4.1), the animal dose (in total number of vector genomes) is either a PAD or NOAEL dose. The morphological factor (Eq. 4.2) accounts for physiological differences between animal and human: the most typically used body or organ metrics are body weight (for intravenously dosed AAV) and organ mass or volume (for AAV directly delivered into an organ with a well-defined boundary such as ocular space, brain parenchyma, or skeletal muscle). The transduction factor takes into consideration potential differences in AAV transfection and transgene expression between animal and human. For AAV gene therapies that either target the liver or use liver as a factory to produce secreted transgene products, at similar vg/(kg body weight) doses, mouse models consistently demonstrated transgene product levels at 20- to 100-fold of those in primate species [41, 54, 60], and nonhuman primates also showed up to 10-fold transgene product levels than human did after normalization by vg/kg dose [41, 48, 54, 60, 61]. Therefore, to achieve a similar level of transgene product in human, the nonclinical PAD must be multiplied by the species-dependent transduction factor to account for interspecies differences in transduction efficiency.

Select examples of dose scaling are listed in Table 4.5.

Table 4.5 Select examples of scaling the total vector genome dose for recombinant AAV gene therapy programs.

Program	Dose scaling basis	Morphological factor	Transduction factor[a]	Notes
AMT-130 (uniQure)	Brain biodistribution of VG in NHP & minipig; Disease biomarker modulation in minipig	Human-to-animal putamen and caudate nucleus volumetric ratios	1	Volumetric analyses applied to both the total dose and dosing volume [43, 46]
Eladocagene exuparvovec	Safety and transgene expression in NHP	Human-to-NHP total brain mass ratio and putamen volumetric ratio	1	
Onasemnogene abeparvovec	Efficacy in SMA mouse model	Body weight	1	Source [47]
Etranacogene dezaparvovec	Activity levels of hFIX in plasma	Body weight	NHP: 2.7	Source [48, 61]
Valoctocogene roxaparvovec	Activity levels of hFVIII in plasma	Body weight	Mouse: 24; NHP: 0.4	Source [28, 50, 51, 52, 62]
Giroctocogene fitelparvovec	Activity levels of hFVIII in plasma	Body weight	Mouse: 22 NHP: 13	Source [41, 54]
Isaralgagene civaparvovec	Activity levels of α-Gal in plasma	Body weight	Mouse: 50–110 NHP: 4	Source [55, 56, 63]

hα-Gal, human α-galactosidase; hFIX, human factor IX; hFVIII, human factor VIII; NHP, nonhuman primates; SMA, spinal muscular atrophy; VG, vector genome.

[a] Higher human-to-animal transduction factor means that human would need a higher AAV dose (after normalization by the morphological factor) to achieve similar transgene expression as the animal models did. Transduction factors were evaluated at peak levels observed in all species.

4.2.3 Mechanistic Approaches to Modeling Gene Therapy

Two levels of mechanistic methods can be implemented for modeling and simulations of AAV gene therapy. First, a platform approach is suitable for modeling AAV BD, as BD is determined by serotype, ROA and dose, and largely unaffected by the actual transgene sequence. Second, a program- or transgene-specific approach is necessary to model the kinetics of the transgene product and to link to downstream effects of biomarkers and clinical assessments of efficacy and safety. Integration of the platform BD module and the transgene product PK/PD would result in a full mechanistic model for a particular AAV program.

4.2.3.1 Modeling and Simulation of AAV Biodistribution

The whole-body and sub-organ level distribution of AAV (Figure 4.1) can be modeled with a system of ordinary differential equations (ODE) used in physiologically-based PK (PBPK) modeling of small molecules and proteins. In this model, the known blood flow rates, organ volumes, and sub-organ compartment volumes are combined with rates of tissue uptake of AAV, intracellular trafficking, and release/formation and loss of episomal DNA. The result is a multispecies PBPK model capable of describing the kinetics of vector genomes in tissues over time [64, 65].

4.2.3.2 Modeling Transgene Product PK and PD of the Transgene Product

The rates of synthesis of the transgene product can either be allometrically scaled from nonclinical data to human as an overall rate of transgene production [63], or mechanistically modeled with rates of transcription, translation, and protein secretion [64]. To model the PK of the secreted transgene protein, it is necessary to either possess a priori knowledge of the protein's clearance and volume of distribution values in human, or project/scale these parameter values separately for human. Once the transgene product kinetics in human has been projected, it can then be integrated with existing PD and/or efficacy models for projection of treatment response over time [63].

4.2.4 Clinical Pharmacology Considerations for Gene Therapy

4.2.4.1 Variability in Transgene Product Levels and/or Treatment Response

Both intra- and inter-individual variabilities should be expected for transgene expression in cohorts administered the same total or per body weight vector genome dose. For intra-individual variability, fluctuations of measurable transgene product levels have been commonly observed [52] and, therefore, when evaluating long-term time course data, averaging the levels over time are typically used to temper the effects of random time-varying fluctuations. On the other hand, moderate to high inter-individual variability in transgene product levels has been observed within patient cohorts, with one example of inter-individual percent coefficient of variation (%CV) at >100% for FVIII levels among 130 subjects with hemophilia A after treatment with valoctocogene roxaparvovec [53]. Known factors contributing to the inter-individual variability in transgene product levels include vector genome counts in target organ, transcription efficiency, and innate expression levels of certain host chaperone proteins that aid the intracellular folding of newly synthesized transgene product [62, 66]. Baseline characteristics and demographics, however, are generally not known to correlate to inter-individual variability in transduction [67].

4.2.4.2 Durability of Transgene Expression and/or Treatment Response

Durability of treatment effect is of particular importance to gene therapy, due to the single-dose nature for recombinant AAV and no proven re-dosing paradigm. Overall, four scenarios of transgene product levels versus time after dose have been observed for multiple AAV gene therapy trials (Figure 4.3). Decline of the transgene product levels can occur at a short time frame after dosing (a few weeks to one year), typically in a precipitous fashion shown in Scenarios II and IV of Figure 4.3, and/or manifest gradually and slowly over a longer period as seen in Scenarios III and IV of Figure 4.3. All four scenarios may manifest within the same dose cohort of subjects receiving a particular recombinant AAV.

The causes of both near- and long-term loss of transgene product levels remain an active field of investigation. Drastic reduction of transgene product levels shortly after the gene therapy dose is most likely due to immune responses (Table 4.5). Prophylactic or timely reactive immunosuppression regimens may mitigate both the incidence and magnitude of these immune responses [68]. In contrast, mechanisms of relatively slower loss of transgene products over a longer period are attributed to multiple potential intrinsic and extrinsic factors (Table 4.6). While it is not possible to modulate rates of host cell turnover, it is rational for subjects administered a gene

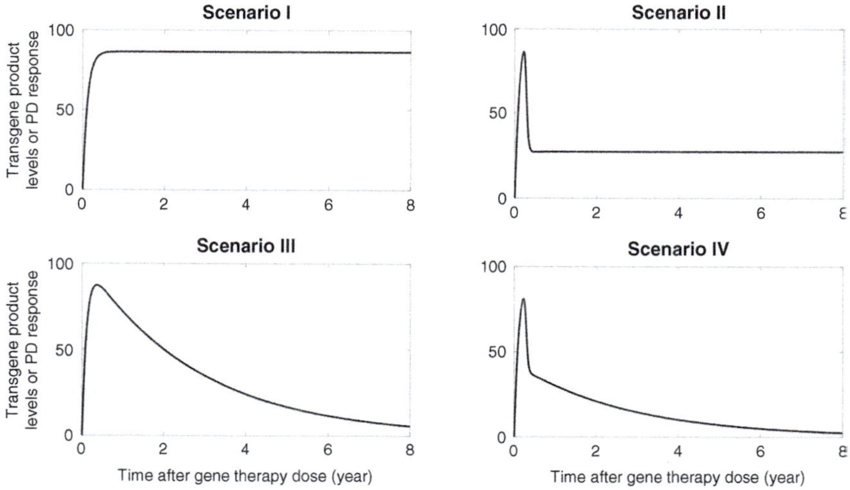

Figure 4.3 Scenarios of transgene product or pharmacodynamic response over time after a single dose of AAV-based gene therapy. The *y*-axis is in linear scale and with an arbitrary unit. PD, pharmacodynamic.

Table 4.6 Potential factors leading to loss of transgene product levels and pharmacodynamic response within different time frames after the gene therapy dosing.

Loss of measurable transgene product levels and/or PD response within...	Potential contributing factors	
	Intrinsic factors	**Extrinsic factors**
Near-term (A few weeks to half or one-year post-dose)	Immune response against transduced host cells Lack of response to the immunosuppression regimen Host cell intracellular stress response	Inadequate immunomodulatory/ immunosuppression regimen
Long-term (Beyond half or one-year post-dose)	Organ growth and host cell turnover Intermittent stress response within the host cells to the transgene product Silencing of the transgene Generation of anti-transgene product-neutralizing antibodies	Lifestyle (smoking, alcohol use) Pathogenic viral infection, drug-induced damage, or chronic disease(s) of the organ(s) hosting the vector genomes

PD, pharmacodynamic.

therapy to avoid, if possible, pathogenic infections and chronic disease to the organ hosting the recombinant AAV vector genomes.

4.2.5 Gaps and Challenges on PK/PD and Clinical Dose Selection

The gaps that remain for PK/PD analyses and clinical dose selection for gene therapy can be roughly grouped into three categories.

4.2.5.1 Interspecies difference in AAV Transduction and Immunogenicity

Nonclinical animal models are crucial for understanding the mechanistic aspects of gene therapy. However, interspecies differences are notable in various aspects of recombinant AAV gene therapy. For transduction and transgene expression from the liver, mouse is known to exhibit much higher transgene production than human does; nonhuman primates typically display similar or slightly higher peak levels of transgene product than human but then decline sharply due to generation of antihuman transgene product immune responses [60]. For immunogenicity, to date, no animal model can duplicate the findings in human on anti-capsid and anti-transgene product T-cell response as well as potential endoplasmic reticulum (ER) stress and transcription silencing [42]; current animal models differ from human in incidence, magnitude, and duration of these events.

4.2.5.2 Availability of Clinical Samples and Bioanalytical Assays

For mechanistic understanding and modeling of gene therapy, it is important to obtain quantitative measurements of vector genome and transgene mRNA levels in the target organ of AAV, as well as transgene product levels in the organ/matrix of interest for the disease [62]. However, this is hampered by the unavailability of biopsied samples due to ethical concerns (skin, skeletal muscle, and liver are the only three solid organs that are routinely biopsied). In addition, due to challenges in developing mass spectrometry-based assays, the absolute amount of intracellular transgene product may be difficult to measure, and thus need to be substituted with either an activity-based assay or downstream biomarker levels as surrogates.

4.2.5.3 Availability of Long-Term Follow-Up Data

To date, only a few clinical trials on recombinant AAV have released follow-up data through or beyond year 5 post-dose [53, 69, 70, 71]. In one case, the transgene product (human factor VIII) levels continuously declined but the efficacy endpoint (reduction of bleeding events) was maintained [53]; in two cases, the transgene product (human factor IX) were stably maintained at low absolute levels [69, 70]; in the last case, the transgene product (SMN1) was not measured but efficacy (survival) was maintained in all 10 subjects in the therapeutic dose cohort [71].

4.3 Summary

Understanding the vector capsid biology (including tissue tropism, secretion/excretion, preexisting immunity) across different species, optimization of transgene expression constructs (e.g. ubiquitous vs. tissue-specific promotors), and clinical dose selection based on exposure-safety or exposure-efficacy findings are important components of the nonclinical development of GT products for which BD analysis of the vector DNA and/or the transgene expression constructs (RNA, protein) is indispensable. Therefore, BD studies usually are integrated in nonclinical pharmacology and toxicology studies.

Shedding analysis is required to protect third parties and the environment in clinical studies. Shedding profiles determined in nonclinical studies can inform clinical sampling regimen (including duration, frequency, and sample types).

Pharmacokinetic/pharmacodynamic analysis of recombinant AAV gene therapy should integrate all available data sources, including BD as measured by the vector genome, transgene product levels, biomarker response, and efficacy endpoints. Clinical dose selection for AAV is informed by both safety and efficacy data in nonclinical models and humans, and should consider both allometric scaling and mechanistic methods.

References

1 ICH guideline S12 on nonclinical biodistribution considerations for gene therapy products 2023. Available from: https://www.fda.gov/media/167605/download.
2 European Medicines Agency (2008). Guideline on the nonclinical studies required before first clinical use of gene therapy medicinal products. EMEA/CHMP/GTWP/125459/2006. Available from: https://www.ema.europa.eu/en/documents/scientific-guideline/guideline-non-clinical-studies-required-first-clinical-use-gene-therapy-medicinal-products_en.pdf.
3 European Medicines Agency Reflection paper on quality, non-clinical and clinical issues related to the development of recombinant adeno- associated viral vectors. EMEA/CHMP/GTWP/587488/2007 Rev. 1. Available from: https://www.ema.europa.eu/en/documents/scientific-guideline/reflection-paper-quality-non-clinical-clinical-issues-related-development-recombinant-adeno_en.pdf.
4 European Medicines Agency (2018). Guideline on the quality, non-clinical and clinical aspects of gene therapy medicinal products. EMA/CAT/80183/2014. Available from: https://www.ema.europa.eu/en/documents/scientific-guideline/guideline-quality-non-clinical-clinical-aspects-gene-therapy-medicinal-products_en.pdf.
5 European Medicines Agency (2019). Guideline on quality, non-clinical and clinical requirements for investigational advanced therapy medicinal products in clinical trials – scientific guideline. Available from: https://www.ema.europa.eu/documents/scientific-guideline/draft-guideline-quality-non-clinical-clinical-requirements-investigational-advanced-therapy_en.pdf.
6 European Medicines Agency (2013). Guideline on the risk-based approach according to annex I, part IV of Directive 2001/83/EC applied to Advanced therapy medicinal products. EMA/CAT/CPWP/686637/2011. Available from: https://www.ema.europa.eu/en/documents/scientific-guideline/guideline-risk-based-approach-according-annex-i-part-iv-directive-2001/83/ec-applied-advanced-therapy-medicinal-products_en.pdf.
7 U.S. Food and Drug Administration (2013). Preclinical assessment of investigational cellular and gene therapy products: guidance for industry. Available from: https://www.fda.gov/media/87564/download.
8 U.S. Food and Drug Administration 2020 Long term follow-up after administration of human gene therapy products: guidance for industry. Available from: https://www.fda.gov/media/113768/download.

9 U.S. Food and Drug Administration (2015). Design and analysis of shedding studies for virus or bacteria-based gene therapy and oncolytic products: guidance for industry. Available from: https://www.fda.gov/files/vaccines%2C%20blood%20%26%20biologics/published/Design-and-Analysis-of-Shedding-Studies-for-Virus-or-Bacteria-Based-Gene-Therapy-and-Oncolytic-Products--Guidance-for-Industry.pdf.

10 International Conference on Harmonisation of Technical Requirements for Registration of Pharmaceuticals for Human Use considerations: General principles to address virus and vector shedding. EMEA/CHMP/ICH/449035/2009. (2009). Available from: https://www.ema.europa.eu/en/documents/scientific-guideline/international-conference-harmonisation-technical-requirements-registration-pharmaceuticals-human-use_en-10.pdf.

11 International Conference on Harmonisation of Technical Requirements for Registration of Pharmaceuticals for Human Use. Final Concept Paper, M6: Guideline on Virus and Gene Therapy Vector Shedding and Transmission. (2009). Available from: https://database.ich.org/sites/default/files/M6_Concept_Paper.pdf.

12 International Pharmaceutical Regulators Programme (2018). Expectations for Biodistribution (BD) Assessments for Gene Therapy (GT) Products. Available from: https://admin.iprp.global/sites/default/files/2018-09/IPRP_GTWG_ReflectionPaper_BD_Final_2018_0713.pdf.

13 Assaf, B.T. and Whiteley, L.O. (2018). Considerations for preclinical safety assessment of adeno-associated virus gene therapy products. *Toxicol. Pathol.* 46 (8): 1020–1027.

14 Korneyenkov, M.A. and Zamyatnin, A.A. Jr. (2021). Next step in gene delivery: modern approaches and further perspectives of AAV Tropism modification. *Pharmaceutics* 13 (5): 750.

15 Bolt, M.W., Whiteley, L.O., Lynch, J.L. et al. (2020). Nonclinical studies that support viral vector-delivered gene therapies: an EFPIA gene therapy working group perspective. *Mol. Ther. Methods Clin. Dev.* 19: 89–98.

16 Wright, J.F. (2014). Product-related impurities in clinical-grade recombinant AAV vectors: characterization and risk assessment. *Biomedicine* 2 (1): 80–97.

17 Saraiva, J., Nobre, R.J., and Pereira de Almeida, L. (2016). Gene therapy for the CNS using AAVs: the impact of systemic delivery by AAV9. *J. Control. Release* 241: 94–109.

18 Samaranch, L., Salegio, E.A., San Sebastian, W. et al. (2012). Adeno-associated virus serotype 9 transduction in the central nervous system of nonhuman primates. *Hum. Gene Ther.* 23 (4): 382–389.

19 Miyake, N., Miyake, K., Yamamoto, M. et al. (2011). Global gene transfer into the CNS across the BBB after neonatal systemic delivery of single-stranded AAV vectors. *Brain Res.* 1389: 19–26.

20 Schenk-Braat, E.A., van Mierlo, M.M., Wagemaker, G. et al. (2007). An inventory of shedding data from clinical gene therapy trials. *J. Gene Med.* 9 (10): 910–921.

21 Gorovits, B., Azadeh, M., Buchlis, G. et al. (2021). Evaluation of the humoral response to adeno-associated virus-based gene therapy modalities using total antibody assays. *AAPS J.* 23 (6): 108.

22 Gorovits, B., Fiscella, M., Havert, M. et al. (2020). Correction to: recommendations for the development of cell-based anti-viral vector neutralizing antibody assays. *AAPS J.* 22 (2): 42.

23 Gorovits, B., Fiscella, M., Havert, M. et al. (2020). Recommendations for the development of cell-based anti-viral vector neutralizing antibody assays. *AAPS J.* 22 (2): 24.

24 Gorovits, B., Marshall, J.C., Smith, J. et al. (2019). Bioanalysis of adeno-associated virus gene therapy therapeutics: regulatory expectations. *Bioanalysis* 11 (21): 2011–2024.

25 Ferla, R., Alliegro, M., Marteau, J.B. et al. (2017). Non-clinical safety and efficacy of an AAV2/8 vector administered intravenously for treatment of mucopolysaccharidosis Type VI. *Mol. Ther. Methods Clin. Dev.* 6: 143–158.

26 Favaro, P., Downey, H.D., Zhou, J.S. et al. (2009). Host and vector-dependent effects on the risk of germline transmission of AAV vectors. *Mol. Ther.* 17 (6): 1022–1030.

27 Nathwani, A.C., Rosales, C., McIntosh, J. et al. (2011). Long-term safety and efficacy following systemic administration of a self-complementary AAV vector encoding human FIX pseudotyped with serotype 5 and 8 capsid proteins. *Mol. Ther.* 19 (5): 876–885.

28 Pasi, K.J., Rangarajan, S., Mitchell, N. et al. (2020). Multiyear Follow-up of AAV5-hFVIII-SQ Gene Therapy for Hemophilia A. *N. Engl. J. Med.* 382 (1): 29–40.

29 Manno, C.S., Pierce, G.F., Arruda, V.R. et al. (2006). Successful transduction of liver in hemophilia by AAV-Factor IX and limitations imposed by the host immune response. *Nat. Med.* 12 (3): 342–347.

30 Nathwani, A.C., Reiss, U.M., Tuddenham, E.G. et al. (2014). Long-term safety and efficacy of factor IX gene therapy in hemophilia B. *N. Engl. J. Med.* 371 (21): 1994–2004.

31 Brunetti-Pierri, N., Ferla, R., Ginocchio, V.M. et al. (2022). Liver-directed adeno-associated virus-mediated gene therapy for mucopolysaccharidosis Type VI. *NEJM Evidence* 1 (7): https://doi.org/10.1056/EVIDoa2200052.

32 Duan, D. and Systemic, A.A.V. (2018). Micro-dystrophin gene therapy for Duchenne muscular dystrophy. *Mol. Ther.* 26 (10): 2337–2356.

33 uniQure Announces Second Quarter 2022 Financial Results and Highlights Recent Company Progress. (2022). Available from: https://www.globenewswire.com/en/news-release/2022/08/08/2493736/0/en/uniQure-Announces-Second-Quarter-2022-Financial-Results-and-Highlights-Recent-Company-Progress.html.

34 Chand, D.H., Zaidman, C., Arya, K. et al. (2021). Thrombotic microangiopathy following onasemnogene Abeparvovec for spinal muscular atrophy: a case series. *J. Pediatr.* 231: 265–268.

35 U.S. Food and Drug Administration (FDA) (2021). Cellular, Tissue, and Gene Therapies Advisory Committee (CTGTAC) meeting #70, transcript for day 2 open session. Available from: https://www.fda.gov/media/154121/download.

36 U.S. Food and Drug Administration (FDA) (2021). Cellular, Tissue, and Gene Therapies Advisory Committee (CTGTAC) meeting #70: Toxicity Risks of Adeno-associated Virus (AAV) Vectors for Gene Therapy (GT), briefing document. Available from: https://www.fda.gov/media/151599/download.

37 Watkins, P.B. (2021). Liver injury due to drugs and viruses: mechanistic similarities and implications for AAV gene therapy. *Clin. Pharmacol. Ther.* 112: 751–753.

38 Chand, D., Mohr, F., McMillan, H. et al. (2021). Hepatotoxicity following administration of onasemnogene abeparvovec (AVXS-101) for the treatment of spinal muscular atrophy. *J. Hepatol.* 74 (3): 560–566.

39 Galletta, F., Cucinotta, U., Marseglia, L. et al. (2022). Hemophagocytic lymphohistiocytosis following gene replacement therapy in a child with type 1 spinal muscular atrophy. *J. Clin. Pharm. Ther.* 47 (9): 1478–1481.

40 Flanigan, K.M., Campbell, K., Viollet, L. et al. (2013). Anti-dystrophin T cell responses in Duchenne muscular dystrophy: prevalence and a glucocorticoid treatment effect. *Hum. Gene Ther.* 24 (9): 797–806.

41 Leavitt, A.D., Konkle, B.A., Stine, K. et al. (2020). Updated follow-up of the Alta study, a phase 1/2 study of Giroctocogene Fitelparvovec (SB-525) gene therapy in adults with severe hemophilia a. *Blood* 136.

42 Ertl, H.C.J. (2019). Preclinical models to assess the immunogenicity of AAV vectors. *Cell. Immunol.* 342: 103722.

43 Evers, M., de Haan, M., Valles-Sanchez, A. et al. (2019). Translating preclinical data to a human equivalent dose for AMT-130 AAV gene therapy for early manifest Huntington's disease. *Mov. Disord.* 34: S7–S8.

44 Rodrigues, F.B. and Wild, E.J. (2020). Huntington's Disease clinical trials corner: April 2020. *J. Huntingtons Dis.* 9 (2): 185–197.

45 (2019). *Safety and Proof-of-Concept (POC) Study With AMT-130 in Adults With Early Manifest Huntington Disease*. ClinicalTrials.gov Identifier: NCT04120493. Available from: https://clinicaltrials.gov/ct2/show/NCT04120493.

46 EMA/CHMP/571076/2022: assessment report for Upstaza, International non-proprietary name: eladocagene exuparvovec (2022). Available from: https://www.ema.europa.eu/en/documents/assessment-report/upstaza-epar-public-assessment-report_en.pdf.

47 U.S. Food and Drug Administration (2019). ZOLGENSMA (onasemnogene abeparvovec-xioi): highlights of prescribing information. Available from: https://www.fda.gov/media/126109/download.

48 Spronck, E.A., Liu, Y.P., Lubelski, J. et al. (2019). Enhanced factor IX activity following administration of AAV5-R338L "Padua" factor IX versus AAV5 WT human factor IX in NHPs. *Mol. Ther. Methods Clin. Dev.* 15: 221–231.

49 uniQure and CSL Behring Announce Primary Endpoint Achieved in HOPE-B Pivotal Trial of Etranacogene Dezaparvovec Gene Therapy in Patients with Hemophilia B. (2021). Available from: https://www.cslbehring.com/newsroom/2021/hope-b-gene-therapy-for-hemophilia-b-topline-results.

50 Bunting, S., Zhang, L., Xie, L. et al. (2018). Gene therapy with BMN 270 results in therapeutic levels of FVIII in mice and primates and normalization of bleeding in hemophilic mice. *Mol. Ther.* 26 (2): 496–509.

51 Long, B.R., Sandza, K., Holcomb, J. et al. (2019). The impact of pre-existing immunity on the non-clinical pharmacodynamics of AAV5-based gene therapy. *Mol. Ther. Methods Clin. Dev.* 13: 440–452.

52 Rangarajan, S., Walsh, L., Lester, W. et al. (2017). AAV5-factor VIII gene transfer in severe hemophilia A. *N. Engl. J. Med.* 377 (26): 2519–2530.

53 Ozelo, M.C., Mahlangu, J., Pasi, K.J. et al. (2022). Valoctocogene Roxaparvovec gene therapy for hemophilia A. *N. Engl. J. Med.* 386 (11): 1013–1025.

54 Meyer, K. (2019). SB-525, a novel gene therapy for treatment of hemophilia A. Presented at the NorCal SOT meeting. Available from: https://www.toxicology.org/groups/rc/NorCal/docs/NorCal-SOT-2019-Fall-Symposium-Meyer-Presentation.pdf.

55 Yasuda, M., Huston, M.W., Pagant, S. et al. (2020). AAV2/6 gene therapy in a murine model of fabry disease results in supraphysiological enzyme activity and effective substrate reduction. *Mol. Ther. Methods Clin. Dev.* 18: 607–619.

56 Sangamo Investor Presentation (May 2022). Available from: https://investor.sangamo.com/static-files/79ddb512-5379-4537-8e55-1e8903e9763d.

57 U.S. Food and Drug Administration (2020). Human gene therapy for hemophilia: guidance for industry. Available from: https://www.fda.gov/media/113799/download.

58 U.S. Food and Drug Administration (2021). Human gene therapy for neurodegenerative diseases: draft guidance for industry. Available from: https://www.fda.gov/media/144886/download.

59 U.S. Food and Drug Administration (2020). Human gene therapy for rare diseases: guidance for industry. Available from: https://www.fda.gov/media/113807/download.

60 Chen, N., Sun, K., Chemuturi, N.V. et al. (2022). The perspective of DMPK on recombinant adeno-associated virus-based gene therapy: past learning, current support, and future contribution. *AAPS J.* 24 (1): 31.

61 Von Drygalski, A., Giermasz, A., Castaman, G. et al. (2019). Etranacogene dezaparvovec (AMT-061 phase 2b): normal/near normal FIX activity and bleed cessation in hemophilia B. *Blood Adv.* 3 (21): 3241–3247.

62 Fong, S., Yates, B., Sihn, C.R. et al. (2022). Interindividual variability in transgene mRNA and protein production following adeno-associated virus gene therapy for hemophilia A. *Nat. Med.* 28 (4): 789–797.

63 Der, K., Wada, R., Huston, M. et al. (2022). ST-920 translational pharmacokinetic–pharmacodynamic (PK/PD) model from mouse to human in Fabry disease. In: *Poster #066 Presented at the 18th Annual WORLDSymposium, February 2022, San Diego, CA, USA*.

64 Narula, J., Luo, H.B., Ko, G. et al. (2022). A quantitative systems pharmacology (QSP) model for design and species-translation of bio-distribution studies of AAV-based gene therapies. *Mol. Ther.* 30 (4): 199–200.

65 Sun, K.F., Zhang, Z.W., Ko, G. et al. (2021). Physiologically based pharmacokinetic modeling for the biodistribution of adeno-associated virus serotype 8 after intravenous administration in mice and non-human primates. *Mol. Ther.* 29 (4): 129–130.

66 Fong, S., Handyside, B., Sihn, C.R. et al. (2020). Induction of ER stress by an AAV5 BDD FVIII construct is dependent on the strength of the hepatic-specific promoter. *Mol. Ther. Methods Clin. Dev.* 18: 620–630.

67 Roctavian: EPAR – Product Information. (2022). Available from: https://www.ema.europa.eu/documents/product-information/roctavian-epar-product-information_en.pdf.

68 Verdera, H.C., Kuranda, K., and Mingozzi, F. (2020). AAV vector immunogenicity in humans: a long journey to successful gene transfer. *Mol. Ther.* 28 (3): 723–746.

69 Nathwani, A.C., Reiss, U., Tuddenham, E. et al. (2018). Adeno-associated mediated gene transfer for hemophilia B:8 year follow up and impact of removing "Empty Viral Particles" on safety and efficacy of gene transfer. *Blood* 132 (Suppl 1): 491.

70 Leebeek, F.W.G., Meijer, K., Coppens, M. et al. (2021). Amt-060 gene therapy in adults with severe or moderate-severe hemophilia B confirm stable fix expression and sustained reductions in bleeding for up to 5 years. *Haemophilia* 27: 41.

71 Mendell, J.R., Al-Zaidy, S.A., Lehman, K.J. et al. (2021). Five-year extension results of the phase 1 START trial of Onasemnogene Abeparvovec in spinal muscular atrophy. *JAMA Neurol.* 78 (7): 834–841.

5

Immunogenicity of AAV Gene Therapy Products

Vibha Jawa[1] *and Bonnie Wu*[2]

[1] *Clinical Pharmacology, Pharmacometrics and Bioanalysis (CPPB), Bristol Myers Squibb, Princeton, NJ, USA*
[2] *Biologics Discovery and Development Sciences, Johnson & Johnson Innovative Medicine, Spring House, PA, USA*

Immune response to recombinant AAV (rAAV) based therapies has been well characterized through preclinical and clinical experience [1]. The immune response can be mediated by product-associated risks that include viral capsids, genome containing the nucleic acid as well as the transgene product. Each of these structural components can induce either an innate or adaptive phase immune response. In addition, the clinical risks can be related to the nature of genetic mutation, disease state, route of administration, and site of injection as well as any standard of care treatments that can contribute to the risks [1].

5.1 Innate and Adaptive Immunity Induced by AAV-Based Gene Therapies

5.1.1 Innate Immune Response

The adeno-associated virus (AAV) vectors were not considered very strong elicitors of innate or adaptive immune response when compared to Adenovirus (Adv) [2]. However, preclinical and clinical observations have shown that the AAV-derived content (capsids, nucleic acid content, translated gene product, etc.) may be recognized as foreign. The membrane bound as well as intracellular pattern recognition receptors (PRRs) on innate immune cells like macrophages and

Drug Development for Gene Therapy: Translational Biomarkers, Bioanalysis, and Companion Diagnostics, First Edition. Edited by Yanmei Lu and Boris Gorovits.

dendritic cells can react with AAV and related content if they have structural motifs/patterns [pathogen associated molecular patterns (PAMPs)] similar to pathogenic organisms [3].

Multiple mechanisms of action have been identified by which rAAV can activate innate immune response. This includes (1) AAV capsid inducing Toll-like receptor 2 (TLR2) present on cell surface or endosomal membrane, (2) AAV packaged genome is hydrolyzed in lysosome or endosome to release viral nucleic acid content triggering a Toll-like receptor 9 (TLR9) mediated response, (3) activation of stimulator of interferon genes (STING)/interferon regulatory factor 3 (IRF3) pathway by mitochondrial DNA that triggers cytosolic sensors leading to upregulation of type I interferons (IFNs), (4) additional pathways like inflammasome formation and viral RNA sensors retinoic acid-inducible gene I (RIG-1) and anti-melanoma differentiation-associated gene-5 (MDA-5) induced IFNs [4, 5]. All pathways lead to release of proinflammatory chemokines like tumor necrosis factor-alpha (TNF-α) and interleukin (IL-6) through engagement of myeloid differentiation primary response 88 (MyD88)/interleukin-1 receptor-associated kinase (IRAK) pathways while inflammasome maturation leads to IL-1β and IL-18 production [6]. An activation of innate immune response through IF-stimulated genes (ISGs) can cause inhibition of viral replication and consequent spreading. In addition, innate phase cytokines and chemokines can prime a T-cell-mediated adaptive immune response that could also induce a long-term memory response [7].

An additional pathway that can induce innate response could be complement-dependent activation triggered by pre-existing anti-AAV antibodies (PEA) complexed with AAV capsids [8, 9]. High-dose AAV complexed with PEA can activate classical complement cascade leading to cell lysis through membrane attack complex (MAC) [10]. Low-dose AAV leads to generation of complement factors C3b and C3d that opsonize AAV and can interact with their respective receptors CR1 and CR3 expressed on macrophages [8]. This leads to an enhanced uptake and antigen presentation which can further recruit T cells. B cells can also uptake C3d opsonized viral particles through CR2 and cross-link with B cells to induce effector and memory B cells (Figure 5.1).

The innate immune response mediated by complement activation (complement proteins C3 and its cleaved products C3b, C3bi, C3d) can be activated by direct interaction with AAV capsid or through pre-existing anti-capsid antibodies complexed with low doses of AAV capsid. The complement cascade can eventually lead to formation of the MAC that can lyse the cell. The anaphylactic fragments C3a and C5a generated during cleavage can induce inflammatory responses. Other innate pathways include engagement and activation of the toll-like receptors TLR 3,7,8 TLR 9 by viral capsid and genome respectively that further prime the adaptive phase immune response.

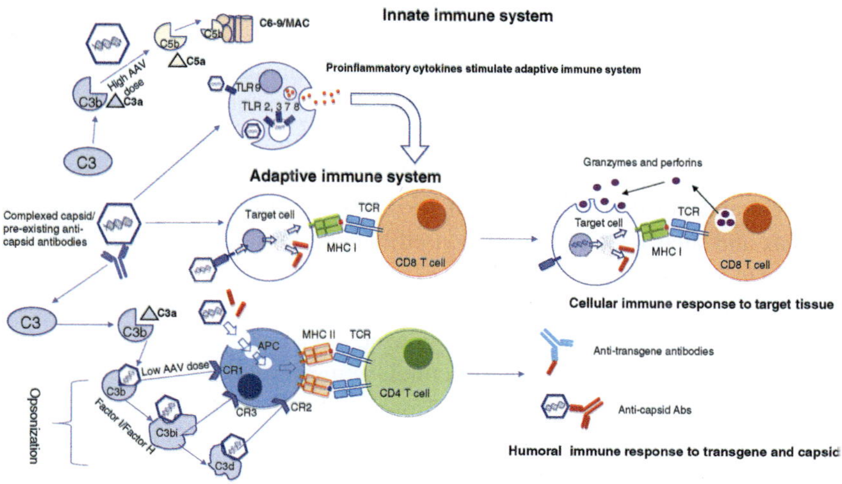

Figure 5.1 AAV-mediated activation of innate and adaptive phase immune responses. APC, antigen-presenting cell (Macrophage/Dendritic cell/B-cell); CR1, complement receptor recognizing complement fragment C3b on macrophages; CR2, complement receptor recognizing complement fragment C3d on B-cells and dendritic cells; CR3, complement receptor recognizing the inactivated C3b (C3bi) on macrophages; and dendritic cells; TCR: T-cell receptor; MHC: major histocompatibility complex; MAC: membrane attack complex.

5.1.2 Adaptive Immune Response

Both humoral and cell-mediated immune responses can occur following the administration of AAV-based gene therapies. The humoral response includes pre-existing antibodies (PEAs) to AAV capsids that can limit the viral transduction as well as treatment emergent antibody response to capsid-specific epitopes [11]. The transgene expression in the targeted cells or tissues can also be associated with an antibody response. Both capsid proteins and transgene product can also elicit MHC Class I and Class II-driven cellular response. MHC Class 1 activates a cytotoxic CD8+ driven T-cell response while MHC Class 2 leads to a CD4+ driven Teffector response.

5.2 Preclinical Immunogenicity Risk Assessment

The immune response to AAV gene therapies (GTs) can occur at any stage of development. Hence, a comprehensive risk assessment is needed that will identify the relevant risk factors. The risk factors can be product, process, and patient derived as summarized in sections below.

5.2.1 Product-related Risk Factors

Product-related risk factors can arise from the vector genome, viral capsid serotype, and transgene product. The capsid-associated risk factors include post-translationally modified capsid-derived peptides and aggregation [12–14]. Similarly, vector genome-related risk factors include unmethylated CpG motifs, self-complementary DNA, as vector DNA and viral dsDNA that can engage PRRs like TLR9 [15–17]. The serotype of capsids and their related transduction efficiencies in different tissues can also contribute to the risk [18–20]. The choice of serotypes can impact clinical outcomes such as safety, vector clearance, treatment eligibility, occurrence of transaminase elevations, activation of capsid-directed cytotoxic T-cell responses, and clinical efficacy. There is a similarity in several structural features responsible for tissue specificity and transduction across serotypes. However, there are distinct features that can also impact gene transfer and immunogenicity. Among the 13 AAV serotypes that have been identified, AAV5 is the most distinct and has the least homology to more common serotypes like AAV2, AAV8, and AAV10. Based on these homologies, the risk of cross-reactive sero-reactivity also varies [21] and selection of capsids where the least pre-existing immunogenicity risk is anticipated can help mitigate any potential immune-mediated vector clearance.

Similarly, serotype tropism can limit the non-specific uptake of the vector and ensures the targeting and concentration in the relevant tissue. While primary receptors influence binding, the secondary receptors support internalization and can influence tropism across the viral vector serotypes [22]. One way to ensure the targeted delivery of the vector is through use of tissue specific promoters as observed in some recent clinical trials where liver-specific promoters were used to treat FVIII and FIX gene defects.

Such strategies can reduce the overall risk of pre-existing antibody cross-reactivity that can lead to elimination of transduced vector and loss of efficacy. Additionally, limited and targeted high level of transgene expression using the tissue-specific serotype can ensure less toxicity due to complement activation and cytotoxic T-cell mediated elimination of off-target cells. There may be differences in capsid-specific T-cell responses based on serotypes. While AAV2, 8, and 10 were associated with elevated liver transaminases, increased capsid-specific cytotoxic T-lymphocyte (CTL) responses, and elimination of transduced gene product in liver cells, AAV5 did not elicit any CTL responses even with or without increase of liver ALT (amino alanine transferase) and AST (aspartate aminotransferase) levels.

5.2.2 Process and Manufacturing-Related Risk Factors

The empty capsid content in the final packaged AAV-GT product can change from lot to lot as part of purification process [23–26]. Additional impurities include the biosynthetic intermediates and incorrectly composed viral particles, AAV encapsidated host DNA fragment from different species like insect cell DNA, nuclease-sensitive nucleic acids, helper components, viruses relevant to cell lines and plasmid DNA, etc. [27, 28]. New analytical characterization approaches have pointed to presence of partially packaged viral genomes even in fully packaged capsids.

5.2.3 Patient-Related Risk Factors

The capsid-derived peptides are presented in the context of Class I and Class II human leukocyte antigen (HLA)/major histocompatibility complex (MHC) that are used interchangeably. The risk of such peptides driving a CTL (CD8+) or Teffector (CD4+) response depends on the HLA alleles of the patients and their affinity to these peptides [29]. The genetic associations with an inflammatory diseased state can also enhance the risk [30].

The transgene-driven protein expression and the consequent immune response could differ based on the nature of gene defects (null vs. point mutation vs. spliced). The engineered protein expressed by transgene has a much higher risk compared to the endogenous protein that is close to self [31]. Based on regional exposure and patient demographics, the pre-existing capsid-specific sero-reactivity can be different. The disease severity and the immune robustness of the patient can also change the risk due to underlying disease that can exacerbate the immune response to the AAV-GT vectors. The highly vascularized nature of site where the GT vector is delivered vs. immune suppressed/immune deficient site can change the risk of immunogenicity. Based on the age of patient (pediatric vs. adult vs. seniors), seroprevalence to the capsid serotypes may differ and can change the risk for reduction of transgene expression due to complement fixing anti-capsid immune complexes. The high dose of AAV genomes has also been associated with adverse events, especially in patients with a severe disease that cannot tolerate the high viral burden [32]. Lastly, the administration through an IV route compared to a local delivery into the subretinal space or liver or muscle may change the risk and is dependent on the immune-privileged state of the site and presence of tissue-associated professional antigen-presenting cells [33]. The administration route will be further discussed in clinical section below.

5.2.4 Nonclinical Assessment of Immunogenicity

Nonclinical studies to assess immune response to rAAV-based GTs can provide an understanding of immune-mediated adverse events and pharmacology as well as biodistribution. These observations also provide a better understanding of the adverse events due to different routes of administration (intra-thecal, subcutaneous, etc.), dose-dependent immune- and geno-toxicities due to viral vector capsids and genomes. Nonclinical species have ranged from non-human primates (NHP), rodents, pigs, and dogs to evaluate pharmacology and toxicity of gene therapies [34]. Animal disease models representing gene defects have also been employed to evaluate pharmacology and efficacy of the gene therapy vectors.

The animal studies do have limited use as they are not predictive of adverse events (AEs) or immunogenicity in humans due to species-specific differences in immune responses [35]. However, the immune response against viral vector components, especially innate immune response, can support interpretation of study observations like impact on safety and efficacy. The aspects of the study design like route and mode of administration, and manufacturing process used to generate IND-enabling studies are also relatable to a certain extent.

5.2.5 Animal Models for Assessing Innate Immunity

The innate immune response can be activated by different components of the adeno-associated virus mediated gene therapy (AAV-GT) product as described earlier. Animal models like murine models with gene defects for metabolic disease, e.g. hemophilia, can be employed to assess impact of innate phase responses on safety and efficacy of AAV-GT. Such mice models can also evaluate the impact of engineered capsids that could reduce the innate and subsequent adaptive immune response [12]. Other PRR-based innate activation mechanisms due to critical quality attributes like presence of empty and partially packaged capsids in the final AAV-GT vector product can also be evaluated using the mouse and NHP models due to homology of the innate mechanisms across species [36]. Some toxicities observed in nonclinical species like loss of dorsal root ganglia following intrathecal delivery of high doses of rAAV-GTs in monkeys, piglets, and mice could be indicative of complement mediate innate phase immune response that propagates to adaptive phase response [37, 38].

5.2.6 Animal Models for Assessing Adaptive Immunity

Both mice and monkeys were utilized as the nonclinical species for assessing adaptive immune response to AAV-GT components. The pre-existing seroreactivity observed in these animals and its impact on transduction of gene of interest and its expression in the target tissue could be translatable to humans and helped drive the enrollment criterion for GT trials in clinic [39]. The pre-existing immunity can also induce complement activation through immune complex formation. The treatment-induced humoral response and associated adverse events could not be predicted through studies conducted in NHP species. The cell-mediated responses in NHP did not reduce transgene expression unlike the clinical experience where there was an elimination of vector transduced cells. The

Teffector and CTL responses associated with NHPs are functionally and phenotypically distinct from the human T cells. Nevertheless, there is value in conducting immune response evaluations in preclinical animal species as it provides insight into the mechanisms behind the adverse events that may be later useful to explain similar consequences in clinic [40, 41]. Even though disease models may be preferred, healthy and immune-competent species are also equally important to be able to understand the impact of immune-mediated adverse events to explain immune-toxicities and be aware of similar observations in clinic.

The preclinical administration of rAAV-GT vectors and the subsequent outcome could be relevant for an immune-privileged site like eye. Multiple studies have demonstrated the value of observations from intravitreal or subretinal administrations that could be kept in consideration while planning a mitigation strategy for clinical trials. Most of the animal studies have not provided any direct evidence for impact of treatment-emergent antibodies on safety or efficacy of GT vectors. However, some studies with monkeys and mice did indicate that pre-existing neutralizing antibodies could impact transgene expression [39, 42]. This observation led to preenrollment criterion to exclude patients with pre-existing antibodies at certain titers.

5.2.7 Impact of Immunogenicity on Animal Selection and Interpretation of Study Results

The NHPs have been considered the most relevant toxicology species to understand disposition (viral vector gene expression) across tissues apart from target where the gene is intended to transduce. Even though ocular delivery should not elicit an immune response due to it being immune privileged, a strong innate and adaptive immune response was observed against the AAV both locally and systemically [43, 44]. The NHP study confirmed the development of an adaptive phase cellular immune response following subretinal injection. The mitigation strategy implemented in clinic was adapted from the NHP study, including immune monitoring and delay in administration between two doses.

5.3 Clinical Manifestation Associated with Immunogenicity

The pre-existing host immunity or treatment-emergent immune responses can induce both humoral and cellular responses to viral capsid and transgene proteins derived from AAV vectors, potentially compromising treatment efficacy and patient safety.

5.3.1 Pre-existing Immunity Against AAV Vector May Compromise Therapeutic Efficacy and Patient Safety

Patients who are exposed to wide-type AAV serotypes early in childhood or during later natural infection usually develop humoral immunity directed against AAV capsid proteins [45, 46]. The pre-existing anti-AAV capsid antibodies can result in the reduction of treatment efficacy and may be associated with patient safety.

Pre-existing neutralizing antibody can block the binding of the AAV vector to target receptor or prevent vector internalization into target cells. This would directly inhibit viral transduction and reduce transgene expression, limiting the therapeutic efficacy. In addition, pre-existing anti-capsid antibody (both non-neutralizing and neutralizing) can also affect efficacy by increasing the clearance of administered AAV vector via opsonization or complement activation [42]. These processes can also trigger safety issues, e.g. inflammation. In addition, complement effector can also cause other complications such as thrombocytopenia, atypical hemolytic uremic syndrome (aHUS, a type of thrombotic microangiopathy), and immune complex deposition, especially in patients treated with systemic AAV gene therapy [47].

Natural exposure to AAV can also produce pre-existing cell-mediated immunity, although this may be less prevalent than pre-existing humoral immunity [47]. It has been reported that cellular responses could eliminate AAV particles or transduced cells, thus affecting the persistence of AAV vector delivered through systemic [48, 49] or intramuscular administration [50]. This can not only reduce treatment efficacy but may also trigger tissue damage, causing potential safety concern. For intramuscular administration, muscle cell damage would lead to the release of biomarkers, e.g. creatine phosphokinase. When AAV vector is administered to liver, hepatic cell damage could result in an elevated level of transaminases in the circulation [37, 51].

5.3.2 Treatment Induced Anti-AAV Capsid Antibodies May Prevent Re-dosing

It has been observed that pre-existing anti-AAV capsid antibodies may increase in titer after administration of AAV vectors. Most seronegative patients can also undergo seroconversion and develop treatment-induced antibodies after administration of AAV vector [52]. Higher dose could further boost the anti-AAV antibody titer [53]. High titers of circulating neutralizing antibodies (Nabs) to AAV2 have been reported to persist up to 9 years following AAV vector administration [53]. These treatment-induced antibodies may prevent readministration of patients using the same or even other AAV vector serotypes.

The post-dosing boosted immunity to AAV capsid protein may be associated with certain serious adverse events (SAEs). The severity of AEs could increase with dose following intravenous administration. Immune complex formation, antibody-dependent cellular cytotoxicity (ADCC), antibody-dependent cellular phagocytosis (ADCP), and opsonization could be accountable for the observed cytotoxicity or inflammation [54]. For example, administration of high doses of AAV9 in children with Duchenne Muscular Dystrophy (DMD) has resulted in acute kidney injury, complement activation, reduced platelets and red blood cells (RBCs), and thrombocytopenia [55].

5.3.3 Antibody Specific to Transgene Protein Could Lead to Toxicity or Unwanted Immunity

Patients may not produce the same endogenous protein as encoded by the AAV vector. For example, cross-reactive immunologic material (CRIM) negative patients carrying a null mutation with infantile Pompe's disease develop an immune response to the enzyme replacement therapy [56]. A similar outcome can occur with gene replacement therapies treating gene defects with a null mutation. In some instances, there could be a pre-existing immunity to the transgene protein if patients were previously treated with a replacement therapy. Under this circumstance, both humoral and cellular response can be elicited against the transgene protein, blocking its activity, and compromising therapeutic efficacy.

Another instance where the transgene protein is homologous to an endogenous counterpart, there is a risk of neutralizing antibodies that can cross-react with endogenous protein and inhibit its physiological function, potentially resulting in more severe clinical consequence, e.g. immunodeficiency syndrome.

If the transgene product is expressed on the cell surface, anti-transgene protein antibodies could directly bind and mediate effector functions, e.g. ADCC, ADCP or CDC (complement-dependent cytotoxicity), leading to cell lysis and tissue destruction. The transgene protein can be presented in the context of Class I MHC on the target cells and leads to activation of cytotoxic CD8+T-cell response driving immunotoxicity and loss of transduced cells and tissue damage.

The transgene product can also be secreted and taken up by antigen-presenting cells in the tissue and presented in the context of Class II MHC which leads to activation of Teffector CD4+ T-cell response.

5.3.4 Risk of Immunogenicity Associated with Different Administration Routes

The route of administration would dictate the immunogenicity risk and its clinical consequence. While systemic administration could induce systemic immune reaction, local administration may result in a lower exposure or activation of Treg cells. Therefore, local, e.g. intravitreal, subretinal, intrathecal, intramuscular, administration may result in a lower risk of immunogenicity than systemic, e.g. intravenous, administration [33, 57, 58].

Delivery of AAV vector to different target tissues may trigger various level of immune response due to differences in the nature and the density of antigen-presenting cells, the cytokine environment, and the existence of tissue/blood barriers.

5.3.4.1 Gene Delivery to the Eye or Central Nervous System

Local administration to an immune-privileged site, such as the eye or central nervous system (CNS) may have a lower risk to induce an immune response. For example, AAV gene therapy for the treatment of retinal diseases usually delivers viral vectors to the subretinal space or vitreous cavity. In addition to the blood–retina barrier, induction of antigen-specific Tregs, the presence of immunosuppressive TGFβ2 cytokine, and a tendency toward Th2 responses (induction of antibodies with no recruitment of complement) would contribute to the ocular immune privilege. The significantly reduced immune response in eyes would protect the inner eye from immunogenic inflammation after administration of first or even repeated dose of AAV vector to the eye [44, 53]. However, inflammation could be detected after elevated level of ADA response was developed to viral capsid following high dose of AAV vector administration [59, 60].

5.3.4.2 Gene Delivery to Liver

AAV gene therapy has also targeted liver for the treatment of a range of genetic and metabolic diseases due to its central function in metabolism, heavy vascularization, and unique immune environment. It has been reported that induction of antigen-specific CD4+CD25+FoxP3+ Tregs and $CD8^+$ Tregs together with apoptosis of reactive T cells would account for the liver-mediated immune tolerance and help establish long-term expression of transgene protein after administration of AAV vector [61, 62]. For example, when all subjects previously undergoing protein replacement therapy were enrolled for the AAV hemophilia B clinical trials, none of the subjects, including CRIM-negative patients, developed immune response to transgene product and the long-term transgene expression was observed [15]. Despite the induced immune tolerance in liver, the cellular response to AAV capsid protein has been reported that led to specific destruction of transduced hepatocytes, transaminase elevation and loss of transgene expression [49, 63].

5.3.4.3 Gene Delivery to Muscle

Muscle is another valuable target tissue for AAV gene therapy to treat diseases related to this organ or to serve as an alternative organ other than liver to produce therapeutic protein in the long term [64]. When compared to liver, muscle presents a unique immune environment that would determine the outcome of AAV gene therapy. Some neuromuscular diseases, e.g. Duchenne muscular dystrophy, are associated with muscle inflammation [65]. This may result in upregulated MHC class I expression and elevated cytotoxic T-cell response to AAV transduced tissue cells, leading to loss of transgene expression [31]. In addition, it has been demonstrated that intramuscular delivery usually results in concentrated transgene expression at the site of injection and thereby could mount more prominent immune response when compared to intravascular delivery [58], potentially reducing the transgene protein production. Nonetheless, sustained transgene expression was documented for subjects administered intramuscularly with AAV1 vector encoding α1-antitrpsin when CD4+CD25+FoxP3+ Tregs were present around the injected site, attenuating the immune response to viral and transgene protein derived from the administered vector [66, 67].

5.3.5 Product- and Process-related Impurity Related Immunogenicity

The product-related impurity includes empty capsid and viral particles with partial genome. The oxidized, deamidated, and aggregated forms of viral particles are also part of the product generated from the cell culture [68]. While empty capsids may be used as a mitigation strategy to quench pre-existing antibodies, empty capsids or capsid protein undergoing biotransformation can also elicit anti-AAV immune response. The anti-capsid humoral response may result in elimination of viral particle through immune complex formation between pre-existing antibodies and capsids. Such immune complexes can also activate complement and impact safety. Cellular response against capsid could additionally be involved in the clearance of AAV-transduced target cells [69]. Both would lead to loss of transgene expression, reduced efficacy, and cytotoxicity. In addition, the AAV genome could potentially recombine with genomic DNA from target cells leading to genotoxicity. The CpG motifs associated with AAV genome may contribute to host immune activation as well as safety risk [11, 68].

5.4 Clinical Mitigation Strategy

The impact of immunogenicity on AAV gene therapy could be mitigated using different strategies to curtail or overcome the immune responses to AAV capsid and transgene protein.

To mitigate the impact of pre-existing antibodies, one important approach is to exclude seropositive patients and enroll patients with undetectable or marginal pre-existing immunity to AAV vector, especially for clinical studies with systemic administration. Therefore, it is important to screen pre-existing anti-AAV antibodies for patient stratification. Either total binding anti-drug antibody ADA assay (TAb) or transduction inhibition assay can be used to measure anti-AAV capsid antibody titer to determine the level of pre-existing immunity and eligibility of patients for clinical trial enrollment. TAb assays are typically immunoassays that are sensitive with adequate robustness. The cell-based transduction inhibition assays can detect not only NAb but also non-antibody factors in the matrix samples that can block the transduction. However, the *in vitro* cell-based transduction inhibition assay may not predict transduction inhibitors *in vivo* and could be susceptible to assay variation. Either assay can be used to exclude patients with antibody titer above a predefined threshold cut-off. The sponsors should consider developing a companion diagnostic test (CDx) for market approval to help determine patient eligibility.

It is also feasible to reduce the pre-existing immunity by engineering AAV vector. Site-directed or random mutagenesis technology can be implemented to deplete the immunogenic motif in the capsid protein or select recombinant mutants with a lower sero-reactivity to pre-existing antibodies in patient circulation without reducing the vector transduction to target cells. Alternatively, AAV vector derived from nonhuman species, e.g. rhesus macaques [70, 71], may have lower seroprevalence and can be used as gene delivery vector. For example, AAVrh74 is a recombinant gene therapy vector from rhesus macaques. It has been reported that 83% of patients with DMD were seronegative for anti-AAVrh74 antibodies [72].

Since an elevated dose of AAV vector may be associated with a higher risk of immunogenicity and potentially elevated incidence of SAEs, a decreased dose can be considered, if possible, during clinical trial planning. Alternatively, a genetically engineered AAV vector could carry mutations that enable a higher transduction efficiency at a lower dose [73].

Other possible solutions to reduce the impact of pre-existing immunity involve removal of anti-capsid antibodies by plasmapheresis or degradation. It has been reported that removal of immunoglobulins in blood from seropositive NHPs via plasmapheresis resulted in a transduction efficiency like that in seronegative animals [74]. To circumvent the limitation of current plasmapheresis procedure that nonspecifically removes all circulating antibodies, many clinical studies consider using AAV-specific plasmapheresis to reduce the pre-existing immunity against AAV gene therapy [75]. Transient degradation of circulating antibodies using Imflidase/IdeS, a bacterial cysteine protease, may provide an alternative approach for efficient AAV transduction of seropositive patients and even readministration of an AAV vector [76, 77].

Immunomodulation is another effective strategy to circumvent the immune response to AAV gene therapy. Transient immunosuppression with immunomodulatory drugs, e.g. rituximab combined with methylprednisolone, prior to AAV vector administration has resulted in reduced immune response to capsid and transgene proteins [16, 78]. Rapamycin is a general immunosuppressant. When combined with prednisolone and coadministered with AAV vector, it can inhibit activation of B and T cells, leading to reduced anti-AAV antibody production [78]. Corticosteroids and other immunomodulatory medications are also available to inhibit innate/adaptive immune cells or T/B-cell production to suppress the immune response to AAV gene therapy [1].

Strategies to mitigate immune response to AAV vector should also involve product optimization and removal of product impurity. Immune reactivity to AAV vector can be reduced by blocking reverse-strand transcription and depletion of CpG motifs. In addition, short non-coding DNA oligonucleotides can be incorporated into the vector genome to inhibit TLR9 activation [16, 78] and lower the risk of CD8+ T-cell response to the transgene product [12, 79]. The risk of immunogenicity could be further mitigated by minimizing the product-related and process-related impurities through optimization of upstream (cell culture) and downstream (purification) manufacturing processes.

Selection of appropriate administration route can help to modulate immune response to AAV gene therapy. It has been known that delivery of AAV vector to liver can introduce the immunotolerance to AAV vector and transgene protein [61, 62]. Likewise, local administration of AAV vector to immune-privileged sites, e.g. eyes or central nervous system, can reduce the risk of immunogenicity and support long-term transgene expression as well [80, 81].

References

1 Yang, T.Y., Braun, M., Lembke, W. et al. (2022). Immunogenicity assessment of AAV-based gene therapies: an IQ consortium industry white paper. *Mol. Ther. Methods Clin. Dev.* 26: 471–494.

2 Zaiss, A.K., Liu, Q., Bowen, G.P. et al. (2002). Differential activation of innate immune responses by adenovirus and adeno-associated virus vectors. *J. Virol.* 76 (9): 4580–4590.

3 Goubau, D., Deddouche, S., and Reis e Sousa, C. (2013). Cytosolic sensing of viruses. *Immunity* 38 (5): 855–869.

4 Ma, Z., Ni, G., and Damania, B. (2018). Innate sensing of DNA Virus genomes. *Annu. Rev. Virol.* 5 (1): 341–362.

5 Kawasaki, T. and Kawai, T. (2014). Toll-like receptor signaling pathways. *Front. Immunol.* 5: 461.

6 Xiao, T.S. (2015). The nucleic acid-sensing inflammasomes. *Immunol. Rev.* 265 (1): 103–111.

7 Ivashkiv, L.B. and Donlin, L.T. (2014). Regulation of type I interferon responses *Nat. Rev. Immunol.* 14 (1): 36–49.

8 Zaiss, A.K., Cotter, M.J., White, L.R. et al. (2008). Complement is an essential component of the immune response to adeno-associated virus vectors. *J. Virol.* 82 (6): 2727–2740.

9 Smith, C.J., Ross, N., Kamal, A. et al. (2022). Pre-existing humoral immunity and complement pathway contribute to immunogenicity of adeno-associated virus (AAV) vector in human blood. *Front. Immunol.* 13: 999021.

10 Merle, N.S., Church, S.E., Fremeaux-Bacchi, V., and Roumenina, L.T. (2015). Complement system part I – molecular mechanisms of activation and regulation. *Front. Immunol.* 6: 262.

11 Calcedo, R., Vandenberghe, L.H., Gao, G. et al. (2009). Worldwide epidemiology of neutralizing antibodies to adeno-associated viruses. *J. Infect. Dis.* 199 (3): 381–390.

12 Faust, S.M., Bell, P., Cutler, B.J. et al. (2013). CpG-depleted adeno-associated virus vectors evade immune detection. *J. Clin. Invest.* 123 (7): 2994–3001.

13 Hosel, M., Broxtermann, M., Janicki, H. et al. (2012). Toll-like receptor 2-mediated innate immune response in human nonparenchymal liver cells toward adeno-associated viral vectors. *Hepatology* 55 (1): 287–297.

14 Xiang, Z., Kurupati, R.K., Li, Y. et al. (2020). The effect of CpG sequences on capsid-specific CD8(+) T cell responses to AAV vector gene transfer. *Mol. Ther.* 28 (3): 771–783.

15 Manno, C.S., Pierce, G.F., Arruda, V.R. et al. (2006). Successful transduction of liver in hemophilia by AAV-Factor IX and limitations imposed by the host immune response. *Nat. Med.* 12 (3): 342–347.

16 Rogers, G.L., Suzuki, M., Zolotukhin, I. et al. (2015). Unique roles of TLR9- and MyD88-dependent and -independent pathways in adaptive immune responses to AAV-mediated gene transfer. *J Innate Immun.* 7 (3): 302–314.

17 Shao, W., Earley, L.F., Chai, Z. et al. (2018). Double-stranded RNA innate immune response activation from long-term adeno-associated virus vector transduction. *JCI Insight* 3 (12): e120474.

18 Balakrishnan, B. and Jayandharan, G.R. (2014). Basic biology of adeno-associated virus (AAV) vectors used in gene therapy. *Curr. Gene Ther.* 14 (2): 86–100.

19 Mays, L.E., Vandenberghe, L.H., Xiao, R. et al. (2009). Adeno-associated virus capsid structure drives CD4-dependent CD8+ T cell response to vector encoded proteins. *J. Immunol.* 182 (10): 6051–6060.

20 Fitzpatrick, Z., Leborgne, C., Barbon, E. et al. (2018). Influence of pre-existing anti-capsid neutralizing and binding antibodies on AAV vector transduction. *Mol. Ther. Methods Clin. Dev.* 9: 119–129.

21 Pipe, S., Leebeek, F.W.G., Ferreira, V. et al. (2019). Clinical considerations for capsid choice in the development of liver-targeted AAV-based gene transfer. *Mol. Ther. Methods Clin. Dev.* 15: 170–178.

22 Wu, Z., Asokan, A., Grieger, J.C. et al. (2006). Single amino acid changes can influence titer, heparin binding, and tissue tropism in different adeno-associated virus serotypes. *J. Virol.* 80 (22): 11393–11397.

23 Grieger, J.C. and Samulski, R.J. (2012). Adeno-associated virus vectorology, manufacturing, and clinical applications. *Methods Enzymol.* 507: 229–254.

24 Wright, J.F., Wellman, J., and High, K.A. (2010). Manufacturing and regulatory strategies for clinical AAV2-hRPE65. *Curr. Gene Ther.* 10 (5): 341–349.

25 Ayuso, E., Mingozzi, F., and Bosch, F. (2010). Production, purification and characterization of adeno-associated vectors. *Curr. Gene Ther.* 10 (6): 423–436.

26 Ayuso, E., Mingozzi, F., Montane, J. et al. (2010). High AAV vector purity results in serotype- and tissue-independent enhancement of transduction efficiency. *Gene Ther.* 17 (4): 503–510.

27 Kishimoto, T.K. and Samulski, R.J. (2022). Addressing high dose AAV toxicity – 'one and done' or 'slower and lower'? *Expert. Opin. Biol. Ther.* 22 (9): 1067–1071.

28 Wright, J.F. (2014). Product-related impurities in clinical-grade recombinant AAV vectors: characterization and risk assessment. *Biomedicine* 2 (1): 80–97.

29 Hui, D.J., Edmonson, S.C., Podsakoff, G.M. et al. (2015). AAV capsid CD8+ T-cell epitopes are highly conserved across AAV serotypes. *Mol. Ther. Methods Clin. Dev.* 2: 15029.

30 Tanaka, T., Narazaki, M., and Kishimoto, T. (2014). IL-6 in inflammation, immunity, and disease. *Cold Spring Harb. Perspect. Biol.* 6 (10): a016295.

31 Mendell, J.R., Campbell, K., Rodino-Klapac, L. et al. (2010). Dystrophin immunity in Duchenne's muscular dystrophy. *N. Engl. J. Med.* 363 (15): 1429–1437.

32 Kumar, S.R.P., Hoffman, B.E., Terhorst, C. et al. (2017). The balance between CD8(+) T cell-mediated clearance of AAV-encoded antigen in the liver and tolerance is dependent on the vector dose. *Mol. Ther.* 25 (4): 880–891.

33 Yiu, G., Chung, S.H., Mollhoff, I.N. et al. (2020). Suprachoroidal and subretinal injections of AAV using transscleral microneedles for retinal gene delivery in nonhuman primates. *Mol. Ther. Methods Clin. Dev.* 16: 179–191.

34 Hinderer, C., Katz, N., Buza, E.L. et al. (2018). Severe toxicity in nonhuman primates and piglets following high-dose intravenous administration of an adeno-associated virus vector expressing human SMN. *Hum. Gene Ther.* 29 (3): 285–298.

35 Ertl, H.C.J. (2019). Preclinical models to assess the immunogenicity of AAV vectors. *Cell. Immunol.* 342: 103722.

36 Armant, M.A. and Fenton, M.J. (2002). Toll-like receptors: a family of pattern-recognition receptors in mammals. *Genome Biol.* 3 (8): REVIEWS3011.

37 Ertl, H.C.J. and High, K.A. (2017). Impact of AAV capsid-specific T-cell responses on design and outcome of clinical gene transfer trials with recombinant adeno-associated viral vectors: an evolving controversy. *Hum. Gene Ther.* 28 (4): 328–337.

38 Mingozzi, F., Anguela, X.M., Pavani, G. et al. (2013). Overcoming pre-existing humoral immunity to AAV using capsid decoys. *Sci. Transl. Med.* 5 (194): 194ra92.

39 Wang, L., Calcedo, R., Bell, P. et al. (2011). Impact of pre-existing immunity on gene transfer to nonhuman primate liver with adeno-associated virus 8 vectors. *Hum. Gene Ther.* 22 (11): 1389–1401.

40 Martino, A.T. and Markusic, D.M. (2020). Immune response mechanisms against AAV vectors in animal models. *Mol. Ther. Methods Clin. Dev.* 17: 198–208.

41 Basner-Tschakarjan, E., Bijjiga, E., and Martino, A.T. (2014). Pre-clinical assessment of immune responses to adeno-associated virus (AAV) vectors. *Front. Immunol.* 5: 28.

42 Davidoff, A.M., Gray, J.T., Ng, C.Y. et al. (2005). Comparison of the ability of adeno-associated viral vectors pseudotyped with serotype 2, 5, and 8 capsid proteins to mediate efficient transduction of the liver in murine and nonhuman primate models. *Mol. Ther.* 11 (6): 875–888.

43 Lai, C.M., Estcourt, M.J., Himbeck, R.P. et al. (2012). Preclinical safety evaluation of subretinal AAV2.sFlt-1 in non-human primates. *Gene Ther.* 19 (10): 999–1009.

44 Amado, D., Mingozzi, F., Hui, D. et al. (2010). Safety and efficacy of subretinal readministration of a viral vector in large animals to treat congenital blindness. *Sci. Transl. Med.* 2 (21): 21ra16.

45 Calcedo, R., Morizono, H., Wang, L. et al. (2011). Adeno-associated virus antibody profiles in newborns, children, and adolescents. *Clin. Vaccine Immunol.* 18 (9): 1586–1588.

46 Li, C., Narkbunnam, N., Samulski, R.J. et al. (2012). Neutralizing antibodies against adeno-associated virus examined prospectively in pediatric patients with hemophilia. *Gene Ther.* 19 (3): 288–294.

47 Vandamme, C., Adjali, O., and Mingozzi, F. (2017). Unraveling the complex story of immune responses to AAV vectors trial after trial. *Hum. Gene Ther.* 28 (11): 1061–1074.

48 Mingozzi, F., Maus, M.V., Hui, D.J. et al. (2007). CD8(+) T-cell responses to adeno-associated virus capsid in humans. *Nat. Med.* 13 (4): 419–422.

49 Nathwani, A.C., Tuddenham, E.G., Rangarajan, S. et al. (2011). Adenovirus-associated virus vector-mediated gene transfer in hemophilia B. *N. Engl. J. Med.* 365 (25): 2357–2365.

50 Mingozzi, F., Meulenberg, J.J., Hui, D.J. et al. (2009). AAV-1-mediated gene transfer to skeletal muscle in humans results in dose-dependent activation of capsid-specific T cells. *Blood* 114 (10): 2077–2086.

51 Ertl, H.C.J. (2021). T cell-mediated immune responses to AAV and AAV vectors. *Front. Immunol.* 12: 666666.

52 Long, B.R., Veron, P., Kuranda, K. et al. (2021). Early phase clinical immunogenicity of valoctocogene Roxaparvovec, an AAV5-mediated gene therapy for Hemophilia A. *Mol. Ther.* 29 (2): 597–610.

53 Mingozzi, F. and High, K.A. (2017). Overcoming the host immune response to adeno-associated virus gene delivery vectors: the race between clearance, tolerance, neutralization, and escape. *Annu. Rev. Virol.* 4 (1): 511–534.

54 Marsic, D., Govindasamy, L., Currlin, S. et al. (2014). Vector design Tour de Force: integrating combinatorial and rational approaches to derive novel adeno-associated virus variants. *Mol. Ther.* 22 (11): 1900–1909.

55 Chand, D.H., Zaidman, C., Arya, K. et al. (2021). Thrombotic microangiopathy following onasemnogene abeparvovec for spinal muscular atrophy: a case series. *J. Pediatr.* 231: 265–268.

56 Corti, M., Liberati, C., Smith, B.K. et al. (2017). Safety of intradiaphragmatic delivery of adeno-associated virus-mediated alpha-glucosidase (rAAV1-CMV-hGAA) gene therapy in children affected by pompe disease. *Hum. Gene Ther. Clin. Dev.* 28 (4): 208–218.

57 Boisgerault, F. and Mingozzi, F. (2015). The skeletal muscle environment and its role in immunity and tolerance to AAV vector-mediated gene transfer. *Curr. Gene Ther.* 15 (4): 381–394.

58 Herzog, R.W., Fields, P.A., Arruda, V.R. et al. (2002). Influence of vector dose on factor IX-specific T and B cell responses in muscle-directed gene therapy. *Hum. Gene Ther.* 13 (11): 1281–1291.

59 Bainbridge, J.W., Mehat, M.S., Sundaram, V. et al. (2015). Long-term effect of gene therapy on Leber's congenital amaurosis. *N. Engl. J. Med.* 372 (20): 1887–1897.

60 Busch, M., Pfeil, J.M., Dahmcke, M. et al. (2022). Anti-drug antibodies to brolucizumab and ranibizumab in serum and vitreous of patients with ocular disease. *Acta Ophthalmol.* 100 (8): 903–910.

61 Franco, L.M., Sun, B., Yang, X. et al. (2005). Evasion of immune responses to introduced human acid alpha-glucosidase by liver-restricted expression in glycogen storage disease type II. *Mol. Ther.* 12 (5): 876–884.

62 Mingozzi, F., Liu, Y.L., Dobrzynski, E. et al. (2003). Induction of immune tolerance to coagulation factor IX antigen by in vivo hepatic gene transfer. *J. Clin. Invest.* 111 (9): 1347–1356.

63 Nathwani, A.C., Reiss, U.M., Tuddenham, E.G. et al. (2014). Long-term safety and efficacy of factor IX gene therapy in hemophilia B. *N. Engl. J. Med.* 371 (21): 1994–2004.

64 Louboutin, J.P., Wang, L., and Wilson, J.M. (2005). Gene transfer into skeletal muscle using novel AAV serotypes. *J Gene Med.* 7 (4): 442–451.

65 Flanigan, K.M., Campbell, K., Viollet, L. et al. (2013). Anti-dystrophin T cell responses in Duchenne muscular dystrophy: prevalence and a glucocorticoid treatment effect. *Hum. Gene Ther.* 24 (9): 797–806.

66 Brantly, M.L., Chulay, J.D., Wang, L. et al. (2009). Sustained transgene expression despite T lymphocyte responses in a clinical trial of rAAV1-AAT gene therapy. *Proc. Natl. Acad. Sci. U. S. A.* 106 (38): 16363–16368.

67 Mueller, C., Chulay, J.D., Trapnell, B.C. et al. (2013). Human Treg responses allow sustained recombinant adeno-associated virus-mediated transgene expression. *J. Clin. Invest.* 123 (12): 5310–5318.

68 Cole, L., Fernandes, D., Hussain, M.T. et al. (2021). Characterization of recombinant adeno-associated viruses (rAAVs) for gene therapy using orthogonal techniques. *Pharmaceutics* 13 (4): 586.

69 Pien, G.C., Basner-Tschakarjan, E., Hui, D.J. et al. (2009). Capsid antigen presentation flags human hepatocytes for destruction after transduction by adeno-associated viral vectors. *J. Clin. Invest.* 119 (6): 1688–1695.

70 Gao, G., Vandenberghe, L.H., Alvira, M.R. et al. (2004). Clades of Adeno-associated viruses are widely disseminated in human tissues. *J. Virol.* 78 (12): 6381–6388.

71 Martin, P.T., Xu, R., Rodino-Klapac, L.R. et al. (2009). Overexpression of Galgt2 in skeletal muscle prevents injury resulting from eccentric contractions in both mdx and wild-type mice. *Am. J. Physiol. Cell Physiol.* 296 (3): C476–C488.

72 Goedeker, N.L., Dharia, S.D., Griffin, D.A. et al. (2023). Evaluation of rAAVrh74 gene therapy vector seroprevalence by measurement of total binding antibodies in patients with Duchenne muscular dystrophy. *Ther. Adv. Neurol. Disord.* 16: 17562864221149781.

73 Zhong, L., Li, B., Mah, C.S. et al. (2008). Next generation of adeno-associated virus 2 vectors: point mutations in tyrosines lead to high-efficiency transduction at lower doses. *Proc. Natl. Acad. Sci. U. S. A.* 105 (22): 7827–7832.

74 Chicoine, L.G., Montgomery, C.L., Bremer, W.G. et al. (2014). Plasmapheresis eliminates the negative impact of AAV antibodies on microdystrophin gene expression following vascular delivery. *Mol. Ther.* 22 (2): 338–347.

75 Bertin, B., Veron, P., Leborgne, C. et al. (2020). Capsid-specific removal of circulating antibodies to adeno-associated virus vectors. *Sci. Rep.* 10 (1): 864.

76 Elmore, Z.C., Oh, D.K., Simon, K.E. et al. (2020). Rescuing AAV gene transfer from neutralizing antibodies with an IgG-degrading enzyme. *JCI Insight* 5 (19): e139881.

77 Leborgne, C., Barbon, E., Alexander, J.M. et al. (2020). IgG-cleaving endopeptidase enables in vivo gene therapy in the presence of anti-AAV neutralizing antibodies. *Nat. Med.* 26 (7): 1096–1101.

78 Mueller, C., Berry, J.D., McKenna-Yasek, D.M. et al. (2020). SOD1 suppression with adeno-associated virus and microRNA in familial ALS. *N. Engl. J. Med.* 383 (2): 151–158.

79 Chan, Y.K., Wang, S.K., Chu, C.J. et al. (2021). Engineering adeno-associated viral vectors to evade innate immune and inflammatory responses. *Sci. Transl. Med.* 13 (580): eabd3438.

80 Shirley, J.L., de Jong, Y.P., Terhorst, C., and Herzog, R.W. (2020). Immune responses to viral gene therapy vectors. *Mol. Ther.* 28 (3): 709–722.

81 Varin, J., Morival, C., Maillard, N. et al. (2021). Risk mitigation of immunogenicity: a key to personalized retinal gene therapy. *Int. J. Mol. Sci.* 22 (23): 12818.

Section III

Bioanalysis for Gene Therapy

6

Bioanalytical Methods to Detect Preexisting and Post-administration Humoral Immune Responses Against AAV Capsid Proteins

Christian Vettermann[1] *and Boris Gorovits*[2]

[1] *Translational Sciences, Assay Strategy & Development, BioMarin Pharmaceutical, Inc., Novato, CA, USA*
[2] *Gorovits BioSolutions, LLC, Andover, MA, USA*

6.1 Introduction

Preexisting and posttreatment humoral anti-Adeno-associated virus (AAV) immunity can significantly impact treatment safety or/and efficacy. The presence of preexisting antibodies may result in insufficient level of transgene protein expression [1] while post-dose immune response is expected and typically is considered as main barrier for repeat administration of AAV-based therapeutics. AAV humoral immune responses to AAV capsid can be assessed by methods detecting either total antibodies (TAb) or neutralizing antibodies (NAb). TAb comprises all immunoglobulins that bind to a particular AAV serotype, including those that can neutralize the AAV vector's ability to transduce cells, i.e. those that are NAb. Due to the nature of the protocols used to detect the presence of AAV NAb, these methods are often referred to as transduction inhibition (TI) methods, which are described in more detail later in this chapter. Some level of correlation between the results from AAV TAb and TI methods can be expected and has been reported [2, 3]. However, this correlation is complex and not always linear. In addition to the difference in mechanisms employed by TAb and TI methods, they can also differ in sensitivity, specificity, and other critical performance parameters.

There is an ongoing debate as to whether preexisting TAb vs. NAb impact study outcomes, with some evidence for AAV5 suggesting that high-sensitivity

Drug Development for Gene Therapy: Translational Biomarkers, Bioanalysis, and Companion Diagnostics, First Edition. Edited by Yanmei Lu and Boris Gorovits.

TI protocols may not only be reporting NAb but also the presence of non-immunoglobulin-based inhibitors without considerable biological significance [2, 4]. Nonetheless, TI methods enjoy wide popularity, sometimes in combination with TAb methods, for pretreatment evaluation of patient eligibility [1]. Contrary to the practice used for protein-based biotherapeutics, a tier-based approach for humoral immunogenicity assessment (first-tier TAb, second-tier Nab only if TAb positive) is not commonly followed when determining eligibility of patients for gene therapy (GTx) [5]. Instead, sponsors select either one or both AAV prescreening methods for pretreatment evaluation.

In this chapter, we present information related to the principle of TAb and TI methods, critical details of the protocols, key performance parameters, and aspects of data interpretation. Finally, value and challenges associated with standardization of TAb and TI methods are briefly discussed.

6.2 Considerations for AAV Total Antibody Assays

6.2.1 Nature of AAV TAb Assay Analyte

A diverse antibody response against capsid proteins occurs after natural exposure to AAV or after therapeutic administration of AAV-based GTx vectors. Such responses vary by epitope specificity, immunoglobulin isotype, and functional characteristics of the antibodies. Because AAV-neutralizing antibodies are a subset of total antibodies, the correlation between TAb and NAb levels is complex, with the general expectation of a higher prevalence of TAb [2, 3]. The TAb to NAb response comparison is further complicated by different assay formats used to detect the two analytes and an oftentimes higher sensitivity of TAb assays as compared to common NAb protocols. Even in the absence of a measurable NAb, detection of TAb can indicate prior exposure to AAV through natural infections.

Based on the definition of TAb responses, all AAV antibody isotypes are expected to be detected in the TAb analytical protocol, with most attention generally paid to the detection of serum IgG and IgM classes. Detection of serum IgE typically falls into a unique category and is only requested based on specific clinical observations, as defined in regulatory guidance [6]. Understanding whether the TAb assay detects both or only one of the IgM and IgG classes of immunoglobulins is viewed as critical and is based on the assay format and reagents used in the method. It is commonly expected that low affinity IgM responses which develop early after exposure will then mature into IgG-based responses. Information about detailed isotype composition of an immune response at a given timepoint could be informative and may facilitate the interpretation of immune-associated toxicities or hypersensitivity reactions [7]. Because a preexisting TAb response is likely to be mature, it is expected to be primarily composed of IgG isotypes, with a smaller fraction of samples containing IgM class antibodies [8, 9].

6.2.2 Primary Analytical Methodologies applied for AAV TAb Detection

Based on the definition of the AAV TAb analyte, assay formats, and analytical platforms used are similar to those applied when detecting anti-drug antibody (ADA) responses to protein therapeutics. Two of the most common protocols are antigen-capture (aka sandwich) and bridging assay formats (Figure 6.1). In the antigen-capture format, the capture reagent, most commonly the AAV vector, defines the specificity of the detected TAb analyte. Antibodies in the sample will bind to the capture reagent on the plate and are detected using, for example, a labeled species-specific anti-immunoglobulin antibody. Isotype specificity of the detector reagent defines whether the method can detect one or more AAV antibody classes and should be carefully evaluated [10]. Alternatively, broadly immunoglobulin-binding proteins A, G, or L can serve as detector reagents, which is typically done in combination to account for differential binding to various antibody isotypes [4]. The detector reagents are labeled with either an enzyme (for enzyme-linked immunosorbent assays, ELISA) or ruthenium (for electrochemiluminescent assays, ECL). Final assay signal generation is triggered by the addition of enzyme-substrate (ELISA) or read buffer (ECL).

The benefit of applying the antigen-capture assay format is the ability to distinguish between IgG and IgM class responses. Understanding isotype composition of antibody responses may be insightful, since it can inform about the status of immune response maturity. Early immune responses are primarily IgM-based with a gradual transition to IgG. Human immunoglobulin isotypes have various ability to engage and activate the complement pathway or trigger various types of

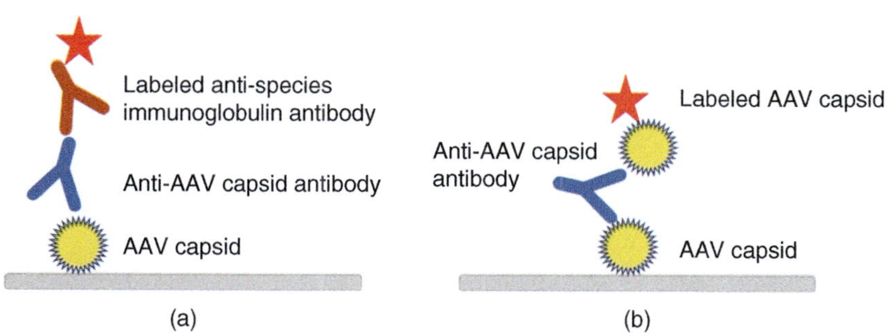

Figure 6.1 Principles of antigen capture (a) and bridging (b) assay formats designed to detect presence of anti-AAV antibodies.

hypersensitivity reactions [7, 11]. The broadly discussed possibility of re-dosing patients previously treated with AAV GTx may greatly depend on the nature and isotype composition of the AAV antibody response.

The potential challenges of the antigen-capture assay format include selecting a detector reagent capable of specific identification of isotypes of interest and the ability to apply a single detector reagent that can detect both species-specific (e.g. human) AAV antibody responses as well as the positive control (PC) used in the assay. A non-human PC antibody reagent is frequently chosen based on the ease of generation and lack of supply concerns. The assay detector reagent therefore should either be cross-reactive to both human antibody and non-human PC reagent or two separate detector reagents need to be applied [12]. This approach may be less favorable as it does not provide information on the integrity of the antihuman detector used for detection of actual sample responses. To monitor performance of the human immunoglobulin detector reagent, additional controls may thus be considered and applied during routine sample testing. For example, a previously tested, confirmed-positive incurred study sample may be used as a surrogate PC.

The TAb bridging assay format offers a robust alternative and has been broadly applied to detect ADA responses against protein-based biotherapeutics [13, 14]. In this format, the multi-valent nature of a typical immunoglobulin is utilized to create a bridge between AAV capture and labeled AAV detector reagents [3]. The bridging assay format is therefore isotype-agnostic and expected to determine the combined presence of IgG, IgM, and other isotypes without distinguishing between them. A subset of monovalent IgG4 antibodies that underwent Fab arm exchange may remain undetectable [15]. Like for the antigen-capture assay format, ELISA and ECL analytical platforms can be applied.

Detection of ADA responses against protein-based biotherapeutics commonly includes assessment of signal specificity, which is conducted in competition-based confirmatory tests [6]. In this assessment, samples are spiked with unlabeled biotherapeutic at a concentration that is sufficient to block antibody binding to the capture reagent used in the assay. A significant reduction in assay signal indicates the presence of biotherapeutic-specific antibodies. While this approach has been implemented for many protein biotherapeutics, it has not found broad application in the AAV GTx field, though exceptions exist [1, 3]. High material cost and the multivalent nature of the capture and detector reagents (AAV capsids), which results in potentially high amounts of unlabeled AAV reagent for successful competition of antibody binding, are two likely reasons.

6.2.3 Tab Assay Critical Reagent Considerations

6.2.3.1 Positive and Negative Control Selection

Like ADA assay protocols, a positive-control AAV antibody reagent is used to determine the quality of TAb assay runs, i.e. it serves as the assay suitability control material. Several commercial AAV antibody reagents are available with various AAV serotype specificities. These monoclonal and polyclonal antibodies can be successfully applied in a proprietary AAV TAb method after it is demonstrated that the commercial antibody reagent sufficiently binds to the AAV serotype of interest. Alternatively, an AAV serotype-specific PC reagent may be developed [16]. The need for serotype-specific AAV PC reagent is debatable since the degree of homology between various AAV serotypes is high, leading to a significant degree of antibody cross-reactivity between serotypes [1]. Hence, even if specificity to a particular AAV serotype was demonstrated for a PC reagent during method validation, it does not rule out the assay's ability to detect antibodies that are cross-reactive to other AAV serotypes in clinical test samples.

Nonetheless, a serotype-specific PC might bind with higher affinity to a particular AAV capsid and thus offer an advantage when characterizing the sensitivity of the method for detecting antibodies to the serotype of interest. While PC reagent is used in characterizing assay sensitivity, precision, selectivity, and robustness, other critical parameters, such as matrix interference and cut-point, are defined based on the analysis of negative control matrix and treatment-naïve individual donor samples.

The TAb assay negative control (NC) is used to calculate plate-specific assay cut-points and further serves as diluent for PC reagent and individual test samples if titered, as well as a general assay suitability control. The NC reagent is commonly generated by combining several individual samples that had shown no or limited AAV antibody reactivity. Selection of individual matrix samples without any AAV reactivity is complicated by high AAV antibody prevalence in treatment-naïve human populations [9, 17, 18]. Screening and confirmatory tests may be required to determine which of the samples can be pooled to generate NC reagent. Alternative solutions include depletion of AAV antibodies from matrix by applying nonspecific (e.g. Protein G-based [19]) or specific (using AAV capsid of interest [20]) affinity pull-down. However, these immune-depletion protocols may potentially produce matrix that is no longer representative of study samples, resulting in a high potential for over-reporting AAV TAb positive responses. Finally, NC matrix needs to match that of the study samples tested for the presence of AAV TAb. The use of surrogate matrix may need to be evaluated in cases where access to natural negative matrix is limited (e.g. aqueous humor) [21].

6.2.3.2 Capture and Detection Reagents

Like in protein biotherapeutic ADA methods, AAV capsid-derived capture reagent is used for TAb assays in both antigen-capture and bridging formats. AAV capsid is also needed as a detector reagent in the bridging format (Figure 6.1). AAV capsid-derived material can be the drug substance or product of interest (i.e. unaltered AAV vector), or surrogate AAV vector encoding another transgene, or a capsid protein preparation representative of investigated AAV serotype. Although using a capsid protein preparation may seem attractive, as it avoids working with transduction-competent AAV vectors, epitopes found on capsid proteins may not be representative of antigenic epitopes on intact AAV capsids. Therefore, it is generally recommended to use intact AAV capsids of the same serotype as the investigated GTx vector. If the AAV vector is identical to a naturally occurring serotype, commercially available AAV preparations may be applied. Similarly, commercially sourced AAV preparations may be used in a confirmatory step, if applied to demonstrate specificity of AAV antibody responses.

6.2.3.3 Sample Testing Strategy

A typical approach to ADA sample analysis may be applied when detecting an AAV TAb. This tier-based approach includes initial screening, confirmatory, and titration testing [14]. For a typical protein biotherapeutic ADA response, the reason to apply a tier-based approach is to reduce the number of samples that are evaluated in the final titration test, since only a fraction of samples is expected to score positive in both screening and confirmatory assays. In contrast, a robust AAV response with high incidence rate, in particular following therapeutic GTx vector administration, is expected. This could support a proposal to directly apply titration analysis and avoid delays related to initial testing of samples in screening and/or confirmatory steps.

6.2.4 Key Assay Qualification/Validation Parameters

6.2.4.1 Assay Sensitivity

Considerations for assay sensitivity that align with expectations for anti-protein biotherapeutic ADA protocols based on current FDA guidance [6] are relevant to AAV TAb assays. Regulatory guidance suggests that at least 100 ng/mL sensitivity based on a suitable PC reagent is expected, and it has been typically relatively easy to achieve this goal [22]. The sensitivity parameter is assessed based on performance of the assay PC that may need to be reconsidered, together with assay conditions, if the desired sensitivity is not reached.

6.2.4.2 Serotype Specificity

Based on high homology between AAV serotypes, it can be expected and has been demonstrated that AAV antibodies detected in GTx treatment-naïve populations are highly cross-reactive [17, 23]. The level of homology between serotypes may determine the degree of cross-reactivity observed [24]. Importantly, a lower degree of cross-reactivity was reported for TI methods, which is likely due to the complexity of the viral transduction process and requirement for high specificity and affinity of immunoglobulins with neutralizing activity [25–28].

Since development of a serotype-monospecific TAb assay may not be possible due to high homology between AAV serotypes, AAV serotype specificity may need to be further characterized when patients are retreated with a different AAV capsid. To understand whether a particular TAb response in an individual is serotype-specific, a confirmatory test can be conducted, whereby excess amount of unlabeled AAV vector for the serotype of interest is added to determine if the assay signal is reduced.

6.2.4.3 Precision

Assay precision is a parameter to describe the degree of variability in assay raw signal, based on results obtained in multiple intra- and inter-run tests [14]. Like protein biopharmaceutical ADA assays, PC samples prepared at a few concentrations in NC matrix (or surrogate) are tested in replicates (typically two per sample) and evaluated on several occasions (typically 3) by two or more analysts. As for regular ADA assays, variability is expressed as coefficient of variation percentage (CV%) and is expected to be less than 20%.

6.2.4.4 Matrix Interference and Selectivity

Impact from sample matrix is evaluated in interference and selectivity assessments. Several (typically 10) individual matrix samples are analyzed unspiked and spiked with PC control reagent at the low PC (LPC) and high PC (HPC) concentration. This enables an understanding of whether inter-individual variations in matrix components interfere with detection and if the assay can reliably detect PC at low concentrations across multiple donors. The test needs to be performed using healthy and disease population samples to understand if any difference in potential interference can be observed. It is understood that for rare diseases obtaining indication-specific matrices may be challenging and alternative solutions, for example, use of sample collected in clinical study, may be required. It is recommended that at least 80% of unspiked samples should score TAb negative and the same percentage of LPC/HPC-spiked samples should score TAb positive. In addition, the potential impact of known interfering factors, such as lipids, hemoglobin and bilirubin, and other matrix components that are expected to be present in the sample (e.g. anti-coagulation factors) has been discussed in regulatory guidance and needs to be considered [6].

6.2.4.5 Assay Cut-Point

TAb assay cut-point is arguably one of the most important parameters to be determined during assay development or validation. Assay cut-point is a threshold value that is used to determine whether a given sample is positive or negative for the presence of AAV TAb, an approach that is very similar to the one used for protein biotherapeutic ADA assays [14, 29]. Cut-point values are assay- and laboratory-specific and are separately defined for screening, confirmatory, and titration protocols. Frequently, screening and titration protocols utilize the same cut-point value. As for the ADA methods, an AAV TAb assay cut-point is determined based on a statistical analysis of assay signals generated for a collection of treatment-naïve individual donor samples that are preferably negative for preexisting AAV immunity. Generally, 50 or more individual samples are tested in 3 or more runs to obtain a sufficient collection of data. The data are analyzed for statistical and biological outliers that are excluded before determining the 95th or 99th percentile of distribution for screening (as well as titration) and confirmatory tests, respectively [14, 29].

Due to preexisting AAV antibodies, it may be challenging to obtain TAb negative treatment-naïve individual donor samples. It may also be difficult to separate true TAb positive vs true TAb negative populations. Conducting a confirmatory test may offer a solution, with an assumption that samples that generate a signal that is inhibited in the presence of unlabeled AAV material by 50% or more contain specific AAV TAb. Results generated for these reactive samples may be removed from the data before performing cut-point data analysis.

6.2.5 TAb Assay Data Interpretation

Information about AAV TAb status of an individual patient may facilitate a decision about AAV treatment eligibility and help understand potential clinical events of interest, including impact on safety or efficacy of the GTx treatment. Application of TAb methods for detection of preexisting AAV immunity has gained popularity since these methods are easy to develop and implement, relatively straightforward for development into a companion diagnostic (CDx) or similar routine test, and relatively robust in their application. If TAb assay results are used as treatment eligibility criteria, a threshold TAb titer for enrolment needs to be pre-defined. The specific treatment eligibility TAb titer threshold may be dependent on disease, route of administration, assay sensitivity, nature of the detected antibodies, and other parameters. It continues to be a subject of ongoing debate. Examples include using the lowest detectable AAV antibody titer, 1 : 1000 titer [30] or no detectable antibody titer [1, 31] as a threshold for eligibility. For some routes of administration (e.g. intrathecal, intravitreal, and subretinal), concerns related to the impact of AAV antibodies may be reduced compared to a systemic, intravenous route of administration. Therefore, the TAb titer thresholds that would be applied when determining patient eligibility would also be different.

6.3 Considerations for Cell-based Transduction Inhibition Assays

6.3.1 Principle and Methodology of Cell-based AAV TI Assays

AAV TI assays are typically performed as cell-based reporter-gene assays [3, 18, 32–34]. A permissible cell line, such as HEK293T/17, is seeded in complete media on white clear-bottom 96 well plates and grown overnight. On the next day, plasma or serum test samples are mixed with a viral AAV reporter gene construct carrying, for example, a CMV promoter-driven luciferase reporter gene (AAV-Luc), in serum-free media and incubated for 30 minutes at room temperature to allow binding of AAV inhibitors to the vector capsid (Figure 6.2).

Following this incubation, samples are added to the cells in replicate wells at a certain multiplicity of infection (MOI), generally between 5000 and 400,000 vg per cell [18, 34]. After a 1-hour incubation at 37 °C, a final concentration of 20 μM etoposide solution in complete media may be added to stimulate DNA repair mechanisms that enhance productive transduction. After 2 days, the cell culture medium is removed and a luciferase substrate, such as Steady-Glo, is added for 10 minutes at ambient temperature. Luminescence is measured, for example, by using a 500-ms integration time on a VICTOR-™X microplate reader. The relative

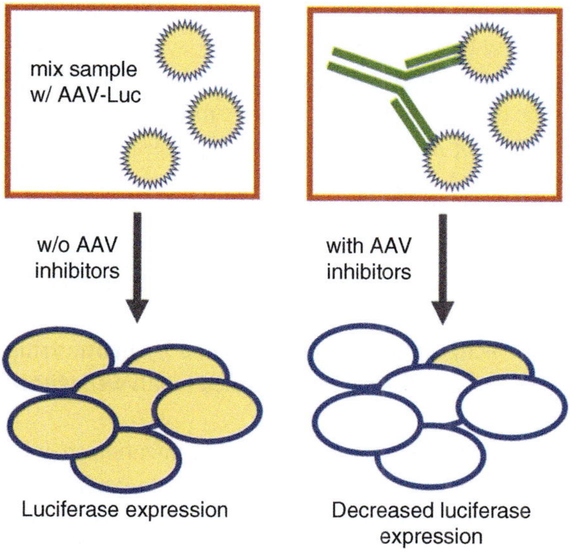

Figure 6.2 Principle of cell-based AAV transduction inhibition (TI) assays.

luminescence of a test sample is then normalized to that of a NC sample, which consists of pooled normal human plasma or serum without any detectable AAV transduction inhibitors. The results can be further converted to percent transduction values relative to NC. Alternatively, test sample results may be expressed as percent inhibition values by subtracting the percent transduction value of a test sample from 100%. Either way, one needs to establish a cut-point below or above which a test sample would be considered positive. Positive test samples are usually reported with a semiquantitative TI titer value as a relative measure of their capacity to inhibit AAV transduction. Titer values are determined by serially diluting a positive test sample to determine the highest sample dilution still detectable in the TI assay. The reciprocal of the highest detectable sample dilution is equal to the numerical neutralizing titer value.

6.3.2 AAV TI Assay Development: Designing for Clinical Relevance

One of the key parameters to optimize during TI assay development is MOI. MOI represents the ratio between the quantity of AAV vector and number of plated cells. Cell lines that require a higher MOI to achieve productive transduction can be thought of as more difficult to transduce. Difficult to transduce cell lines can pose challenges for developing sensitive AAV TI assays, since higher test sample titers will be required to neutralize the higher quantity of AAV vector used in the assay, resulting in limited detectability of lower titers. Cell lines commonly used for TI assays are HEK293, HeLa, and HuH7. Depending on AAV serotype, a popular choice is HEK293 and its derivatives, such as HEK293T/17 and 2V6.11 [3, 35].

To optimize MOI, AAV vector titrations are performed, similar to how drug-response curves are needed to develop cell-based anti-protein therapeutic NAb assays. These titrations are performed to establish the dynamic range, within which a meaningful decrease in AAV vector concentration generates a distinctly measurable change in assay signal. The goal is to ensure that – at the selected MOI – the assay signal, such as measured luminescence, is sufficiently robust yet not fully saturated. Otherwise, the assay would remain unresponsive to a meaningful decrease in effective AAV vector concentration mediated by neutralizing factors in a test sample. In conclusion, MOIs should be selected to fall well within the dynamic range of the assay.

Dynamic ranges can be broad, and this is where strategic optimization may be employed to maximize clinical relevance of a cell-based TI assay. The lower the MOI for a given number of plated cells, the more sensitive the assay will be, since a lower quantity of AAV vector will need to be neutralized by a particular volume of test sample. The question is how sensitive does the assay need to be? To answer this with clinical efficacy in mind, it can be helpful to determine the volume of test sample (plasma or serum) that is incubated together with a particular quantity of AAV vector in the assay, based on the selected MOI and number of plated cells. The AAV vector-to-sample ratio can then be compared to the maximal AAV vector plasma concentration anticipated in study participants, based on intended clinical dose level. For example, an adult clinical trial participant who is administered an AAV vector dose level of 6 E12 vector genome per kilogram (vg/kg) will receive a total vector dose of 4.2 E14 vg, assuming average body weight of 70 kg. This would lead to a theoretical maximal plasma concentration of 1.4 E11 vg/mL, assuming average total plasma volume of 3000 mL. In a cell-based TI assay, 3.36 E09 vg AAV vector may be incubated with 42 microliters of undiluted plasma test sample, resulting in a vector-to-sample ratio of 3.36 E09 vg/0.042 mL = 8 E10 vg/mL, which is about two times less than theoretical maximal plasma concentration of 1.4 E11 vg/mL.

To be relevant for clinical efficacy evaluations, the vector-to-sample ratio in a TI assay should be equal to or lower than the maximal plasma concentration of the AAV GTx theoretically achieved in study participants, as illustrated in the example above. Otherwise, the neutralizing potential associated with a particular volume of plasma or serum could completely neutralize clinically administered GTx quantities, while barely inhibiting transduction of the cell line. Therefore, cell lines requiring exceedingly high MOIs may not be suitable for use in TI assays. In conclusion, to develop a TI assay with clinically relevant sensitivity, it is useful to concurrently optimize cell density per well, MOI, and test sample volume, while comparing vector-to-sample ratio *in vitro* to anticipated GTx plasma levels *in vivo*. Similar considerations have been successfully applied to interpolate clinical relevance of conventional anti-biologics NAb assays [36].

6.3.3 Key Assay Validation Parameters

This section has been written in consideration of available regulatory guidance for immunogenicity assays [6, 37] and industry white papers [38, 39].

6.3.3.1 Screening and Titer Cut-Points

Test samples that contain AAV inhibitors are expected to decrease transduction in a cell-based TI assay compared to NC. To adjudicate a sample as positive or negative for AAV inhibitors, a cut-point is required to define a meaningful decrease in transduction. This analytical cut-point is established in reference to cell-based assay procedures; it does not directly translate to the anticipated impact on clinical efficacy. This will be discussed further below, refer to: Data Interpretation: Determining a Meaningful Titer Threshold for Clinical Efficacy.

Two main strategies exist to establish TI assay cut-points: (1) using a threshold of 50% transduction at a predetermined sample dilution compared to a NC sample [35]; (2) using a statistical cut-point derived from treatment-naïve, AAV inhibitor-negative individual donor samples [3, 40], as recommended by regulatory guidance [6].

A 50% transduction cut-point is commonly used in antiviral and vaccine research to report neutralizing titers against target viruses, and thus this strategy is oftentimes applied for measuring AAV inhibitors. However, this approach could miss less potent but potentially impactful inhibitors that only partially decrease transduction to 60–70%.

Whether reduced detectability of less potent AAV inhibitors poses a significant hurdle for clinical data interpretation will depend on the assay setup, for example, on how much lower the vector-to-sample ratio is compared to the theoretical maximal vector plasma concentration in patients: If the vector-to-sample ratio is 10 times lower than the anticipated vector plasma concentration, then using an analytical cut-point of 50% transduction could be justified, simply because the quantity of AAV vector neutralized in vitro would translate to as little as 5% of the maximal clinical vector concentration anticipated in the tested sample volume. It will also be critical when implementing a 50% transduction cut-point to account for assay variability: Variability needs to be empirically evaluated and may be caused by different reagent lots, composition of individual sample matrices, passage number of cells, and operators.

Statistical cut-points have been discussed elsewhere [6, 14, 39] and are commonly used for detection of anti-drug antibodies to therapeutic proteins. This bioanalytical strategy has also been successfully implemented for clinical AAV TI assays [3, 40]. Statistical cut-points are generally considered to be more objective, even though they can be difficult to establish due to a varying degree of preexisting AAV antibodies in healthy humans and patient populations alike [3, 41]. A high number of antibody-positive samples may confound the statistical analysis of data generated during cut-point establishment. To mitigate cut-point inflation resulting from using samples with preexisting AAV antibodies, samples may be prescreened prior to the cut-point experiments to exclude those associated with less than 50% transduction, indicative of preexisting AAV immunity [3]. To be more rigorous, prescreening could include a confirmatory assay (described below) and any sample with more twofold increase in transduction following immunoglobulin depletion would be considered AAV NAb-positive and excluded from cut-point experiments. These approaches may require an initial analysis of a larger number of commercial samples to find 50 individual samples without preexisting AAV immunity. Alternative methods to deal with antibody-positive samples and analyze data generated during cut-point establishment are discussed in the literature [29, 42–45].

6.3.3.2 Limit of Detection

To characterize the analytical sensitivity of a cell-based AAV TI assay, a limit of detection (LOD) can be determined using a surrogate positive-control AAV antibody. Both polyclonal and monoclonal antibodies may serve as PC, with polyclonal antibodies better reflecting the nature of humoral immune responses in future study samples. The LOD is defined as the concentration of a positive-control antibody in neat (undiluted) sample matrix that is detectable at the cut-point of the TI assay (Figure 6.3). Serial 1 : 2 or 1 : 3 dilutions of a positive-control antibody are typically performed and tested. Dilutions expected to fall above and below cut-point should be included to accurately interpolate the LOD. The LOD may be interpolated using either a fitted curve, as shown in Figure 6.3, or a linear curve interval between the two antibody concentrations whose transduction values flank the cut-point.

Regulatory agencies recommend a detection limit of 100 ng/mL or less for the positive-control antibody in conventional anti-protein therapeutic TAb assays and acknowledge that cell-based NAb assays may not always meet this mark [6]. Meeting this specification for AAV TI assay will depend on the type and affinity of the positive-control antibody and hence does not necessarily reflect the level of AAV inhibitors in study samples. Furthermore, it is presently unknown what level of AAV inhibitors in study participants will have a considerable impact on GTx efficacy or safety. This level may also depend on GTx vector-specific factors, such as AAV serotype and route of administration. In conclusion, a clinically relevant LOD target value cannot be specified for AAV TI assays at this time.

Therefore, suitable assay sensitivity may need to be ascertained through alternative avenues, such as (1) consideration of the MOI and vector-to-sample ratio, as discussed above; (2) adequate resolution of AAV inhibitor titers in a large number of individual subjects with pre-existing immunity, which should range from low to high titers; (3) empirical verification of predicted dose-neutralization curves, as described in the last paragraph; (4) correlation with efficacy and/or safety signals in nonclinical studies.

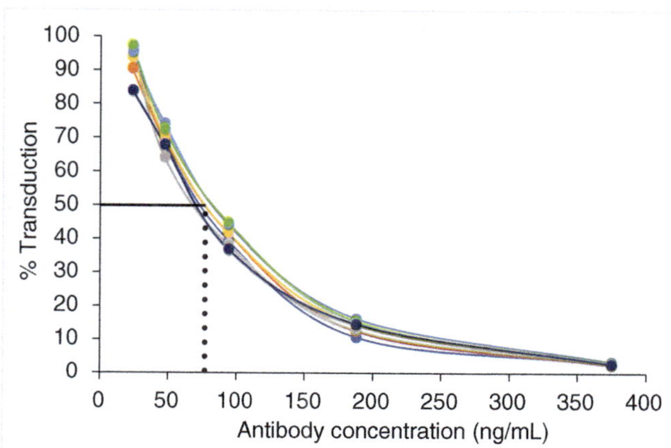

Figure 6.3 LOD interpolation at the assay cutpoint set for 50% transduction.

6.3.3.3 Precision

Precision of cell-based TI assays is generally lower than that of ligand-binding assays, resulting in higher coefficients of variance (CV). Intra-replicate, intra-assay, and inter-assay precision of TI assays can be assessed using low (LPC) and high (HPC) PC AAV antibody-spiked samples. CVs can be evaluated for raw luminescence signals or normalized transduction values and should generally remain below 30%. Inter-assay CV is also expected to remain below 30%, even though this limit could be elevated to 50%, if such higher variance does not affect the reproducibility of qualitative results (negative/positive) for low-positive samples.

Titer precision is also important to consider, which can be assessed as the range of titer values measured repeatedly for the same sample. Titer precision is primarily dependent on the dilution scheme applied to the test samples and will typically fall within ±1 of the mean $\log_2$(titer) derived from repeated measurements if a 1 : 2 dilution scheme is used. This corresponds to the last detectable sample dilution shifting one step upwards or downwards.

6.3.3.4 Specificity

Specificity for AAV TI assay addresses: (1) detection of neutralizing AAV antibodies versus non-antibody-based neutralizing factors (antibody specificity); (2) no detection of AAV inhibitors that exclusively react with other AAV serotypes (serotype specificity); (3) detection of AAV inhibitors to both reporter gene construct and GTx (GTx specificity). For (1), refer to section: Confirmatory Steps to Ensure Specific Detection of Neutralizing AAV Antibodies. Serotype specificity (2) can be assessed by testing samples spiked with monoclonal antibodies that exhibit exclusive reactivity against related AAV serotypes, which should remain undetectable. This assessment is, however, of limited value and merely confirms that the reporter gene construct is not contaminated with another AAV serotype. Clinical test samples typically contain polyclonal mixtures of AAV antibodies that broadly cross-react with multiple AAV serotypes [41].

The underlying question of (3) is whether the reporter gene construct adequately represents the GTx. Even though reporter gene construct and GTx have the same AAV serotype, they may have been produced by different manufacturing processes and differ in the size of packaged DNA, both of which could impact capsid antigenicity. One way to confirm GTx specificity in TI assays is to spike excess GTx into a sample containing AAV inhibitors. The GTx will bind to AAV inhibitors, preventing them from neutralizing the reporter gene construct, thus leading to increased transduction.

6.3.3.5 Confirmatory Steps to Ensure Specific Detection of Neutralizing AAV Antibodies

While TI assays detect AAV inhibitors, these assays do not routinely distinguish between neutralizing antibodies (NAbs) and non-antibody-based neutralizing factors [3]. This could obfuscate GTx patient eligibility when investigating preexisting AAV immunity, since some non-antibody-based neutralizing factors decrease transduction in vitro but are not impactful in vivo, as demonstrated for an AAV5 GTx in nonhuman primates [4].

The nature of non-antibody-based neutralizing factors in plasma is heterogeneous and poorly understood [46–49]. To ensure specific detection of NAbs in AAV TI assays, a confirmatory procedure may be implemented that depletes antibodies using resins coupled with immunoglobulin-binding proteins, such as protein A, G, and L [33, 40, 50]. Samples that screened positive for AAV inhibitors are retested following immunoglobulin depletion, which restores transduction if and only if NAbs were present. A mock depletion of the same sample is performed in parallel to account for non-specific depletion by unconjugated resin. Transduction following protein AGL depletion can be normalized to that observed following mock depletion, establishing an AGL/mock ratio [40]. For regulated studies, a statistical confirmatory cutpoint can be derived for the AGL/mock ratio; sensitivity, precision, and selectivity of the confirmatory assay could also be validated.

Alternative strategies to confirm the specificity of AAV TI assays include depletion of test samples using resins conjugated with the AAV vector of interest [20, 51], or the addition of empty or generic AAV vectors of the same serotype to absorb and functionally deplete AAV inhibitors in test samples [52]. Even though these strategies confirm the interaction of AAV inhibitors with a particular AAV serotype of interest, they do not immediately provide for a molecular classification of the involved neutralizing factors as antibodies, since any AAV inhibitor would be physically or functionally depleted.

6.3.3.6 Selectivity/Matrix Interference

Selectivity assessments for AAV TI assays are performed as described for TAb assays. At least 80% of spiked donor samples should test positive, while at least 80% of unspiked donor samples should test negative. A related assessment is interference from individual matrix components. For AAV TI assays, it is not advisable to use endogenously lipemic, hemolytic, and icteric test samples for interference studies, since results may be confounded by preexisting AAV inhibitors, falsely indicating additional interference. It may be more conclusive to add interfering factors under investigation (lipids, hemoglobin, or bilirubin) exogenously to individual or pooled samples previously confirmed to be negative for AAV inhibitors.

6.3.3.7 Stability

Stability studies can be performed by spiking low and high concentrations of positive-control AAV antibodies into negative test samples. Stability of non-antibody-based neutralizing factors could alternatively be assessed by retesting a positive donor sample over time. Neutralizing AAV antibodies are the most relevant AAV inhibitors, and antibodies are generally stable if stored at −70 °C or colder [53, 54]. Thus, there is limited value in performing long-term stability studies at low temperatures. Short-term and freeze/thaw stability assessments may be more applicable; for example, samples may be stored at higher temperatures, such as −20 °C, 4 °C, or ambient temperature, given logistical constraints, or be subject to repeated freeze–thaws to allow for retesting.

6.3.4 Sample Testing Strategy and Monitoring Assay Performance

Samples from GTx treatment-naïve subjects can be positive for AAV inhibitors but seroprevalence typically does not reach 100% [3, 41]. This could justify tiered testing for pre-dose samples: initial screening at minimally required dilution (MRD), followed by titering of screen-positive samples only. Though for rare diseases, where sample batches are small, screening and titer assays are typically performed simultaneously to enable timely eligibility decisions. Screen-positive samples may also be tested in a confirmatory assay, if deemed suitable for the serotype of interest.

In contrast, samples collected following systemic GTx administration have high AAV TI titers due to the formation of neutralizing AAV antibodies, usually with an incidence rate of 100% and persisting for at least 15 years [55–58]. Therefore, in most clinical scenarios, post-dose samples could be titered directly. While post-dose samples might also contain non-antibody-based neutralizing factors, their titers will be negligible compared to those from treatment-emergent neutralizing AAV antibodies. Hence, confirmatory assays for post-dose samples may not be necessary or be restricted to a subset of samples to demonstrate that neutralizing AAV antibodies have indeed formed as expected.

TI assay performance can be monitored long-term by trending luminescence signals and/or transduction/inhibition values for LPC and HPC, as well as titer values for HPC. Appropriate trending ranges may be established based on results from at least 10 assay runs.

6.3.5 Data Interpretation: Preexisting TI Titer and Clinical Efficacy

Preexisting AAV immunity detected in TI assays does not always correlate with decreased GTx efficacy. Lack of biological impact has been linked to low TI titers [2, 59] as well as non-antibody-based neutralizing factors, which are typically present at low TI titers [3, 4]. This raises the question of whether there is a clinically meaningful correlation between preexisting TI titers and the likelihood of clinical success, which could inform GTx patient eligibility.

For TI data interpretation, it can be useful to determine the theoretical neutralizing capacity of plasma/serum in relation to a particular TI titer, based on assay procedures. For example, in a cell-based TI assay, 3.36 E+09 vg of an AAV reporter gene construct may be incubated with 42 µL of a plasma test sample. Assuming a linear dynamic range, a test sample that shows 50% transduction at MRD will have neutralized 50% of the AAV reporter gene construct, which is equal to 0.5×3.36 E+09 vg = 1.68 E+09 vg (Table 6.1).

In comparison, an average human adult has an estimated total plasma volume of about 3000 mL, and thus the theoretical total neutralizing capacity of plasma is derived as (1.68 E+09 vg/0.042 mL) × 3000 mL = 1.20 E+14 vg, assuming linearity and complete mixing. Plasma samples with higher TI titers that can be diluted beyond the MRD to achieve 50% transduction in the assay will have higher neutralizing capacity, which can be theoretically extrapolated by multiplying the previously calculated quantities of neutralized reporter gene construct with the term "TI titer/MRD." An example of these calculations is shown in Table 6.1.

If plotted as a double-logarithmic graph, there will be a linear relationship between TI titer (x) and the theoretical total neutralizing capacity of plasma (y), the latter of which is expressed as the quantity of AAV vector neutralized by 3000 mL plasma. In the example, this relationship is represented by the equation $y = 6$ E+13 $\times x$ (Figure 6.4).

Using the theoretical total neutralizing capacity of plasma for various TI titers, dose neutralization curves can be derived for different GTx dose levels. For example, at the 6 E+12 vg/kg dose level, an adult with 70 kg body weight will receive a total vector dose of 4.2 E+14 vg. From the graph and table above, for a test sample

Table 6.1 Theoretical neutralizing capacity of plasma volumes *in vitro* and *in vivo*.

	Theoretical neutralizing capacity (vg)	
TI Titer	**42 µL plasma (*in vitro*)**	**3000 mL plasma (*in vivo*)**
0	0	0
2 (MRD)	1.68 E+09	1.20 E+14
4	3.36 E+09	2.40 E+14
8	6.72 E+09	4.80 E+14
16	1.34 E+10	9.60 E+14
32	2.69 E+10	1.92 E+15
64	5.38 E+10	3.84 E+15
128	1.08 E+11	7.68 E+15
256	2.15 E+11	1.54 E+16
512	4.30 E+11	3.07 E+16

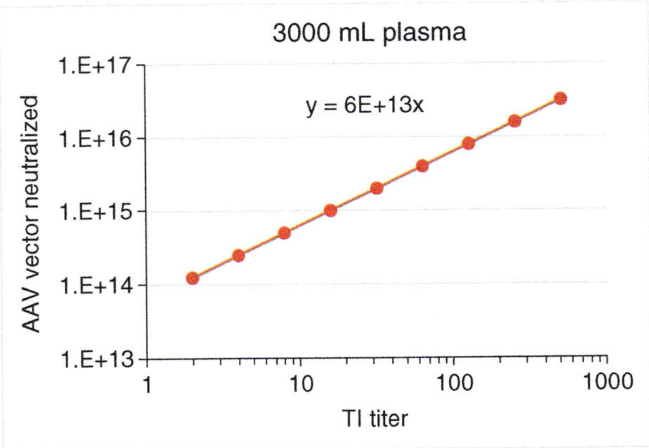

Figure 6.4 Theoretical relationship between TI titer and total AAV dose neutralized.

with a TI titer equal to the MRD of 2, the theoretical total neutralizing capacity of plasma was 1.20 E+14 vg, which corresponds to 29% of the total vector dose received by this patient (Table 6.2). In other words, at least 71% of the total vector dose are expected to remain active. This calculation can be repeated for the theoretical total neutralizing capacity of plasma for higher TI titers at the same dose level and expanded to higher dose levels, resulting in dose neutralization curves that estimate at which TI titer complete dose neutralization (0% active clinical dose) would theoretically occur (Figure 6.5).

While theoretical dose neutralization curves are unlikely to accurately predict TI titer limits associated with individual treatment success, they can provide benchmarks for empirical interrogation of GTx efficacy in the presence of preexisting AAV immunity [4, 60]. The current clinical praxis of weight-based AAV-GTx dosing may also not be fully compatible with using average assumptions for body weight and total plasma volume. The real-world challenges of determining patient eligibility, and tentatively adjusting vector dose level based on preexisting TI titers, may require more sophisticated modeling than the one discussed above, while also carefully considering infusion-associated reactions and safety in dose escalation studies conducted to overcome preexisting TI titers. Nonetheless, theoretical dose neutralization curves can confirm that a cell-based TI assay has sufficient analytical sensitivity to detect AAV inhibitors that are predicted to neutralize a substantial portion of the clinical dose.

AAV antibodies may have clinical impact apart from dose neutralization, such as accelerated capsid clearance, modified capsid tropism, or complement activation; corresponding orthogonal methods should therefore be implemented as needed.

Table 6.2 Theoretical relationship between TI titer and percentage dose neutralization.

	Theoretical percentage of total clinical vector dose neutralized by 3000 mL plasma[a]			
TI titer	**6 E+12 (vg/kg)**	**2 E+13 (vg/kg)**	**6 E+13 (vg/kg)**	**2 E+14 (vg/kg)**
0	0%	0%	0%	0%
2	29%	9%	3%	1%
4	57%	17%	6%	2%
8	114%	34%	11%	3%
16	229%	69%	23%	7%
32	457%	137%	46%	14%
64	914%	274%	91%	27%
128	1829%	549%	183%	55%
256	3657%	1097%	366%	110%

[a] The total clinical vector dose was calculated for a 70-kg adult patient as 4.2 E+14 vg (at 6 E+12 vg/kg dose level), 1.4 E+15 vg (at 2 E+13 vg/kg dose level), 4.2 E+15 vg (at 6 E+13 vg/kg), and 1.4 E+16 vg (at 2 E+14 vg/kg).

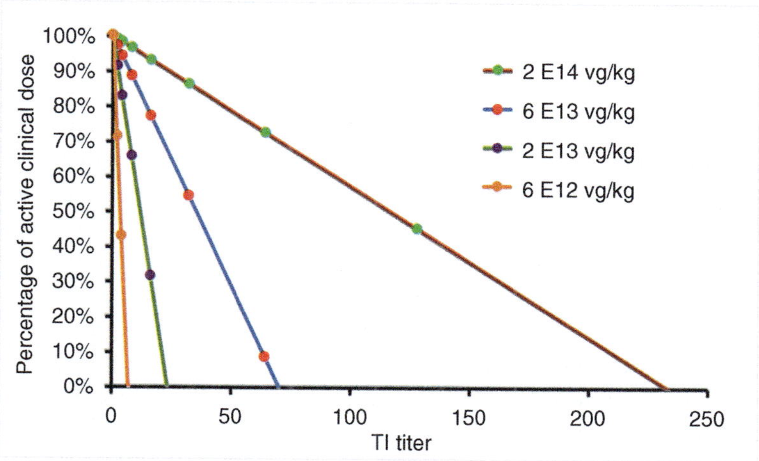

Figure 6.5 Theoretical dose neutralization curves were obtained by plotting the difference between 100% and the theoretical percentage of the total clinical dose neutralized by 3000 mL plasma from Table 6.2.

6.3.6 Value and Challenges of Standardizing TAb and TI Assays

There are ongoing discussions on how to harmonize AAV TAb and TI assays across sponsors. Potential benefits could be increased transparency of data used to inform GTx patient eligibility and comparability of antibody titers and efficacy across clinical trials. Presently, AAV antibody titers from different sponsors are not comparable, due to different assay formats, technologies, cut-points, titer schemes, and algorithms (endpoint vs. interpolated titer; endpoint titer is the reciprocal of the highest sample dilution empirically detectable, interpolated titer is the reciprocal of the mathematically derived sample dilution at the assay cut-point using curve fit or linear regression). Moreover, TAb and TI assays generally do not use reference standards against which antibody concentrations in test samples would be calibrated, since no single reference standard would adequately represent the unique mixture of AAV antibodies in individual samples. This idiosyncrasy distinguishes immunogenicity assays from pharmacokinetic or biomarker assays, for which standardization can be achieved by using universally accepted and widely accessible primary and secondary reference standards [61].

A first step toward AAV TAb and TI assay standardization could be regulatory guidance detailing minimally required assay performance characteristics and best practices for bioanalytical method validation. A second step could be to establish a panel of positive polyclonal reference samples, and/or a panel of representative monoclonal AAV antibodies isolated from human subjects. While these reference samples or antibodies would not serve as calibration standards, they could be tested and titered in assays across sponsors and compared to real-world clinical samples [62, 63]. This would allow for comparison of analytical sensitivity across sponsors and thus provide a better understanding of numerical titer measurements.

A third step toward harmonization could be to gain alignment on whether AAV TAb or TI assays are more clinically meaningful for detecting preexisting AAV immunity. Agreements across industry and academia on analytical assay formats have been achieved previously, for example, for detecting coagulation factor VIII inhibitors (Nijmegen-Bethesda Assay, [64–69], or for detecting NAbs to interferon beta [70]. For detecting preexisting AAV immunity, the choice of assay would need to reflect not only its predictive value with regard to clinical GTx efficacy but also the feasibility of commercialization as a CDx. Cell-based TI assays are intrinsically more complex and thus more challenging to develop into a globally viable CDx. Hence, even though TI titers can adequately describe the neutralizing potential of AAV antibodies, as opposed to TAb titers that represent a combined measure of binding strength and concentration, TAb assays could be a less expensive and more practical solution for large-scale real-world CDx implementation. For more information, see Chapter 15, Chapter 16, and Chapter 17.

References

1 Long BR, Veron P, Kuranda K, et al. Early phase clinical immunogenicity of valoctocogene roxaparvovec, an AAV5-mediated gene therapy for hemophilia A. *Mol. Ther.* 2021;29(2):597-610. doi: https://doi.org/10.1016/j.ymthe.2020.12.008.

2 Majowicz A, Nijmeijer B, Lampen MH, et al. Therapeutic hFIX activity achieved after single AAV5-hFIX treatment in hemophilia B patients and NHPs with pre-existing anti-AAV5 NABs. *Mol. Ther. Methods Clin. Dev.* 2019;14:27-36. doi: https://doi.org/10.1016/j.omtm.2019.05.009.

3 Falese L, Sandza K, Yates B, et al. Strategy to detect pre-existing immunity to AAV gene therapy. *Gene Ther.* 2017;24(12):768-78. doi: https://doi.org/10.1038/gt.2017.95.

4 Long BR, Sandza K, Holcomb J, et al. The impact of pre-existing immunity on the non-clinical pharmacodynamics of AAV5-based gene therapy. *Mol. Ther. Methods Clin. Dev.* 2019;13:440-52. doi: https://doi.org/10.1016/j.omtm.2019.03.006.

5 Gorovits B, Azadeh M, Buchlis G, et al. Evaluation of the humoral response to adeno-associated virus-based gene therapy modalities using total antibody assays. *AAPS J.* 2021;23(6):108. doi: https://doi.org/10.1208/s12248-021-00628-3.

6 FDA. (2019). Immunogenicity testing of therapeutic protein products – developing and validating assays for anti-drug antibody detection. Guidance for industry. https://www.fda.gov/media/119788/download (accessed July 2019).

7 Gorovits B. Current considerations for immunoglobulin isotype characterization of antibody response against biotherapeutics. *AAPS J.* 2020;22(6):144. doi: https://doi.org/10.1208/s12248-020-00530-4.

8 Erles K, Sebökovà P, Schlehofer JR. Update on the prevalence of serum antibodies (IgG and IgM) to adeno-associated virus (AAV). *J. Med. Virol.* 1999;59(3):406-11. doi: https://doi.org/10.1002/(sici)1096-9071(199911)59:3<406::aid-jmv22>3.0.co;2-n.

9 Kruzik A, Fetahagic D, Hartlieb B, et al. Prevalence of anti-adeno-associated virus immune responses in international cohorts of healthy donors. *Mol. Ther. Methods Clin. Dev.* 2019;14:126-33. doi: https://doi.org/10.1016/j.omtm.2019.05.014.

10 Murphy SL, Li H, Mingozzi F, et al. Diverse IgG subclass responses to adeno-associated virus infection and vector administration. *J. Med. Virol.* 2009;81(1):65-74. doi: https://doi.org/10.1002/jmv.21360.

11 Vidarsson G, Dekkers G, Rispens T. IgG subclasses and allotypes: from structure to effector functions. *Front. Immunol.* 2014;5:520. doi: https://doi.org/10.3389/fimmu.2014.00520.

12 Wang Y, Smith JF, Araya MM, et al. Development of a highly specific anti-drug antibody assay in support of a nanoparticle-based therapeutic. *AAPS J.* 2020;22(4):81. doi: https://doi.org/10.1208/s12248-020-00462-z.

13 Patterson M, Beausang LA, Rup B, et al. A bridging assay for detection and characterization of anti-drug antibodies to dostarlimab, a new anti-PD-1 therapeutic monoclonal antibody. *AAPS Open* 2021;7(1):11. doi: https://doi.org/10.1186/s41120-021-00045-y.

14 Shankar G, Devanarayan V, Amaravadi L, et al. Recommendations for the validation of immunoassays used for detection of host antibodies against biotechnology products. *J. Pharm. Biomed. Anal.* 2008;48(5):1267-81. doi: https://doi.org/10.1016/j.jpba.2008.09.020.

15 Vultaggio A, Nencini F, Carraresi A, et al. IgG4 anti-infliximab in treated patients: clinical impact and temporal evolution. *Allergy* 2018;73(11):2172-81. doi: https://doi.org/10.1111/all.13471.

16 Tseng YS, Vliet KV, Rao L, et al. Generation and characterization of anti-Adeno-associated virus serotype 8 (AAV8) and anti-AAV9 monoclonal antibodies. *J. Virol. Methods* 2016;236:105-10. doi: https://doi.org/10.1016/j.jviromet.2016.07.009.

17 Calcedo R, Vandenberghe LH, Gao G, et al. Worldwide epidemiology of neutralizing antibodies to adeno-associated viruses. *J. Infect. Dis.* 2009;199(3):381-90. doi: https://doi.org/10.1086/595830.

18 Boutin S, Monteilhet V, Veron P, et al. Prevalence of serum IgG and neutralizing factors against adeno-associated virus (AAV) types 1, 2, 5, 6, 8, and 9 in the healthy population: implications for gene therapy using AAV vectors. *Hum. Gene Ther.* 2010;21(6):704-12. doi: https://doi.org/10.1089/hum.2009.182.

19 Salas D, Kwikkers KL, Zabaleta N, et al. Immunoadsorption enables successful rAAV5-mediated repeated hepatic gene delivery in nonhuman primates. *Blood Adv.* 2019;3(17):2632-41. doi: https://doi.org/10.1182/bloodadvances.2019000380.

20 Bertin B, Veron P, Leborgne C, et al. Capsid-specific removal of circulating antibodies to adeno-associated virus vectors. *Sci. Rep.* 2020;10(1):864. doi: https://doi.org/10.1038/s41598-020-57893-z.

21 Azadeh M, Sondag P, Wang Y, et al. Quality controls in ligand binding assays: recommendations and best practices for preparation, qualification, maintenance of lot to lot consistency, and prevention of assay drift. *AAPS J.* 2019;21(5):89. doi: https://doi.org/10.1208/s12248-019-0354-6.

22 Song S, Yang L, Trepicchio WL, Wyant T. Understanding the supersensitive anti-drug antibody assay: unexpected high anti-drug antibody incidence and its clinical relevance. *J Immunol Res* 2016;2016:3072586. doi: https://doi.org/10.1155/2016/3072586.

23 Li C, Narkbunnam N, Samulski RJ, et al. Neutralizing antibodies against adeno-associated virus examined prospectively in pediatric patients with hemophilia. *Gene Ther.* 2012;19(3):288-94. doi: https://doi.org/10.1038/gt.2011.90.

24 Gao G, Vandenberghe LH, Wilson JM. New recombinant serotypes of AAV vectors. *Curr. Gene Ther.* 2005;5(3):285-97. doi: https://doi.org/10.2174/1566523054065057.

25 Gao GP, Alvira MR, Wang L, Novel adeno-associated viruses from rhesus monkeys as vectors for human gene therapy. *Proc. Natl. Acad. Sci. U. S. A.* 2002;99(18):11854-9. doi: https://doi.org/10.1073/pnas.182412299.

26 Peden CS, Burger C, Muzyczka N, Mandel RJ. Circulating anti-wild-type adeno-associated virus type 2 (AAV2) antibodies inhibit recombinant AAV2 (rAAV2)-mediated, but not rAAV5-mediated, gene transfer in the brain. *J. Virol.* 2004;78(12):6344-59. doi: https://doi.org/10.1128/jvi.78.12.6344-6359.2004.

27 Grimm D, Kay MA. From virus evolution to vector revolution: use of naturally occurring serotypes of adeno-associated virus (AAV) as novel vectors for human gene therapy. *Curr. Gene Ther.* 2003;3(4):281-304. doi: https://doi.org/10.2174/1566523034578285.

28 Rivière C, Danos O, Douar AM. Long-term expression and repeated administration of AAV type 1, 2 and 5 vectors in skeletal muscle of immunocompetent adult mice. *Gene Ther.* 2006;13(17):1300-8. doi: https://doi.org/10.1038/sj.gt.3302766.

29 Devanarayan V, Smith WC, Brunelle RL, et al. Recommendations for systematic statistical computation of immunogenicity cut points. *AAPS J.* 2017;19(5):1487-98. doi: https://doi.org/10.1208/s12248-017-0107-3.

30 Spark T. (2017). LUXTURNA package insert. https://www.fda.gov/media/109906/download (accessed 2019).

31 EMA. (2022). Valoctocogene roxaparvovec. https://www.ema.europa.eu/en/medicines/human/EPAR/roctavian-0 (Accessed 30 October 2022).

32 Andreas Jungmann, O.M. and Rapti, K. (2017). Cell-based measurement of neutralizing antibodies against adeno-associated virus (AAV). In: *Cardiac Gene Therapy: Methods and Protocols, Methods in Molecular Biology* (ed. K. Ishikawa). New York: Springer Science+Business Media.

33 Rapti K, Louis-Jeune V, Kohlbrenner E, et al. Neutralizing antibodies against AAV serotypes 1, 2, 6, and 9 in sera of commonly used animal models. *Mol. Ther.* 2012;20(1):73-83. doi: https://doi.org/10.1038/mt.2011.177.

34 Grimm D, Lee JS, Wang L, et al. In vitro and in vivo gene therapy vector evolution via multispecies interbreeding and retargeting of adeno-associated viruses. *J. Virol.* 2008;82(12):5887-911. doi: https://doi.org/10.1128/jvi.00254-08.

35 Meliani A, Leborgne C, Triffault S, et al. Determination of anti-adeno-associated virus vector neutralizing antibody titer with an in vitro reporter system. *Hum. Gene Ther. Methods* 2015;26(2):45-53. doi: https://doi.org/10.1089/hgtb.2015.037.

36 Kaliyaperumal A, Pennucci J, Nagatani J, et al. A method to quantitate the neutralizing capacity of anti-therapeutic protein antibodies in serum and their correlation to clinical impact. *J. Pharm. Biomed. Anal.* 2015;102:176-83. doi: https://doi.org/10.1016/j.jpba.2014.09.009.

37 EMA. (2017). Guideline on immunogenicity assessment of therapeutic proteins. https://www.ema.europa.eu/en/documents/scientific-guideline/guideline-immunogenicity-assessment-therapeutic-proteins-revision-1_en.pdf (accessed 2019).

38 Gorovits B, Fiscella M, Havert M, et al. Recommendations for the development of cell-based anti-viral vector neutralizing antibody assays. *AAPS J.* 2020;22(2):24. doi: https://doi.org/10.1208/s12248-019-0403-1.

39 Gupta S, Devanarayan V, Finco D, et al. Recommendations for the validation of cell-based assays used for the detection of neutralizing antibody immune responses elicited against biological therapeutics. *J. Pharm. Biomed. Anal.* 2011;55(5):878-88. doi: https://doi.org/10.1016/j.jpba.2011.03.038.

40 Kasprzyk T, Triffault S, Long BR, et al. Confirmatory detection of neutralizing antibodies to AAV gene therapy using a cell-based transduction inhibition assay. *Mol. Ther. Methods Clin. Dev.* 2022;24:222-9. doi: https://doi.org/10.1016/j.omtm.2022.01.004.

41 Klamroth R, Hayes G, Andreeva T, et al. Global seroprevalence of pre-existing immunity against AAV5 and other AAV serotypes in people with hemophilia A. *Hum. Gene Ther.* 2022;33(7–8):432-41. doi: https://doi.org/10.1089/hum.2021.287.

42 Kumar SC, DelCarpini JA, Qu Q, Mitigation of pre-existing antibodies to a biotherapeutic in non-clinical species when establishing anti-drug antibody assay cutpoint. *AAPS J.* 2017;19(1):313-9. doi: https://doi.org/10.1208/s12248-016-0011-2.

43 Mikulskis A, Yeung D, Chen W, et al. Novel data analysis methods to overcome cut point challenges and enable comprehensive assessment of antidrug binding activity in confirmatory assays. *J. Immunol. Methods* 2013;392(1–2):38-48. doi: https://doi.org/10.1016/j.jim.2013.03.008.

44 Schaarschmidt F, Hofmann M, Jaki T, et al. Statistical approaches for the determination of cut points in anti-drug antibody bioassays. *J. Immunol. Methods* 2015;418:84-100. doi: https://doi.org/10.1016/j.jim.2015.02.004.

45 Jaki T, Lawo JP, Wolfsegger MJ, A formal comparison of different methods for establishing cut points to distinguish positive and negative samples in immunoassays. *J. Pharm. Biomed. Anal.* 2011;55(5):1148-56. doi: https://doi.org/10.1016/j.jpba.2011.04.006.

46 Berry G, Murlidharan G, Asokan A. 30. Modulation of intracellular calcium enhances AAV transduction in the CNS. *Mol. Ther.* 2016;24:S14. doi: https://doi.org/10.1016/S1525-0016(16)32839-8.

47 Hirosue S, Senn K, Clément N, et al. Effect of inhibition of dynein function and microtubule-altering drugs on AAV2 transduction. *Virology* 2007;367(1):10-8. doi: https://doi.org/10.1016/j.virol.2007.05.009.

48 Virella-Lowell I, Poirier A, Chesnut KA, et al. Inhibition of recombinant adeno-associated virus (rAAV) transduction by bronchial secretions from cystic fibrosis patients. *Gene Ther.* 2000;7(20):1783-9. doi: https://doi.org/10.1038/sj.gt.3301268.

49 Nonnenmacher M, Weber T. Intracellular transport of recombinant adeno-associated virus vectors. *Gene Ther.* 2012;19(6):649-58. doi: https://doi.org/10.1038/gt.2012.6.

50 Sanchez S, Barger T, Zhou L, et al. Strategy to confirm the presence of anti-erythropoietin neutralizing antibodies in human serum. *J. Pharm. Biomed. Anal.* 2011;55(5):1265-74. doi: https://doi.org/10.1016/j.jpba.2011.03.029.

51 Orlowski A, Katz MG, Gubara SM, et al. Successful transduction with AAV vectors after selective depletion of anti-AAV antibodies by immunoadsorption. *Mol. Ther. Methods Clin. Dev.* 2020;16:192-203. doi: https://doi.org/10.1016/j.omtm.2020.01.004.

52 Mingozzi F, High KA. Overcoming the host immune response to adeno-associated virus gene delivery vectors: the race between clearance, tolerance, neutralization, and escape. *Annu. Rev. Virol.* 2017;4(1):511-34. doi: https://doi.org/10.1146/annurev-virology-101416-041936.

53 Michaut L, Laurent N, Kentsch K, et al. Stability of anti-immunotherapeutic antibodies in frozen human serum samples. *Bioanalysis* 2014;6(10):1395-407. doi: https://doi.org/10.4155/bio.14.97.

54 Hendriks J, Stals C, Versteilen A, et al. Stability studies of binding and functional anti-vaccine antibodies. *Bioanalysis* 2014;6(10):1385-93. doi: https://doi.org/10.4155/bio.14.96.

55 George LA, Sullivan SK, Giermasz A, et al. Hemophilia B gene therapy with a high-specific-activity factor IX variant. *N. Engl. J. Med.* 2017;377(23):2215-27. doi: https://doi.org/10.1056/NEJMoa1708538.

56 Nathwani AC, Tuddenham EG, Rangarajan S, et al. Adenovirus-associated virus vector-mediated gene transfer in hemophilia B. *N. Engl. J. Med.* 2011;365(25):2357-65. doi: https://doi.org/10.1056/NEJMoa1108046.

57 Manno CS, Chew AJ, Hutchison S, et al. AAV-mediated factor IX gene transfer to skeletal muscle in patients with severe hemophilia B. *Blood* 2003;101(8):2963-72. doi: https://doi.org/10.1182/blood-2002-10-3296.

58 George LA, Ragni MV, Rasko JEJ, et al. Long-term follow-up of the first in human intravascular delivery of AAV for gene transfer: AAV2-hFIX16 for severe hemophilia B. *Mol. Ther.* 2020;28(9):2073-82. doi: https://doi.org/10.1016/j.ymthe.2020.06.001.

59 Aronson SJ, Veron P, Collaud F, et al. Prevalence and relevance of pre-existing anti-adeno-associated virus immunity in the context of gene therapy for Crigler–Najjar syndrome. *Hum. Gene Ther.* 2019;30(10):1297-305. doi: https://doi.org/10.1089/hum.2019.143.

60 Meadows AS, Pineda RJ, Goodchild L, Threshold for pre-existing antibody levels limiting transduction efficiency of systemic rAAV9 gene delivery: relevance for translation. *Mol. Ther.* 2019;13:453-62. doi: https://doi.org/10.1016/j.omtm.2019.04.004.

61 Hubbard A.R. (2005). JS: calibration of the proposed WHO 1st international standard for blood coagulation factor V in human plasma. https://www.who.int/bloodproducts/publications/05-2007%20BCF%20V%20in%20plasma.pdf (accessed 24 March 2021).

62 Mytych DT, Barger TE, King C, et al. Development and characterization of a human antibody reference panel against erythropoietin suitable for the standardization of ESA immunogenicity testing. *J. Immunol. Methods* 2012;382(1-2):129-41. doi: https://doi.org/10.1016/j.jim.2012.05.013.

63 Wadhwa M, Mytych DT, Bird C, et al. Establishment of the first WHO erythropoietin antibody reference panel: report of an international collaborative study. *J. Immunol. Methods* 2016;435:32-42. doi: https://doi.org/10.1016/j.jim.2016.05.005.

64 Blanchette VS, Key NS, Ljung LR, et al. Definitions in hemophilia: communication from the SSC of the ISTH. *J. Thromb. Haemost.* 2014; 12(11):1935-9. doi: https://doi.org/10.1111/jth.12672.

65 Verbruggen, B., Novakova, I., Wessels, H. et al. (1995). The Nijmegen modification of the Bethesda assay for factor VIII:C inhibitors: improved specificity and reliability. *Thromb. Haemost.* 73 (2): 247–251.

66 Giles, A.R., Verbruggen, B., Rivard, G.E. et al. (1998). A detailed comparison of the performance of the standard versus the Nijmegen modification of the Bethesda assay in detecting factor VIII:C inhibitors in the haemophilia A population of Canada. Association of Hemophilia Centre Directors of Canada. Factor VIII/IX Subcommittee of Scientific and Standardization Committee of International Society on Thrombosis and Haemostasis. *Thromb. Haemost.* 79 (4): 872–875.

67 Collins PW, Chalmers E, Hart DP, et al. Diagnosis and treatment of factor VIII and IX inhibitors in congenital haemophilia: (4th edition). UK Haemophilia Centre Doctors Organization. *Br. J. Haematol.* 2013;160(2):153-70. doi: https://doi.org/10.1111/bjh.12091.

68 Miller CH, Rice AS, Boylan B, et al. Comparison of clot-based, chromogenic and fluorescence assays for measurement of factor VIII inhibitors in the US Hemophilia Inhibitor Research Study. *J. Thromb. Haemost.* 2013;11(7):1300-9. doi: https://doi.org/10.1111/jth.12259.

69 EMA. (2018). Clinical investigation of recombinant and human plasma-derived factor VIII products – scientific guideline. https://www.ema.europa.eu/en/clinical-investigation-recombinant-human-plasma-derived-factor-viii-products-scientific-guideline (accessed November 2022).

70 Wadhwa M, Subramanyam M, Goelz S, et al. Use of a standardized MxA protein measurement-based assay for validation of assays for the assessment of neutralizing antibodies against interferon-β. *J. Interf. Cytokine Res.* 2013;33(11):660-71. doi: https://doi.org/10.1089/jir.2012.0079.

7

Bioanalytical Methods to Study Biodistribution and Shedding of AAV-Based Gene Therapy Vectors

Christian Vettermann and Russell Soon

Translational Sciences, Assay Strategy & Development, BioMarin Pharmaceutical, Inc., Novato, CA, USA

7.1 Introduction

Pharmacokinetic studies monitor the presence and fate of a therapeutic following administration to animals or clinical trial participants. For adeno-associated virus (AAV)-based gene therapies (GTx), pharmacokinetic studies typically address tissue biodistribution, intracellular vector processing, and vector shedding. Given their invasive nature, biodistribution and vector processing studies are typically conducted non-clinically, and only limited data exist in humans [1, 2]. In contrast, while vector shedding studies may be conducted non-clinically, they are more routinely implemented in clinical GTx trials. Regulatory expectations for nonclinical biodistribution and clinical shedding studies are covered in Chapter 4. In this chapter, we focus on bioanalytical methods to detect GTx vectors in tissues and biological fluids.

Polymerase Chain Reaction (PCR)-based methods to detect GTx vectors are generally more sensitive than cell-based or ligand-binding assays [3]. In addition to PCR being highly specific, it can broadly detect both encapsidated and non-encapsidated (i.e. processed) vector DNA. Regulatory guidance for qualification or validation of PCR-based methods is limited [4, 5]. Thus, bioanalytical practices are often rooted in regulatory guidances for validation of non-PCR-based methods. These guidances focus on methodologies that have dominated the field so far, namely ligand-binding and chromatographic assays [6], and therefore do not fully address the specific needs of PCR-based methods. Consequently, while these

Drug Development for Gene Therapy: Translational Biomarkers, Bioanalysis, and Companion Diagnostics, First Edition. Edited by Yanmei Lu and Boris Gorovits.

guidances may serve as reference points, specific recommendations require consideration of context-of-use and fit-for-purpose aspects for PCR methods, as the field acquires the critical mass necessary for establishing consensus and specific regulatory guidance.

Some industry white papers and regulatory reflection papers on PCR have emerged within the past few years [7–9]. These publications started to delve into the details underpinning critical areas of set-up, validation, and sample testing for PCR-based methods in gene and cell therapy development. A valuable summary of regulatory and technical aspects for PCR-based biodistribution and shedding methods was recently provided [10]. In this chapter, we will discuss the application of real-time (q) PCR and digital PCR to bioanalytical assessments that support GTx development. Our focus is on areas where practical use of these methods still requires clarification, or where strategic and tactical considerations come into play in choosing one PCR platform over the other. While another book chapter will focus in-depth on reverse transcription PCR methods for the detection and measurement of GTx vector RNA transcripts, this chapter will focus on detection of GTx vector DNA.

7.2 Choice of Platform: qPCR vs. Digital PCR

Two technology platforms exist for PCR-based methods that differ in instrumentation, measurement principle, and applicability to translational or clinical questions (Table 7.1). On the one hand, quantitative polymerase chain reaction (qPCR) has been a standard molecular biology technique since its invention in the 1980s. On the other hand, digital PCR is a relative newcomer to the world of quantitative assays, and even newer to the bioanalytical suite of assays used in drug development. Understanding the principles of each technology should help inform platform selection within the intended contexts of use.

In qPCR, a sample containing target DNA sequence is mixed with DNA polymerase and primers plus probe, and subjected to several reaction cycles at set temperatures for specific durations to amplify the DNA and generate a fluorescent signal in real time. The fluorescent signal increases as the target DNA sequence amplifies with each thermal cycle until a given detection threshold is reached. The cycle number at which this occurs is designated Cq or Ct value and is proportional to the starting amount of target DNA in the sample. The Cq or Ct value from test samples with an unknown concentration of target DNA is then interpolated against a calibration curve of reference GTx drug material, vector DNA, or surrogate vector DNA, generating a result reported as vector genome (VG) copy number. Sample volumes in qPCR reactions range from 20 to 200 μL, and real-time measurements are carried out on qPCR instruments, such as Roche

Table 7.1 Characteristics of qPCR and ddPCR assays.

qPCR	ddPCR
Standard real-time PCR instruments, such as Roche LightCycler systems. Validated methods require system to be CFR Part 11 compliant.	Microfluidic chamber (e.g. BioMark digital PCR, Fluidigm), micro-well chip (e.g. QuantStudio3D digital PCR, Life Technologies), droplet-digital (e.g. QX200 and QXONE, Bio-Rad), or Crystal™ digital-based (Naica, Stilla) instruments. Validated methods require system to be CFR Part 11 compliant.
Real-time fluorescence measurement.	End-point fluorescence measurement.
Calibration curve prepared from GTx reference material, vector DNA, or surrogate material.	No calibration curve required, absolute quantification through application of Poisson statistics to number of positive/negative reaction partitions.
Unit of measurement: single-stranded or double-stranded copies of target DNA per volume.	Unit of measurement: molecules of target DNA per volume.
Dynamic range of about 8–9 logs.	Dynamic range of about 4–5 logs.
Limited precision for lower template amounts.	Higher precision for lower template amounts.
Tolerance for PCR inhibitors can vary with reagents, sensitivity may be negatively impacted in interfering matrix components.	High tolerance for PCR inhibitors due to reaction partitioning and end-point measurement, greater sensitivity in the presence of interfering matrices.
Sequences with low PCR efficiency difficult to quantify accurately.	Quantification less dependent on PCR efficiency, advantageous for sequences with secondary structures, such as ITRs.
Multiple-amplicon reactions possible, but no read-out of connectivity between amplicons.	Multiple-amplicon reactions possible ('drop-phase'), read-out of linked amplicons possible with systems allowing for multiple detectors per reaction partition.
Generally lower cost ($25,000–$50,000 for instrument, $2 reagent cost per sample)[a]	Generally higher cost ($80,000-$100,000 fcr instrument, $5 reagent cost per sample)[a]
Average run-time is around 2 h per plate, option to run 384-well plate offers increased throughput over 96-well plate.	Average run-time is around 5 h per plate, current instruments limited to 96-well plate.
Data analysis includes calibration curve and quality control performance, followed by sample interpolation.	Data analysis simplified since calibration curve is not required, and quality controls and samples absolutely quantified.

[a] Basu, A.S. (2017). Digital assays part I: partitioning statistics and digital PCR. *SLAS Technol. Transl. Life Sci. Innov.* 22 (4): 369–386. doi:10.1177/2472630317705680.

LightCycler. Conventional qPCR is well-established, and a variety of commercial master-mixes and instruments are available to provide optimal PCR conditions, even for difficult samples or matrices that may otherwise impact PCR efficiency, accuracy, or precision of quantification.

The concept of digital PCR was introduced in the 1990s [11, 12] and found immediate application in the detection of rare genetic targets and mutations. The breadth and depth of the platform's application have grown steadily in both academia and industry, as the technology is being continually refined and greater technical understanding facilitates increasingly varied applications. At its core, digital PCR enables absolute quantification of a nucleic acid target through endpoint-measured PCR amplification; the platform does not require interpolation against a standard curve to generate a quantitative measurement. In contrast to qPCR, where an analog signal is generated in real time in a microliter scaled (typically 20 μL) PCR reaction, a digital PCR reaction employs the same thermal-cycling and detection chemistry, but the amplification reaction is partitioned into tens of thousands of smaller reactions on the nano or femtoliter scale. At sufficiently low target template concentrations, partitioning a sample into a high number of reactions enables each partition to generate a binary (digital) endpoint signal depending on the presence or absence of a single copy of target DNA in the partition, following a Poisson distribution. This partitioning of a sample allows for advantages over qPCR, namely absolute quantification of target DNA molecules, greater tolerance to PCR inhibitors, improved precision, and potentially greater sensitivity. In essence, an absolute count of target amplicon copies is derived from applying Poisson statistics to the number of negative and positive partitions in a sample.

While various commercialized digital PCR platforms exist on the market, the differences between them boil down to how the reaction is partitioned. Microfluidic chamber (e.g. BioMark dPCR, Fluidigm), micro-well chip (e.g. QuantStudio3D dPCR, Life Technologies), droplet-digital (e.g. QX200 and QXONE, Bio-Rad), and Crystal™ digital-based (Naica, Stilla) systems generate digital PCR data using the same underlying principle: PCR reaction partitioning to achieve a digital readout. Bio-Rad's droplet digital PCR (ddPCR) platform is presently widely used and therefore the focus of this chapter.

The difference in measurement principles between qPCR and ddPCR if used in single-amplicon mode can be illustrated by the following theoretical examples: A monomeric circular episome of an AAV GTx vector would be detected as one copy of vector DNA using qPCR and as one molecule of vector DNA using ddPCR. However, a concatemeric circular episome consisting of two vector copies fused head-to-tail at the ITRs would be detected as two copies of vector DNA using qPCR but as one molecule of vector DNA using ddPCR. The reason behind this theoretical discrepancy is that the two covalently joined vector copies in a circular episome would be co-partitioned in ddPCR into the same droplet and counted as one positive event after amplification.

Similarly, if using an appropriate calibration standard, qPCR can resolve the number of single-stranded VGs (the DNA packaging form in AAV capsids), while ddPCR indistinguishably counts both single-stranded and double-stranded copies of vector DNA, due to co-partitioning of both individual strands of a double-stranded DNA molecule. In summary, there may be a tendency to under-estimate VG copies using single-amplicon ddPCR compared to using single-amplicon qPCR, if higher-order molecular aggregates exist. This line of thought may warrant further consideration when ddPCR is used to compare the presence of vector DNA in longitudinal GTx biodistribution studies since the fraction of double-stranded concatemeric vector DNA is expected to increase in transduced tissues over the first few weeks following GTx administration.

ddPCR has clear advantages over qPCR if used for structural characterization of vector DNA. The ability of ddPCR to accurately quantify DNA sequences with substantial secondary structure, such as ITR fusions, makes it the method of choice for estimating the number of fully circularized AAV vector episomes in transduced tissue. In contrast, qPCR measurements of ITR fusions are oftentimes unreliable due to lower PCR amplification efficiency, a circumstance that does not affect ddPCR as long as positive droplets remain clearly separated from negative droplets.

In addition, ddPCR offers the unique feature to establish the contiguity (linkage) between two or more target sequences, when ddPCR is used in multi-amplicon mode ("drop-phase") [2]. This feature can be utilized to confirm the length of detected vector DNA fragments and ensure that adequate DNA repair processes have occurred when 5′ and 3′ truncated single strands of an over-sized VG are packaged in AAV capsids and need to be assembled in transduced target cells. The same linkage analysis can be applied to characterize transgene-derived mRNA in reverse transcription ddPCR (RT-ddPCR). As an example, multiplex RT-ddPCR could quantify two distinct amplicons at the 5′ and 3′ ends of a vector transcript. Measuring the number of "linked" (double-positive) vector transcripts provides an indication of productive vector expression and proper splicing.

Program-specific considerations also influence which PCR platform is most appropriate to detect vector DNA. While there may not always be an immanent reason to choose one over the other in nonclinical biodistribution or clinical vector-shedding studies, it may be helpful to use the same platform in both types of studies for a particular program (Part 3 of the 2020 White Paper on Recent Issues in Bioanalysis). This ensures better comparability of results since data interpretations remain unaffected by potential differences in sample concentrations that may be attributable to platform differences.

Study-specific details, such as anticipated sample testing load, also deserve consideration: Technologies for improving ddPCR throughput are still nascent, and hence the reagent cost of ddPCR on a per sample basis currently remains about twice as high as that of qPCR. Another consideration could be the anticipated peak vector concentration in test samples: Given qPCR's greater dynamic range, the method may prove more efficient than ddPCR from a labor and cost perspective when testing samples, since some shedding matrices or target tissues are likely to have high vector concentrations requiring dilution. Samples with higher concentrations would require fewer dilutions to quantitate in qPCR, translating into less time and lower consumable costs over the duration of a study. This aspect might also contribute to overall qPCR data accuracy since fewer sample dilutions translate into fewer opportunities for analyst errors or instrument malfunctions.

Other context-of-use considerations include availability of a well-characterized reference calibration DNA standard for qPCR: if unavailable, ddPCR may be preferable since this platform relies on absolute quantification and hence does not require interpolation from a calibration standard. Finally, very large template DNA fragments can negatively affect droplet generation in ddPCR, thus requiring enzymatic pre-digestion of samples, which increases sample preparation time and cost. All these considerations should be taken into account when deciding on the most suitable PCR platform for GTx studies.

7.3 Aspects of Method Development

Once a PCR platform has been selected, the development of the bioanalytical assay can begin. Many aspects of method development were covered in-depth recently [10]. Consequently, the aim of this chapter is to highlight only a few notable points and frame them within the context of planned applications and logistical considerations. While this section will discuss method development, subsequent sections will delve deeper into assay parameters that are critical for method characterization and validation.

With knowledge of pharmacokinetic study design, model species, and test sample matrices, the first step in PCR-based method development is sourcing of appropriate reference material that serves as a calibration standard (qPCR) and as a positive control (qPCR and ddPCR). Ideally, the structure/form of the reference material should resemble that of the GTx vector DNA in future test samples. If surrogate reference materials are used, comparable PCR efficiency and colinearity may be evaluated between the chosen material and the anticipated form of the GTx vector. Early in assay development, synthetic, commercially manufactured, single-stranded, or double-stranded DNA fragments or bacterial plasmids are often used as reference material. These synthetic materials should be characterized thoroughly since commercial products show varying degrees of quality. For example, a commercially purchased synthetic DNA fragment product that contains only 80% DNA molecules with complete fidelity to the ordered sequence would result in an immediate negative 20% relative error (RE) from the nominal concentration measured by UV absorbance or nucleic acid-intercalating dyes. This potential impact on accuracy precedes any manipulation of the material, and hence the initial negative bias would be additive to any subsequent error introduced during preparation of reference material stocks or during in-assay dilutions. One potential solution could be to use the known purity of DNA reference material to compensate and adjust reported nominal concentrations.

ddPCR is also increasingly being used to characterize reference materials because absolute quantification allows for selective measurement of functionally intact sequences that can be amplified in a PCR reaction [13–17]. This experimental approach to assigning nominal reference material concentrations is largely unaffected by the presence nontarget or low-fidelity sequences that may interfere in spectrophotometric or fluorometric methods. The advent of ddPCR has also led to a better understanding of potential impact from passive adsorption to plasticware, which may result in loss of nucleic acids during preparation of reference material stocks or when performing bioanalytical assays [18]. While passive-adsorption loss of an encapsidated genome or naked nucleic acid template may not have a quantitatively significant impact on highly concentrated stock solutions, under-recovery of template DNA may become more evident at low-concentration calibrators or assay controls. Addition of surfactants like Pluronic® F-68 can increase accurate quantification of encapsidated VGs [19], while addition of carrier nucleic acids can mitigate passive adsorptive loss of non-encapsidated vector DNA templates [20]. Moreover, manufacturers of ddPCR instruments recommend addition of carrier nucleic acids, such as tRNA or polyA, to mitigate loss of DNA template due to non-specific binding to plasticware [21].

Another focus of early method development is the selection of sample matrix into which the reference material will be spiked. In pharmacokinetic studies, GTx vector DNA can be measured in target as well as off-target tissues and various biofluids. Therefore, tissues and biofluids from naïve subjects of the test species should be acquired and gDNA extracted for use as matrix into which reference material is spiked. The extraction method employed for a specific tissue or biofluid during development should be selected with consideration of future testing needs, such as suitability for study samples and throughput. Other considerations for nucleic acid extraction are discussed below.

Un-spiked and spiked calibrator and/or controls are then used to demonstrate specific, precise, and accurate detection of target vector DNA within endogenous gDNA. Some biofluids (e.g. urine, saliva, CSF) may not yield a high concentration of gDNA since they are acellular in nature. Thus, it may be helpful to characterize typical nucleic acid yield early in development and decide how to best report assay sensitivity and other method parameters that are typically expressed as a quantity of VG per microgram of gDNA.

Given that GTx biodistribution and shedding studies often measure VGs in different tissues and biofluids, assay development may also need to demonstrate equivalency of gDNA extracted from each source. Validation of bioanalytical assays can be costly and time-intensive, so while it may be ideal to validate every parameter using gDNA from each matrix, scientific rigor needs to be balanced against logistical feasibility, return of investment, and assay development timelines. Early development experiments can establish if a given parameter requires evaluation in gDNA extracted from each matrix, or whether DNA from a representative matrix or pooled matrix can be leveraged to limit the number of validation experiments. These decisions should be guided by the assay's context of use and scientific rationale.

Another key aspect for assay development is the design of PCR amplification primer and probes. Since the mechanics of PCR amplification do not change between qPCR and ddPCR, we will not delve into specific recommendations for *in silico* primer and probe design that have been covered elsewhere [10]. In addition, the steady increase in PCR applications has resulted in development and constant improvement of software tools that simplify PCR reagent design and allow one to predict the performance of those reagents under various reaction conditions (e.g. PrimerQuest by Integrated DNA Technology, Geneious by Dotmatics, and the public software Primer3).

Even though *in silico* tools may be used for designing PCR assays, a myriad of experimental and biological variables necessitates empirical screening of PCR amplicon, primer, and probe candidates, usually followed by optimizing PCR conditions. Performing a thermal-gradient experiment to optimize ddPCR annealing temperature can have a large impact on amplification efficiency and resolution of droplet clusters (Figure 7.1). Empirical verification of *in silico* predicted performance becomes especially important in multiplexed PCR assays where the target amplicon may be amplified together with invariant genes for normalization of results. Design considerations may also extend beyond amplification of a specific sequence. For example, if using amplicon linkage capability offered by RT-ddPCR, the method employed for RT priming becomes critical – one needs to ensure that RT is primed so that both targeted amplicons are reverse transcribed into one contiguous first strand of cDNA. This would be possible by using Oligo(dT) to prime from the poly(A) tail of a transcript, or by using a sequence-specific primer that is downstream of the most 3′ amplicon, but it would not be possible by using random hexamers for RT priming.

Once a PCR-based method format has been established, its performance should be further characterized and/or validated. The following sections will highlight

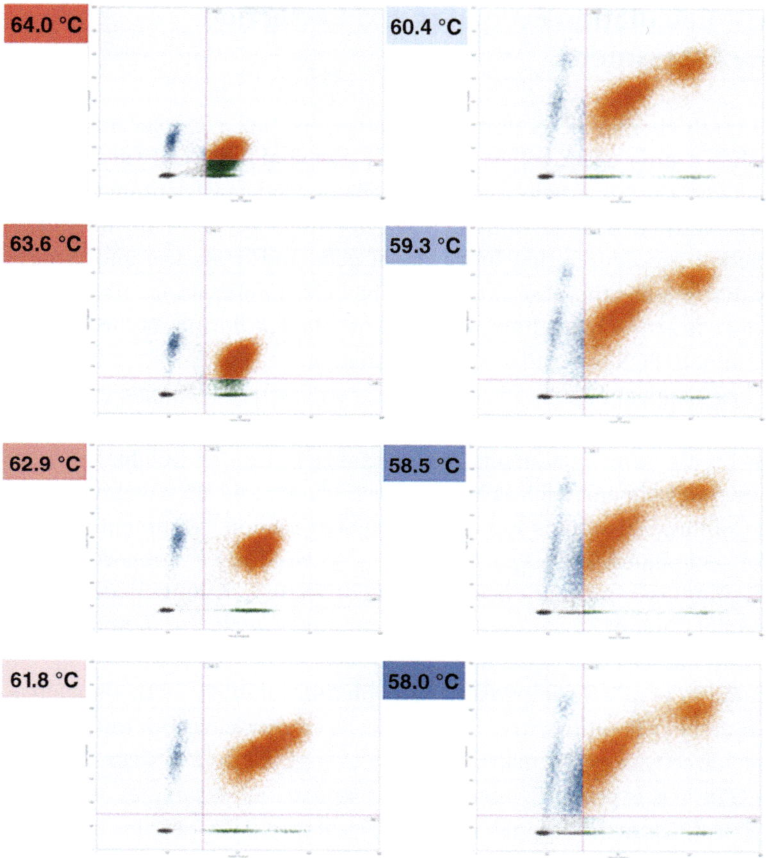

Figure 7.1 Optimizing the annealing temperature for thermal cycling in a duplexed ddPCR assay. Changes in annealing temperature improve spatial resolution of droplet clusters and reduce ddPCR "rain" or the appearance of multiple double-positive droplet populations due to different reaction efficiencies. This is visible as better fluorescent signal separation between double-positive (orange), single-positive (blue and green), and double-negative (gray) droplet populations. Two double-positive droplet populations arise at lower annealing temperatures, likely due to varying reaction efficiencies arising from partial inhibition. Increasing the annealing temperature to 62.9 °C from the default of 60.0 °C resulted in one defined double-positive population, which permitted more accurate thresholding and quantification of target.

key bioanalytical parameters to assess, such as extraction efficiency, sensitivity, specificity, standard curve performance, range of quantification, linearity, precision, accuracy, selectivity/matrix interference, and stability of extracted DNA samples and original biological specimens. Strategic and tactical considerations related to these assessments or how they are affected by critical method development choices will also be described.

7.4 Back-Calculation Formulas and Extraction Efficiency Assessments

PCR-based methods may be performed directly with biological specimens, for example, when testing saliva, cell lysate, or cerebrospinal fluid (CSF) at appropriate dilutions [22, 23]. Nonetheless, to ensure reliable and consistent method performance in pharmacokinetic studies, it is oftentimes advisable to implement DNA extraction procedures that break down cells and/or tissues and remove impurities that might otherwise interfere with PCR amplification. Hence, efficiency of the extraction procedure needs to be evaluated during method validation, in addition to PCR performance characteristics.

Prior to assessing extraction efficiency, one should select a suitable extraction kit/method for each matrix. While many organ and tissue samples may be extracted using the same commercial kit, specialized kits are available for more difficult clinical specimens, such as human feces where rapid DNA degradation is a concern. Commercial manual kits or automated extraction instruments employ a variety of techniques, such as DNA binding to solid-phase supports (silica-coated spin columns or magnetic beads). Forms of vector DNA anticipated in biological specimen, type of biological fluid or tissue, and sample throughput should also factor into the selection of extraction methods.

One peculiarity of qPCR compared to other pharmacokinetic methods, such as LC/MS, is that the calibration curve is not typically co-extracted with test samples but prepared directly in gDNA matrix or buffer. Consequently, PCR-based quantification of GTx VGs is primarily obtained for extracted DNA samples, while the concentration of VGs in the original (neat) biological specimens remains unknown and may need to be extrapolated. Similarly, absolute quantification by ddPCR is obtained only for extracted DNA samples. Hence, primary PCR data may require back-calculation to determine quantities of VGs per milliliter or milligram of biological specimens. Establishing back-calculation formulas is also essential when assessing extraction efficiency during method validation.

To derive back-calculation formulas, volume changes and sample losses during the extraction process are considered while working backwards from primary quantities measured in an extracted PCR test sample, as shown in Figure 7.2. In the example shown, the copy number of VGs detected in 5 μL of an extracted PCR test sample derived from a semen specimen is first multiplied by 24 to obtain the number of VGs in total elution volume of 120 μL. An identical number of VGs would then be expected in 200 μL pre-extraction solution, assuming all vector DNA was adsorbed by and eluted from the DNA purification column (100% extraction efficiency), resulting in a multiplication factor of 1. Due to sample dead volumes on the automated extraction instrument, only 200 μL of the initially prepared 300 μL pre-extraction solution (consisting of 75 μL semen samples plus

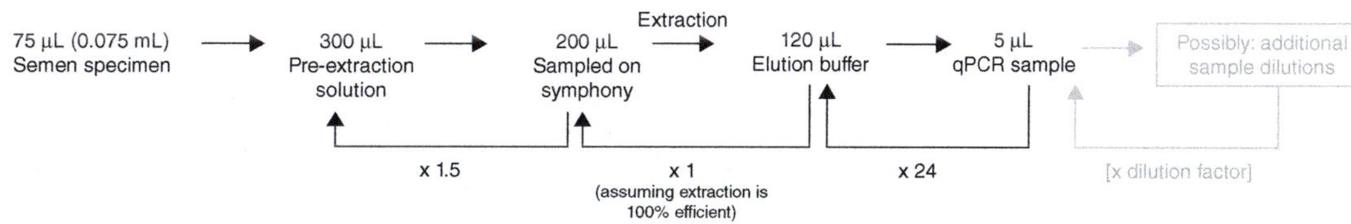

copies/mL semen = [copies (in 5µL qPCR sample) x 24 x 1.5]/0.075 mL semen

e.g. [75 copies(in 5 µL qPCR sample) x 24 x 1.5]/0.075 mL = 36000 copies/mL semen.

Figure 7.2 Exemplary formula to back-calculate copy numbers of vector genomes per mL semen.

225 μL lysis buffer) are taken up during the process. Thus, a correction factor of 1.5 is introduced to determine the copy number of VGs in 75 μL (or 0.075 mL) of the original semen specimen. Back-calculated VG quantities are then normalized to 0.075 mL to yield the concentration of VGs per mL semen. A back-calculation formula to report copies of VGs per milligram of a tissue specimen is derived analogously in Figure 7.3.

Extraction efficiency is defined as the percentage yield of extracted vector DNA normalized to input vector DNA in the original biological specimen. To determine extraction efficiency, a treatment-naïve biological specimen is spiked with vector DNA, or intact vector capsids, at one or more concentrations, followed by extraction and PCR quantification. In practice, vector DNA is oftentimes represented by a surrogate reference material, e.g. double-stranded bacterial plasmid DNA or a restriction enzyme-derived plasmid DNA fragment that contains the GTx vector sequence.

Whether vector DNA or intact vector capsids are used as spike material may depend on study context. For nonclinical GTx biodistribution studies, the use of vector DNA or surrogate vector DNA may better represent biological specimens, since transduced cells in tissues collected several weeks after GTx administration are expected to contain episomal double-stranded vector DNA rather than single-stranded vector DNA packaged in intact AAV capsids that were present in dosing solutions. For clinical shedding studies, vector DNA may still be packaged in capsids, and hence it is important to establish that encapsidated vector DNA can be efficiently extracted from biological specimens. Encapsidated vector DNA may be potentially transduction-competent and its efficient detection is crucial to the key objectives of shedding studies, which are to assess the potential risk of horizontal transmission and the potential risk of release into the environment.

Using the back-calculation formula, a nominal spike quantity is chosen based on its anticipated target concentration in the extracted sample, assuming 100% recovery. An example of an extraction efficiency assessment is shown in Table 7.2. The target concentration in the extracted PCR test samples was chosen to be 500 copies/reaction, which is 1 log above the LLOQ of 50 copies/reaction. Consequently, each biological specimen was spiked with a nominal quantity of 1.35E+04 copies of VGs, based on the back-calculation formula (nominally spiked copies = copies (in 5 μL PCR sample) × 24 × 1.125). It is important to determine extraction efficiencies for specimen containing relatively low vector copy numbers, as extraction efficiency typically decreases with decreased vector DNA input. As a negative control, biological specimens were also left unspiked.

All samples were extracted and tested in qPCR, and compared to the theoretically expected vector quantities representing complete spike recovery. In the example shown in Table 7.2, the spike working solution was also re-quantified

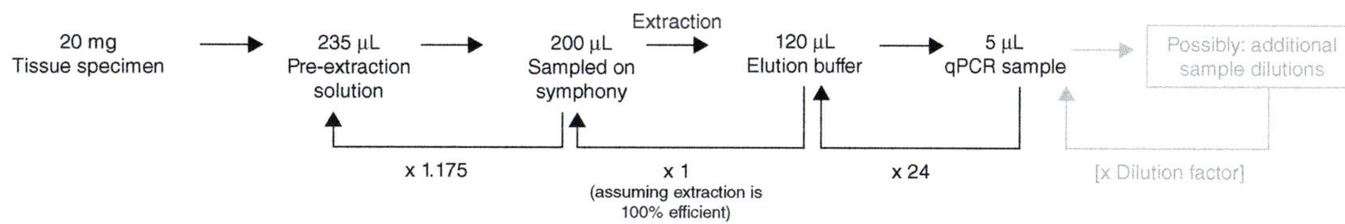

copies/mg tissue = [copies (in 5µL qPCR sample) x 24 x 1.175]/20 mg tissue

e.g. [75 copies(in 5 µL qPCR sample) x 24 x 1.175]/20 mg = 106 copies/mg tissue.

Figure 7.3 Formula to back-calculate copy numbers of vector genomes per mg tissue.

Table 7.2 Example of extraction efficiency assessment during PCR method validation.

				Mean Qty	StDev	% CV	StDev	% CV	% Recovery
Sample name	Ct	Mean Ct	Qty Copies/5 µL	Copies/5 µL	of replicate Qty	of replicate Qty	of replicate Ct	of replicate Ct	Relative to re-quantified spike material
EPC	32.01	32.06	262.3	254.3	11.3	4.45	0.1	0.20	59.59
	32.11		246.3						
Blood 25 µL unspiked	U	NA	0	0	NA	NA	NA	NA	0.00
	U		0						
Blood 50 µL unspiked	U	NA	0	0	NA	NA	NA	NA	0.00
	U		0						
Blood 25 µL spiked	31.62	31.78	344.2	309.8	48.7	15.72	0.2	0.73	72.60
	31.94		275.3						
Blood 50 µL spiked	31.23	31.56	446.3	366.5	112.9	30.81	0.5	1.45	85.88
	31.88		286.6						

	Copies per 5 µL spike solution	Expected copies per qPCR reaction
Theoretical spike	1.35 E+04	500
Re-quantified spike material (Target 1.35 E+04)	1.15 E+04	426.7

Ct, cycle threshold; Qty, quantity; StDev, standard deviation; CV; coefficient of variation; EPC, extraction positive control (spiked buffer); U, undetermined; NA, not applicable.

alongside spiked extraction efficiency validation samples to account for minor variations that may occur during the preparation of the spike working solution. Therefore, the detected copies of vector DNA in the extracted spiked specimen were normalized to 426.72 copies instead of the nominal 500 copies/reaction, based on re-quantified concentration of the spike working solution at approximately 1.15E+04 copies/5 μL.

Extraction efficiency may be evaluated on three different occasions (days) and ideally by a different analyst on each occasion. The range and average of observed extraction efficiencies may be reported to describe expected variability of the procedure. Each specimen type and each volume or mass used for extraction needs its own assessment, since these parameters may impact vector DNA recovery. Consequently, if specimen quantities are limiting during regulated sample analysis, this may require additional validation of the extraction method.

Extraction efficiency assessments for PCR-based methods are informational only since there is no regulatory method validation guidance from health authorities yet. The goal is to achieve relatively *consistent* recovery for each specimen, since variable recovery within the same matrix would confound longitudinal monitoring and comparisons between individuals or dose levels. Variable recovery would also confound assessments of biological specimen stability. As a general rule, if extraction efficiency drops below 10%, then even minor variations in recovery of vector DNA may lead to high variability in the PCR data: For example, if extraction efficiency varied between 3% and 9% of vector quantities present in a biological specimen, then quantities recovered in extracted PCR test samples would vary by a factor of up to 3 times (= 9% divided by 3%) for the same input amount. This would significantly increase data variability, without even considering additional variability introduced by PCR. In contrast, if extraction efficiency varied between 13% and 19%, or between 23% and 29%, then vector quantities recovered in PCR test samples would vary by less than 1.5 times, thus facilitating a more precise assessment. In addition, extraction efficiencies of less than 10% would necessitate a correction factor when back-calculating sample results, since the assumption of complete vector DNA recovery would under-estimate true vector quantities in biological specimens by at least 10 times.

7.5 Sensitivity Requirements

Regulatory guidance on sensitivity of PCR-based methods exists only for nonclinical GTx biodistribution studies: “The assay should have a demonstrated limit of quantification of ≤50 copies/μg gDNA, so that your assay can detect this limit with

95% confidence" [4]. While the guidance refers to the lower limit of quantification (LLOQ) in the first part of the sentence, the criterion referenced in the second part specifies a reliable limit of detection (LOD). This duality is notable because limits of quantification are usually based on acceptable precision and accuracy, and not detectability. Hence, the guidance is usually interpreted as providing that PCR-based vector DNA detection methods must have a demonstrated LOD at which 95% of samples remain reliably detectable and that is equal to or lower than the LLOQ, whereby the LLOQ can be as high as 50 copies per microgram gDNA. Precision or accuracy need not be maintained at the LOD if it is lower than 50 copies/microgram.

In the absence of guidance for the sensitivity of clinical shedding assays, one may apply a similar criterion for method validation purposes, considering that PCR procedures in nonclinical and clinical assays are oftentimes identical even if DNA extraction procedures differ. One characteristic of clinical shedding specimens can be a lower gDNA content, for example, in plasma, urine, saliva, and feces. Hence, it may suffice to assess LOD in buffer rather than matrix containing 1 microgram of gDNA. LOD would then be reported as vector copies per reaction, or vector copies per tested volume of extracted sample, instead of vector copies per microgram gDNA.

To demonstrate reliable detection with 95% confidence for qPCR, a twofold dilution series of vector DNA or intact vector capsids is prepared in buffer or matrix containing 1 μg gDNA, oftentimes starting at LLOQ of 50 copies per reaction as the highest concentration. This dilution series is tested across 10 replicates for each concentration level. The results are analyzed by logistic regression, whereby each replicate is given a binary value of either "not detected" or "detected," depending on whether there was no increase in fluorescence after 40 PCR cycles or whether there was an increase resulting in a measurable Ct or Cp value.

The percentage of detected replicates at each concentration forms the empirical basis for logistic regression that interpolates the concentration of vector DNA at which 95% of the replicates would theoretically be detected. It may be helpful to perform at least two sensitivity assessments over a period of 2 days and either pool the data or average the LOD values derived by separate logistic regression. A reliable LOD value with 95% detectability derived by logistic regression is typically reported together with a 90% confidence interval, whereby the upper confidence limit should ideally remain at or below 50 copies per reaction, i.e. below the desired LLOQ.

The experiments to establish a reliable LOD with 95% detectability for ddPCR can be similar to those for qPCR but will differ in how replicate results are designated as "not detected" or "detected." In qPCR, a cycle count of 40 is often applied as the threshold by which a target template is considered detected or not, based on collective experience with typical qPCR performance. Since ddPCR does not rely on cycle count for quantifying vector copies, a threshold for "not detected" vs. "detected" may need to be established empirically prior to applying logistic regression. While there is no consensus on how to establish this threshold for ddPCR assays, we oftentimes first characterize several replicates of a negative template control (NTC) that contains the same amount of gDNA per reaction as would be loaded with test samples. We then determine the maximum number of false-positive droplets that may be observed occasionally in the absence of any specific DNA template, using an appropriate statistical distribution limit (e.g. 95th percentile). This approach focuses on the raw signal generated by the instrument (i.e. positive droplets) to characterize background noise, rather than using low fractional copy numbers generated after application of Poisson statistics to positive droplet counts (typically less than 5 per reaction). Once the false-positive droplet limit has been determined for NTC, it is applied as the threshold for designating sample replicate results as "not detected" or "detected" in LOD experiments to enable logistic regression.

Without clear regulatory guidance or bioanalytical consensus on LOD determinations, other approaches may be taken as long as they are based on scientific rationale that supports context of use.

7.6 Specificity Requirements

Assessment of specificity determines the potential for detecting structurally similar analytes of exogenous or endogenous origin that might lead to false-positive results. In addition, it includes a verification that positive samples are accurately quantified despite the presence of similar analytes. Hence, PCR primers and probes need to be specific for the target sequence and not generate off-target amplicons. Specificity can be evaluated *in silico* using NCBI BLAST to query oligo sequences against a species-specific nucleotide database. Primer and probes with the lowest percentage of matched sequence coverage and identity are then further evaluated empirically for nonspecific amplicons and target amplification efficiency in gDNA matrix.

This evaluation is performed in two ways, with and without spikes of a positive control template into up to 1 µg of gDNA reaction input. If 1 µg reaction input is not possible for low DNA-content shedding matrices, then input may be guided by conservative expectations for clinical samples. Firstly, absence of amplification in unspiked gDNA demonstrates that primers and probe do not generate nonspecific amplicons. Secondly, spikes of positive control DNA are tested at various concentrations in the presence or absence of gDNA, ideally extracted from a biological specimen that has a high DNA content. A decrease in PCR reaction efficiency for a spiked sample containing gDNA relative to an identically spiked sample without gDNA suggests that primers or probe anneal to a nonspecific gDNA sequence.

The sequestration of oligo reagents consequently decreases PCR amplification efficiency of specific target template.

Assessing specificity early during PCR method development facilitates rapid evaluation of other parameters. Confirming specificity by routinely testing gDNA-containing NTC samples can support previously established sensitivity thresholds and monitor for potential contamination.

7.7 Standard Curve Performance, Colinearity, Precision, and Accuracy

Given the fundamental differences in quantification between qPCR and ddPCR, method development and validation diverge most strikingly in the parameters relevant for quantifying the target sequence. The most obvious difference is that qPCR requires a standard/calibration curve against which unknown samples are interpolated, whereas ddPCR does not require a standard curve but quantifies absolutely.

Consequently, qPCR method development includes optimization of standard curve performance, followed by validation. Standard curves for qPCR typically meet acceptance criteria for reaction efficiency between 90% and 110% (corresponding to a slope of −3.60 and −3.10)[1] and a coefficient $R^2 \geq 0.980$. Furthermore, back-calculated standard concentrations typically have Ct CV ≤ 3.00% (precision) and RE between −50% and +100% (accuracy). More narrow precision and wider accuracy criteria compared to ligand binding assays can be justified by the doubling nature of PCR reactions. A change in Ct value of 1 indicates a doubling in the amount of target DNA, therefore the standard deviation must be small compared to the mean Ct for the amount of amplification product to not differ significantly between replicate reactions. Therefore, if CVs are evaluated for Ct values rather than back-calculated VG concentrations, they are oftentimes set as ≤3.00%.

In contrast, back-calculated VG concentrations should be used to evaluate accuracy of quantification relative to nominal input. Due to the exponential nature of PCR, sample quantifications can easily vary within 1 Ct cycle, translating into target DNA levels within one-half of (−50%) or double (+100%) the nominal concentration. Therefore, a target RE range of −50% to +100% for VG quantities may be adequate. A similar acceptance criterion is used to specify accuracy required for antibody titer measurements, which is ±1 $\log_2$(titer) if a 1 : 2 dilution scheme was implemented. This criterion corresponds to the last detectable sample dilution shifting one dilution step up or down. Using the ±1 Ct cycle criterion for PCR

1 PCR efficiency is calculated as: $E = 10^{(-1/S)} - 1$, where E: theoretical PCR efficiency, S: slope of standard curve plotted with Ct values on y-axis and log(quantity) on x-axis.

methods ensures that raw Ct values greater than 10 (i.e. those within the core measurement range) do not differ more than 10% from expectations. Moreover, the decline of vector concentrations in clinical shedding studies is typically graphed using a $\log_{10}$ scale; hence, the use of a two-fold RE range for measured quantities appears to be reasonably accurate.

While ddPCR does not require a standard curve, linearity of assay response in addition to precision and accuracy still needs to be verified using spiked quality controls. Acceptance criteria from bioanalytical validation of ligand binding assays can be adopted in the absence of PCR-specific regulatory guidance. Thus, quality controls at the limits of quantification may be allowed to have up to 25% CV and up to ±25% RE, while quality controls within these limits are allowed up to 20% CV and up to ±20% RE, if following BMV guidance. Acceptance criteria determined through method development and validation should be fit-for-purpose and appropriate for the assay's context of use. Linearity is evaluated in the same fashion as for qPCR, but performed with a dilution series of spiked controls, and typically only the $R^2 \geq 0.980$ acceptance criteria is applied. Due to the greater independence of ddPCR quantification from PCR reaction efficiency, assessing efficiency is not a critical component for validation. While ddPCR reaction efficiency is maximized during method development, as long as positive droplets can be adequately distinguished from negative droplets with low "rain" (droplets with varying reaction efficiencies due to inhibition), quantification is not significantly impacted. As such, assessment of precision and accuracy is the most critical parameter in ddPCR method validations.

Colinearity is another validation parameter for PCR assays on either platform. Colinearity assesses if a surrogate positive control material (e.g. synthetic DNA or linearized plasmid fragment) amplifies comparably to the GTx vector. In the simplest experiment, positive control template and GTx vector are identically diluted serially and the slopes and coefficient $R^2 \geq 0.980$ are compared. If both dilution series meet acceptance criteria and the curves are reasonably close and parallel, the surrogate positive control material is suitable as a calibrator or quality control. Using more easily procurable surrogate positive control material could help conserve GTx products when early manufacturing lots may be limited in quantity.

7.8 Selectivity Assessment and Matrix Interference

Selectivity of ligand-binding assays is assessed by evaluating the degree and nature of interference from components in sample matrix and its effect on analyte quantification. Taking plasma as an example, one would assess the impact of normal, hemolyzed, and lipemic plasma, in addition to possible disease-state or study-driven alterations in plasma composition. Since the analyte is measured directly in matrixed sample, or in some minimally required dilution (MRD) thereof, all possible types of interfering matrix components could be present in study samples. By contrast, when testing DNA extracted from biological specimens, most potentially interfering matrix components have been stripped away from the analyte. Nonetheless, individual post-extraction samples may contain a unique mix of co-extracted matrix contaminants that could still interfere with downstream PCR. There may also be carry-over of residual extraction reagents. Therefore, selectivity is still an important component of method characterization for bioanalytical PCR assays.

Selectivity can be evaluated during method development to assess all matrices but may be limited to target tissue or a key biofluid during validation. Off-target tissues and other biofluids may be included in validation if they are critical to interpret study data. Since composition of biological specimens is variable, DNA extractions may also vary in their final composition of potentially interfering factors. Selectivity can be assessed, for example, using 10 individual donors from which gDNA is extracted and tested either unspiked or spiked with vector DNA near the LLOQ. At least 80% of the unspiked samples should yield a result below the LOD, while at least 80% of the spiked samples should yield a positive result within established precision and accuracy criteria.

7.9 Sample Stability Considerations

Depending on method procedures, stability may be evaluated for both vector DNA in purified PCR test samples following extraction (if performed) and vector DNA in non-extracted biological specimens (tissues, organs, biofluids). The duration and design of stability studies depend on study logistics, sample collection and testing schedule, number of anticipated freeze/thaws, and storage conditions. Collection of specimens in clinical trials may precede sample testing by only a few weeks or months. In contrast, specimens in nonclinical GTx biodistribution studies may be held for years until analysis is triggered.

Biological specimens for PCR-based testing of vector DNA are typically stored frozen at −80 °C, but study-specific circumstances may necessitate storage at higher temperatures for varying amounts of time. For example, at-home collection of clinical shedding specimen may require temporary storage at −20 or 4 °C. Extracted DNA samples may also be stored differently than biological specimens during short-term storage at test sites. Retests of extracted DNA samples or retests requiring re-extraction of DNA from biological specimens could also trigger assessment of freeze–thaw stability for all sample types involved.

To assess stability prior to a study, biological specimens from treatment-naïve individuals could be spiked with surrogate vector DNA or intact GTx capsids and stored for repeated PCR testing over a defined time period. Spiked vector quantities could be selected analogously to vector quantities used in extraction efficiency assessments, targeting at least one level within 1 log of the LLOQ, for example,

500 copies/reaction. This spike-in approach is particularly useful for liquid specimens, such as urine, saliva, blood, plasma, and semen, which are monitored in clinical vector shedding studies. While feasible, this approach faces obvious difficulties with solid samples. These difficulties may be overcome by preparing multiple specimen portions to accommodate single extractions at each stability time point, and by adapting testing procedures to lyse stability specimens directly in their original containers.

To assess stability concurrently to or after a study, biological specimens from GTx-treated individuals can be extracted and tested using a primary PCR test that serves as baseline assessment, followed by multiple secondary extractions and PCR retests of the same specimen (or aliquots/portions thereof) at defined stability intervals. This would not require spiking any vector material and resembles incurred sample reanalysis (ISR), which is sometimes performed for validation of other pharmacokinetic methods. This retest approach is useful for tissue specimen in biodistribution studies, where spiked surrogate vector DNA or intact GTx vector capsids may not fully recapitulate the biological conditions that affect episomal vector DNA within the nuclei of transduced cells. In particular, linearized surrogate vector DNA spiked onto tissue specimens may represent more of a worst-case scenario for analyte stability, since circularly closed vector episomes within cells would be expected to be more stable than open-ended surrogates that could be exposed to extracellular DNases in tissue samples. The obvious disadvantage of using secondary PCR tests of incurred specimens for stability studies is that the results cannot inform study design, and thus sample storage conditions and duration are decided at risk. Another consideration is limited sample quantity in small animal studies and the need to plan for collecting multiple specimen aliquots/portions to facilitate evaluation after various storage intervals without incurring freeze–thaws.

As for extracted DNA stability, most PCR workflows aim to test shortly after DNA samples have been extracted from biological specimens. In addition, the stability of purified DNA frozen at low temperature is relatively well established, and thus long-term stability studies may not be needed. Considering the possibility of PCR retests following a failed initial run, freeze–thaw assessments combined with short-term stability periods (e.g. 1 month) appear to be more useful. This evaluation can be performed pre-study using spiked surrogate vector DNA, or post-study by repeated PCR testing of extracted DNA samples after the primary test was completed on study.

In-assay short-term stability assessments differ between qPCR and ddPCR. In qPCR, one would typically proceed from sample preparation to amplification and generate results in real time. Using Bio-Rad's QX200 ddPCR system, it may be possible to "pause" the run for a period of time following thermal-cycling of the generated droplets (e.g. overnight). This allows for greater analyst flexibility and increased throughput, since test plates can be prepared and thermal-cycled on Day 1 and read out on Day 2, with runs staggered by one day. Furthermore, holding droplets at 12 °C for a minimum of 4 hours can increase assay sensitivity, since more droplets will meet the instrument's droplet quality metrics [24]. A higher number of acceptable droplets translates into a higher probability that low abundance targets will be detected.

Stability acceptance criteria for PCR-based methods vary. Measured GTx vector quantities in stability samples can be compared back to baseline samples to reveal any trends of degrading sample integrity. Alternatively, a comparison to nominal spike levels can be made. The latter approach lends itself more easily to extracted DNA samples spiked with known amounts of vector, while the former approach is more applicable for biological specimens, where extraction variability and unknown vector quantities preclude the use of nominal concentrations.

When comparing results from stability time points to those from baseline samples, a threshold needs to be established for the biological specimen to be considered stable. In other words, what relative difference from baseline results indicates sample instability? No consensus exists but reasonable acceptance ranges may fall between 0.5 and 0.2 times baseline value. To justify a threshold, the variability of both DNA extraction efficiency and PCR measurement needs to be considered. For example, if extraction efficiency varies by a factor of 2, then halving of PCR-measured vector quantities would not be conclusive for specimen instability. Assuming a potential negative bias of −50% for both extraction efficiency and PCR measurements, one would obtain a theoretical lower limit for overall analytical variability of $0.5 \times 0.5 = 0.25$. Thus, after adding a small margin of tolerance, a value of 0.2 times baseline could be a reasonable threshold, below which biological specimens are considered instable.

7.10 Data Reporting Formats, Acceptance Criteria, and Trending

Depending on study context, PCR data may be reported in various formats: In nonclinical biodistribution studies, VG copies per microgram of gDNA are commonly used, and VG copies per mg tissue may also be reported. The former unit can be converted to VG copies per cell by multiplying with the approximate mass of a diploid genome (6 pg in humans/mice, 5 pg in NHP). Potentially higher degrees of ploidy, for example in liver and heart cells [25–27], may need to be considered as appropriate. In shedding studies, reporting as VG copies per mL biological fluid or mg solid specimen is oftentimes more clinically relevant, since volumes and masses are more easily visualized. For clinical specimens with sufficient DNA content, such as whole blood and seminal fluid, reporting as VG copies per microgram gDNA could additionally be provided.

Exemplary acceptance criteria for regulated PCR sample testing are provided in Tables 7.3 and 7.4. The standard curve and low-level quality control (QC) within 1 log of the LLOQ are routinely included, together with various negative controls that monitor for potential cross-contamination. Medium and high QCs can be added to increase confidence in accurate and precise sample quantification.

The range of acceptable accuracy for qPCR-based methods is typically broader (−50% to +100% relative error, RE) than that of other pharmacokinetic methods, as discussed above. When performing qPCR, it can be useful to verify that no sample-mediated inhibition occurred, in particular for samples without detectable increase in fluorescence that would be reported as negative. Sample-mediated inhibition can be assessed by spiking a low quantity of vector DNA (50–500 copies/reaction) into a designated test sample replicate and comparing with a spike volume control (SVC) sample. The SVC sample contains the same number of vector DNA

Table 7.3 Exemplary standard curve, QC, and run acceptance criteria for PCR.

Parameter	Acceptance criteria qPCR sample analysis	Acceptance criteria dPCR sample analysis
Standard Curve Regression	The standard curve should have an $R^2 \geq 0.98$ and a slope between −3.1 and −3.6 (corresponding to 90–110% amplification efficiency).	Not applicable, absolute quantification mode alleviates the need for a standard curve
Precision	The CV of replicate Cq or Ct values for each calibration point and quality control (QC) sample should be ≤3%. OR: The CV of interpolated VG copies for each standard curve point and QC samples should be ≤30%, except for LLOQ-QC where CV should be ≤50%. Outlier exclusion: 1 out of 3 replicates may be omitted, if the Cq or Ct differences between the omitted replicate and each of the other two replicates are greater than 1.	The CV of measured VG copies for QC samples should be ≤30%, except for LLOQ-QC where CV should be ≤50%. Outlier exclusion: 1 out of 3 replicates may be omitted, using an appropriate statistical method (e.g. Median Absolute Deviation, MAD).
Accuracy	The RE for each standard calibration point and QC sample should be within −50% and +100% of the nominal value.	The RE for QC samples should be within ±30% of the nominal value, except for LLOQ-QC where RE should be within ±50% of the nominal value.
Negative controls	Negative extraction controls, no-template controls, and sentinel controls should not show any increase in fluorescence OR: remain below the LOD.	Negative extraction controls, no-template controls, and sentinel controls should not have VG copies greater than LOD.

copies/reactions in the same spike volume added to a non-interfering matrix or plain buffer. If the Cq or Ct value of the spiked test sample replicate is significantly higher than that of the SVC, sample-mediated inhibition has occurred and adequate remediation procedures may be needed, as detailed in Table 7.4.

Table 7.4 Exemplary test sample acceptance criteria for PCR.

Parameter	Acceptance criteria qPCR sample analysis	Acceptance criteria dPCR sample analysis
Sample-mediated inhibition	Samples should be non-inhibitory. Samples are considered inhibitory if the spiked replicate exhibits a Cq or Ct value that is greater than 1.73 + the mean Cq or Ct value of the spike volume control (SVC). Inhibitory samples shall be retested. If a sample remains inhibitory upon retest, it should be reextracted from biological specimens. If re-extraction is not feasible, the sample should be reported as not determinable (ND).	Samples should be non-inhibitory. Samples are considered inhibitory if the spiked replicate exhibits measured VG copies that are more than 30% lower than those in the SVC. Inhibitory samples shall be retested. If a sample remains inhibitory upon retest, it should be reextracted from biological specimens. If re-extraction is not feasible, the sample should be reported as ND.
Precision	Samples should be tested in triplicate or duplicate. The CV of replicate Cq or Ct values for a sample above the LLOQ should be $\leq$3%. OR: the CV of interpolated VG copies for a sample above the LLOQ should be $\leq$30%. Outlier exclusion: 1 out of 3 replicates may be omitted, if the Cq or Ct differences between the omitted replicate and each of the other two replicates are greater than 1.	Samples should be tested in triplicate or duplicate. Samples with mean VG copies above 2 times the LLOQ must have CV $\leq$30%. Samples with mean VG copies between 2 times the LLOQ and the LLOQ must have CV $\leq$50%. Outlier exclusion: 1 out of 3 replicates may be omitted, using appropriate statistical methods (e.g., Median Absolute Deviation, MAD).
Sample result determination	Samples are considered positive if the level of fluorescence after 40 cycles rises above the run-specific threshold and is indicative of true amplification. Otherwise, samples are considered negative. OR: Samples are considered positive if interpolated mean VG copies fall at or above the LOD. Otherwise, samples are considered negative.	Samples are considered positive if mean VG copies fall at or above the LOD. Otherwise, samples are considered negative.

Parameter	Acceptance criteria qPCR sample analysis	Acceptance criteria dPCR sample analysis
Reporting ranges	Positive samples should be quantified and reported with a numerical result if they fall within the limits of quantification (LLOQ, ULOQ) established in assay validation. Samples above the ULOQ should be diluted adequately and retested. Positive samples below the LLOQ should be reported as falling below the lower limit of quantification (BLQ). Negative samples should be reported as 0 VG copies or below the limit of detection (BLD).	Positive samples should be quantified and reported with a numerical result if they fall within the limits of quantification (LLOQ, ULOQ) established in assay validation. Samples above the ULOQ should be diluted adequately and retested. Positive samples below the LLOQ should be reported as falling BLQ. Negative samples should be reported as 0 VG copies or BLD.

Routine clinical vector shedding assays should be monitored for consistent performance. For qPCR methods, the Cq or Ct values of the standard curve could be trended and should typically fall within two standard deviations from the mean. VG copies for QCs can also be trended, which may be the most appropriate way to monitor ddPCR performance over time. The proportion of droplets that pass internal instrument quality checks may also be monitored in ddPCR since a significant and prolonged change in accepted droplets used for quantification could impact assay performance.

7.11 Immunocapture qPCR: An Ultra-Sensitive Method to Detect Intact AAV Capsids

The main objectives of clinical vector shedding studies are to evaluate the potential risk of horizontal transmission and release into the environment. AAV-based GTx are generally designed to have minimal risk, given that they are replication-incompetent, non-pathogenic, encode nontoxic gene products, have weak or moderate promoters/enhancers, do not carry antibiotic resistance genes, and rarely integrate into the host genome. Nonetheless, there is a desire to limit any unintended exposure or release. Monitoring vector shedding can inform the duration of potential pre-cautionary measures and is therefore mandated by regulatory agencies during clinical studies.

If standard PCR-based methods are used to monitor shedding, encapsidated and non-encapsidated vector DNA cannot be distinguished. This distinction, however, could be informative, since the potential risk associated with each of these two forms of vector DNA is not equal. On the one hand, encapsidated vector DNA is contained in structurally intact capsid particles that may have the capability to enter cells, i.e. this form of vector DNA is potentially transduction competent. On the other hand, non-encapsidated vector DNA could stem from low amounts of degraded capsid particles or emerge during uncoating within transduced cells and be released following cell death. No meaningful transduction competence would therefore be expected for non-encapsidated vector DNA, which decreases any potential risk.

To measure intact GTx vector capsids, cell-based transduction assays or ELISA-based immunoassays may be used [3, 28]. In clinical practice, however, the limited sensitivity of these methods poses a significant challenge to detecting low residual levels of GTx capsids in shedding matrices or blood [29, 30]. Recently, a more sensitive ELISA-based method named VIRELISA was described, which employs a recombinant high-affinity AAV receptor (AAVR) to capture and detect AAV2 capsids [31]. With an estimated sensitivity of detecting as low as 5E+05 vg/mL of AAV2 capsids in buffered aqueous solution, this method may warrant further investigation, even though the sensitivity for matrixed clinical samples and other AAV serotypes remains unclear.

To overcome the hurdles imposed by limited sensitivity of existing methods, an ultra-sensitive assay for detecting structurally intact, and thus potentially transduction-competent capsids has recently been developed for an AAV5-based GTx [3]. This two-step methodology, named immunocapture-qPCR (iqPCR), features a monoclonal antibody (ADK5a/b) that specifically recognizes a conformational epitope on AAV5 capsids and facilitates magnetic bead-mediated pull-down, followed by capsid denaturation and PCR amplification of released vector DNA. The major advantage of iqPCR is the quantum leap in analytical sensitivity for detecting AAV capsids across various clinical matrices. IqPCR exceeds the sensitivity of both cell-based methods and advanced immunoassay platforms (including single-molecule counting, or SMC-Erenna), and approximates detection limits of standard PCR-based methods. Compared to VIRELISA, iqPCR is at least 40 times more sensitive.

Another advantage of iqPCR over ELISA-based methods is its dually assured specificity for intact GTx capsids, (1) by using a conformation-specific antibody for a particular AAV serotype and (2) by ensuring that captured capsids contain specific vector DNA. ELISA-based methods, such as VIRELISA, detect AAV capsids regardless of whether they are full ("heavy") or empty ("light"), or regardless of whether they contain vector DNA or a wild-type AAV genomes. Hence, one may expect potential interference from natural AAV infections for ELISA-based capsid detection methods if used in clinical studies.

A notable limitation of iqPCR (as well as ELISA-based capsid detection methods) is interference from AAV antibodies. These antibodies may compete for epitopes with the monoclonal antibody reagent used to capture capsids, and consequently iqPCR has a low tolerance to AAV antibodies present in plasma or semen [3]. This interference is of lesser concern within the first few days after GTx administration, since patients have usually been pre-screened to have low or no pre-existing humoral immunity to AAV. While treatment-emergent AAV antibodies can impact capsid quantification in iqPCR, they are generally also highly neutralizing [32–34]. Hence, AAV capsid/antibody complexes that remain undetectable by iqPCR would also be less likely to mediate horizontal transmission or environmental impact, as compared to "free" transduction-competent AAV capsids.

IqPCR has been successfully implemented in clinical GTx studies and used to describe clearance kinetics of transduction-competent "free" GTx vector in human biological fluids [3]. This helped inform the duration of precautionary measures deemed necessary to limit unintended exposure or release. In summary, PCR-based methods to detect vector DNA in tissues and biological fluids are essential components of bioanalytical portfolios that support the development of novel AAV GTx.

References

1 Gil-Farina, I., Fronza, R., Kaeppel, C. et al. (2016). Recombinant AAV integration is not associated with hepatic genotoxicity in nonhuman primates and patients. *Mol. Ther.* 24 (6): 1100–1105.

2 Pasi, K.J., Rangarajan, S., Mitchell, N. et al. (2020). Multiyear follow-up of AAV5-hFVIII-SQ gene therapy for hemophilia A. *N. Engl. J. Med.* 382 (1): 29–40.

3 Sandza, K., Clark, A., Koziol, E. et al. (2022). Ultra-sensitive AAV capsid detection by immunocapture-based qPCR following factor VIII gene transfer. *Gene Ther.* 29 (1–2): 94–105.

4 United States Food and Drug Administration. (2020). Long term follow-up after administration of human gene therapy products draft guidance for industry. January [cited 22 March 2023], Available from: https://www.govinfo.gov/content/pkg/FR-2020-01-30/pdf/2020-01710.pdf.

5 United States Food and Drug Administration. (2015). Design and analysis of shedding studies for virus or bacteria-based gene therapy and oncolytic products: guidance for industry. August [cited 22 March 2023], Available from: https://www.fda.gov/files/vaccines%2C%20blood%20%26%20biologics/published/Design-and-Analysis-of-Shedding-Studies-for-Virus-or-Bacteria-Based-Gene-Therapy-and-Oncolytic-Products--Guidance-for-Industry.pdf.

6 United States Food and Drug Administration. (2018). Bioanalytical method validation guidance for industry. May [cited 22 March 2023], Available from: https://www.fda.gov/files/drugs/published/Bioanalytical-Method-Validation-Guidance-for-Industry.pdf.

7 Piccoli, S., Mehta, D., Vitaliti, A. et al. (2019). 2019 white paper on recent issues in bioanalysis: FDA immunogenicity guidance, gene therapy, critical reagents, biomarkers and flow cytometry validation (part 3-recommendations on 2019 FDA immunogenicity guidance, gene therapy bioanalytical challenges, strategies for critical reagent management, biomarker assay validation, flow cytometry validation & CLSI H62). *Bioanalysis* 11: 2207–2244.

8 Corsaro, B., Yang, T.Y., Murphy, R. et al. (2021). 2020 white paper on recent issues in bioanalysis: vaccine assay validation, qPCR assay validation, QC for CAR-T flow cytometry, NAb assay harmonization and ELISpot validation (part 3 – recommendations on immunogenicity assay strategies, NAb assays, biosimilars and FDA/EMA immunogenicity guidance/guideline, gene & cell therapy and vaccine assays). *Bioanalysis* 13: 415–463.

9 Gene Therapy Working Group. (2018). Expectations for biodistribution (BD) assessments for gene therapy (GT) products. Inernational pharmaceutical regulators programme. [Internet], [cited 2023 Mar 2022], Available from: https://admin.iprp.global/sites/default/files/2018-09/IPRP_GTWG_ReflectionPaper_BD_Final_2018_0713.pdf.

10 Ma, H., Bell, K.N., and Loker, R.N. (2021). qPCR and qRT-PCR analysis: regulatory points to consider when conducting biodistribution and vector shedding studies. *Mol. Ther. Methods Clin. Dev.* 20: 152–168.

11 Sykes, P.J., Neoh, S.H., Brisco, M.J. et al. (1992). Quantification of targets for PCR by use of limiting dilution. *Biotechniques* 13 (3): 444–449.

12 Vogelstein, B. and Kinzler, K.W. (1999). Digital PCR. *Proc. Natl. Acad. Sci.* 96 (16): 9236–9241. Available from: https://doi.org/10.1073/pnas.96.16.9236.

13 Furuta-Hanawa, B., Yamaguchi, T., and Uchida, E. (2019). Two-dimensional droplet digital PCR as a tool for titration and integrity evaluation of recombinant adeno-associated viral vectors. *Hum. Gene Ther. Methods* 30 (4): 127–136.

14 Gimpel, A.L., Katsikis, G., Sha, S. et al. (2021). Analytical methods for process and product characterization of recombinant adeno-associated virus-based gene therapies. *Mol. Ther. Methods Clin. Dev.* 20: 740–754.

15 Dobnik, D., Kogovšek, P., Jakomin, T. et al. (2019). Accurate quantification and characterization of adeno-associated viral vectors. *Front. Microbiol.* 10 (July): 1570.

16 Lock, M., Alvira, M.R., Chen, S.J., and Wilson, J.M. (2014). Absolute determination of single-stranded and self-complementary adeno-associated viral vector genome titers by droplet digital PCR. *Hum. Gene Ther. Methods* 25 (2): 115–125.

17 Sivaganesan, M., Varma, M., Siefring, S., and Haugland, R. (2018). Quantification of plasmid DNA standards for U.S. EPA fecal indicator bacteria qPCR methods by droplet digital PCR analysis. *J. Microbiol. Methods* 152: 135–142.

18 Gaillard, C. and Strauss, F. (1998). Avoiding adsorption of DNA to polypropylene tubes and denaturation of short DNA fragments. *Tech. Tips Online* 3 (1): 1–3.

19 Sanmiguel, J., Gao, G., and Vandenberghe, L.H. (2019). Quantitative and digital droplet-based AAV genome titration. *Methods Mol. Biol.* 1950: 51–83.

20 Wang, Y., Keith, M., Leyme, A. et al. (2019). Monitoring long-term DNA storage via absolute copy number quantification by ddPCR. *Anal. Biochem.* 583: 113363.

21 Bio-Rad Laboratories Inc. (2014). ddPCR Library Quantification Kit for Ion Torrent – 10033630. [Internet], [cited 2023 March 2022], Available from: https://www.bio-rad.com/webroot/web/pdf/lsr/literature/10033630.pdf.

22 Deiana, M., Mori, A., Piubelli, C. et al. (2020). Assessment of the direct quantification of SARS-CoV-2 by droplet digital PCR. *Sci. Rep.* 10 (1): 18764.

23 Vasudevan, H.N., Xu, P., Servellita, V. et al. (2021). Digital droplet PCR accurately quantifies SARS-CoV-2 viral load from crude lysate without nucleic acid purification. *Sci. Rep.* 11 (1): 780.

24 Rowlands, V., Rutkowski, A.J., Meuser, E. et al. (2019). Optimisation of robust singleplex and multiplex droplet digital PCR assays for high confidence mutation detection in circulating tumour DNA. *Sci. Rep.* 9 (1): 12620.

25 Donne, R., Saroul-Aïnama, M., Cordier, P. et al. (2020). Polyploidy in liver development, homeostasis and disease. *Nat. Rev. Gastroenterol. Hepatol.* 17: 391–405.

26 Donne, R., Sangouard, F., Celton-Morizur, S., and Desdouets, C. (2021). Hepatocyte polyploidy: driver or gatekeeper of chronic liver diseases. *Cancers* 13: 5151.

27 Derks, W. and Bergmann, O. (2020). Polyploidy in cardiomyocytes: roadblock to heart regeneration? *Circ. Res.* 126: 552–565.

28 Grimm, D., Kern, A., Pawlita, M. et al. (1999). Titration of AAV-2 particles via a novel capsid ELISA: packaging of genomes can limit production of recombinant AAV-2. *Gene Ther.* 6 (7): 1322–1330.

29 Nathwani, A.C., Tuddenham, E.G.D., Rangarajan, S. et al. (2011). Adenovirus-associated virus vector-mediated gene transfer in hemophilia B. *N. Engl. J. Med.* 365 (25): 2357–2365.

30 Rangarajan, S., Walsh, L., Lester, W. et al. (2017). AAV5-factor VIII gene transfer in severe hemophilia A. *N. Engl. J. Med.* 377 (26): 2519–2530.

31 Cui, M., Lu, Y., Tang, C. et al. (2019). A generic method for fast and sensitive detection of adeno-associated viruses using modified AAV receptor recombinant proteins. *Molecules* 24 (21): 3973.

32 Long, B.R., Veron, P., Kuranda, K. et al. (2021). Early phase clinical immunogenicity of valoctocogene roxaparvovec, an AAV5-mediated gene therapy for hemophilia A. *Mol. Ther.* 29 (2): 597–610.

33 D'Avola, D., López-Franco, E., Sangro, B. et al. (2016). Phase I open label liver-directed gene therapy clinical trial for acute intermittent porphyria. *J. Hepatol.* 65 (4): 776–783.

34 George, L.A., Ragni, M.V., Rasko, J.E.J. et al. (2020). Long-term follow-up of the first in human intravascular delivery of AAV for gene transfer: AAV2-hFIX16 for severe hemophilia B. *Mol. Ther.* 28 (9): 2073–2082.

8

Transgene mRNA Expression Analysis

Venkata Vepachedu and Hsing-Yin Liu

Molecular Biology, Johnson and Johnson Innovative Medicine, Janssen Pharmaceuticals, Spring House, PA, USA

8.1 Purpose of Measuring Transgene mRNA

One important parameter in measuring the efficacy of traditional small-molecule medicine is to follow the drug concentration to build a pharmacokinetic (PK) profile. The possible drug metabolites are captured carefully to assess when and where the functional drug and its metabolites are therapeutic. The pharmacodynamics (PD) studies are employed to correlate the relationship between drug concentration and its biologic effects. GTx is different from small molecule medicine due to the complexity of the "drug" itself. In an AAV-mediated GTx, for example, the direct drug is the DNA packed into the AAV capsid. This packed DNA encodes the therapeutic payload, including cDNA replacement/gene addition, exon-skipping, RNA degradation, or genome editing. [1]. When combined with possible systemic or target-tissue-specific deliveries by modifying the AAV capsid, AAV-mediated GTx is a very attractive platform for future medicine.

The additional advantages of AAV-mediated GTx are the rare incident of chromosomal integration and the ability to infect both dividing and nondividing cells. However, these can also be seen as disadvantages. AAVs enter the host cells and stay as an episomal form without further replication [2]. This might not be a problem if the target cells are nondividing cells. If the target cells are rapidly dividing cells, a number of these episomal AAV DNA will be diluted quickly and lose its therapeutic efficacy. Hence, monitoring the persistence of AAV genome and its products is critical for the PK profile.

Drug Development for Gene Therapy: Translational Biomarkers, Bioanalysis, and Companion Diagnostics, First Edition. Edited by Yanmei Lu and Boris Gorovits.

Despite an elaborate design and engineering process before packing the recombinant DNA into the AAV capsids, there is no guarantee that the intended gene product will be expressed and functional. In addition, the expression levels of the intended drug product are critical for it to be therapeutic. As mentioned earlier, AAV-mediated GTx is flexible to deliver many different drug modalities: (1) gene replacement or addition (~90%) [3] that produces full-length or truncated-but-functional replacement proteins; (2) small interference RNAs (siRNAs) or microRNAs (miRNAs) to reduce a target messenger RNA (mRNA)[3], and (3) CRISPR-Cas9/guide RNA (gRNA) components to perform gene-editing in the target cells [4]. Depending on the MOA, the AAV-mediate GTx can result in increased or decreased levels of targeted gene product (protein), which often can be captured at the RNA level as well.

For AAV-mediated GTx, the PK profile includes the tracking of the drug (AAV vector genome), the payload (the direct transgene RNA), and final drug product (transgene product, both the target mRNA species and the translated protein). The final protein product is the driving force for a successful PD profile [5], and it will be discussed in separate chapter of this book. Currently, the measurement of transgene mRNA is performed with quantitative RT-PCR (RT-qPCR or RT-dPCR) and in situ hybridization (ISH), both will be discussed in this chapter.

8.1.1 Transgene Encodes Therapeutic Protein Entity

The AAV vector has an insert size limitation of ~4.7 kb [2], which translates into approximately 170 kDa of protein size. Regardless of the MOA of the specific AAV therapeutics, it is critical to quantify the direct transcription product of the AAV recombinant DNA during the development and preclinical stage. Absolute RT-qPCR is commonly used to quantify the expression levels of the target transgene mRNA. One must keep in mind that the transgene mRNA could be the direct or indirect product of the AAV-mediated GTx depending on the modality type the AAV cassette carries (Figure 8.1).

If the replacement protein carried by an AAV-mediated GTx is a full-length endogenous protein, the recombinant DNA sequence in the AAV cassette is likely significantly different from the endogenous transcript due to codon optimization. If so, the RT-qPCR assay development strategy is straightforward to differentiate the recombinant from the endogenous transcript (Figure 8.1a). Depending on the program, it might be also important to monitor the expression levels of both endogenous mRNA (if any) and transgene mRNA.

Not all AAV-mediated protein replacement therapies are designed to produce full-length proteins. To treat diseases like Duchenne muscular dystrophy (DMD), a truncated protein produced by antisense oligo-directed exon-skipping mRNA is sufficient to restore the dystrophin (DMD gene product) function. In this case, it

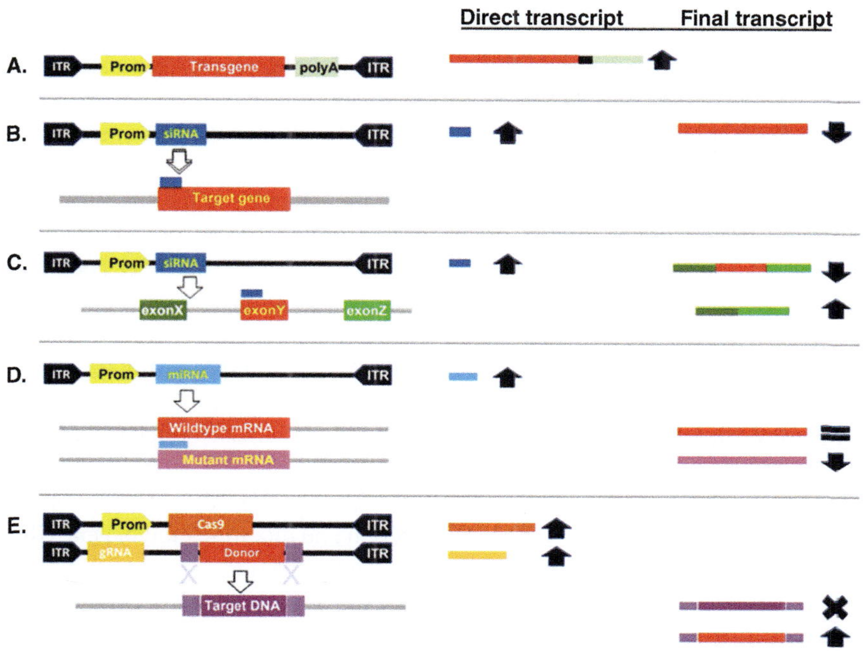

Figure 8.1 The AAV-mediated GTx may deliver direct transcript from the AAV vector and then the final transcripts mediated by the direct transcript. See text for detail.

is critical to monitor both exon-skipped transgene mRNA and antisense oligos (which will be discussed in Section 8.1.2) [6] to further demonstrate the MOA as well as the accuracy of the exon-skipping product (Figure 8.1c).

Alternatively, an AAV carrying artificial miRNA targeting mutant Huntingtin (mHTT) mRNA results in decreased levels of mHTT mRNA, but not wild-type HTT (wtHTT) is currently on clinical trial for treating Huntington disease (HD). The key pathology of HD is caused by the aggregation of mHTT protein in the brain. This specific AAV5-miHTT binds to the mutant exon1 of the mRNA to inhibit it from being translated into the toxic mHTT (Figure 8.1d). Because wtHTT is an essential protein, it is important to monitor the expression levels of both wild-type and mutant HTT mRNA during the development stage [7]. For AAV platform that carries antisense oligonucleotides to mediate the gene silencing, carefully monitoring the reduction of target mRNA is obviously the key (Figure 8.1b). Lastly, dual vector systems to deliver CRISPR-Cas9/gRNA gene editing therapy are on the rise [8]. Depending on the MOA of specific GTx, quantifications of RNA species derived from gene editing components (the Cas9 nuclease mRNA and gRNA), as well as the homologous recombination template DNA, may be required (Figure 8.1e).

8.1.2 Transgene Encodes Other Entities

As mentioned previously, in addition to delivering mRNA that generates gene replacement protein, the AAV-mediated GTx can also deliver oligonucleotides that produce specific-splice variant mRNA, or gene editing machinery, including the nuclease and the gRNA.

While it is critical to track the final gene product quantity of the specific-splice variant mRNA or edited mRNA, the expression levels of these oligonucleotides and the mRNA encoding the editing nuclease are also important per Food and Drug Administration (FDA) guidance [9]. For unconventional GTx like these, it is essential to monitor all AAV-generated products: the direct RNA species (mRNA, siRNA, miRNA gRNA, etc.), and the final RNA product (spliced mRNA, exon-skipped mRNA, edited mRNA, etc.), during the development, preclinical, nonclinical, and clinical stages, to ensure intended and efficacious gene products are generated in the target tissue, as well as monitoring the off-target toxicity associated with these RNA species. The techniques to quantify oligonucleotides will be covered in Section 8.2.1, and the detection and quantification of gene editing at the genome level will be discussed in separate chapter.

8.2 Technologies to Quantify Transgene Expression in Tissues

8.2.1 RT-qPCR or RT-dPCR

The discovery of critical reagent thermostable DNA polymerase gave a breakthrough in the PCR from a strenuous frequent addition of Klenow fragment to a continuous reaction and made it possible for universal application [10]. Development of programmable thermostats transformed the process of PCR from a three water bath to a single dry bath reaction [11]. Subsequently, real-time PCR machines hit the market in 1996 [11] allowing the detection of amplification in real time up to 48 samples in one run for detection and quantification of gene or sequence of interest, i.e. quantitative PCR or qPCR. By reverse transcription as the first step to make cDNA and then by qPCR of cDNA, i.e. RT-qPCR, levels of RNA or transgene expression are quantified. The terminology used for qPCR and RT-qPCR has been discussed briefly in a separate chapter. As per Minimum Information for Quantitative PCR Experiments (MIQE) [12] guidelines, we used the terms qPCR for DNA and RT-qPCR for mRNA quantitation by real time PCR. RT-qPCR is the gold-standard technique for either absolute or relative quantification of transgene expression levels in mammalian or human tissues (or body fluids such as saliva, tears, or serum). Thus, except for the conversion of RNA to cDNA by using a reverse transcriptase (RT step), the technology of qPCR and RT-qPCR remained the same. The development of single-step RT-qPCR and efficient reverse transcriptases have made the RT-qPCR process more easy and simpler [13]. Digital PCR (dPCR) is also a quantitative PCR method but a novel advancement over qPCR termed by some as next-generation qPCR technology that involves absolute quantitation of nucleic acid target sequences without a standard curve. In brief, the main difference between real-time based qPCR and dPCR is that in qPCR the amplification reaction is monitored throughout the process and quantification is based on the fluorescent signal analysis at the exponential phase, whereas in dPCR the sample is distributed in partitions and signals are measured at the end to quantify based on the number of positive partitions over the total [14]. The details and a comparative discussion will be performed while discussing the quantification strategies. Factors affecting the accurate quantification of transgene expression include the quality of RNA, assay design, and analysis. Hence, the technologies/methods available (Figure 8.2) and in use for

1) Extraction/purification
2) Quantification and quality check
3) Reverse-transcription and qPCR assay
4) Analysis of the raw data

will be discussed.

8.2.1.1 RNA Extraction (Separate vs. DNA/RNA Co-extraction), Quality Testing, and Quantification

Total RNA extraction of large number of samples has become easier to perform with the development of a single-step extraction method using guanidium thiocyanate–phenol–chloroform extraction technique under acidic conditions [15]. This also led to the production of several kits, including column-based kits globally. A study was performed on the efficiency of six RNA extraction methods using commercially available kits, RNAqueous kit, micro-to-midi total RNA purification system, NucleoSpin RNA II, GenElute mammalian total RNA kit, RNeasy mini kit, and TRIzol LS reagent, were evaluated on blood and seven tissues showed acceptable performance of all methods for RT-qPCR [16]. A Trizol-based method showed slightly better performance in this study. However, successful extraction and purification of either DNA or RNA depends on the quality of homogenization of the tissue for the release of RNA and removal of contaminating DNA. Currently, high-quality RNA extraction kits are available from various vendors, such as ThermoFisher scientific and Qiagen, for blood and other tissues or prokaryotes. However, gene/cell therapy projects (AAV vectors or LNP) involving PK/PD studies require the extraction of RNA from large number of samples that require high-throughput platforms and kits for the same. New kits and technologies or platforms have been

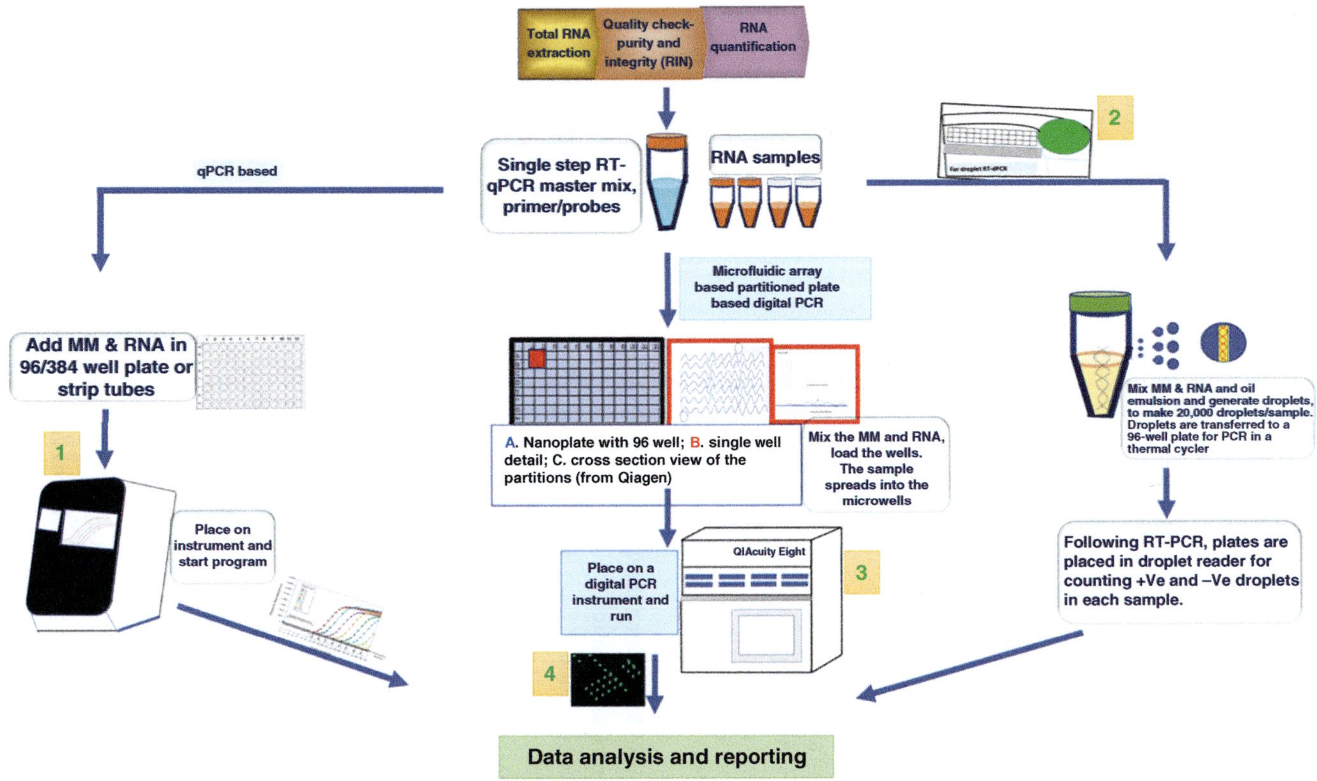

Figure 8.2 Schematic of RT-qPCR/RT-dPCR for gene expression analysis in GTx pharmacokinetic studies. All cartoon pictures are drawn on PowerPoint.

introduced for high-throughput extraction of DNA/RNA with greater simplicity, cost saving, and maintaining quality and yield. The filter plate and nucleic acid extraction spin column offer for high- and low-throughput nucleic acid purification [17]. The introduction of magnetic bead-based procedure improved the high-throughput extractions as filter plate or spin-column based protocols may lead to clogging when used for tissues. New platforms along with kits to use have been introduced for semi-automated extraction platforms for 24, 48, or 96 samples in a single run. Some of the popular automated instruments are Roche Magnapure, Qiagen-Qiacube, ThermoFisher – Kingfisher and Promega-Maxwell.

However, they supply separate kits for DNA or RNA for different tissues/blood. The quality and yield vary between different kits for each sample type. However, the yield of the automated extraction kits also depends on the extent of sample (tissue) homogenization either by manual or high-throughput instruments such as a Tissue lyser or Omniruptor. The selection of proper lysis protocol and homogenizer is essential. DNase treatment is very crucial in RNA purification and some automated instruments/kits tend to show slightly higher DNA contamination from specific tissues [18]. For most of the gene/cell therapy projects the available automated instruments and kits are useful in the extraction of quality RNA with high RIN (RNA Integrity Number), a user-independent algorithm for the assignment of integrity numbers to an RNA sample using a scale of 1–10.

8.2.1.2 Co-extraction of DNA and RNA from same Sample

Blood/tissue collection limitations from mice [19, 20], low concentration of GTx drug target in clinical samples (shedding) and for better comparable results [21] some protocols combining the DNA and RNA kits have been designed and tested [22]. In this, the lysis buffer which is usually discarded after binding DNA to resin was processed for RNA extraction. Also, there are kits developed for the co-extraction of DNA and RNA from the same sample using magnetic bead separation and precipitation at different stages. Comparative assessment of those kits was performed from environmental samples and biological tissues [23]. It was found that though the yields vary, the quality of the RNA is acceptable for sequencing or RT-qPCR. Simultaneous extraction of DNA and RNA from Ambion's ToTally RNA kit from tumor, spleen, and other tissues seems to show quality results and is also compatible with Kingfisher automated systems [24]. However, co-extraction of DNA/RNA is not yet widely used in gene/cell therapy studies for PK/PD assays probably due to higher quantities needed for validated quantitative/relative quantitative assays per non-GLP and GLP requirements. As the GxP studies for biodistribution require processing of large number of samples, automation-friendly kits for co-extraction will encourage wide usage.

8.2.1.3 Quantification and Quality Testing of total RNA in Purified Extracts

For academic or diagnostic project purposes, purified DNA or RNA can directly go for target detection or quantification by qPCR/RT-qPCR [25]. However, AAV-based or other GTx/CTx studies have stringent validation requirements for PK analysis. To meet this, the samples need to have equal total RNA concentration to avoid any matrix effect. Hence, the guidelines for PK analysis in AAV or other GTx studies suggest quantitating the total RNA extractions from samples and normalize them before performing drug or target RNA quantitation by RT-qPCR [26]. For total RNA quantification, the instruments Nanodrop/Denovix/UV–VIS spectrophotometer [26], Qubit, Agilent TapeStation, and Promega Glomax, etc. are popularly used in different labs globally (Table 8.1). However, specificity of Nanodrop or Denovix is very less, compared to the other three instruments which are fluorescent based (Qubit & Glomax or electrophoresis based – TapeStation). Because, both DNA/RNA, nucleotides, and several aromatic compounds also absorb at 260/280 nm. In addition, Agilent TapeStation and Promega Glomax offer high-throughput technologies more suitable for the analysis of large number of samples. Nanodrop is also used to test the quality based on A260/A280 nm absorbance as explained in the next section.

The easiest and routine way to confirm the purity is by taking absorbance at 260/280 nm using a spectrophotometer or Nanodrop [27]. As RNA is relatively labile and prone to degradation, additional care is taken to preserve and confirm its integrity. This includes optimizing extraction conditions and confirmation of RNA quality in the samples – random or individual as preferred. Because mRNA is expressed in low quantities, RNA integrity in eukaryotic cells is commonly determined by quantifying the more abundant 18S and 28S ribosomal RNA (rRNA) transcripts [28]. The quality of RNA is proportional to the percentage of full-length transcripts. In the earlier days, the quality used to be assessed by agarose or capillary electrophoresis [29]. However, this is a bit subjective. Using

Table 8.1 Commonly used instruments for total RNA quality check and quantification in GTx studies.

Type	Model	Manufacturer
Spectrophotometer	DeNovix	Denovix
	NanoDrop	ThermoFisher
Fluorescent-based	Qubit	Promega
	Glomax (high throughput)	Promega
Electrophoresis-based	Bioanalyzer (chip-based)	Agilent
	TapeStation (high throughput)	Agilent

microcapillary electrophoresis, Agilent technologies have developed a user-independent, automated procedure for standardization of RNA quality control that allows the calculation of an RIN [30]. This is widely applied in research and clinical studies to confirm the quality of RNA before going for RT-qPCR. The Agilent 5200 Fragment Analyzer system and Agilent ProSize data analysis software use an RNA quality number (RQN) as a quality metric indicator. ProSize considers the entire electropherogram to give a RQN number [31]. However, in another technical note they published in the recent Agilent platforms total electropherogram is taken into consideration to provide RIN or RQN and both values are interchangeable [32]. A method to measure spatial RNA integrity number (sRIN) in situ from a single tissue section suitable for clinical samples has been developed as a new tool [33].

8.2.1.4 Quantification Using DNA vs. RNA Standards

Quantification of target RNA by RT-qPCR requires a standard curve and needs preparation of standards. However, the RT-dPCR is gaining more preference due to some advantages especially as this does not need a standard curve. The term "digital PCR" was coined by Vogelstein and Kinzler in 1999 [34]. However, in the case of digital PCRs, they are of several varieties depending on the manufacturer. Bio-Rad's digital droplet PCR became very popular. The general parameters for real-time RT-qPCR guidelines were provided in MIQE 2009 [12]. For digital qPCR also MIQE guidelines were published initially in 2013 and an update was provided in 2020 [35, 36]. Though RT-dPCR has several advantages over RT-qPCR as listed in Table 8.2, there is a need to test the reliability of the absolute copy number output and differentiation of positive and negative reaction units.

Some of the commercially available digital PCR systems are listed in Table 8.3.

8.2.1.5 Assay Qualification/Validation and Report

Due to the lack of specific regulations from FDA, European Medicines Agency (EMA), or other regulatory agencies, the guidelines from different forums, white papers, and MIQE [12, 33] are followed for assay qualification/validation and reporting (see Chapter 10. Some of the points were discussed in separate chapter and will be briefly touched on here. Assay qualification of RT-**q**PCR or RT-**d**PCR assays for clinical trials is required to meet same acceptance criteria for assay qualification and validation in general except that while performing the assay, RT-dPCR does not require a standard curve. In either way, the RNA extraction method, and its stability are equally important. The general criteria recommended and followed for validation and acceptance criteria for RT-qPCR and RT-dPCR are listed in Table 8.4.

How extensive a method needs to be validated depends on how much uncertainty exists in the study samples and regulatory needs. Though a standard curve is not needed, RT-dPCR also requires calibrators/quality controls with the range expected in the study samples [37].

Table 8.2 Comparative overview of RT-qPCR vs. RT-dPCR.

Item	Real Time	Digital
Principle	Quantification in real time as it occurs after every cycle. Data collected during exponential (log) phase – PCR product formed is equal to the amount of template.	Partitioning of the PCR reaction into thousands of individual reaction vessels prior to amplification. Microfluidics technology is used for digitization. Endpoint data acquisition.
Quantification	Relative quantification with a reference gene or absolute quantification with a set of standards.	Absolute quantification possible without a standard curve.
Quantification	**Ct Value** is the PCR cycle number at which a specific sample reaches fluorescent intensity above the set threshold background. A standard curve is constructed with the Ct values of a set of standards of known copy numbers and used in downstream quantitation or detection in samples. No post-PCR processing required.	After PCR, the target copy number is calculated based on the fraction of positive/negative microreaction wells/droplets. A *Poisson distribution co-efficient* is used to correct the copies per droplet using the equation = $-\ln(1-p)$, where p = fraction of positive droplets
Precision	Good at higher copy number in the range.	Higher precision can be obtained with a greater number of replicates and is better than real-time RT-qPCR.
PCR inhibitors	No option to specially reduce PCR inhibitor other than including a positive control.	Tolerance to PCR inhibitors is more due to digitalization.
Cost	Platforms are already installed in the labs worldwide and less expenditure.	Initial startup slightly more expensive.
Upper limit of quantitation (ULOQ)	Upper limit of quantification is 10^8 copies in general.	Upper limit of quantitation with current platforms is 10^5 copies as of 2022.
Lower limit of quantitation (LLOQ)	Varies (3 and above copies/reaction) depending on the target.	Reproducibility is higher at LLOQ.
Multiplexing	Higher capability	Limited
Samples/plate	96 or 384 well format	96 well format maximum
Run time	Can go to <90 min	2 h or higher

Table 8.3 Currently available digital PCR systems.

Type	Model	Manufacturer
Non-droplet	QIAcuity	QIAGEN
	QuantStudio 3D	ThermoFisher
	QuantStudio Absolute Q	ThermoFisher
	BiomarkX	Fluidigm
Digital droplet	QX-200/QX-600/QXOne	Bio-Rad
	RainDance	Bio-Rad
	Naica	Stilla

Table 8.4 Recommendations for RT-qPCR/RT-dPCR method validation for regulated AAV gene therapy PK studies.

Molecular reagent / parameter for validation	Recommendations for preparation	Suggested acceptance criteria/plan for validation
Number of steps – cDNA/PCR	One-step RT-qPCR/RT-dPCR is preferred due to reduced hands-on time or steps with labile RNA samples, less contamination, and human or technical errors. Chances of more variability in RT step.	Recommended to validate all assays with at least two lots of kits/reagents and have a protocol for lot-to-lot variation.
Preparation of Reference materials/ Standard calibrators/ Quality Controls	As standard calibrator/controls, a synthetic or in vitro transcribed (IVT) RNA of the transgene/gene of interest (GOI) can be used. Using modified nucleotides for RNA is acceptable and beneficial to increase the stability of the RNA.	Calibrators for standard curve are made fresh before the assay every time. Controls are made, evaluated, and stored at −70 °C.
	LLOQ and ULOQ standard points (range of quantification) are determined based on the performance of the standard curve that is confirmed in development and to be verified by accuracy and precision in the method validation.	No template controls (NTCs) should be below LOD. Calculation of LOD is determined by an appropriate serial dilution starting from copies less than or equal to LLOQ in the respective total RNA matrix and calculated by statistical determination.

(Continued)

Table 8.4 (Continued)

Molecular reagent / parameter for validation	Recommendations for preparation	Suggested acceptance criteria/plan for validation
Standard/ calibration curve	For standard curve, 6–8 non-zero calibrators (in triplicates) spiked into the total RNA from surrogate tissue needs to be prepared. A standard curve is not required for **RT-dPCR** but calibrators in the range of expected sample concentration are recommended to check the efficiency. The standard curve is used to evaluate the ***linear range, specificity, sensitivity, and robustness*** of the RT-qPCR assay by evaluating the LOD, LLOQ, ULOQ, linearity, and amplification efficiency (E). For **RT-dPCR** – gene specific reference/calibrators are suggested in the predicted working range of the study. Acceptance criteria can be same as RT-qPCR calibrators.	(a) Range of Efficiency (**E**) should be 90%–110.0% and R^2 should be ≥0.980. (b) At least 75% of the calibrators should be within the criteria. (c) At >LLOQ, CV and RE should be ≤ +25%. At LLOQ, the CV and RE should be ≤ +45%. (d) For RT-dPCR, the **uncertainty** contribution of the threshold setting and assigned **partition volume** needs to be corrected/updated in the absolute value calculations for accuracy during method development.
Quality Controls-Accuracy	Controls (Positive) – at ULOQ, HQC, MQC, LQC, LLOQ will be run on each plate.	%RE and %CV should be within ±25% for controls >LLOQ. For LLOQ control they can be within ±45%. QCs >50% at each level and overall >67% should meet the above criteria. Same criteria apply for inter-assay run.
Selectivity/ specificity	To be determined with spiked and naïve total RNA samples in the relevant tissues. If the AAV genome-positive (DNA) data is available, the positive tissues may be selected. Same criteria can be set for RT-dPCR also.	%RE and %CV should be ≤45% for a minimum of 3 out of 4 spiked RNA copy numbers at LLOQ level and ±25% for 3X LLOQ or above. At least 3 out of 4 of the unspiked and '0' copy spike of RNA samples should be BLOD or UND.
Extraction recovery	Full transcript of the RNA (GOI) is required. Either the drug-infected cell lysate or full-length transcript (surrogate) obtained by other means should be acceptable. Appropriate level of QCs should be used.	Recovery ≥20% is expected from 75% of the sample types tested.

Molecular reagent / parameter for validation	Recommendations for preparation	Suggested acceptance criteria/plan for validation
Stability	Due to the labile nature of the RNA, this parameter is considered important, specifically for Bioanalysis. Freeze–thaw and storage stability need to be tested depending on the expected number of samples and the predicted storage duration and conditions. Different QC levels need to be tested.	Results should be within 25% of the spiked copy number and %CV. At least 67% of QC's should meet the criteria.

Note: As PK of drug transcript levels are not measured in majority of clinical studies for shedding unless the drug is RNA virus, the above parameters/criteria are mainly focused on nonclinical GLP studies. HQC/MQC/LQC stands for **H**igh/**M**iddle/**L**ow copy number **Q**uality **C**ontrols, respectively. %CV (Coefficient of Variation): Is a measure of random error; the closeness of agreement between repeated individual measurement of an analyte in a single sample. %CV is calculated as (Standard Deviation (SD)/Mean) × 100. Relative error (RE) is a measure of the uncertainty of measurement compared to the size of the measurement. Absolute Error (AE) = True Value (TV) minus Measured Value (MV). RE = AE/TV. % RE = RE × 100. In this context, value refers to copy number.

8.2.1.6 Reporting

The raw data of the digital PCR provides the number of positive/negative droplets for each sample. Conversion of this to copies per μL will be performed using manufacturer's software. However, the volume of each reaction/droplet needs to be calculated carefully to avoid errors. After performing the data analysis, for PK studies the transcript copy number is expressed either as drug target copy number per ug of Total RNA isolated from the sample. If preferred the copy number maybe calculated using a reference gene. In addition, the reporting format can be modified depending on the needs of the study. MIQE guidelines[12, 34, 35] listed the information to be reported for the publication of qPCR/RT-qPCR or dPCR data. Details for the following points are considered essential, (1) experimental design; (2) sample; (3) nucleic acid extraction; (4) storage; stability and quality assessment details; (5) For RNA: RIN number; method used for it, reverse transcription conditions; (6) Details of the standards used, amplicon location/length; (7) qPCR or dPCR instrument details; assay validation and software used for analysis. Almost all items are common to both RT-qPCR and RT-dPCR. Those specific for digital are related to the partitioning of the sample. They are, (1) number of partitions measured; (2) partition volume; (3) copies per partition, and (d) dPCR analysis program details.

However, these guidelines are mostly for publication. For clinical or nonclinical studies of AAV GTx, the requirements vary depending on whether it is regulated or not. More rigorous acceptance criteria and parameters are required for the validation and data reporting for regulatory studies as mentioned in Chapter 10. For AAV-based GTx programs, the transcript levels are reported as copies/μg of total RNA or normalized to a reference gene such as albumin. For a typical AAV-based GTx study, a standard curve with at least 6 non-zero points ranging from 10^8 copies to 0 is made. At least 3 sets of QCs – in that range namely HQC, MQC, and LQC are prepared, validated, and stored. For regulated studies, sensitivity (LOD and LLOQ), specificity, linearity/dilutional linearity (R^2 and Error), precision with %CV at each level, accuracy and reproducibility are validated. Inhibition is tested using control, such as SPUD. SPUD consists of potato (*Solanum tuberosum*) phyB gene (Gen-Bank Y14572), which encodes for a species-specific regulatory photoreceptor that is involved in the pathway responsible for the purple coloration of potato root. The total gene sequence is of 4907 bp in size. The upstream non-translated region of phyB gene between nucleotides 449 and 549 was found to lack homology with any other known sequence which ensures that this stretch can be used for qPCR control in the presence of all samples except from potato [38]. Synthetic sequences are also successfully used as controls [39]. Recognizing false positives: synthetic oligonucleotide controls for environmental DNA surveillance. Inhibition observed will be reported.

Recently, a capillary-based flow cytometric setup was described for sequence-specific counting of individual RNA molecules by tagging with a number of fluorescent oligo probes, eliminating the need for reverse transcription and PCR [40]. In addition to the vast advances in the quantification methods, improvements in the high-throughput extraction platforms and clear regulations for clinical studies are warranted for consistent analysis and reporting.

8.2.2 In Situ Hybridization (ISH)

Commonly used RT-qPCR or RT-dPCR techniques quantify the transgene mRNA expression in a "collective" manner because the RNA is extracted from a chunk of tissues. However, unless the target cells are in the blood circulatory system, the efficacy of an AAV-mediated GTx falls on the successful delivery of the AAV gene product at the cellular level.

Similar to immunohistochemistry, which is used to localize specific protein in a tissue section, ISH is used to locate sequence-specific nucleotide fragments in a tissue section. ISH is a powerful tool to detect specific DNA or RNA species from histological sections under light microscope or observe detailed DNA or RNA organization under electronic microscope.

8.2.2.1 Values of ISH for Discovery Studies

In situ hybridization provides a direct approach to surveil the biodistribution of the AAV vector as well as the gene product at the cellular level, to further ensure the AAV-mediated GTx is delivered to the target cells. The technologies for ISH have greatly advanced in the past decade. Not only have the probe-labeling expanded from chromogenic to more sensitive fluorescent type, the capability of multiplex assays also makes the direct correlation of multiple target sequences in the same section image possible.

Using commercially available ISH techniques such as RNAscope, Zhao et al. were able to follow the kinetic distribution patterns of AAV9-GFP in various tissues at the cellular level after injections into the mouse [41]. In the example of muscle tissue, the AAV genome DNA was found mostly localized in the interstitial space 2-hour post i.v. injection, and slowly moved toward the cytoplasm of sarcolemma over the next 22 hours. Started in 5 days post injection, the AAV genome DNA was found mostly in myonuclei, and the staining became barely visible 4 weeks post injection. In contrast, the GFP mRNA became readily visible 2 weeks post injection, and the signal persisted up to 24 weeks post injection, the last time point of this study. This is an excellent example to demonstrate that ISH can provide detailed kinetics characterizations of AAV-mediated GTx throughout all stages of clinical program development.

Another great example is shown in Wang et al., where the SABER-FISH technology was used to fluorescently label DNA or RNA probes [42]. After using a limiting dilution analysis to confirm that one DNA punctum with SABER-FISH represents a single AAV genome, the detailed biodistributions of the subretinal-injected AAV8-GFP in different layers of retina were further traced. For both rod and cone photoreceptor cells, the majority of GFP mRNA-positive cells were AAV genome-positive, with an average of 2.1 and 2.5 genomes per cell, respectively. However, in the retinal pigment epithelium (RPE) cells, the majority of the GFP mRNA-positive cells were also AAV genome-positive, but with an average of 29.9 genome per cell. Interestingly, some GFP mRNA-negative cone and RPE cells were actually AAV genome-positive, suggesting that not all AAV-infected cells express the transgenes. These levels of understanding cannot be reached with other techniques like qPCR unless it is single-cell qPCR.

Both examples showed strong evidence that the high-resolution characterizations of the biodistribution analysis can only be performed with ISH assays. Additionally, some of the ISH technologies discussed in separate chapter also allow for fluorescence in situ hybridization (FISH) and immunohistochemistry (IHC) performed on the same section to image RNA and protein, which significantly empowers the biodistribution studies of the GTx.

The riboprobes for one RNA species usually cover a couple of hundreds of bases long for RNA ISH. The need to detect small RNA oligoes, such as siRNA, miRNA, and gRNA, became more prominent after the discovery of siRNA [43]. Novel technologies like different locked nucleotide probes and an additional signal amplification step [44] make small RNA-ISH a readily available tool for biodistribution study for AAV-mediated GTx that carries small RNA payload [45].

8.2.2.2 Semi-quantitative, Tissue Fixation, Probe to Reference

Classic Procedure

The classic step-by-step mRNA-ISH is listed in Hunter et al. [46], and one example of miRNA-ISH is listed by Yin [47]. Four basic steps for ISH: tissue section preparation, probe preparation, hybridization and detection, and signal quantification, are outlined in Table 8.5.

Tissue Section Preparation

Additional precautions need to be taken if an RNA-ISH will be performed in a tissue section, for example, all reagents should be made in DEPC-water to maintain the RNA integrity. The most commonly used fixative for tissues is 4% paraformaldehyde, which is a cross-linking agent that maintains cellular and subcellular structures. Other fixatives such as ethanol or methanol can also be used depending on what structures or antigens are to be visualized later. The fixed tissues then can be embedded in paraffin (wax) and then sectioned with microtome, or dipped into OCT solution, frozen with liquid nitrogen, and then sectioned with cryostat.

For paraffin-embedded sections, dewaxing and a series of stepwise rehydration need to be performed before RNA-ISH [48]; for frozen sections, the slides only need to be brought to room temperature and dry. Because the tissues are usually fixed with formaldehyde-based solution, a proteinase treatment is necessary to open the tissue and cell membranes, allowing the riboprobe to find the targeted RNA species. Additionally, a DNase might be needed if the riboprobe might also bind to a DNA template.

Probe Preparation

The most common protocol to labeling riboprobe is to clone the desired probe sequence (usually the antisense strand that can hybridize with the mRNA) in downstream of a bacteriophage promoter and then run an in vitro transcription with digoxigenin (DIG)-UTP, followed by the signal detection with anti-DIG-conjugated reporter dye. The signal strength is directly proportional to the numbers of uracil (U) in the riboprobe. After the in vitro transcription reaction, the riboprobe is purified, ethanol-precipitated, and resuspended in proper solution [46].

Many new technologies are available to make riboprobes with higher signal amplification and easier to multiplex. A few examples are described below.

Table 8.5 Outlines of RNA-ISH procedure.

Tissue section preparation	Fixation	• Formaldehyde-based fixatives • Better made in DEPC-water
	Embedding and sectioning	• Dehydration, and embed the tissue in paraffin, and sectioned with microtome. • Embed the tissue in OCT, freeze, and sectioned with cryostat.
	Get ready for hybridization	• For paraffin-embedded tissue slides dewaxing and rehydration are necessary. • For frozen section slides, thaw at room temperature and dry.
Probe preparation	DIG-UTP-labeled riboprobe	• In vitro transcription with DIG-UTP
	Commercially available ready-to-use riboprobes	• Purchase customized probes
	Probes that need to be amplified by the user	• Order or design customized oligos • Amplify the probe according to the protocol
Hybridization	Blocking	• Denhardt's solution-based • SSC-based
	Denaturation Hybridization Detection	• Should be riboprobe-specific
Signal quantification	Manual counting Automatic counting (with proper software)	• Dot-counting if one punctum equals one target • If the counts cover the whole cell range. • Mostly is qualitative (semi-quantitative)

See text for detail.

Commercially available RNAscope probe [49] is a mix of 20 pairs of proprietarily-made riboprobes for each target. The signal amplification step is performed after the hybridization step. Each riboprobe comes with a ready-to-use tube format. SABER-FISH technology was developed by professors at Harvard University and the Wyss Institute [50]. Users will design sequence-specific probes and mix with PER (primer exchange reaction) primers provided on the manuscript to generate PER concatemers. A pool of PER concatemers hybridize to the target RNA sequence as well as dye-labeled oligos to achieve signal amplification. Similar to

SABER-FISH, HCR-FISH, or hybridization chain reaction-FISH, uses hairpin hybridization to incorporate dye-conjugated dye and amplify the signal for mRNA detection [51].

One important probe that should be made is the proper negative control probe. This could be a sense probe that should not bind to the transgene RNA or probes for background control (e.g. a bacterial *dapB* probe was recommended for RNAscope) [49].

Hybridization and Detection

Same principles for Northern hybridization also apply to RNA-ISH. A blocking step is essential for reducing nonspecific binding of the riboprobe(s). The blocking solution such as Denhardt's solutions, SSC (saline-sodium citrate)-based buffer, with the addition of salmon sperm DNA or tRNA are often used [50]. The riboprobes and the samples can go through a denaturation step at 75–100 °C for 5–10 minutes [46, 52, 53] before hybridization. However, the actual hybridization conditions should depend on which riboprobe technology is used to ensure specific target binding.

The signal detection method is also riboprobe technology dependent, followed by mounting with anti-fade mounting media if fluorescent dyes are used. Some mounting media also serve as curing mounts for long-term storage, which means re-hybridization is impossible. A good variety of mounting media are commercially available, users need to pay attention to the compatibility to the dye used (fluorescent spectrum), refractive index (after curing if it is a curing mount), with or without nucleus staining (some already has DAPI), and if curing is needed.

Signal Quantification

It is always a big challenge to quantify image-based assays under the microscope; the thickness of a histological section adds on the additional complexity because the highest signal found in one section does not always represent the whole signal in the target cell. However, compared to protein immunofluorescent microscopy, ISH type of signals is often localized in the cytoplasmic space or in the nucleus, also some technology allows for 1 punctum staining for a target-binding [42], the quantification of RNA-ISH could become a dot-counting work [42, 51, 54], kind of similar to counting digital PCR signals.

One important point to consider is that the human cell is sized from 7.5 to 150 microns, but the histological sections are usually under 10 microns [55], so several sections might be needed to cover the RNA-ISH counts for each cell, if accurate quantification is important. Unless the RNA-ISH technique employed is properly validated to confirm one dot per RNA strand, and the image covers the whole cell range, most of the time, RNA-ISH is still considered a qualitative assay.

8.3 Summary

The biodistribution and expression of transgene mRNA is part of the PK profile for the AAV-mediated GTx. It is important to establish the relationship of the AAV genome copy number vs transgene expression copy number, as well as to demonstrate the transgene mRNA is expressed in the target cells at the early discovery stage. The RT-qPCR/RT-dPCR is the most used and reliable method to quantify the expression levels of AAV-mediated transgene mRNA species, including the direct and final transgene product. The transgene mRNA expression in non-targeted tissues also needs to be monitored as part of the toxicity study. Successful transgene mRNA quantification depends on parameters such as clean RNA extraction from tissues, optimized primers and probe set, proper standard curves, and fit-for-purpose method assessment/validation process, for both nonclinical and clinical studies. The pros and cons of using RT-qPCR or RT-dPCR were also discussed in this chapter.

ISH usually serves as a semiquantitative observation to confirm the on-target transgene mRNA expression at the cellular levels, as it could be critical for certain disease treatment. While the precise quantification of ISH assays might be difficult, the qualitative information obtained is often sufficient for biodistribution studies.

References

1. Mendell, J.R., Al-Zaidy, S.A., Rodino-Klapac, L.R. et al. (2021). Current clinical applications of in vivo gene therapy with AAVs. *Mol. Ther.* 29 (2): 464–488.
2. Colella, P., Ronzitti, G., and Mingozzi, F. (2018). Emerging issues in AAV-mediated in vivo gene therapy. *Mol. Ther. Methods Clin. Dev.* 8: 87–104.
3. Au, H.K.E., Isalan, M., and Mielcarek, M. (2021). Gene therapy advances: a meta-analysis of AAV usage in clinical settings. *Front. Med.* 8: 809118.
4. Moreno, A.M., Fu, X., Zhu, J. et al. (2018). In situ gene therapy via AAV-CRISPR-Cas9-mediated targeted gene regulation. *Mol. Ther.* 26 (7): 1818–1827.
5. Chen, N., Sun, K., Chemuturi, N.V. et al. (2022). The perspective of DMPK on recombinant adeno-associated virus-based gene therapy: past learning, current support, and future contribution. *AAPS J.* 24 (1): 31.
6. Verheul, R.C., van Deutekom, J.C., and Datson, N.A. (2016). Digital droplet PCR for the absolute quantification of exon skipping induced by antisense oligonucleotides in (pre-)clinical development for Duchenne muscular dystrophy. *PLoS One* 11 (9): e0162467.

7 Keskin, S., Brouwers, C.C., Sogorb-Gonzalez, M. et al. (2019). AAV5-miHTT lowers Huntingtin mRNA and protein without off-target effects in patient-derived neuronal cultures and astrocytes. *Mol. Ther. Methods Clin. Dev.* 15: 275–284.

8 Dasgupta, I., Flotte, T.R., and Keeler, A.M. (2021). CRISPR/Cas-dependent and nuclease-free in vivo therapeutic gene editing. *Hum. Gene Ther.* 32 (5-6): 275–293.

9 FDA. (2022). Human-gene-therapy-incorporating-human-genome-editing_draft guidance for industry.

10 Saiki, R.K., Gelfand, D.H., Stoffel, S. et al. (1988). Primer-directed enzymatic amplification of DNA with a thermostable DNA polymerase. *Science (New York, N.Y.)* 239 (4839): 487–491.

11 Campos, J. (2015). The cutting edges of contemporary diagnostics. *Culture* 2 (2): 34–41.

12 Bustin, S.A., Benes, V., Garson, J.A. et al. (2009). The MIQE guidelines: minimum information for publication of quantitative real-time PCR experiments. *Clin. Chem.* 55 (4): 611–622.

13 Raja, S., Luketich, J.D., Kelly, L.A. et al. (2000). Increased sensitivity of one-tube, quantitative RT-PCR. *Biotechniques* 29 (4): 702. 4, 6.

14 Quan, P.L., Sauzade, M., and Brouzes, E. (2018). dPCR: a technology review. *Sensors (Basel)* 18 (4): 1271.

15 Chomczynski, P. and Sacchi, N. (1987). Single-step method of RNA isolation by acid guanidinium thiocyanate-phenol-chloroform extraction. *Anal. Biochem.* 162 (1): 156–159.

16 Deng, M.Y., Wang, H., Ward, G.B. et al. (2005). Comparison of six RNA extraction methods for the detection of classical swine fever virus by real-time and conventional reverse transcription-PCR. *J. Vet. Diagn. Investig.* 17 (6): 574–578.

17 Euler, L. (2021). Understanding the new technologies for RNA isolation/extraction. *Gen. Eng. News* 41 (2).

18 Sellin Jeffries, M.K., Kiss, A.J., Smith, A.W. et al. (2014). A comparison of commercially-available automated and manual extraction kits for the isolation of total RNA from small tissue samples. *BMC Biotechnol.* 14: 94.

19 Blood Collection from Animals. Available from: https://www.nal.usda.gov/animal-health-and-welfare/animal-use-alternatives/blood-collection-animals.

20 Raabe, B.M., Artwohl, J.E., Purcell, J.E. et al. (2011). Effects of weekly blood collection in C57BL/6 mice. *J. Am. Assoc. Lab. Anim. Sci.* 50 (5): 680–685.

21 Lim, N.Y., Roco, C.A., and Frostegård, Å. (2016). Transparent DNA/RNA co-extraction workflow protocol suitable for inhibitor-rich environmental samples that focuses on complete DNA removal for transcriptomic analyses. *Front. Microbiol.* 7: 1588.

22 Bowden, A., Fleming, R., and Harbison, S. (2011). A method for DNA and RNA co-extraction for use on forensic samples using the Promega DNA IQ™ system. *Forensic Sci. Int. Genet.* 5 (1): 64–68.

23 Wang, S., Shanthan, G., Bouzga, M.M. et al. (2021). Evaluating the performance of five up-to-date DNA/RNA co-extraction methods for forensic application. *Forensic Sci. Int.* 328: 110996.

24 Simultaneous extraction of RNA and DNA. Available from: https://www.thermofisher.com/us/en/home/references/ambion-tech-support/rna-isolation/tech-notes/simultaneous-extraction-of-rna-and-dna.html.

25 Summay of available potocols fo COVID diagnosis WHO. Available from: https://www.who.int/docs/default-source/coronaviruse/whoinhouseassays.pdf.

26 C. F. Barbas III, D. R Burton, J. K. Scott, and G. J. Silverman. Quantitation of DNA and RNA 2007. Available from: http://cshprotocols.cshlp.org/content/2007/11/pdb.ip47.full.

27 NanoDrop spectrophotometers; assessment of nucleic acid purity. T042. Rev 1/11.

28 Sambrook, J., Fritsch, E., and Maniatis, T. (1989). *Molecular Cloning, a Laboratory Manual.* Cold Spring Harbor Laboratory Press.

29 Imbeaud, S., Graudens, E., Boulanger, V. et al. (2005). Towards standardization of RNA quality assessment using user-independent classifiers of microcapillary electrophoresis traces. *Nucleic Acids Res.* 33 (6): e56.

30 Schroeder, A., Mueller, O., Stocker, S. et al. (2006). The RIN: an RNA integrity number for assigning integrity values to RNA measurements. *BMC Mol. Biol.* 7: 3.

31 Quality analysis of eukaryotic total RNA with the Agilent 5200 fragment analyzer system (Application Note 5994-0519EN). Agilent Technologies.

32 Comparison of RIN and RQN for the Agilent 2100 bioanalyzer and the fragment analyzer systems technical overview. Agilent Technologies.

33 Kvastad, L., Carlberg, K., Larsson, L. et al. (2021). The spatial RNA integrity number assay for in situ evaluation of transcriptome quality. *Commun. Biol.* 4 (1): 57.

34 Vogelstein, B. and Kinzler, K.W. (1999). Digital PCR. *Proc. Natl. Acad. Sci. U. S. A.* 96 (16): 9236–9241.

35 Huggett, J.F., Foy, C.A., Benes, V. et al. (2013). The digital MIQE guidelines: minimum information for publication of quantitative digital PCR experiments. *Clin. Chem.* 59 (6): 892–902.

36 Huggett, J.F. (2020). The digital MIQE guidelines update: minimum information for publication of quantitative digital PCR experiments for 2020. *Clin. Chem.* 66 (8): 1012–1029.

37 Deprez, L., Corbisier, P., Kortekaas, A.M. et al. (2016). Validation of a digital PCR method for quantification of DNA copy number concentrations by using a certified reference material. *Biomol. Detect. Quantif.* 9: 29–39.

38 Nolan, T., Hands, R.E., Ogunkolade, W., and Bustin, S.A. (2006). SPUD: a quantitative PCR assay for the detection of inhibitors in nucleic acid preparations. *Anal. Biochem.* 351: 308–310.

39 Wilson, C.C., Wozney, K.M., and Smith, C.M. (2016). *Methods Ecol. Evol.* 7: 23–29.

40 Yoo, H.B., Park, S.R., Hong, K.S. et al. (2022). Precise RNA quantification by counting individual RNA molecules using high-sensitivity capillary flow cytometry. *Anal. Chem.* 94 (3): 1752–1759.

41 Zhao, J., Yue, Y., Patel, A. et al. (2020). High-resolution histological landscape of AAV DNA distribution in cellular compartments and tissues following local and systemic injection. *Mol. Ther. Methods Clin. Dev.* 18: 856–868.

42 Wang, S.K., Lapan, S.W., Hong, C.M. et al. (2020). In situ detection of adeno-associated viral vector genomes with SABER-FISH. *Mol. Ther. Methods Clin. Dev.* 19: 376–386.

43 Andrew Fire, S.X., Montgomery, M.K., Kostas, S.A. et al. (1998). Potent and specific genetic interference by double-strandedRNA in Caenorhabditiselegans. *Nature* 391: 806–811.

44 Urbanek, M.O., Nawrocka, A.U., and Krzyzosiak, W.J. (2015). Small RNA detection by in situ hybridization methods. *Int. J. Mol. Sci.* 16 (6): 13259–13286.

45 Suhy, D.A., Kao, S.C., Mao, T. et al. (2012). Safe, long-term hepatic expression of anti-HCV shRNA in a nonhuman primate model. *Mol. Ther.* 20 (9): 1737–1749.

46 Hunter, J.E., Gurda, B.L., Yoon, S.Y. et al. (2019). In situ hybridization for detection of AAV-mediated gene expression. *Methods Mol. Biol.* 1950: 107–122.

47 Yin, V.P. (2018). In situ detection of microRNA expression with RNAscope probes. In: *RNA Detection Methods in Molecular Biology*, vol. 1649 (ed. I. Gaspar). New York, NY: Humana Press.

48 Redmayne, N. and Chavez, S.L. (2019). Optimizing tissue preservation for high-resolution confocal imaging of single-molecule RNA-FISH. *Curr. Protoc. Mol. Biol.* 129 (1): e107.

49 Wang, F., Flanagan, J., Su, N. et al. (2012). RNAscope: a novel in situ RNA analysis platform for formalin-fixed, paraffin-embedded tissues. *J. Mol. Diagn.* 14 (1): 22–29.

50 Kishi, J.Y., Lapan, S.W., Beliveau, B.J. et al. (2019). SABER amplifies FISH: enhanced multiplexed imaging of RNA and DNA in cells and tissues. *Nat. Methods* 16 (6): 533–544.

51 Lovely, A.M., Duerr, T.J., Stein, D.F. et al. (2023). Hybridization chain reaction fluorescence in situ hybridization (HCR-FISH) in Ambystoma mexicanum tissue. *Methods Mol. Biol.* 2562: 109–122.

52 Young, A.P., Jackson, D.J., and Wyeth, R.C. (2020). A technical review and guide to RNA fluorescence in situ hybridization. *PeerJ* 8: e8806.

53 Lee, D., Xiong, S., and Xiong, W.C. (2013). General introduction to in situ hybridization protocol using nonradioactively labeled probes to detect mRNAs on tissue sections. *Methods Mol. Biol.* 1018: 165–174.

54 Bondarenko, G., Sorden, S.D., Christian, B.J. et al. (2021). Semiquantitative methods for GFP immunohistochemistry and in situ hybridization to evaluate AAV transduction of mouse retinal cells following subretinal injection. *Toxicol. Pathol.* 49 (3): 537–543.

55 Khan, Y.S. and Farhana, A. (2022). *Histology, Cell.* Statepearls Publishing.

9

Quantification of Transgene Protein Expression and Biochemical Function

Robert Dodge[1] *and Liching Cao*[2]

[1] *Department of BioMedical Research, Novartis, East Hanover, NJ, USA*
[2] *Biomarker and BioAnalytical Sciences, Sangamo Therapeutics, Richmond, CA, USA*

9.1 Introduction

Many transgene proteins in gene replacement therapy are from protein classes not typically encountered in traditional bioanalysis of soluble biomarkers or targets of biotherapeutics. Those classes include intracellular proteins located in subcellular compartments or structural proteins, or membrane proteins which require new methods and platforms to be developed to high quantitative standard in addition to traditional methods like Enzyme-Linked Immunosorbent Assays (ELISA). This chapter aims to discuss the current industrial practices and challenges encountered during method development and qualification/validation of transgene protein quantitation in terms of expression and biochemical function.

Discovery and early development gene therapy studies are carried out in animals. Because of that, there are many more tools available for obtaining samples and assessing protein expression and biodistribution that may not be available in the clinical setting. Most notably, assessing expression in a variety of tissues may only be possible in animals except in the very limited cases where planning and consent are available to test human tissue for transgene expression retrieved during a human autopsy [1].

Because of the challenges with quantification of transgene protein expression as presented in this chapter, most gene therapy biodistribution studies are carried out by assessing transgene DNA levels, or sometimes transgene mRNA. However,

Drug Development for Gene Therapy: Translational Biomarkers, Bioanalysis, and Companion Diagnostics, First Edition. Edited by Yanmei Lu and Boris Gorovits.

the ability to test protein expression in animal models has some advantages. For example, detecting transgene DNA in the brain may be useful and predictive; however, proof of expression of a functional transgene protein may be valuable and more conclusive of effective entry of a functional transgene DNA into cells. And while the presence of transgene-translated mRNA in a cell or tissue is generally predictive of transgene protein expression, determination of functional transgene protein will always be the most effective pharmacodynamic marker indicating success of a gene therapy treatment.

9.2 Transgene Protein Concentration Determination

9.2.1 Human Transgene in Preclinical Species

As many gene therapy transgene proteins are wild-type proteins, designed to be a correction for genetic mutations that exist in the disease state, or cause the increased expression of an already endogenous expressed protein, the ability to distinguish the transgene protein of interest over the background of existing protein in the animal. In animal models, this may be especially true of a human transgene in the gene therapy therapeutic has considerable homology to the animal protein. This creates a challenge for the bioanalytically determination of transgene protein expression measurements [2]. In addition, as most transgene protein bioanalytical work is done in animal models, where functional and homologous transgene protein is likely expressed absent a knockout or genetically modified animal and so differentiating the human gene therapy transgene product from the endogenous animal protein is challenging. Typically, the ability to detect small variations in two otherwise equivalent proteins, with specific antibodies in a ligand binding assay (LBA) is challenging. While it is possible to generate an antibody that would distinguish, for example, between a genetically mutated protein and transgene with the wild-type sequence, access to these specific antibodies or generating these antibodies de novo may not be possible within timeframe needed for testing. Hence, the major bioanalytical methodology for assessing expression of a transgene when an endogenous or very similar animal protein exists is mass spectrometry (MS) assay protein formats as discussed further in the chapter. With MS, the ability to distinguish differences between two similar proteins differing only slightly in amino acid sequence and thus mass is possible and straightforward. Because of this, most transgene protein assessments of human gene therapy therapeutics are carried out with MS platforms.

9.2.2 Human Transgene Assessment for Intracellular Proteins

While most early development and discovery studies are done in animals or even specific animal models, ultimately the most interesting and scientifically relevant pharmacodynamic measurements are in humans. As described in this chapter, measurement of gene therapy therapeutic transgene proteins is highly dependent on the type of transgene product, from bioanalytical assay technique to limitations on obtaining relevant samples.

Measurement of intracellular proteins can be illustrated by the case study with the assessment of the intracellular transgene survival motor neuron (SMN) protein in the gene therapy therapeutic ZOLGENSMA® (onasemnogene abeparvovec-xioi). This work and strategies for transgene assessment provides a comprehensive understanding of transgene protein assessment for intracellular proteins.

Spinal muscular atrophy (SMA) is a neuromuscular disease that affects the part of the nervous system that controls voluntary muscle movement. Unlike many other neuromuscular diseases, there is a clear understanding of the specific genetic cause of SMA and that results directly from bi-allelic mutations in the survival motor neuron 1 (SMN1) gene. Therefore, direct measurement and relevant concentration levels in various solid and liquid tissues of SMN transgene protein that Zolgensma gene therapy provides are of scientific interest.

Development of an electrochemiluminescence (ECL) immunoassay for detection and quantification of SMN protein has been reported using standard sandwich immunoassay format with anti-SMN antibodies used for capture and detection reagents [3]. As will be relevant for any transgene protein expressed, stability *ex vivo* will be important to assess to ensure accurate interpretation of results. This is important because the environment of a sample *ex vivo* will be very different than the intracellular environment. In addition, many intracellular proteins may not have a long half-life, and so maintaining the protein *ex vivo* in stable form prior to measurements of functional assessments can be challenging. In the work of Zaworski et al., stability was carefully assessed and it was noted that room temperature, 4 and −20 °C even was noted to be limited in whole blood. These challenges with *ex vivo* stability are not unusual in a protein expressed *in vivo* intracellularly, especially in this case where measurements were made in blood, which would not be the native environment for this transgene protein and thus stability in storage in blood may be expected to be limited. Although careful fractionation of whole blood showed that red cell or platelet levels of SMN were the primary source of SMN protein in whole blood.

Once a sensitive and robust assay, validated for stability, specificity, parallelism, and accuracy is available, investigation on correlation between accessible matrix (whole blood in this case) and disease state or genotype may be carried out. If there is such a correlation, there may be the ability for trans-gene protein expression levels to be evaluated in standard matrix like whole blood if the protein is soluble and functional even if the native environment is primarily intracellular. In the case of SMA, it was determined that there was a correlation between disease severity, genotype, and levels of circulating SMN protein in whole blood [4]. Other work also showed, in general, correlation between mRNA levels and protein levels for most patients, but not all types of patients when analyzed based on time of disease onset [5].

In addition, use of the SMN protein immunoassay to assess SMN protein levels after treatment with Zolgensma did not show an obvious increase in the levels of SMN protein in whole blood. The correlation between mRNA levels, measured transgene protein concentrations and response to gene therapy treatment may not be straightforward. Therefore, even when measurements of transgene intracellular proteins are available, it is necessary to examine all data from DNA, mRNA, and protein levels assays as well as efficacy levels before using transgene protein expression data for a conclusion.

9.2.3 Human Transgene Protein Assessment for Non-secreted Proteins

Perhaps the largest area of interest and most developed field of non-secreted gene therapy transgene proteins are in ocular gene therapy products [6]. The approval of LUXTURNA™ (voretigene neparvovec-rzyl) in 2017 is evidence of the viability of this type of therapeutic. Luxterna treats a specific disorder in a class of genetic hereditary retinal dystrophies that are associated with progressive visual dysfunction and due to mutations in any one of more than 220 different genes. Specially, Luxterna delivers a gene that codes for the enzyme RPE65. While RPE65 is not specifically a transmembrane protein, it is expressed in the retina pigment epithelium and is not freely available to measure the expression in the circulating blood. Because of the challenge of measuring the RPE65 transgene-expressed product, all clinical pharmacokinetic results focused only on the measurement of the vector gene using PCR [7].

Even in cell lines or animal expression studies, because of the challenge of detecting a membrane-associated protein, most studies of expression of RPE65 are inferred by levels of mRNA present as opposed to actual detection of protein transgene [8]. Thus, detection and quantification of the RPE65 protein has not been done using traditional bioanalytical methodologies associated with soluble transgene expression. However, direct detection of the membrane-associated enzyme RPE65 can be carried out in a semiquantitative manner. For example, RPE65 protein can be detected from harvested mouse retinas using Western blots [9]. In this work, retinas were homogenized with detergent and extracts were electrophoresed on SDS-PAGE gels, transferred to nitrocellulose and detected with anti-RPE65 antibodies. Relative levels of protein were determined based on the relative intensity of the bands on the Western blot relative.

Assessing levels of protein expressed using Western blot imagining and relative interest can be a very useful technique for assessing membrane-associated proteins in animal studies, but is it subject to several caveats. First, while semiquantitative Western blot analysis within a study or experiment can be very informative, due to lack of universal standards and potential technique variation between labs, comparison of expression levels in different studies, or between laboratories is challenging. Thus, Western blot quantification is best suited to a specific study and comparison of conditions rather than universal quantification of transgene protein levels. Second, Western blot analysis relies extensively on having specific antibodies to the gene of interest. While for RPE65, because of the overall specific interest and disease state, there are sufficient useful antibodies commercially available, it may not always be the case for all membrane-associated gene therapy expressed transgene proteins.

Related to the antibody specificity, as in the case with Luxterna, which is a treatment of patients with confirmed biallelic RPE65 mutation-associated retinal dystrophy, the mutated RPE65 protein is present in the disease state. And the antibodies used in Western blot imaging will likely not be available that are specific enough to mutated epitope so that they can distinguish between mutated non-functional RPE65 protein and the wild-type functional RPE65 enzyme. Thus, the Western blot method of quantification of transgene protein products may be of limited utility in assessing expression levels of transgene after treatment even in animal models.

Relative expression levels can also be determined with immuno-histochemistry (IHC) staining [10]. Relevant sections of the eye can be isolated and fixed using standard IHC techniques, and then stained with anti-RPE65 antibody and detected with secondary labeled antibody. As in Western blot analysis, relative quantitation of a membrane-associated protein is primarily semi-quantitative and also subject to the caveats of antibody availability, specificity to mutated vs. wild-type RPE65 protein and variation between experiments and between laboratories. However, an advantage of the IHC methodology may enable investigation into expression patterns within tissue at a higher resolution than the Western blot, which generally involves homogenization of bulk tissue, resulting in loss of specific location expression information.

For the case of RPE65, which is a membrane-associated enzyme, as in secreted enzymes, it may still be possible to assess enzyme activity. As in the case of Western blot and IHC methodologies, assessing differences between nonfunctional mutated RPE65 in the disease state and functional RPE65, resulting from transgene expression after gene therapy treatment is not generally possible. However, enzyme activity assays have been carried out with RPE65 expressed in the eye [9]. The substrate for RPE65 enzyme has been identified as retinyl esters and in vitro methodologies have been developed to measure RPE65 enzyme activity and have been established and shown in RPE65 enzyme knock-out mice [11]. While the extraction-homogenization procedure for all enzymes may not be amenable to retaining enzyme activity, RPE65 enzyme activity is possible to be measured after extraction. For RPE65 transgene expression evaluation, the ability to distinguish between no or low levels of enzyme activity and increased activity in the presence of functional RPE65 is clear. As in Western blot and IHC, variability between experiments and between labs makes this method semi-quantitative. However, the key advantage of direct detection of enzyme activity allows one to access the general success of functional protein expression in animals for an enzyme transgene product delivered using gene therapy techniques.

9.2.4 Human Transgene Protein Assessment for Secreted Proteins

Assessment of protein levels for intracellular proteins or membrane-bound or associated proteins has been discussed. Primarily, assessment of proteins in these cases in human studies may be semi-quantitative, or for example, in the case of SMN protein, measured in whole bold as a potential surrogate to intracellular concentrations in applicable cell types. In addition, non-secreted protein concentration or location assessment may be applicable in early discovering animal work for biodistribution and potentially assisting benefit if an animal model is available.

For secreted proteins, that is those proteins we expect to exist extra-cellularly, for example, in blood, the clinical applications for determining protein concentration levels may be much greater. The largest class of secreted proteins, those naturally or expected to be present in the blood, are enzymes and are the most common gene therapy therapeutics in development. And of the enzyme gene therapy disease targets, hemophilia is highly targeted [12]. Hemophilia is a bleeding disorder in which the blood does not properly clot. This cause of the disease is primarily genetic, leading to low levels of Factor VIII or Factor IX. The levels of clotting factor present are typically directly correlated to severity of the disease. Therefore, for gene therapy therapeutics that are designed to increase levels of Factor VIII or Factor IX, quantitation of the amount of protein present during treatment will be a valuable pharmacodynamic marker. Hence, for these secreted transgenes, measurement either concentration or as more typically by enzyme activity is an important pharmacodynamic biomarker.

EtranaDez (AMT-061) is a gene therapy using AAV5 to deliver an enhanced functionality Factor IX gene [13] to liver cells. Because of the enhanced functionality of the transgene and direct correlation of enzyme activity levels to disease severity, measurement of transgene enzyme activity is a key aspect of clinical development. In clinical studies, the Factor IX activity levels is a secondary outcome of clinical trial after the primary outcome, which is a decrease in annualized bleeding rate (Clinicaltrails.gov NCT03569891, [14]). As described in detail later in this chapter, enzyme activity assays can be quantitative and provide direct levels of functional protein levels and thus can be used to assess clinical pharmacodynamics.

9.2.5 Human Transgene Protein Assessment for Expressed Therapeutics

The gene therapy transgene proteins discussed so far have been endogenous or naturally occurring proteins that the gene therapy therapeutic was designed to provide or enhance levels of. Another class of gene therapy transgene product is a designed therapeutic that is not a naturally occurring protein. This type of gene therapy therapeutic would potentially replace repeated doses of a therapeutic protein and instead, have patient's cells express the therapeutic *in vivo*. The utility of understanding the concentration of the transgene protein *in vivo* in this class of gene therapy transgenes is clear because the pharmacokinetics of the original protein therapeutics delivered with traditional methodologies (e.g. Injections at intervals) will be well understood and a therapeutic window (safety and efficacy) based on protein therapeutic concentrations over time will have been established. Therefore, the ability to measure gene therapy-expressed transgene protein concentrations to verify similar expression levels will be especially important for this class of gene therapy therapeutics.

RGX-314 is an AAV8 vector designed to deliver the transgene for a soluble anti-VEGF Fab and provide continuous anti-VEGF therapy. Approved anti-VEGF therapies to treat eye diseases, such as diabetic macular edema, myopic degeneration, and wet age-related macular degeneration, for example, Avastin, Lucentis, Eylea. RGX-314 delivers the gene sequence to produce the transgene protein with the sequence of Eylea. The pharmacokinetics of Eylea are described in the drug label [15] as well as in the literature [16]. Standard bioanalytical assays either validated or qualified can be used to quantitate the transgene protein therapeutic in plasma as well as other matrices that is possible to collect in human clinical trials, such as aqueous humor. The phase 2 clinical trial for RGX-314 includes a secondary outcome for the concentration of the RGX-314 transgene product in serum and aqueous humor (Clinicaltriais.gov NCT04514653, [17]).

9.2.6 Transgene Protein Assay Format Considerations

Prior to initiation of method development for an assay to quantitate transgene protein from gene therapy, several factors should be considered. The most important consideration will be the type of protein. For example, is the protein an enzyme, membrane-bound or membrane-associated, or secreted and expected to circulate in blood extra-cellularly. Choosing the platform will depend on those factors as well as availability of matrix. So an understanding of the assay will be needed for animal studies where a variety of tissues will be available vs. human clinical assessment, where typically blood is readily available and other matrices may be limited. Finally, understanding the purpose of the assay measurement will be critical in understanding platforms and sensitivity. For assessment of short-term increases in transgene levels, options will be varied. Although if one wants to compare long-term durability of expression, a protein assay for a transgene protein expressed in liver cell hepatocytes with lifetime of several hundred days may be different than the requirement to assess transgene protein expression levels in neurons with potentially a cell lifetime as long as the lifespan of the individual [18].

9.2.6.1 Immunoassays

The most common methodologies for quantitation of proteins are LBA. Formats such as ELISA. LBAs are precise, accurate and allow, in general, for very specific detection of a particular protein (Findlay and editor). In addition, the ability to amplify the signal with ELISA or ImmunoPCR LBA platforms, or detect very low levels of analyte using luminescence techniques (e.g. ElectroChemical Luminescence) makes immunoassays the most sensitive platform for protein quantitation. Immunoassays are well suited for the detection and quantification of transgene protein levels in biological matrices.

However, two key challenges can exist while using immunoassays for gene therapy transgene protein detection. First, while specificity of an immunoassay is typically high, the ability to distinguish between endogenous levels of a protein and transgene-expressed protein if they are very homologous may not be possible. Of course, no method can distinguish between identical proteins, but in some cases, the transgene protein sequence may be slightly different than the endogenous protein [2] and the ability to distinguish between them critical for understanding success of the gene therapy treatment.

The second challenge is that immunoassays may be subject to background interference from the biological matrices. While liquid matrices such as serum, cerebral spinal fluid, and aqueous or vitreous humor in general to not have significant interference that cannot be resolved [19], tissue homogenates from transgene protein biodistribution studies may result in sufficient interference to limit the limit of quantitation of an immunoassay [20].

9.2.6.2 Mass Spectrometry Assays

MS, coupled with liquid chromatography (LC) separation is an ideal platform for quantifying low molecular weight molecules. And the platform, coupled with protein digestion to peptides, allows this platform to be used to quantify proteins [21]. In general, LC–MS platforms are not as sensitive as immunoassays because of extensive background in biological matrices. However, use of an immunoaffinity step to isolate the protein of interest may allow the LC–MS methodology to achieve similar sensitivities as immunoassays [22]. In general, the specific application where LC–MS can excel over immunoassays for quantifying transgene proteins from gene therapy therapeutics is when two closely related proteins by amino acid sequence exist, one endogenous and one coming from the gene therapy transgene. For example, in the case of Factor IX protein, a modified gene and hence transgene protein may have higher potency, but with only slightly different amino acid sequence [13]. In this case, it may not be possible to use an immunoassay to distinguish between the wild-type and gene therapy transgene protein with an immunoassay; however, the two proteins will differ by mass, and so MS could quantify both the wild-type endogenous and a transgene protein if that was of interest.

9.2.6.3 Semiquantitative Assay Formats

Both immunoassays and MS methods are quantitative. The ability to prepare standards and quantitate unknown concentrations in biological matrices with total error well under 40% is standard and typical for these two platforms. However, immunoassay and MS platforms may not be sufficiently sensitive enough to detect all types of transgenes. The LC–MS and immunoassays platforms described are best suited for quantifying soluble proteins and so transmembrane proteins or membrane-associated proteins and enzymes may not be able to be detected with these platforms at sensitivities required. Transmembrane proteins or membrane-associated proteins and enzymes are not likely to be able to be detected without extensive pretreatment of samples. For membrane-bound or -associated transgene proteins, it may be sufficient to use semiquantitative methodologies to assess transgene protein biodistribution or relative expression levels.

Semiquantitative methods, techniques such as Western blotting and IHC are typical formats used in gene therapy transgene protein expression determinations. And the use of Western blots for quantification can nearly quantitative when appropriate controls, standards are used [23]. In addition, as access to extensive tissue samples is usually only available in Discovery or early preclinical studies, Western blots for transgene protein quantification can be rapid to implement, easy to carryout, and provide sufficient information to move programs toward clinical development.

IHC is potentially the most useful technique to not only determine transgene protein levels in various tissues but also provide information on transgene protein levels within a cell [24]. Early discovery transgene protein biodistribution studies may employ IHC as the primary technique for protein detection. Quantification of transgene protein by computer-supported image analysis can also be used; however, standardization, normalization, and within laboratory comparisons may be challenging [25]. Thus, IHC is useful and informative to show success of gene therapy transgene protein expression, and although it may not be absolutely quantitative, it can likely provide information to move a product forward to clinical development.

9.3 Transgene Protein Activity Determination

Many investigational gene therapy programs are targeting genetic diseases and aim to provide a normal copy of the gene to restore the function of the protein. A variety of bioanalytical approaches to measure the expression in terms of protein concentration have been described in earlier sections. However, when transgene protein is an active enzyme or cofactor for other enzymes, the biochemical function of the protein in the form of protein activity measurement is more relevant and the preferred approach. Enzyme activity assays measure the catalytic potential of the enzyme by either measuring the consumption of substrate or production of product over time with enzyme activity defined as moles of substrate converted per time unit [26], for example, nmol/hour/mL of plasma or nmol/hour/mg of protein. For coagulation factors such as factor VIII activity, the assay measures factor activity by either using a one-stage clotting assay or a chromogenic substrate assay. In this case, a unified international unit is used for activity measurement and the factor concentration is expressed in international units per milliliter (IU/mL) of human plasma, whereby 1 IU/mL is equivalent to 100% factor activity [27]. Several factors, such as temperature, pH, enzyme or factor concentration, substrate concentration, and the presence of any inhibitors or activators, can impact enzymatic reactions and the individual enzyme or cofactor dictates the assay conditions. This also leads to different measured activity values in different laboratories because of the slight differences in the assay conditions. From a bioanalytical perspective, a functional protein activity assay is challenging to develop, validate, and monitor as it can be impacted by many factors. This section will focus on addressing protein activity method development challenges and providing guidance on method optimization, validation, and the use of the methods to support sample testing. To limit the scope, the discussion will focus on using lysosomal storage enzymes as an example; however, the presented information can be applied to other protein activity assays, including protein activity for pharmacodynamic biomarkers.

9.3.1 Method Development Considerations

9.3.1.1 Enzyme Kinetics, the Initial Rate of Reaction, and Substrate Concentration

Most enzyme reactions except for allosteric enzymes involving a single substrate often follow Michaelis–Menten kinetics. The kinetics model takes the form of an equation:

$$E + S \leftrightarrow [ES] \rightarrow E + P$$

Here, E is the enzyme, S is the substrate, ES is the enzyme–substrate complex, and P is the product. The enzyme interacts with the substrate to form an enzyme–substrate complex and leads to the generation of the product and the release of the enzyme. The Michaelis–Menten equation as shown below shows how the initial rate of reaction (V_0) depends on the substrate concentration $[S]$.

$$V_0 = \frac{V_{max} * [S]}{K_m + [S]}$$

Here, V_{max} represents the maximum rate at saturating substrate concentration for a given enzyme concentration. K_m is the Michaelis constant. At half V_{max}, K_m is equal to the $[S]$ Figure 9.1 [28].

Michaelis–Menten kinetics only applies to the initial velocity, therefore, the reaction conditions must be chosen under the initial rate conditions such that the activity is proportional to the concentration of enzyme over the entire dynamic range. One underlying assumption is that substrate is not the limiting factor, and the substrate concentration is sufficient to remain almost unchanged during the reaction. During the initial method development stage, it is essential to perform experiments to determine the initial rate of reaction and generate a substrate saturation curve to estimate the K_m and V_{max}. The initial rate window can be determined by generating a reaction progress curve via mixing the enzyme and its substrate and measuring the time course of product generation. A given enzyme may have a different reaction progress curve in different matrices and time-course experiments will need to be determined in the matrix of interest, Figure 9.2 [29]. A substrate saturation curve can then be generated by varying the substrate concentration in reactions occurring within the initial rate time window to determine

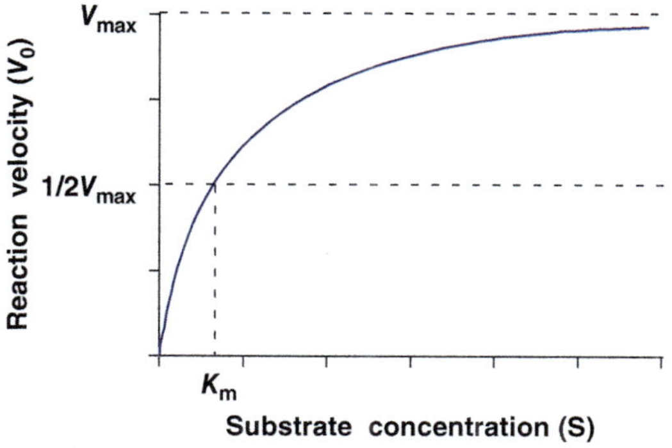

Figure 9.1 Michaelis–Menten Equation. "V_0" denotes reaction velocity and "$[S]$" denotes substrate concentration. V_{max} is the highest reaction rate that can be attained for a given enzyme concentration. K_m is the concentration of substrate at which a half-maximal reaction rate occurs.

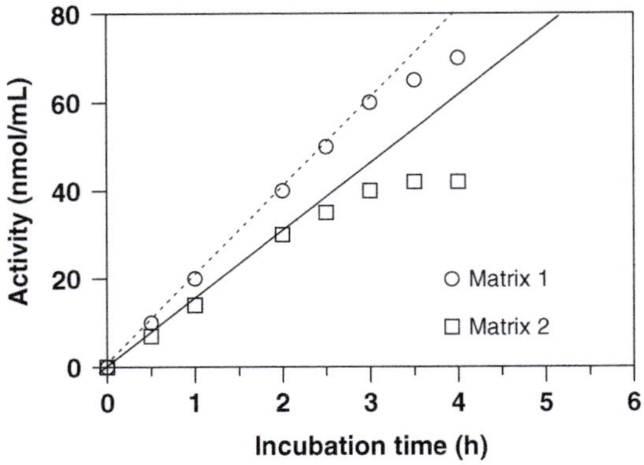

Figure 9.2 Time course of an enzyme activity expressed in nmol/mL of the matrix. Matrix 1 is represented by an open circle with a dashed line fitted through the linear range of the reaction. Within 3 hours, the reaction is linear. Matrix 2 is represented by an open square with a solid line fitted through the linear range of the reaction. Within 2 hours, the reaction is linear.

the substrate concentration for further method development. Ideally, it is best to conduct an enzyme assay under substrate-saturating conditions. However, some substrates are quite costly like those for lysosomal storage enzymes, thus an alternative approach is to conduct the assay using a final concentration of K_m as shown in Ou et al. [30] to reduce overall cost. It is important to note that the velocity should remain relatively stable throughout the whole incubation time when using non-saturating substrate concentrations.

9.3.1.2 Reference Standard

The source of reference standards for assay development is one of the biggest challenges for protein activity assay development. The World Health Organization (WHO) established biological reference materials (called International Standards) as global reference standards for various antibodies, antigens, blood products and related substances, cytokines, growth factors, endocrinological substances, and *in vitro* diagnostic devices [31]; however, WHO international standards are not available for all biological materials such as lysosomal storage enzymes.

For lysosomal storage enzymes, 4-Methylumbelliferone (4MU) derived artificial substrates have been widely used for measuring lysosomal enzyme activity [4, 7]. The enzymatic reaction occurs at acidic pH 4.5–5.5 and the substrate is hydrolyzed by the enzyme to release 4MU, which is highly fluorogenic at pH 10–11. The measured fluorescence intensity is proportional to the enzyme function. One common reagent that is used as the reference standard for activity calculation is the free fluorophore, 4MU [29, 32]. Since the reference standard (4MU) fluorescence is independent of the enzymatic reaction, for quantitative enzyme activity measurement the use of recombinant enzyme is recommended during method development and as additional controls to monitor the assay performance.

During the early development phase when the assay is not well-controlled, a non-parallel relationship between enzyme activity and 4MU and a shift of x-intercept of enzyme curves are often observed (Figure 9.3a). The use of recombinant enzyme can help to characterize the assay, to use as a tool for assay optimization to ensure consistency and parallel relationship between enzyme reaction and 4MU standard curve, and to help define the 4MU quantifiable range (Figure 9.3b). It is not

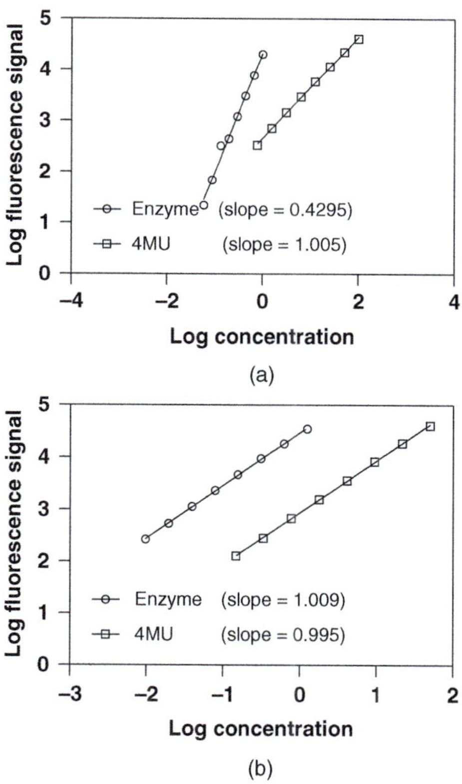

Figure 9.3 Enzyme activity. Lysosomal storage enzyme activity in relation to 4MU for activity calculation. A recombinant enzyme can be used during method development to optimize assay and to define the enzyme activity quantifiable range. (3A) Nonparallel relationship between enzyme activity and 4MU. (3B) Parallel relationship between enzyme activity and 4MU. The enzyme is represented by an open circle with a solid line fitted through the enzyme curve. 4MU is represented by an open square with a solid line fitted through the 4MU.

Table 9.1 Fluorescent background.

	Vendor 1 (substrate only)	Vendor 2 (substrate only)	Assay diluent	Human plasma at MRD 6 with vendor 1 substrate	Human plasma at MRD 10 with vendor 1 substrate
Fluorescent signal	2106	1462	389	1884	1952

It is necessary to evaluate the fluorescence signal of all reagents used in the enzymatic assay. The table illustrates the fluorescent background of substrate 4-Methylumbelliferyl α-D-galactopyranoside used in a-Gal A activity assay from Sigma (vendor 1), Cayman (vendor 2), assay diluent, the minimum required dilution (MRD) 6 and MRD 10 plasma diluted in assay diluent containing substrate from vendor 1.

recommended to assign an activity unit per enzyme concentration and use the recombinant enzyme as the reference standard for activity calculation as the production of commercially available enzymes is not well controlled and hence the activity of each enzyme can vary from lot to lot. However, if assigning an activity unit per enzyme lot helps to reduce assay variability, an extensive characterization of a respective lot of enzymes should be implemented and a lot-to-lot recombinant enzyme bridging strategy should be established.

When generating the reference standard curve, it is essential to prepare the standard curve containing similar reaction compositions as test samples as matrix and substrate may have a different background than assay diluent. If matrix effects are not an issue, standards can be prepared in buffered assay diluent. However, it is necessary to include substrate in the standard curve preparation as substrate tends to have a different fluorescence background as assay diluent and background can also vary from vendor to vendor and lot to lot. Higher substrate background may lead to an overestimation of the enzyme activity, Table 9.1. Additional factors to consider for the standard curve are whether to institute background subtraction and the curve fit of choice. Background subtraction is typically applied. Linear and log-log linear fit are two commonly used curve fit models for enzyme activity assay. Though the provided examples are specific to lysosomal storage enzymes, the same principle applies to all quantitative enzyme assays regarding the use of recombinant protein, curve fit, and ensuring the reference standard contains similar reaction components as test samples.

9.3.1.3 Sample Processing

Enzyme activity assay can be measured in different types of matrices, for example, plasma, serum, leukocyte, tissues, etc. Sample collection and sample processing are important factors that can affect enzyme activity, but these factors are often overlooked during method development. The types of blood collection tubes and downstream processing conditions such as centrifugation speed, storage temperature, and duration between collection to plasma/serum separation should be evaluated early on to determine the most suitable and reproducible conditions [33]. The evaluation should include different donors and the reproducibility of the measured enzyme activity should trump the conditions that give the highest measured enzyme activity, Figure 9.4. If the matrix for enzyme assay is leukocyte, it is important to ensure laboratories are well-trained and qualified for leukocyte preparation. Hemolysis can have an impact on downstream enzyme activity and protein concentration determination for leukocyte lysate. Additionally, the different means of transportation and shipping times should be evaluated to ensure good quality of leukocytes can be generated at the processing laboratory.

During sample preparation for the activity assay measurement, the stability of the enzyme under assay conditions should be evaluated and monitored. For enzyme activity in leukocyte and tissue lysates, heat generation during the leukocyte sonication or tissue homogenization process may inactivate the enzyme. It is important to evaluate the impact of temperature changes and derive a process to best preserve the enzyme function.

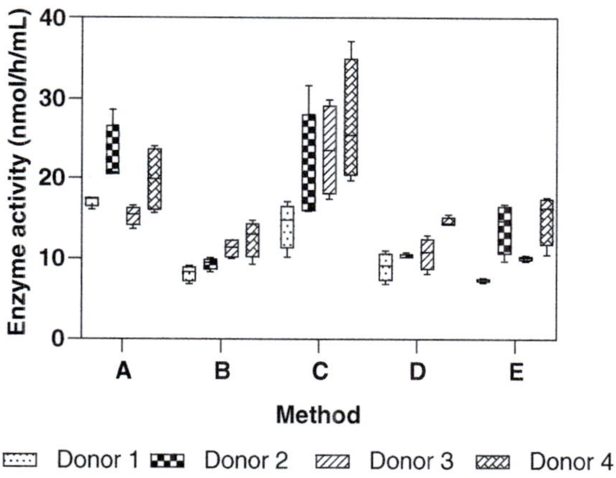

Figure 9.4 Plasma sample processing comparison. Four donors (represented by different filled patterns in the legend) with whole blood collected in K_2EDTA were subjected to different plasma preparation conditions. A: spin within 8 hours at 2400 revolutions per minute (rpm) for 10 minutes; B: spin within 8 hours at 3400 rpm for 15 minutes; C: spin within 24 hours at 2400 rpm for 10 minutes; D: spin within 24 hours at 3400 rpm for 15 minutes. E: spin within 24 hours at 2465 rpm for 20 minutes. Each sample was subjected to four independent measurements. Conditions B, D, and E were comparable with better reproducibility. Conditions A and C yielded higher enzyme activity but poor precision.

9.3.1.4 Buffers and Incubation Temperature

Assay buffer is enzyme-dependent with slightly different compositions reported in different literature for a given enzyme [29, 34]. During the method development, it is necessary to further optimize the buffer compositions and buffer pH using the method described in the literature as a guide in the tested matrix. Lysosomal storage enzymes are naturally found in the lysosomes, pH optimization is critical for this type of enzymatic assay. Additionally, some enzymatic assays include the addition of surfactants, such as Tween-20 and Triton X-100, to improve assay reproducibility. The impact of changes in surfactant percentage should be evaluated as some assays are sensitive to small variations in surfactant amount. For example, Triton X-100 may help to improve performance consistency due to ease of pipetting, but the window of optimal Triton X-100 percentage may be very narrow, Figure 9.5. Since the enzyme activity is measured by fluorescence signal and correlated to 4MU for activity calculation, the preparation of assay buffer will be critical as it can impact the overall calculated enzyme activity. In addition to buffer composition and pH, the reaction temperature is also an important factor. Most enzymatic assay occurs at body temperature (37 °C). The impact of temperature changes with respect to enzyme activity should also be evaluated during development.

9.3.1.5 Assay Dynamic Range, Minimum Required Dilution, Matrix Interference, and Parallelism

Many gene therapy programs are targeting genetic diseases that lack a functioning protein. These patients have little to no active protein present in their system. Therefore, a quantitative and sensitive assay is likely needed to support drug

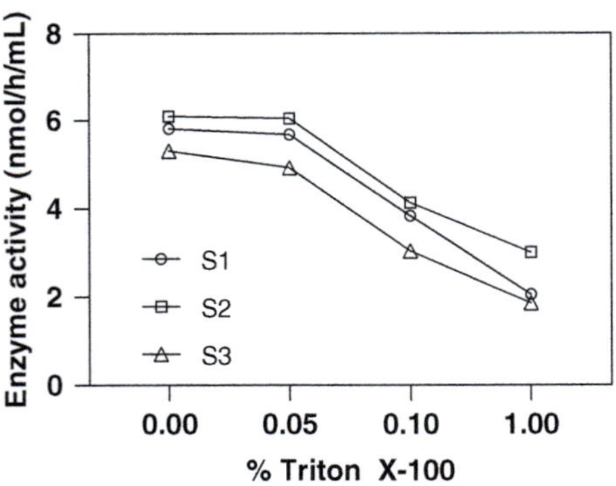

Figure 9.5 Impact of surfactants. Three donors represented by S1 (open circle), S2 (open square), and S3 (open square) were tested in assay diluent containing different Triton-X %. A significant shift in enzyme activity when the activity was measured in 0.05% vs. 0.1% Triton X-100.

development. In terms of assay dynamic range, published data can help to define the initial target assay range and the required assay sensitivity during method development. Some enzymes may require many folds over healthy donors' normal range to achieve cross-correction whereby the extracellular enzymes are taken up by the enzyme-deficient cells. An assay with a wider dynamic range will be helpful in this scenario.

Assay matrices are the trickiest component for assay development and careful consideration of this variable in terms of the MRD, matrix interference, and parallelism should be evaluated during development. If matrix effects are not an issue, standards can be prepared in buffered assay diluent like the majority of biomarker assays instead of the matrix of test samples. To apply this approach, it is crucial to demonstrate that the protein activity–response relationship in the sample matrix is parallel to that of the buffered assay diluent through parallelism evaluation. Parallelism experiments are performed through serial dilutions of a high-activity sample in the buffered assay diluent using multiple individual matrix lots (≥3 lots during development). The consistency of the back-calculated in-range concentrations in a dilution series not exceeding 30% coefficient of variation (CV) or a relative bias approach by comparing serially diluted samples to the first few dilutions to assess acceptable %Bias (20–30%) can be used to assess parallelism. The latter is the preferred method. Additionally, parallelism experiments also help to define the MRD of the assay whereby the lowest dilution factor alleviates the matrix interference sufficiently and can achieve the required assay sensitivity. In cases when parallelism cannot be performed due to a lack of high protein activity samples, dilutional linearity should be tested in a similar manner to parallelism.

For some protein activity assays, a matrix assay diluent is needed. As reported in the literature, lysosomal sulfatases are strongly inhibited by sulfates, phosphates, and metal ions [29]. To reduce the inhibitory effects, one option is to perform high sample dilutions to reduce the inhibitory effect, but this reduces the assay sensitivity as well. An alternative approach is to use a heat-inactivated matrix with enzyme deactivated as the background matrix for the standard curve and sample diluent. Though an inhibitory effect is present, the use of matrix assay diluent will help to reduce the variability due to matrix interference. MRD can be determined based on the targeted assay sensitivity and parallelism evaluation in the matrix assay diluent.

9.3.1.6 Specificity and Selectivity

The nonspecific activity of another enzyme or some matrix components can result in a reduction or enhancement of the overall response, therefore, evaluation of specificity and selectivity are essential during assay development. The demonstration of specificity and selectivity should be evaluated in the context of the assay as it can be difficult to demonstrate absolute specificity and selectivity for enzyme activity assays. For example, the lysosomal enzyme α-N-acetylgalactosaminidase has nonspecific activity toward the artificial substrate used for α-galactosidase A activity assay. To ensure specificity, N-acetylgalactosamine (GalNAc) inhibitor is included in the α-galactosidase A activity assay to eliminate any potential contribution from N-acetylgalactosaminidase [35]. The optimal concentration of the inhibitor should be determined during assay optimization to ensure the generated activity is specific to the enzyme of interest. In some other cases, a second enzyme is needed to cleave the intermediate substrates generated by the enzyme of interest, like the two steps iduronidate-2-sulfatase enzymatic reaction. Understanding the specificity of the enzymes involved in the reaction is needed. In this case, testing the ability of the first enzyme to generate the intermediate substrate is an important factor. To perform enzyme specificity evaluations, heat-inactivated plasma with and without the addition of recombinant enzymes can be a viable approach.

As for selectivity, this parameter measures the ability of the assay to discriminate the analyte unequivocally in the presence of components that may be expected to be present in the sample. To perform enzyme selectivity evaluations, recombinant enzymes at a minimum double the endogenous activity can be used to spike into individual donors (>5 donors during development) for this evaluation.

9.3.1.7 Quality Controls (QCs)

When a well-characterized enzyme or protein reference standard is available, quality controls (QCs) generation is straightforward and can follow the approaches used for the majority of biomarker assays. When a reference standard is based on free fluorophore like 4MU, which is independent of the enzyme activity, the selection of the QC levels will be based on enzyme activity rather than the concentration of the enzyme. The use of recombinant enzymes can help to define the range of quantitation and to set the upper limit of quantitation (ULOQ) and lower limit of quantitation (LLOQ). It should be noted the raw responses of the recombinant enzyme should fall within the raw response of the 4MU standard curve, Figure 9.6. Based on the enzyme activity of ULOQ and LLOQ, the remaining 3 QCs can be set at about 75–85% of ULOQ for high-quality control (HQC), at about 3 times the LLOQ for low-quality control (LQC) and middle-quality control (MQC) at mid-range of the standard curve. Sometimes it may be helpful to set two LLOQs and finalize the selection during method validation. It may not be possible to obtain donor samples with enzyme activity that can span either range of quantitation, a combination of spiked QC with the recombinant enzyme in assay diluent and healthy donor samples can be used to establish the QCs for the assay. There are different ways QCs can be generated.

a) If no matrix interference is observed at the selected MRD and in a rare matrix like cerebrospinal fluid, spiking recombinant enzyme into assay diluent can be used to establish all five levels of QC samples but the inclusion of individual donors (>3 donors) is important during validation to assess assay precision and stability in the tested matrix.
b) If no matrix interference is observed at the selected MRD, a combination of recombinant enzyme spiked into assay diluent and healthy donors is another option. For example, ULOQ, HQC, and LLOQ are generated using recombinant enzymes in assay diluent with MQC and LQC from two individual donors.
c) If matrix interference is observed and the assay is conducted at a fixed percent matrix with high endogenous enzyme activity, spiking recombinant enzyme into the heat-inactivated matrix is an option. The inclusion of individual donors (>3 donors) to assess assay precision and stability in the tested matrix will be important during validation.

Once QCs are established, samples should be tested for reproducibility assessment and the targeted enzyme activity range for each QC level prior to method validation.

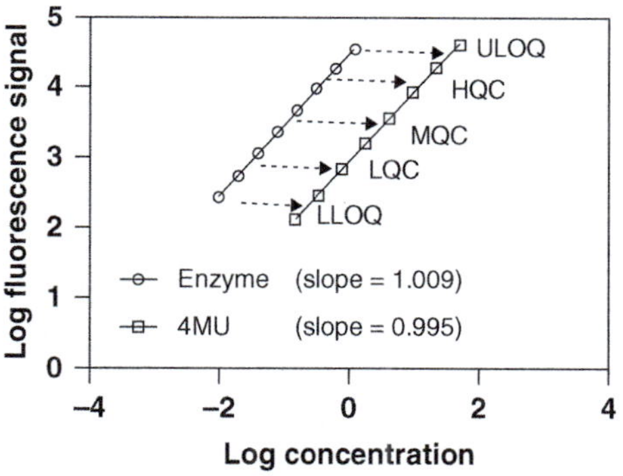

Figure 9.6 Enzyme activity response. Lysosomal storage enzyme activity in relation to 4MU for activity calculation. A recombinant enzyme can be used during method development to define the enzyme activity quantifiable range and define the ULOQ and LLOQ of the assay. The remaining QCs (HQC, MQC, and LQC) can be set accordingly. The enzyme is represented by an open circle with a solid line fitted through the enzyme curve. 4MU is represented by an open square with a solid line fitted through the 4MU.

9.3.2 Method Validation

The intended use of the biomarker data drives the level of rigor of the assay validation and is relative to the stage of drug development. Typically, proof-of-concept studies that occur earlier in drug development will require the least rigor and a well-characterized assay should be sufficient. When biomarker data are submitted for regulatory decision-making for approval, safety, or labeling, the bioanalytical methods should be fully validated as indicated in FDA guidance [10]. The published FDA and EMA [36, 37] method validation guidance for the industry can be used as a guide to validate the method. Since enzymatic assay is a biomarker assay, fit-for-purpose assay acceptance per the individual assay can be adopted. Parameters such as standard curve, range of quantification, precision (there is no accuracy evaluation for gene therapy enzyme activity assays unless a WHO reference standard is available), dilution linearity, parallelism, specificity, selectivity, matrix interference (hemolysis, lipemia), short-term and long-term stability using recombinant protein in assay diluent and individual donors. Stability using recombinant protein in assay diluent and donors serves a different purpose. One is to track the stability of recombinant protein as quality control samples and the individual donors track the stability of the endogenous enzyme. In addition to the recommended parameters in the guidance, the establishment of healthy donor ranges for both male and female populations with a minimum of 30 donors for each gender should be included as well. The mean donor range can then be calculated from the obtained data and used in clinical studies to compare transgene-expressed enzyme activity relative to the mean normal for the respective enzyme.

To support assay acceptance for future testing, enzyme activity ranges for each QC level can be calculated using a variety of approaches. Using a bias of ±20–35% of the measured mean to establish the range is an option. Alternatively, the acceptable activity range for each QC level can be calculated using mean ± 2SD (95% confidence interval), mean ± 2.5SD (97% confidence interval), mean ± 3SD (99.7% confidence interval), or mean ± 3.3SD (99.9% confidence interval). Which SD to select will depend on the assay variability and the selected SD should not yield a bias greater than 50% as compared to the established mean. For plate acceptance during sample testing, the standard approach of using ≥67% of QCs and ≥50% of QCs per level meeting the established enzyme activity range acceptance criteria can be applied.

9.4 Summary

Quantification of gene therapy transgene protein expression levels and function is an active and evolving area of bioanalytical science. And the methodologies and platforms available for assessing protein expression are very dependent on phase of development, whether in animals where obtaining relevant samples is typically easier to human clinical studies where the results are the most relevant to therapeutic development, but where relevant matrices are sometimes more difficult to routinely obtain. In addition, the techniques and platforms use to quantitate gene therapy transgene proteins are very dependent on the class and type of protein expressed. Gene therapy products that are expressed as soluble proteins in standard matrices, such as blood, do not typically require new methodologies or platforms although functional assessment to ensure active protein is still specific to the protein product itself and can be challenging.

Gene therapy therapeutics however are novel in their ability to also produce transgene proteins that are expressed intracellularly, are membrane-associated enzymes or even transmembrane proteins; classes of protein products that are not typically possible to be delivered to patients with traditional biologic therapeutics. Hence, for these novel classes of transgene proteins, novel bioanalytical methods and platforms, as well as potentially surrogate matrices analysis are likely necessary as highlighted in the case studies in this chapter.

Finally, perhaps the most targeted class of proteins by gene therapy therapeutics are enzymes. For this class of proteins, it is critical to understand not just the presence of absence of the enzyme after gene therapy treatment, but to carry out accurate and quantitative functional assessment. As described in this chapter, expansive and novel methodologies and techniques can be developed enabling accurate and robust transgene protein activity measurements. Thus, using the bioanalytical techniques as described can result in validated methodologies providing robust data to assess patient treatment and progress the therapeutic product development.

References

1 Zolgensma Package Insert https://www.fda.gov/media/126109/download (accessed 29 October 2022).

2 Sleczka, B.G., Levesque, P.C., Adam, L.P. et al. (2020). LC/MS/MS-based quantitation of pig and human S100A1 protein in cardiac tissues: application to gene therapy. *Anal. Biochem.* 602: 113766.

3 Zaworski, P., von Herrmann, K.M., Taylor, S. et al. (2016). SMN protein can be reliably measured in whole blood with an electrochemiluminescence (ECL) immunoassay: implications for clinical trials. *PLoS One* 11: e0150640.

4 Alves, C.R.R., Zhang, R., Johnstone, A.J. et al. (2020). Whole blood survival motor neuron protein levels correlate with severity of denervation in spinal muscular atrophy. *Muscle Nerve* 62: 351–357.

5 Czech, C., Tang, W., Bugawan, T. et al. (2015). Biomarker for spinal muscular atrophy: expression of SMN in peripheral blood of SMA patients and healthy controls. *PLoS One* 10: e0139950.

6 Rodrigues, G.A., Shalaev, E., Karami, T.K. et al. (2019). Pharmaceutical development of AAV-based gene therapy products for the eye. *Pharm. Res.* 36: 29.

7 Package Insert. Luxterna. https://www.fda.gov/media/109906/download (accessed 29 October 2022).

8 Boulanger, A. and Redmond, T.M. (2002). Expression and promoter activation of the Rpe65 gene in retinal pigment epithelium cell lines. *Curr. Eye Res.* 24: 368–375.

9 Chucair-Elliott, A.J., Elliott, M.H., Wang, J. et al. (2012). Leukemia inhibitory factor coordinates the down-regulation of the visual cycle in the retina and retinal-pigmented epithelium. *J. Biol. Chem.* 287: 24092–24102.

10 Machalińska, A., Kawa, M.P., Pius-Sadowska, E. et al. (2013). Endogenous regeneration of damaged retinal pigment epithelium following low dose sodium iodate administration: an insight into the role of glial cells in retinal repair. *Exp. Eye Res.* 112: 68–78.

11 Moiseyev, G., Crouch, R.K., Goletz, P. et al. (2003). Retinyl esters are the substrate for isomerohydrolase. *Biochemistry* 42: 2229–2238.

12 Bolous, N.S., Bhatt, N., Bhakta, N. et al. (2022). Gene therapy and hemophilia: where do we go from here? *J. Blood Med.* 13: 559–580.

13 Simioni, P., Tormene, D., Tognin, G. et al. (2009). X-linked thrombophilia with a mutant factor IX (factor IX Padua). *N. Engl. J. Med.* 361: 1671–1675.

14 HOPE-B: Trial of AMT-061 in severe or moderately severe hemophilia B patients. https://clinicaltrials.gov/ct2/show/NCT03569891 (accessed 29 October 2022).

15 EYLEA Package insert https://www.accessdata.fda.gov/drugsatfda_docs/label/2011/125387lbl.pdf (accessed 29 October 2022)

16 Avery, R.L., Castellarin, A.A., Steinle, N.C. et al. (2014). Systemic pharmacokinetics following intravitreal injections of ranibizumab, bevacizumab or aflibercept in patients with neovascular AMD. *Br. J. Ophthalmol.* 98: 1636–1641.

17 ClinicalTrials.gov. NCT04514653 RGX-314 gene therapy administered in the suprachoroidal space for participants with neovascular age-related macular degeneration (nAMD) (AAVIATE). https://clinicaltrials.gov/ct2/show/NCT04514653 (accessed 29 October 2022).

18 Muhuri, M., Levy, D.I., Schulz, M. et al. (2022). Durability of transgene expression after rAAV gene therapy. *Mol. Ther.* 30: 1364–1380.

19 Gorovits, B., McNally, J., Fiorotti, C. et al. (2014). Protein-based matrix interferences in ligand-binding assays. *Bioanalysis* 6: 1131–1140.

20 Sugimoto, H., Chen, S., and Qian, M.G. (2020). Pharmacokinetic characterization and tissue distribution of fusion protein therapeutics by orthogonal bioanalytical assays and minimal PBPK modeling. *Molecules* 25: 535.

21 Lipton, M.S. and Paša-Tolic, L. (ed.) *Mass Spectrometry of Proteins and Peptides: Methods and Protocols*, 2e. Springer.

22 Palandra, J., Psychogios, N., and Neubert, H. (2022). Application of immunoaffinity mass spectrometry (IA-MS) for protein biomarker quantification. *Methods Mol. Biol.* 2466: 111–119.

23 Barkovits, K., Pfeiffer, K., Eggers, B. et al. (2021). Protein quantification using the "Rapid Western Blot" approach. *Methods Mol. Biol.* 2228: 29–39.

24 Ramos-Vara, J.A. (2017). Principles and methods of immunohistochemistry. *Methods Mol. Biol.* 1641: 115–128.

25 Braun, M., Kirsten, R., Rupp, N.J. et al. (2013). Quantification of protein expression in cells and cellular subcompartments on immunohistochemical sections using a computer supported image analysis system. *Histol. Histopathol* 28: 605–610.

26 Bisswanger, H. (2014). Enzyme assays. *Perspect. Sci.* 1: 41–55.

27 Raut, S. and Hubbard, A.R. (2010). International reference standards in coagulation. *Biologicals* 38: 423–429.

28 Segel, I.H. (2013). Enzyme kinetics. In: *Encyclopedia of Biological Chemistry*, 216–220. Elsevier.

29 Voznyi Ya, V., Keulemans, J.L., and van Diggelen, O.P. (2001). A fluorimetric enzyme assay for the diagnosis of MPS II (Hunter disease). *J. Inherit. Metab. Dis.* 24: 675–680.

30 Ou, L., Herzog, T.L., Wilmot, C.M. et al. (2014). Standardization of α-L-iduronidase enzyme assay with Michaelis–Menten kinetics. *Mol. Genet. Metab.* 111: 113–115.

31 WHO. Expert Committee on biological standardization sixtieth report. http://www.who.int/bookorders.

32 Wenger, D.A. and Luzi, P. (2020). The lysosomal diseases testing laboratory: a review of the past 47 years. *JIMD Rep.* 54: 61–67.

33 Burin, M., Dutra-Filho, C., Brum, J. et al. (2000). Effect of collection, transport, processing and storage of blood specimens on the activity of lysosomal enzymes in plasma and leukocytes. *Braz. J. Med. Biol. Res.* 33: 1003–1013.

34 Kim, C., Seo, J., Chung, Y. et al. (2017). Comparative study of idursulfase beta and idursulfase in vitro and in vivo. *J. Hum. Genet.* 62: 167–174.

35 Andrade, J., Waters, P.J., Singh, R.S. et al. (2008). Screening for fabry disease in patients with chronic kidney disease: limitations of plasma α-galactosidase assay as a screening test. *Clin. J. Am. Soc. Nephrol.* 3: 139–145.

36 FDA, CDER. (2018). Bioanalytical method validation guidance for industry biopharmaceutics bioanalytical method validation guidance for industry biopharmaceutics contains nonbinding recommendations. http://www.fda.gov/Drugs/GuidanceComplianceRegulatoryInformation/Guidances/default.htmand/orhttp://www.fda.gov/AnimalVeterinary/GuidanceComplianceEnforcement/GuidanceforIndustry/default.htm.

37 EMA guideline of method validation. https://www.ema.europa.eu/en/documents/scientific-guideline/guideline-bioanalytical-method-validation_en.pdf (accessed 29 October 2022).

10

Substrate and Distal Pharmacodynamic Biomarker Measurements for Gene Therapy

Liching Cao[1], Kai Wang[2], John Lin[3], and Venkata Vepachedu[4]

[1] *Biomarker and Bioanalytical Sciences, Sangamo Therapeutics, Richmond, CA, USA*
[2] *Immunoassay Bioanalysis and Biomarker, GlaxoSmithKline, Collegeville, PA, USA*
[3] *Bioanalytical and Biologics Services, Frontage Laboratories, Exton, PA, USA*
[4] *Preclinical Sciences and Translational Safety, Johnson and Johnson Innovative Medicine, Spring House, PA, USA*

10.1 Introduction

The analysis of pharmacodynamic (PD) biomarkers plays an essential role in drug development. It is used in preclinical and clinical studies to provide information on the pharmacologic effects of a drug on its target and has become a critical component of decision-making processes in drug development. Most clinical research phase studies follow a typical series of studies, starting with Phase 1 studies primarily in healthy volunteers to test the safety of the investigational drugs [1]. In contrast, gene therapy studies often combine Phases 1 and 2 due to the use of recombinant adeno-associated virus (AAV) where redosing is not possible currently for gene delivery and in disease populations. Because of the inclusion of patients in early-phase clinical trials, PD biomarkers are often adopted early on to demonstrate proof of mechanism of action, assess pharmacological response, confirm target engagement, and provide evidence of clinical benefit. In some diseases, the specific PD biomarker can also serve as a surrogate endpoint for approval, for example, plasma phenylalanine for Phenylketonuria and complete/near-complete clearance of globotriaosylceramide (GL-3) inclusions in biopsied renal peritubular capillaries for Fabry disease [2]. Therefore, the ability to measure a PD biomarker for its intended purpose with specificity, relative accuracy,

Drug Development for Gene Therapy: Translational Biomarkers, Bioanalysis, and Companion Diagnostics, First Edition. Edited by Yanmei Lu and Boris Gorovits.

precision, and sensitivity is essential for making critical study decisions, supporting regulatory submissions, and facilitating regulatory approval of new drugs.

The bioanalytical methods for the detection of PD biomarkers are broad and utilize a wide variety of bioanalytical technologies to generate data for the intended use. Technologies such as liquid chromatography with tandem mass spectrometry (LC-MS/MS), histologic and imaging techniques, functional protein activity monitoring tools, a spectrum of immunological assays, and molecular techniques like reverse transcription-quantitative polymerase chain reaction (RT-qPCR) and ribonucleic acid (RNA) sequencing for messenger ribonucleic acid (mRNA) detection/quantitation of downstream target expression are commonly used for method development. In terms of data outputs, a PD biomarker whether it is used to support an endpoint or as an exploratory biomarker typically falls under one of the four categories, definitive quantitative, relative quantitative, semi-quantitative, and qualitative [3]. The intended use of the PD biomarker will guide the data outputs and whether a qualitative or quantitative method is needed. The majority of PD biomarker assays that utilize reference standards to determine analyte concentration or activity are considered relative quantitative assays because the reference standards are likely synthetic or recombinant materials, not well characterized, or not fully representative of the endogenous form (e.g. glycosaminoglycans (GAGs) measurement using LC-MS/MS for mucopolysaccharidosis type I lysosomal storage disease). Data generated from molecular biology techniques (RNAseq, NanoString) and histological approaches to quantify PD biomarkers can be relatively quantitative, semi-quantitative, and qualitative as these measurements typically do not employ the use of a calibration standard. For example, for target engagement evaluation, housekeeping genes are typically used for normalization to evaluate the relative gene expression and a semi-quantitative scoring scale is commonly used for histological assays. However, some data from histological evaluations can be considered definitive quantitative as data are presented as the absolute count of the PD biomarker within the evaluated tissues. Because of the diverse data outputs, it is not possible to have one recommendation of parameters to be included for PD biomarker assays during method qualification or validation. The different data types and different stages of the drug development cycle will guide the method qualification or validation strategies and will require differential consideration and plan on the performance evaluation parameters to be included during the method qualification or validation.

This chapter will attempt to describe the current industrial practices and challenges encountered during method development and qualification/validation. Additionally, the chapter will provide recommendations or solutions to challenges encountered under the specific context of the substrate and distal PD biomarker measurements for gene therapy. The gene therapy field continues to evolve, and new bioanalytical technologies will be needed to support these new modalities.

More integrated technologies will likely be adopted, e.g. omics and big data analysis in preclinical models, clinical trials, and precision medicine. Lastly, the chapter will also address the use of current regulatory guidance in method validation and address regulatory guidance gaps associated with PD biomarker assays.

10.2 Technologies to Quantify Substrate and Distal PD Biomarker

10.2.1 Liquid Chromatography/Tandem Mass Spectrometry (LC-MS/MS)

10.2.1.1 Method Development Challenges and Resolutions

The advancement of biopharmaceutical and medicinal technologies with the aim to constantly improve the quality and extend the longevity of human life, revolutionary science, and technology in the research and development of cutting-edge innovative technologies are in the pace of exploring fashion [4–6]. Biomarkers, sometimes called biomolecules (including cytokines, chemokines, and growth factors) are important indicators of physiological or pathological processes and play a key role in clinical decision-making and are major targeted analytes for quantitation in drug development and translational medicine [7–10]. Because of the complexity, dynamic nature, and interactive variability of physiological and/or pathological conditions among human subjects, precise, accurate, and quantitative measurement of targeted endogenous analytes has been challenging. Biomarkers exist in various forms, including small molecule biomarkers, large molecule protein biomarkers, and metal ions. Small molecule biomarkers are often quantitated using LC-MS platform. As a mature technology, LC-MS/MS is a powerful tool for small molecule biomarker quantitation. However, biomarker quantitation using the LC-MS platform faces some common challenges that can also be found in the developmental stage for drug entities and metabolite methods. The most common challenge is the stability concerns of the biomarkers in biological matrices. Biomarkers that experience stability issues should be carefully investigated before setting up the assay, to determine the fundamental cause of the stability issues. Biomarkers such as methylcobalamin (Vitamin B12), coproporphyrin I (CP-I), and III (CP-III) are known to be unstable under exposure to light [11, 12]. Developing and implementing proper handling procedures for study samples during sample collection, storage, and processing is highly recommended. During the sample collection and storage, amber tubes must be used, and throughout the sample collection and sample analysis workflow, only yellow light should be used to protect the analytes from degradation. If stability issues arise from the enzyme activity, acidifying the study samples by adding citric acid or Sorenson's phosphate buffer should be considered. For stability issues that result from nonspecific binding, especially in low-protein bio-matrices such as urine or cerebrospinal fluid (CSF), introducing external protein sources such as bovine albumin (BSA), or surfactants such as Triton-X, Tween-20, or Tween-80 can help overcome the nonspecific binding. The surfactants can be harmful to the performance of the LC-MS instruments and hence are not recommended when other approaches are available. When oxidation of the compound is predicted, ascorbic acid (Vitamin C) can be added. Biomarkers are endogenous in nature. Unlike the stability behavior of drug entities or their metabolites, biomarker concentration sometimes increases rather than decreases, when demonstrating their stability problems. One good example is some lipids such as lysophosphatidic acids (LPAs) [13, 14]. When testing these biomarkers in biological matrices, their stability should be carefully evaluated. To prevent the concentration from falsely elevating, the samples should be collected in chilled conditions with shortened sample collection and processing time. The bioanalytical sample processing should be performed as fast as possible, samples should be stored in chilled conditions when taken out of the freezer and returned to the freezer immediately after samples are pipetted. Under rare conditions, when the biomarkers are extremely unstable after sample collection, the samples can be processed at the clinical site. For small molecule biomarker sample collection, adding organic reagents to the biological samples can help stabilize selected biomarkers. Some stabilizers can be pre-added to the sample collection tubes. For example, the BD™ designed P800 blood collection system is designed to accurately measure and stabilize metabolic markers, such as glucagon-like peptide-1 (GLP-1), Glucagon, and gastric inhibitory polypeptide (GIP). Complement biomarkers are known for their unstable natures due to the complement activation process. For such types of biomarkers, rigorous sample preservation must be done. Regardless of the strategies chosen to stabilize the samples, the procedures must be carefully designed, developed, and validated before implementation, to ensure the data quality for biomarkers. Clear instructions should be addressed in the lab manual to cover sample collection, sample processing, and sample shipment. Large molecule biomarker targets, on the other hand, employ immunoassays as the gold standard. Various types of technologies and platforms are available under this category, such as enzyme-linked immunoassays (ELISAs), electrochemiluminescence (ECL) assays on Meso scale discovery (MSD), and single molecule array (SIMOA) assays using Quanterix HD-1/HD-X instruments. The details of these assays are discussed in a separate section.

The second challenge of biomarker quantitation, regardless of the platform and techniques, often results from the lack of analyte-free matrices. To remedy this, a surrogate matrix or a surrogate analyte may be used. Herein, we discuss the strategies of protein biomarker analysis using LC-MS, and also compare the

LC-MS technique with immunoassay platforms. Intact protein biomarker quantification using LC-MS can be challenging. There are efforts to explore the extent of the possibility of quantitating intact protein biomarkers and the intact protein quantitation using LC-MS and achieved success. However, this process requires high-resolution mass spectrometry (HR-MS) and specific sample clean-up techniques. Instead, quantitating specific constituent peptides, called signature peptides, draws much more success. The signature peptides are selected after tryptic digestion and searched through open-source software platforms such as Skyline, followed by further comparison and identification through basic local alignment search tool for protein (BLASTP), an online software by the National Center for Biotechnology Information (NCBI). Compared to immunoassays, LC-MS/MS is superior in terms of specificity, versatility, and reduced matrix effect. However, immunoassays are still preferred for protein biomarker analysis as the most sensitive and reliable approach. For multiplexing capabilities, both LC-MS/MS and immunoassays have demonstrated mature and reliable capabilities of such. A surrogate matrix, often a buffer with appropriate pH, can be used in biomarker assays, to prepare calibration standards. QC (Quality Control) samples should be prepared in the authentic matrix, if possible, unless the endogenous concentration is too high to prepare one or more levels of the QC samples. A majority of kit-based immunoassays use buffer to prepare calibration standards. When a surrogate matrix approach is used, appropriate parallelism testing should be conducted to demonstrate the correlation between the authentic matrix and the surrogate matrix. For LC-MS, when using this surrogate matrix approach, a stable isotope-labeled internal standard (SIL-IS) is recommended to correct the possible absolute matrix effect with the internal standard normalized matrix effect, yielding more accurate concentration results [15]. For protein biomarker bioanalysis, using the stable isotope-labeled protein as an internal standard is possible, but not very common [16]. The stable isotope labeled signature peptide instead can be used as the internal standard. The labeled peptide should be added prior to sample processing to track the whole sample process whenever possible.

The "surrogate analyte" approach can be used for biomarker analysis when the authentic analyte reference material or the authentic and recombinant protein cannot be obtained. For small molecule biomarkers, typical surrogate analyte features in a structure analog or stable isotope labeled analytes. Choosing the surrogate analyte approach allows the investigators to retrieve important data quickly in the exploratory phase of drug development. However, it is not typically used under regulated settings, unless the equivalency between the surrogate analyte and authentic analyte is fully investigated and validated. The surrogate analyte approach can also be applied to protein biomarkers when appropriate protein biomarker materials are not available from commercial or in-house sources. In this case, the same protein from a different species might be chosen [17]. In doing so, the signature peptide after tryptic digestion might be the same, or slightly different in selected amino acids. A signature peptide within ± 2 Da of molecular weight from the signature peptide of the analytical targeted protein might be chosen as the surrogate analyte, and since a typical Sciex mass spectrometer does not have the capability to differentiate such small mass-to-charge ratio, the same multiple reaction monitoring (MRM) transition can be used. However, if the molecular weight is equal to or greater than 3 Daltons on a Sciex mass spectrometer, different MRM transitions have to be monitored which creates additional variables in the methodology but is still feasible.

Glycosaminoglycans, including heparan sulfate (HS), dermatan sulfate (DS), keratan sulfate (KS), and chondroitin sulfate (CS), are unbranched linear sulfated anionic polysaccharides, composed of a series of structurally different disaccharides. When a large molecule analytical target such as glycosaminoglycans is to be quantitated, neither top-down intact analysis on a large-molecule level nor bottom-up analysis is feasible, due to their highly heterogeneous structures, without a consistent sequence or molecular weight. In this case, a different analytical strategy must be adopted. The quantitation of HS can be achieved by quantitating individual disaccharides after chemical or enzymatic digestion. A similar approach can be taken for the quantitation of DS. The disaccharides to be used in quantitation are carefully investigated in healthy and diseased matrices and the most abundant and/or clinically meaningful disaccharides are selected for quantitation. Individual or total disaccharide concentrations can be reported using this approach. An alternative approach for data processing can be utilized by adding up the individual peak area of different disaccharides of the same sample, which represents the total HS or DS. This approach is useful in reporting the total HS and DS but lacks information on individual disaccharide concentrations. Based on the purpose of the biomarker quantitation, different strategies can be considered and adopted.

Even though the LC-MS platform provides advanced tools for biomarker quantitation, compared with immunoassays, the sensitivity of the LC-MS technique limits its application in the large molecule biomarker fields. Immunoassays usually allow the detection of protein biomarkers at pg/mL levels. Using high-sensitive platforms such as Quanterix SIMOA or MSD (S-plex assays), the sensitivity can reach low fg/mL levels. However, typical LC-MS methods are only capable of detecting biomarkers at ng/mL to μg/mL levels. Recently, a few highly sensitive protein biomarker quantitation methods using LC-MS were also reported using nano-flow HPLC coupled with a Q-Evacuative mass spectrometer. These methods also feature high selectivity and provide an alternative option for protein biomarker quantitation. Secondly, after validation is complete, the immunoassay workflows are usually more concise and have a higher throughput than LC-MS assays. After a sample batch is prepared, the LC-MS method requires samples to be quantitated by sequential injections and is more time-consuming in data acquisition than immunoassays, most of which only take a few seconds to minutes for data acquisition. On the other hand, the LC-MS platform is superior to immunoassays with respect to assay selectivity, assay linearity, and reduced matrix effect. According to the physiological or pathological processes, the observed biomarker response change may be an increase or decrease and the percent change may be great or very slight. The normal range can vary both within a patient and between patients.

Choosing the appropriate platform is important and each platform has its unique advantages and disadvantages. Investigators should carefully evaluate the platforms, reagents, and context of use for the biomarkers to develop the most fit-for-purpose methodologies.

10.2.1.2 Method Validation by LC-MS/MS

To determine the extent of biomarker method validation needed, it is necessary to establish how the data will be used in the context of the entire project being supported. The associated method validation parameters and their stringency will be defined based on:

- Intended use of the biomarker data,
- The importance of the data with respect to the conclusion of the study,
- The type of study in which it is placed, and
- How the data are used in a regulated submission.

Some questions that may help define the current validity of the biomarker, which will drive the degree of method validation in return, for example, include:

- Is this biomarker part of a panel of biomarkers whose significance in this indication is not yet fully understood?
- Is there data in the literature describing this biomarker for the current indication or drug class?
- Are references available?
- Is this a biologically qualified biomarker for the context of use (i.e. animal model, efficacy, safety endpoint)?
- Is this biomarker understood in terms of:
  - Anticipated individual physiologic/biologic variability
  - Subtle or robust changes in values depending on the disease state or intended indication
- Is there a baseline level in the matrix? Does it vary with disease state?
- What is known about the target (e.g. identity, possibility of cross-reactivities, and stability in matrix, etc.)?

An example of the tiered approach to defining the intended use of the biomarker data is summarized in Table 10.1.

Table 10.1 Biomarker purpose and suggested fit-for-purpose validation extent.

Stage of drug program	Biomarker qualification	Method	Examples where is will be useful
Discovery	Exploratory	Exploratory	To screen for possible makers; to help rank drug candidates.
Proof of Concept	Selection	Qualified	Make decisions on the biomarker's utility; useful in discovery studies but not for safety or submission studies.
Established markers	Further corroboration	Validated	Safety Studies; submission studies
Surrogate endpoint	Surrogacy	Validated	Can be used in place of traditional endpoints, including primary, secondary, and exploratory endpoints.

Biomarker validation, including LC-MS biomarker validation, should follow US 2018 Guidance: Bioanalytical method validation guidance for industry, European Medicines Agency (EMA) guidance: Guideline on bioanalytical method validation, and ICH M10 guidance. However, the criteria for accuracy and precision for protein biomarker analysis is generally following that of immunoassays, where accuracy should be within $100 \pm 20\%$ for non-LLOQ calibration standards or QCs, and $100 \pm 25\%$ for lowest calibration standard and LLOQ samples; and percent CV should be within 20% for non-LLOQ intra- and inter- QC precision and 25% for LLOQ intra- and inter-precision. In terms of validation parameters, since the selectivity cannot be evaluated due to the endogenous nature of the biomarkers in biological matrices, a matrix sample screening run, including healthy or diseased matrices, from at least 25 different donors should be performed. For recovery and matrix effect evaluation, if the endogenous concentration is too high to perform at one or more concentration levels (e.g. QC-Low and/or QC-Mid concentration levels), the stable isotope-labeled reference material can be used instead as an analyte to perform the matrix effect testing. For protein or other large molecule bioanalysis involving chemical or enzymatic digestion, the digestion efficiency should be evaluated whenever possible.

10.2.2 Histology

Histologic assessment of tissue biomarkers can range from histochemistry such as hematoxylin and eosin (H&E) to in situ hybridization (ISH) for nucleic acid and immunohistochemistry (IHC) for protein detection. It has traditionally been used in drug development for *in vivo* diagnosis, evaluating morphological changes, evaluating response to therapy, nonclinical safety assessment, and basic research [18]. However, there are cases where histology evaluation was used as a surrogate endpoint to reasonably predict clinical benefit to support drug approval. Histological reduction of the accumulated GL-3 inclusion burden in biopsied kidney interstitial capillaries (KIC) was used as a surrogate endpoint [2] to support the approval of Fabrazyme® and Galafold® in the United States, Fabrazyme for patients with confirmed Fabry disease and Galafold to treat Fabry patients with amenable GLA gene variants. For both approved drugs, the reduction of GL-3 inclusions in kidney KIC was evaluated on renal tissues that were fixed, embedded, and stained using similar procedures [19, 20]; however, each approved drug used a different method for GL-3 inclusion evaluation and scoring. The Fabrazyme Scoring System (FSS) based on a semiquantitative scoring scale of 0 to 3 evaluated on fresh glass slides under light microscopy by three independent renal pathologists in a blinded manner (0 = none/trace, 1 = mild, 2 = moderate, 3 = severe accumulation) was used in Fabrazyme trials to support the traditional approval of the therapy [2, 19]. Whereas Galafold, also evaluated by three independent renal pathologists in a blinded manner, used the quantitative Barisoni Lipid Inclusion Scoring System (BLISS) method based on KIC GL-3 inclusions per renal peritubular capillaries using whole-slide digital (WSI) images to support the accelerated approval of the chaperon therapy [2, 20].

Traditionally, the interpretation of the histology results is performed by observation using light or immunofluorescence microscopy. However, the digital WSI workflow has become a trend in modern pathology practice and is increasingly adopted in clinical diagnostic and clinical trials [20–23]. Additionally, the incorporation of artificial intelligence and machine-learning techniques for computation image analysis also takes the histology field to the next level. The incorporation of these new capabilities will enable information extraction and quantitative analysis that is not possible using the conventional microscopy method [22]. There are advantages and disadvantages to either approach. For conventional microscopy, pathologists regularly evaluate multiple focus planes; however, individual practice could potentially introduce variations, bias, and errors. Although Digital WSI offers many advantages over the conventional microscopy approach, it does have some limitations. These include the lack of Z-dimensional focus (depth on an image) unless using the Z-stacking approach to produce a composite multiplane image, the imperfect control of the scanner and scanning software that can contribute to imaging artifacts, the need to use microscopy scanner with high objective lens (e.g. ×40, ×100) results in the large file size, which will require a substantial network bandwidth to handle these data sets, and complicated workflow to securely store, share images and data [21, 22]. Additionally, the individual pathologist bias may still exist with WSI if images are evaluated by an individual pathologist only.

From clinical trials support perspective, whether it is through traditional microscopy-based with manual evaluation or scoring by an individual pathologist through a digital platform, there are many method developments, validation, logistical, and operation challenges when it comes to using histology in clinical trials to support endpoints evaluation. Testing logistics of histology evaluations may involve many laboratories from sample collection, fixation, staining, and image scanning to data outputs and on-boarding experienced pathologists. Each clinical site may have different procedures for sample collection, processing, and fixation. It is crucial to understand variations in procedures used by different clinical sites and implement a strategy to best unify these procedures. During method development, the preanalytical variables, such as tissue collection, fixation, processing, and embedding, should be evaluated carefully as differences in the procedure applied will likely impact the results [24]. Improper tissue handling and mechanical manipulation during tissue processing may introduce artifacts or impact RNA integrity for ISH assessments. Inappropriate fixation or prolonged formalin-fixation may impact staining results or cause excessive protein cross-linking and reduce the biomarker availability for antibody binding for IHC evaluations. During the analytical histology phase, the sectioning and staining procedures should be optimized and evaluated for consistency. Utilizing the same laboratory to perform the analytical phase will help to reduce variability. If a digital platform is used, the selection of the scanner, image and monitor resolutions, data analysis and interpretation of results, and reporting of data process will need to be vetted and planned out early [21].

The draft Food and Drug Administration (FDA) guidance for the development of Fabry disease indicated a standardized and validated method conducted by experienced pathologists in a blinded and systematic manner should be used if the surrogate endpoint is based on histological assessment [25]. However, guidance on how to systematically validate this type of histology evaluation is still lacking. FDA guidance of histopathology to support biomarker qualification for nonclinical studies [18] and the white paper from the College of American Pathologists and Laboratory Quality Center for diagnostic purposes [23] can be used as a general guide for method setup and validation. During validation, the validation study design should closely mimic the clinical environment and the specific technology for the intended use. Demonstration of accuracy, intra- and inter-observer agreement and the concordance between digital and glass slides are some recommended parameters if a digital platform is the selected method [23]. Additionally, the data analysis, data reporting, and the software used for the analysis will also need to be evaluated and incorporated as part of the validation if applicable.

10.2.3 Functional Activity and Immunoassays

Functional protein activity and immunoassays have also been used in drug development for PD biomarkers measurement. Some examples of PD biomarkers using protein activity are chitotriosidase for Gaucher disease and serum ceruloplasmin activity for Wilson disease [26–28]. Method development and validation challenges and solutions for functional activity were discussed extensively in Chapter 9 and the presented information can be applied to this application.

Immunoassays are still considered the gold standard for biomarker testing, especially in the field of large molecule protein biomarker testing. Over the past two decades, immunoassays and technologies have developed in various aspects, and recent innovations have advanced in many directions: sensitivity, multiplexing, robustness, reproducibility, time, and cost-effectiveness. It covers a wide spectrum of immunoassay quantitation platforms, including traditional ELISAs and novel platforms, such as Quanterix SIMOA, MSD, Ella™ Automated Immunoassay System (ELLA), Luminex, Quansys Q-view imaging system, SMC, and O-Link. Table 10.2 lists current platforms that are commonly applied for quantitative immunoassays of biomarkers. In this table, a comprehensive cross-platform and cross-assay evaluation was performed, using different technology platforms, and immunoassays of different analytes, by comparing a set of common assay parameters: precision, sensitivity, parallelism, frequency of endogenous analyte detection (FEAD), and data correlation between platforms.

Immunoassays are trending in two main directions: high sensitivity and multiplexing. The driving force for these two directions is the ultra-low amounts of a lot of newly discovered and important biomarkers (picomolar to femtomolar range) in the body, their dynamic secretion processes, and short half-lives. Moreover, to understand the mechanism of drug functionality and to lend support to the proof of concept of the biomarker, more than one biomarker needs to be quantitated at the same time and these biomarkers can be selected in return to facilitate the disease diagnosis and treatment, patient selection, companion diagnosis, etc. All these highlight the value of biological specimens from disease and healthy populations and the urgent need for sensitive methods that are capable of multiplexing. Currently, the Quanterix SIMOA demonstrates superior sensitivity due to their unique technology. Competitors like MSD S-plex assays and the SMC Pro assays also showcase the ability for high-sensitivity testing and quantitation. There are various platforms capable of supporting multiplexing, such as Luminex assays, ELLA assays, MSD V-plex and U-plex assays, and O-link assays. Most of these multiplexing assays demonstrate stable, reliable performance. However, when looking to combine the sensitivity and the multiplexing capabilities, there is still a gap in the technology. It is challenging to find a multiplexing assay that demonstrates ultra-high sensitivity in many fields, such as Alzheimer's biomarkers. Various companies continue their efforts in building high-performing instruments capable of multiplexing without compromising sensitivity and throughput.

10.2.3.1 Method Validation of Immunoassay

Biomarker validation using immunoassay techniques follows FDA, EMA, and ICH M10 guidelines for immunoassays. Biomarker validation is deemed as fit-for-purpose, and the extent of the validation should be appropriate for the intent of

Table 10.2 Current spectrum of technologies and platforms for quantitative immunoassays of biomarkers.

Platform	Vendor	Sensitivity	multiplexity	Typical Sample Volume	Rep[a]	Range[b]	Automation	Duration	Assay Cost
SIMOA HD-X Analyzer	Quanterix	sub pg/mL	Up to 6	1–100 μL	1–2	>4 logs	Yes	1–3 h/2–4 h	+++
	Description: Bead-based technology isolates each bead-immunocomplex to allow for enzyme bound to the capture antibody to produce sufficient fluorescence in each well to be detectable, even when just a single molecule is present. As a result, the analyzer produces a map of the array—each of which contains more than 200,000 wells—where each well can be digitally analyzed as either "on" (containing a target molecule) or "off." The proprietary SIMOA algorithm then converts the measurement into a concentration.								
MSD S600	MSD	sub pg/mL	Up to 10	25–50 μL	1–2	4 logs	No	2–6 h/2–5 min	++
	Description: Combining electrochemiluminescence and multiarray technology for the detection of multiple proteins in a single sample.								
Ella	Protein Simple	pg/mL	Up to 8	25 μL	2–3	4–5 logs	Yes	1–2 h/~1.5 h	++
	Description: Fully automated cartridge-based system to perform single analyte, multianalyte, and multiplexed immunoassays with the specificity of a traditional single-plex ELISA.								
Q-View Imager LS	(Quidel)	pg/mL	Up to 18	50 μL	1–2	3–4 logs	No	4–5hr/5–6min	++
	Description: A multiplex ELISA based on chemiluminescence. Each well of the 96-well plate contains a defined array of printed capture microspots. Detection of the array is performed by the Q-View Imager and data are reported in different individual formats.								

Platform	Vendor	Sensitivity	multiplexity	Typical Sample Volume	Rep[a]	Range[b]	Automation	Duration	Assay Cost
Luminex	R&D Systems	pg/mL	Up to 50	50 μL	1–2	5 logs	No	4–5 h/ 5–6 min	++
	Description: Luminex multiplex assays utilize color-coded superparamagnetic beads coated with analyte-specific antibodies. Beads recognizing different target analytes are mixed together and incubated with the sample. Captured analytes are subsequently detected using a cocktail of biotinylated detection antibodies and a streptavidin-phycoerythrin conjugate.								
ELISA	Various commerical sources	pg/mL	None, single analyte only	25–100 μL	1–2	3–4 logs	No	1–5 h/1–5 min	+
	Description: Traditional enzyme-linked immunosorbent assay (home-brewed or kits).								
SMCxPRO	Sigma-Aldrich	sub pg/mL	Up to 12	25–75 μL	1–2	>4 logs	No	3–4 h/1–2 h	+++
	Description: Bead-based assay. Signals collected from analyte fluorochromes free-floating in suspension using a rotating laser, CCD camera optics, and individual photons are counted by Avalanche Photodiode (APD).								
Olink	Olink	sub pg/mL	Up to 1920	1 μL	1–2	>4 logs	No	1–1.5 day/–4 h	+++
	Description: Unique technology called proximity extension assay (PEA) technology.								

[a] Rep: Replicate
[b] Range: Dynamic Range

the study that the biomarker supports. Fit-for-purpose method validation ranges from a simple method qualification, a including one precision and accuracy (P&A) run to test accuracy and precision, to a full validation covering all required parameters defined in the guidance.

Elements of method qualification and method validation are usually assessed on a case-by-case basis and depend on many factors, including the importance of the study in terms of decision-making and whether the biomarker data will be used to support submission. Documentation is always expected, even if the earlier stages of method development are less complete in terms of validation.

Once the biomarker is positioned in its importance to the study protocol, the next step should be to collect the available information about the proposed method:

- Are there any proprietary or commercial kits available?
- How is this study to be used and how is the sample collected? Will the sample be used fresh or stored?
- Platform
- What are the species (and dynamic range and sensitivity required?) to be researched on: rat/dog/monkey/human?
- What is the matrix type: urine/plasma/serum/CSF/other?
- What are the sample volume requirements: microliter/nanoliter?
- Are critical reagents available, supplied consistently, and meet the standard for the context of use?
- Is there a WHO or other recognized or commercially available reference standard?
- What is the sensitivity required: ng/mL or pg/mL or fg/mL?
- What is the nature of throughput required (approximate samples per study/ submission)?

Once the method is established and the quantitation range is identified, a proposal for method qualification/validation can be developed.

- Partial Validations (bridging studies)
  The stage of validation and the extent of the modifications made to the current method determines the degree of testing required. For some modifications, e.g. change in the reagent lot, reagent "bridging" is necessary to verify the new lot acceptability. For other modifications, e.g. change in platform, critical reagent, matrix, etc., at least three assays must be performed to verify the validity of the method and should include at least recovery, P&A, and linearity.
- Cross-Site Verification
  Optimally cross-site verification should be conducted using both spiked QCs and incurred (study) samples. The method transfer should follow an established SOP.
- Commercial Kits
  Never assume a kit meets the criteria described in the package insert. The kit should follow the stringencies described above, depending on its intended use.

Table 10.3 Comparison of method validation parameters between fully validated regulated bioanalytical assays support clinical studies against full validation under CLIA regulations.

Method Validation Requirement	Regulated Bioanalysis	CLIA
Dynamic Range, Sensitivity and Calibration Standards	✓	✓
Dilution Linearity and Parallelism	✓	✓
Intra- and Inter-run Accuracy and Precision, Spike Recovery	✓	✓
Matrix Selectivity	✓	✓
Endogenous Screening	✓	✓
Stability (Short and Long Term)	✓	✓
Inter-lot Bridging	✓	✓
Drug Interference Test, Hemolysis, Lipemic Effect	✓	✓
QA Audit	✓	✓
Normal Population Range Determination	✗	✓
Personnel Competency Test	✗	✓
Lab Accreditation and Certification per Regulations	✗	✓
CLIA Director Review and Signature on Validation Protocols and Reports	✗	✓
External Proficiency Test	✗	✓

Moreover, when biomarker data is used for physicians to make the diagnosis decision, a CLIA validation should be performed in a CLIA-certified laboratory. The details of the comparison for parameters between full validation of regulated bioanalytical methods against full validation under CLIA regulations are listed in Table 10.3.

10.2.4 mRNA Detection of Downstream Target Expression as a PD Biomarker

Transcription is a highly regulated process in both prokaryotes and eukaryotes. Hence, specific transcript or transcriptome analysis, either qualitative or quantitative is preferentially performed to assess health conditions or a therapeutic intervention by monitoring variations in PD biomarker levels. Though Northern blotting and DNA microarrays also are in use, PCR-based molecular methods are still preferred for the detection and quantitation of target expression. Real-time PCR is considered by many as the gold standard in nucleic acid quantification because of its accuracy and sensitivity [29, 30]. Real-time PCR is termed qPCR when used for quantitation. For the quantitation of target gene (mRNA) expression, the RNA is first reverse transcribed to its complementary DNA using reverse transcriptase and then quantitated by qPCR (RT-qPCR or qRT-PCR) [31]. DNA microarray technology was popular a decade ago due to the surge of omics-level studies. This found applications in studying expression levels throughout the genome for transcriptome profiling and biomarker discovery [32, 33]. After the human genome project, the importance and the business of sequencing has been realized, resulting in the development of Next Generation Sequencing (NGS) technologies of which RNAseq is a part. Advances in NGS technologies and RNAseq, shifted the focus preferring sequencing-based methods over microarrays [34, 35]. RNA sequencing using NGS can detect both known and novel transcripts. The NanoString nCounter gene expression system captures and counts individual mRNA transcripts by direct measurement without amplification [36]

10.2.4.1 RT-qPCR for Relative Gene Expression Analysis

RT-qPCR was originally developed from real-time PCR to amplify distinct nucleic acid sequences for the detection and relative quantification of mRNA levels in human monocyte-derived macrophages [37]. In some publications, the term RT-PCR was used for real-time PCR. In this chapter, RT is used for reverse transcription and qPCR for quantitative (Real-Time) PCR, i.e. RT-qPCR. If the technique is used only for detection, the term reverse transcription Real-Time PCR (rRT-PCR) was also used by some groups, such as CDC [38]. There are tremendous advancements in the design of primers and probes (Minor Grove Binding), master mixes with mutated enzymes, such as hot-start DNA polymerase, reverse transcriptase, and efficient real-time PCR instruments, that can detect multiple fluorophores (multiplex RT-qPCR). This made the qPCR or RT-qPCR a routine method of choice to quantitate PD biomarker analysis in several gene and cell therapy studies. The availability of kits for single-step RT-qPCR reduced the time and improved efficiency to a greater extent [39]. In addition, there are several tools available by the vendors, such as IDT-DNA or ThermoFisher Scientific, for the design of primers and probes for real-time PCR (https://www.thermofisher.com/order/catalog/product/4316034?SID=srch-hj-4316034 and https://www.idtdna.com/pages/products/qpcr-and-pcr/gene-expression/primetime-qpcr-probes). Quantitation of mRNA in a sample by RT-qPCR can be performed either as absolute quantification or relative quantification. As in the case of any analyte quantitation, for absolute quantification of mRNA also, serially diluted standards that produce a linear relationship between Ct values and initial amounts of total RNA or cDNA is used allowing the determination of the concentration of unknowns based on their Ct values assuming that all standards and samples have approximately equal amplification efficiencies. In the case of single-step RT-qPCR reactions, the standards, and any QC must be RNA only [40]. In addition, the range of the standard curve should cover all sample concentrations and in the range of accurately quantifiable and detectable levels specific for both the instrument and assay. However, the digital PCR technology allows us to perform absolute quantitation without the need of a standard curve as it provides a direct number rather than a Ct value. The main disadvantage of these systems is extended sample preparation time and the upper limit of quantitation (ULOQ) is only 10^5 copies/reaction (https://www.thermofisher.com/order/catalog/product/4316034?SID=srch-hj-4316034) [41].

Relative Quantitation: Relative quantitation of gene expression analysis, which will be discussed more in this chapter also uses regular real-time PCR instrumentation and the same general reagents for amplification. In relative quantitation, variations in target gene expression are measured based on the levels of a reference gene used as either internal or external sample [42]. The results are expressed as a target/reference ratio, for accurate expression. In several biodistribution studies, the data are expressed as the mRNA copy number per unit of total RNA in the sample [29, 43]. Several mathematical models, for example, geNorm, geNorm Kits, REST-2009-standalone application software, Bestkeeper applet, NormFinder, and "R," Genevestigator are available to calculate the mean normalized gene expression from relative quantitation assays. As the levels are expressed in comparison to a selected reference gene, standard curve preparation and calculation of copy number are not required. Relative quantitation is also used in microarray technology to study the variations in transcriptome levels [33]. The main factor and issue in the relative quantitation of target gene is the selection of the reference gene(s) and normalization of the data. In clinical trials, relative quantification by RT-qPCR has been successfully used when there is availability of untreated or normal groups [44, 45]. An ideal reference gene is one whose expression is, (1) constitutive, (2) unaffected by the drug target and experimental conditions, (3) unaltered by the developmental stage, and (4) equal in different tissue or cell types [43]. Amplification efficiency is another important factor that affects relative quantitation. A reference gene is completely different from the target gene. So, unless by chance, it is not possible to have the same amplification efficiency as target gene mRNA. This mandates the requirement of a correction factor. A simple equation to calculate the amplification efficiency from the data derived using a standard curve, $e = 10^{-1/\text{slope}}$ [46], where: e = theoretical efficiency, Slope = the slope of the standard curve, plotted with the y-axis as Ct and the x-axis as log(quantity). Alternatively, several data analysis models have been developed that enable the calculation of PCR amplification efficiencies from individual amplification plots, such as data analysis for real-time PCR (DART-PCR), LinReg, and the real-time PCR Miner algorithms. Some of the reference genes found to be useful in different studies are listed in Table 10.4.

Table 10.4 Frequently used reference genes in RT-qPCR-based relative quantitation studies.

Sl. No.	Gene	Name of the Protein	One Example of Successful Study	Doi
1	18SrRNA	18S Ribosomal RNA gene	Mutat Res. 2008 Jan 8;649(1–2):126–34	10.1016/j.mrgentox.2007.08.005
2	ABL1	Tyrosine-protein kinase ABL1	PLoS One. 2019 Mar 21;14(3):e0214305	10.1371/journal.pone.0214305
3	ACTB	Beta-actin	Asian Pac J Cancer Prev. 2017 Aug 27;18(8):2269–2274	10.22034/APJCP.2017
4	ALB	Albumin	Methods Mol Biol. 2014;1087:315–23	10.1007/978-1-62703-670-2_25
5	B2M	Beta-2-Microglobulin	Ann Hematol. 2018 Mar;97(3):467–474	10.1007/s00277-017-3210-8
6	BCR	Breakpoint cluster region protein	Leukemia. 2016 Sep;30(9):1844–52	10.1038/leu.2016.90
7	CD20	Cluster of differentiate 20 (MS4A1-membrane-spanning 4A gene family)	J Mol Diagn. 2002 Nov;4(4):201–8	10.1016/S1525-1578(10)60704-0
8	CD71	Transferrin receptor protein 1 (TfR1), also known as cluster of differentiation 71	Mutat Res. 2008 Jan 8;649(1–2):126–34	10.1016/j.mrgentox.2007.08.005
9	CD8A	Cluster of differentiation 8a	Br J Haematol. 2000 Nov;111(2):618–25	10.1046/j.1365-2141.2000.02352.x
10	CYCS	Cytochrome-C, Somatic	Mutat Res. 2008 Jan 8;649(1–2):126–34	10.1016/j.mrgentox.2007.08.005
11	G6PD	Glucose-6-phosphate dehydrogenase	Methods Mol Biol. 2020;2065:153–173. 10.1007/978-1-4939-9833-3_12	10.1007/978-1-4939-9833-3_12
12	GAPDH	Glyceraldehyde 3-phosphate dehydrogenase	Blood. 2003 Oct 15;102(8):2763–7	10.1182/blood-2002-10-2998
13	GUSB	Beta-glucuronidase (β-glucuronidase)	Molecular Monitoring of Chronic Myeloid Leukemia. Methods Mol Biol. 2020;2065:153–173	10.1007/978-1-4939-9833-3_12
14	HMBS	Hydroxymethylbilane synthase	Exp Mol Pathol. 2014 Dec;97(3):433–9	10.1016/j.yexmp.2014.09.014
15	HPRT1	Hypoxanthine phosphoribosyltransferase 1	Ann Clin Lab Sci. 2015 Winter;45(1): 64-72	PMID: 25696013

Sl. No.	Gene	Name of the Protein	One Example of Successful Study	Doi
16	KLF4	Kruppel-like factor 4 (Transcription Factor)	Electrophoresis. 2010 Jan;31(3):528–34	10.1002/elps.200900362
17	KMT2C	Lysine methyltransferase 2C	Cancer Detect Prev. 2001;25(5):454–69	PMID: 11718452
18	MRPL19	Mitochondrial ribosomal protein L19	Br J Haematol. 2010 Nov;151(3):232–8	10.1111/j.1365-2141.2010.08363.x
19	MTIF2	Mitochondrial translational initiation factor 2	J Clin Pathol. 2006 Oct;59(10):1059–65	10.1136/jcp.2005.031161
20	NPM1	Nucleophosmin 1	Br J Haematol. 2009 Feb;144(4):517–23	10.1111/j.1365-2141.2008.07488.x
21	PCNA	Proliferating cell nuclear antigen	Cancer Biol Ther. 2004 Oct;3(10):989–94	10.4161/cbt.3.10.1137
22	PGGT1B	geranylgeranyl transferase type-1 subunit beta	Genes (Basel). 2019 May 17;10(5):376	10.3390/genes10050376
23	PGK1	Phosphoglycerate kinase	Mutat Res. 2008 Jan 8;649(1-2):126–34	10.1016/j.mrgentox.2007.08.005
24	PGM1	Phosphoglucomutase-1	Iran Biomed J. 2010 Jul;14(3):77–82	PMID: 21079657
25	PPIA	Peptidylprolyl isomerase A	Leukemia. 2003 Apr;17(4):789–95	10.1038/sj.leu.2402880
26	PRKG1	Protein kinase CGMP-dependent-1	Diagn Mol Pathol. 2009 Dec;18(4):243–9	10.1097/PDM.0b013e3181a06f42
27	PSMB6	Proteasome 20S subunit beta 6	Genes (Basel). 2019 May 17;10(5):376	10.3390/genes10050376
28	PSMC4	Proteasome 26S subunit, ATPase, 4	Sci Rep. 2015 Jul 21;5:12435	10.1038/srep12435
29	RARA	Retinoic acid receptor alpha	Clin Chem Lab Med. 2000 Feb;38(2):171–7	10.1515/CCLM.2000.026
30	RPL13A	Ribosomal protein L13a	Eur Respir J. 2015 Jul;46(1):207–18	10.1183/09031936.00119914
31	RPL30	Ribosomal subunit protein, Large 30	Sci Rep. 2019 Mar 15;9(1):4616	10.1038/s41598-019-40851-9
32	RPL37A	Ribosomal protein L37a	Clin Chem. 2010 Dec;56(12):1845–53	10.1373/clinchem.2010.151233

(Continued)

Table 10.4 (Continued)

Sl. No.	Gene	Name of the Protein	One Example of Successful Study	Doi
33	RPLP0	Ribosomal protein lateral stalk subunit P0	Exp Hematol Oncol. 2018 Feb 20;7:5	10.1186/s40164-018-0097-6
34	RPLP2	60S acidic ribosomal protein P2	Mutat Res. 2008 Jan 8;649(1–2):126–34	10.1016/j.mrgentox.2007.08.005
35	SDHA	Succinate dehydrogenase complex, subunit A (SDHA)	Technol Cancer Res Treat. 2018 17:1533033818802318	10.1177/1533033818802318
36	SF3A1	Splicing factor 3a subunit 1	J Invest Dermatol. 2008 Feb;128(2):465–72	10.1038/sj.jid.5701013
37	SOX5	SRY-box transcription factor 5	Electrophoresis. 2010 Jan;31(3):528–34	10.1002/elps.200900362
38	STAT3	Signal transducer and activator of transcription 3	Biotechniques. 2005 May;38(5):785–92	10.2144/05385MT01
39	TBP	TATA box-binding protein	Methods Mol Biol. 2020;2065:153–173	10.1007/978-1-4939-9833-3_12
40	TERT	Telomerase reverse transcriptase	Genes Chromosomes Cancer. 2008 Dec;47(12):1067–75	10.1002/gcc.20601
41	UBC	Ubiquitin-C	Ann Clin Lab Sci. 2015 Winter;45(1):64–72	PMID: 25696013
42	UBQLN2	Ubiquilin-2	Genes (Basel). 2019 May 17;10(5):376	10.3390/genes10050376
43	UQCR2	Ubiquinol-cytochrome c reductase core protein 2	Genes (Basel). 2019 May 17;10(5):376	10.3390/genes10050376
44	YWHAZ	Tyrosine 3-monooxygenase/tryptophan 5-monooxygenase activation protein, zeta polypeptide	Ann Clin Lab Sci. 2015 Winter;45(1):64–72	PMID: 25696013

Methods popularly used in different studies for relative quantification of transcript mRNA are,

a) Standard curve method – copy numbers quantitated using a standard curve and relative quantitation calculated in comparison to a selected calibrator sample [42].
b) The comparative Ct method – a mathematical model that calculates changes to the relative fold difference between an experimental and calibrator sample, with no standard curve required [43].
c) Pfaffl model – calculates relative gene expression data while accounting for differences in primer efficiencies using the equation:

$$\text{Gene Expression Ratio} = (\text{EGOI})^{\Delta \text{Ct GOI}} / (\text{EHKG})^{\Delta \text{Ct HKG}}$$

where E = Efficiency, GOI = Gene of Interest, HKG = House Keeping Gene or Reference gene. The term housekeeping gene has been discontinued [44].

The d) Q-gene method [45], e) Liu and Saint method [46], and f) Amplification plot method [47] are other methods used to calculate amplification efficiencies and normalize the data. However, it was found from several studies that the reference genes cannot be considered universal even in broadly similar conditions. It is necessary to confirm that the selected reference gene is expressed to the same level in the selected tissues under experimental conditions [47]. However, the new tools available from different vendors or institutes to design the primer/probe sets allow to have them good efficiency. For example, a) IDT DNA Inc. (https://www.idtdna.com/pages/tools/), b) ThermoFisher Scientific Inc (https://www.thermofisher.com/us/en/home/life-science/oligonucleotides-primers-probes-genes/custom-dna-oligos/oligo-design-tools.html), c) National Center for Bioinformatics, NIH (https://www.ncbi.nlm.nih.gov/tools/primer-blast/). Relative quantitation is easier as it does not require standard curve preparation, but it is applicable only when variations in levels of the target are under study for a PD biomarker. To reduce the time and effort for method development, testing the efficiency of three or more sets of primer/probes has been suggested for biodistribution studies[48]. Specific regulatory guidance for qPCR is warranted for uniformity in the qPCR-based clinical studies.

10.2.4.2 RNA-seq

Rapid advances in sequencing technologies and reagents simplified the sequencing methods. No wonder, over 13 million SARS-CoV2 whole genome sequences are uploaded to GISAID by September 2022 (https://gisaid.org). There are several PD studies where in RNA-seq has been used successfully to monitor the levels of known PD biomarkers and also to identify any novel genes affected. In a US FDA coordinated study RNA-seq showed agreement with RT-qPCR at relative gene expression measurement levels [41]. The analysis of RNA-seq data are complex. Unless there is a good pipeline developed for data analysis, quality check, and quantitation, the results could be erroneous. Designing an appropriate pipeline with reference sequences is essential for quality sequence data. With an appropriate pipeline, the RNA-seq was used successfully in several AAV-based gene therapy studies to find out the changes in tropism [49]. In addition, the minimum number of copies of the target RNA required to get high-quality sequence also causes issues in RNAseq. A general flowchart for RNA-seq analysis [50] was provided for successful processing of the samples. However, RNA-seq possesses the potential to check multiple targets in a single run and is very useful in discovery or genomic studies.

10.2.4.3 Nanostring Technology

This technique has been developed by integrating the ligand-protein interaction, primer binding, and fluorescence. It consists of two oligonucleotide probes of 35–50 bases in length. The two probes are complementary to the target mRNA separated by a gap of some bases. One probe is biotinylated and is called, "Capture Probe". The other one is called, "Reporter Probe" which is linked to a unique fluorescent barcode. In brief, both probes are hybridized to the target mRNA. The probe-target complex is purified using a streptavidin-coated surface by the binding of the biotinylated capture probe, excess probe is removed by washing. The complexes are aligned, and the reporter probes are scanned and counted using a barcode imaging device, Figure 10.1. High level of multiplexing is possible by using different reporter barcodes. The number of counts is equal to the number of target RNA molecules. Normalization of the data with internal controls and reference genes is performed to produce highly precise relative counts. Nanostring technology is considered best suitable for relative quantification rather than absolute quantification [31].

In a single reaction, relative quantitation of hundreds of target genes can be performed. Several comparative studies of RNA-seq and Nanostring technology in relative quantitation of gene expression showed that both of them are equally effective and complementary [51]. Nanostring-based detection and quantitation are also found to be very sensitive in the rAAV biodistribution studies with a titer of [52] $1–2\times10^{10}$ genomes, ~10–40 times less virus than what is typically used for intravenous or peripheral administration for targeting the spinal cord.

If the task is only the detection of biomarkers, all three techniques – RT-qPCR, RNA-Seq, and Nanostring work well with limitations specific to each. For absolute quantification, RT-qPCR is still the method of choice either by regular real-time PCR platform or a digital PCR platform. Parallel studies performed with

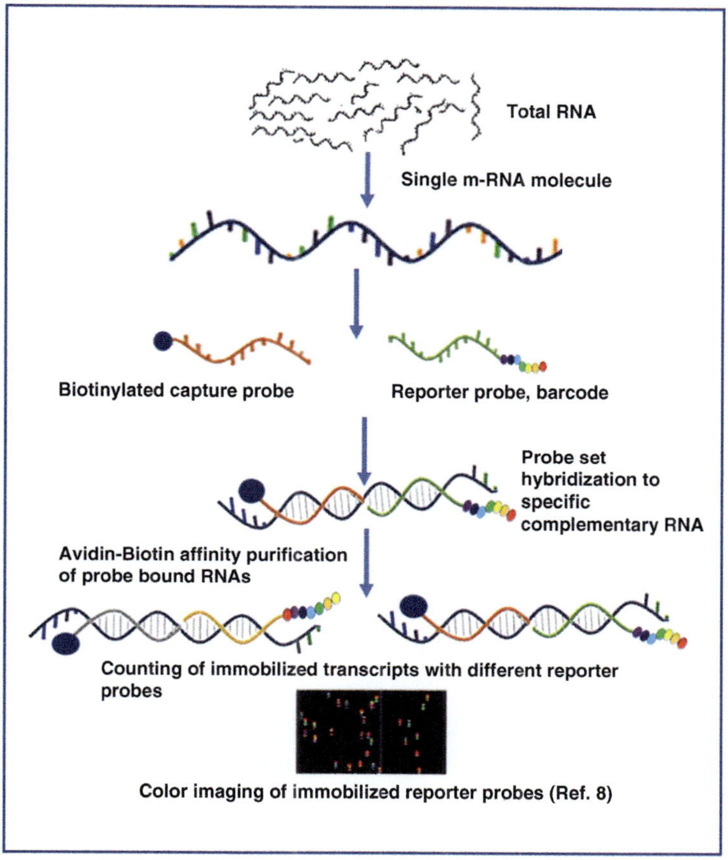

Figure 10.1 Principle of NanoString nCounter for RNA relative quantification.

RT-qPCR, RNA-seq, and Nanostring showed that they all qualify for relative quantitation [50, 53, 54]. A comparison of the three techniques for detection and relative quantitation has been made in Table 10.5.

10.2.4.4 Regulatory Considerations for RNA Quantitation in GLP Studies

As mentioned earlier, due to the low cost of the reagents, sensitivity, precision qRT-PCR is still the method of choice for measuring transcript levels. Efforts to have a consensus on experimental design and data analysis of real-time PCR-based studies were made for regular publications and other research work for both absolute and relative quantitation [42, 43, 46, 47, 54–56]. For cell and gene therapy studies, FDA, USA (FDA) and EMA have released bioanalytical guidance documents for clinical and nonclinical studies [57, 58]. FDA guidance recommends the

Table 10.5 Comparison between RT-qPCR – RNA-seq and Nanostring.

Parameters	RT-qPCR	RNA-Seq	Nanostring
Instrument operation/Sample preparation	Simple and established in many labs, familiar platform.	Needs more skills than real-time PCR and Nanostring, increasingly used where transcriptome/ multitarget analysis is required.	Simple, platform is still not as popular to replace RT-qPCR.
Sample quantity	Minimal (Less than femtograms, <5 copies of target)	Minimum 25 ng Total RNA	Minimum 25 ng Total RNA
Prior Knowledge of Target Sequence	Essential	Not essential	Essential
Sample Throughput	Very High	Medium	High
Number of Targets per run (Multiplexing)	One to five	One to hundreds	One to several
Novel transcript/ Sequence Discovery Power	Nil	High	Nil
Pharmacokinetics Application - Target gene Transcript	Very Useful	Not Likely, as the target number is 1 or a few	Useful
Pharmacodynamic Studies	Good for one to a limited number of marker gene expression studies.	Very useful for larger number to transcriptome level studies in variation in gene expression.	Good for one to a large number of gene expression studies possible.
Absolute/Relative Quantification	Works for both	Works for both, sensitivity lesser than RT-qPCR.	Good for relative quantification.
QC Review and Data Analysis	Simple	Needs pipeline development and bioinformatics knowledge.	Simple

LLOQ of a qPCR assay to be at least 50 copies per μg of host gDNA with consideration given to the sample size used for analysis relative to the size of the tissue in its entirety [59, 60]. However, many of them are easily applicable to techniques such as ligand-binding assays (LBA) but are difficult to apply for qPCR-based assays due to their distinctive characteristics [55]. The EMA has released guidelines for PCR assays but FDA has not yet required validation of qPCR/qRT-PCR assays [61]. However, pharma industries and contract research organizations are trying to harmonize divergent global practices for method development, qualification/validation, and sample analysis through publications and conferences [62]. Recently, different serotypes of AAV have emerged as preferred viral vectors for gene therapy applications due to several advantages such as broad tropism, nonpathogenic nature, low immunogenicity, and prolonged period of transgene expression [63]. Quantitative biodistribution of the viral vector and expression of the target gene (PK) is essential to draw conclusions on the effect and relationship between the dosage vs. PD marker levels [49, 64]. Extensive discussion and recommendations for "best practice" for qPCR/qRT-PCR assay design during method development, validation for gene therapy sample analysis with a special focus on AAV-based clinical and nonclinical studies were made recently [48, 65]. These practices are followed for biodistribution analysis in AAV-based gene therapy studies for vector transgene or its mRNA quantitation by qPCR or RT-qPCR in both nonclinical tissue distribution and shedding in clinical samples, respectively. Several white papers are published by various institutes and scientific forums on the regulatory guidelines. One of them discussed the upcoming digital PCR in comparison to regular real-time PCR. The steps/workflow of the guidelines in practice and recommendations are provided in Figure 10.2.

Recently, the use of digital PCR with (ddPCR) or without droplet is also being used as it does not require a standard curve for absolute quantification. In view of the changing instrumentation and reagents, there is a need to regularly update the specific steps per the requirement of the method. Also, quantitative RNA-seq-based methods are being put into use in AAV-based transcriptome levels [49]. Due to the complexity of the sequencing methodology, quality control, and analysis pipelines, it is very difficult to form regulatory guidelines for RNA-seq-based quantitation of the target gene, its product or PD biomarkers [66]. Signal amplification-based branched-DNA assay technology is also considered as a viable alternative to RT-qPCR but recent advances in the probe design and technology made real-time PCR a more sensitive technique [67]. In addition, unless there is a need for transcriptome analysis or large-scale biomarker level changes, RT-qPCR by either regular or digital platforms serves the needs for current gene therapy projects, including the preferred AAV-based platforms. Though the suggestions and guidelines from forums and labs are available, there remains a need for FDA/health authorities to refine the best practices and provide unambiguous guidelines for qPCR in regulated bioanalysis.

10.2.5 Single-cell Analysis

Another application of histology to support gene therapy programs is target engagement assessment in nonclinical studies to understand the mechanism of

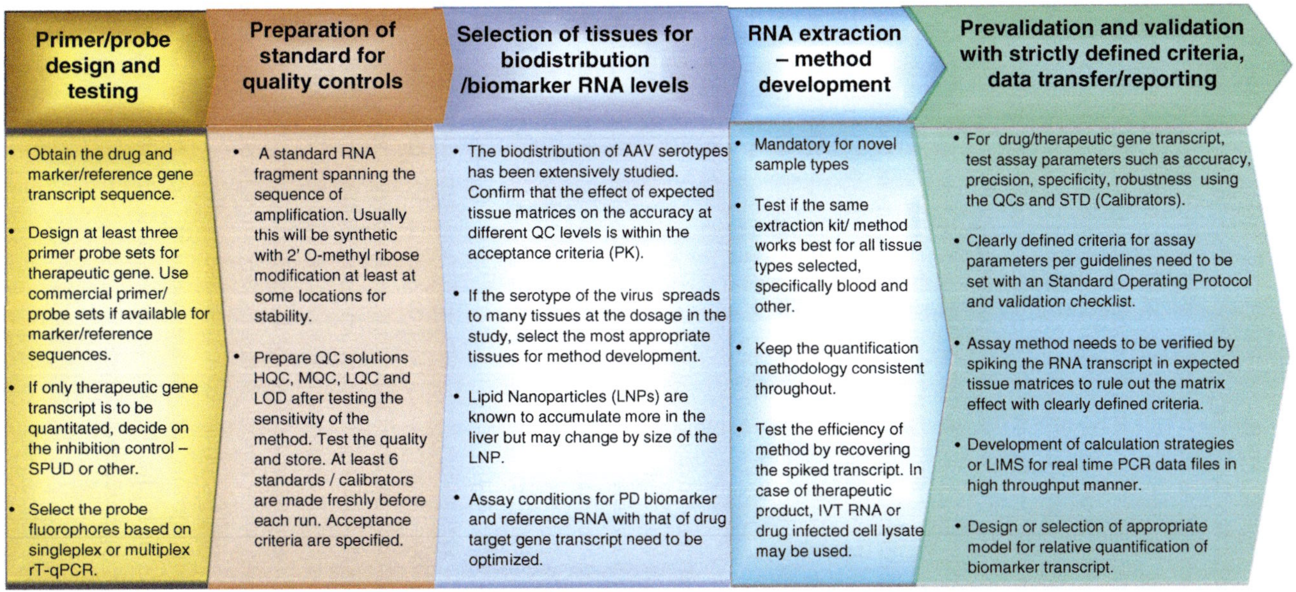

Figure 10.2 Guidelines for steps/workflow in practice for quantitation of AAV-based drug target transcripts (mRNA) by RT-qPCR.

action and the regulation of expression and to identify biomarkers and correlative endpoints of a potential new drug. Analysis of RNA and protein provides invaluable information of gene and protein expression within the spatial and morphological tissue context. Although *in situ* RNA detection techniques have been around for years, the widely used *in situ* RNA ISH approach is the RNAscope ISH based on ACD patented signal amplification and background suppression technology [68]. RNAscope can be used for both conventional chromogenic dyes for bright-field microscopy or fluorescence microscopy applications with multiplexing capability. The technology employs a probe design strategy of two independent double Z probes hybridizing to the target sequence in tandem for signal amplification to occur. On average, ~20 double Z target probe pairs within a 1 kb target region are designed to specifically hybridize to the target molecule to ensure selective amplification of target-specific signals [68, 69]. During development phase, it is crucial to evaluate probe specificity in the selected tissue type in addition to assay reproducibility. Preanalytical variables as stated previously also apply as well. Since the ISH and IHC analysis share similar preanalytical workflows, this allows the possibility to combine in situ hybridization and immunohistochemistry on the same slide. A combination of multiplex ISH and IHC also allows for the co-localization of gene and protein evaluation at the single-cell level and to understand cell-by-cell expression profiles, which is a powerful tool for understanding target engagement.

10.3 Summary

Biomarker analysis has drawn more and more attention from industrial professionals, academic researchers, and regulatory agencies. Biomarker method covers a wide range of analytical targets, technologies, and platforms; hence, understanding the context of substrate and distal biomarker measurements for gene therapy is critical to determine the extent of method development, and method validation work. In the current chapter, we discussed a variety of methodologies for biomarker quantitation for gene therapy, including LC-MS, histology and imaging, functional activity and immunoassays, and mRNA detection assays. The general overview, challenges for method development and validation, and respective regulatory guidelines were discussed, including selected examples and case studies. The intention of this chapter is to provide guidance for biomarker quantitation, method selection, development and validation for genomic therapy, standardize practices, and eventually lead to more reliable biomarker data during drug development.

References

1 FDA. The Drug Development Process. https://www.fda.gov/patients/learn-about-drug-and-device-approvals/drug-development-process (accessed 21 November 2023).

2 FDA (2021). Research C for DE. Table of Surrogate Endpoints That Were the Basis of Drug Approval or Licensure. *FDA*. 1–19.

3 Lee, J.W., Devanarayan, V., Barrett, Y.C. et al. (2006). Fit-for-purpose method development and validation for successful biomarker measurement. *Pharm. Res.* 23: 312–328.

4 Strimbu, K. and Tavel, J.A. (2010). What are biomarkers? *Curr. Opin. HIV AIDS* 5: 463–466.

5 Chiswick, E.L., Duffy, E., Japp, B. et al. (2012). Detection and quantification of cytokines and other biomarkers. *Methods Mol. Biol.* 15–30.

6 Liu, C., Chu, D., Kalantar-Zadeh, K. et al. (2021). Cytokines: from clinical significance to quantification. *Adv. Sci.* 8: 2004433.

7 Ganau, M., Bosco, A., Palma, A. et al. (2015). A DNA-based nano-immunoassay for the label-free detection of glial fibrillary acidic protein in multicell lysates. *Nanomedicine* 11: 293–300.

8 Yeung, D., Ciotti, S., Purushothama, S. et al. (2016). Evaluation of highly sensitive immunoassay technologies for quantitative measurements of sub-pg/mL levels of cytokines in human serum. *J. Immunol. Methods* 437: 53–63.

9 Lasseter, H.C., Provost, A.C., Chaby, L.E. et al. (2020). Cross-platform comparison of highly sensitive immunoassay technologies for cytokine markers: Platform performance in post-traumatic stress disorder and Parkinson's disease. *Cytokine X* 2: 100027.

10 Collings, F.B. and Vaidya, V.S. (2008). Novel technologies for the discovery and quantitation of biomarkers of toxicity. *Toxicology* 245: 167–174.

11 Hotta, K., Wang, Y., and Mano, Y. (2020). A sensitive bioanalytical assay for methylcobalamin, an endogenous and light-labile substance, in human plasma by liquid chromatography with tandem mass spectrometry and its application to a pharmacokinetic study. *J. Pharm. Biomed. Anal.* 191: 113621.

12 Njumbe Ediage, E., Dillen, L., Vroman, A. et al. (2018). Development of an LC–MS method to quantify coproporphyrin I and III as endogenous biomarkers for drug transporter-mediated drug-drug interactions. *J. Chromatogr. B* 1073: 80–89.

13 Onorato, J.M., Shipkova, P., Minnich, A. et al. (2014). Challenges in accurate quantitation of lysophosphatidic acids in human biofluids. *J. Lipid Res.* 55: 1784–1796.

14 Kano, K., Matsumoto, H., Kono, N. et al. (2021). Suppressing postcollection lysophosphatidic acid metabolism improves the precision of plasma LPA quantification. *J. Lipid Res.* 62: 100029.

15 Rigo-Bonnin, R. and Canalias, F. (2019). Traceability of immunosuppressant's mass concentration results obtained using different commercial calibrators. *Clin. Biochem.* 63: 113–120.

16 Becker, G.W. (2008). Stable isotopic labeling of proteins for quantitative proteomic applications. *Brief Funct. Genomic Proteomic.* 7: 371–382.

17 Gao, X., Lin, H., Krantz, C. et al. (2016). Quantitative analysis of factor P (Properdin) in monkey serum using immunoaffinity capturing in combination with LC–MS/MS. *Bioanalysis* 8: 425–438.

18 CDER. Considerations for Use of Histopathology and Its Associated Methodologies to Support Biomarker Qualification. *FDA Guidance for Industry*, May. https://www.fda.gov/regulatory-information/search-fda-guidance-documents/considerations-use-histopathology-and-its-associated-methodologies-support-biomarker-qualification (accessed 21 November 2023).

19 Thurberg, B.L., Rennke, H., Colvin, R.B. et al. (2002). Globotriaosylceramide accumulation in the fabry kidney is cleared from multiple cell types after enzyme replacement therapy. *Kidney Int.* 62: 1933–1946.

20 Barisoni, L., Jennette, J.C., Colvin, R. et al. (2012). Novel quantitative method to evaluate globotriaosylceramide inclusions in renal peritubular capillaries by virtual microscopy in patients with fabry disease. *Arch. Pathol. Lab. Med.* 136: 816–824.

21 Lara, H., Li, Z., Abels, E. et al. (2021). Quantitative image analysis for tissue biomarker use: a white paper from the digital pathology association. *Appl. Immunohistochem. Mol. Morphol.* 29: 479–493.

22 Barisoni, L., Lafata, K.J., Hewitt, S.M. et al. (2020). Digital pathology and computational image analysis in nephropathology. *Nat. Rev. Nephrol.* 16: 669–685.

23 Pantanowitz, L., Sinard, J.H., Henricks, W.H. et al. (2013). Validating whole slice imaging for diagnostic purposes in Pathology: Guideline from the College of American pathologists Pathology and Laboratory Quality Center. *Arch. Pathol. Lab. Med.* 137: 1710–1722.

24 Rolls, G. (2022). *BETTER HISTOLOGY*, https://www.leicabiosystems.com/us/knowledge-pathway/101-steps-to-better-histology-a-practical-guide-to-good-histology-practice/ (accessed 24 October).

25 U.S. Department of Health and Human Services, Food and Drug Administration, Center for Drug Evaluation and Research, et al. Fabry Disease: Developing Drugs for Treatment Guidance for Industry.

26 Raskovalova, T., Deegan, P.B., Mistry, P.K. et al. (2020). Accuracy of chitotriosidase activity and CCL18 concentration in assessing type I Gaucher disease severity. A systematic review with meta-analysis of individual participant data. *Haematologica* 106: 437–445.

27 Hollak, C.E., van Weely, S., van Oers, M.H. et al. (1994). Marked elevation of plasma chitotriosidase activity. A novel hallmark of Gaucher disease. *J. Clin. Invest.* 93: 1288–1292.

28 EASL Clinical Practice Guidelines (2012). Wilson's disease. *J. Hepatol.* 56: 671–685.

29 Bustin, S. (2000). Absolute quantification of mRNA using real-time reverse transcription polymerase chain reaction assays. *J. Mol. Endocrinol.* 25: 169–193.

30 Ho-Pun-Cheung, A., Cellier, D., and Lopez-Crapez, E. Considerations for normalisation of RT-qPCR in oncology. *Ann. Biol. Clin. (Paris)* 66: 121–129.

31 Adams, G. (2020). A beginner's guide to RT-PCR, qPCR and RT-qPCR. *Biochem. (Lond)* 42: 48–53.

32 Yang, X., Kui, L., Tang, M. et al. (2020). High-throughput transcriptome profiling in drug and biomarker discovery. *Front. Genet.* 11: https://doi.org/10.3389/fgene.2020.00019.

33 Reichlen, M.J., Vepachedu, V.R., Murakami, K.S. et al. (2012). MreA functions in the global regulation of methanogenic pathways in *Methanosarcina acetivorans. mBio* 3: https://doi.org/10.1128/mBio.00189-12.

34 Roberts, J.P. (2019). Microarrays still have their uses and fans. BioCompare, October 10, 2019 https://www.biocompare.com/Editorial-Articles/518117-Microarrays-Still-Have-Their-Uses-and-Fans/#:~:text=Microarrays%20still%20play%20a%20significant,approved%2C%E2%80%9D%20points%20out%20Head (accessed 27 September 2023).

35 Bumgarner R. Overview of DNA microarrays: types, applications, and their future. *Curr. Protoc. Mol. Biol.*; 101. Epub ahead of print January 2013. https://doi.org/10.1002/0471142727.mb2201s101.

36 Geiss, G.K., Bumgarner, R.E., Birditt, B. et al. (2008). Direct multiplexed measurement of gene expression with color-coded probe pairs. *Nat. Biotechnol.* 26: 317–325.

37 Wang, A.M., Doyle, M.V., and Mark, D.F. (1989). Quantitation of mRNA by the polymerase chain reaction. *Proc. Natl. Acad. Sci. U.S.A.* 86: 9717–9721.

38 Real-time RT-PCR Primers and Probes. Updated June 6, 2020.

39 VanGuilder, H.D., Vrana, K.E., and Freeman, W.M. (2008). Twenty-five years of quantitative PCR for gene expression analysis. *Biotechniques* 44: 619–626.

40 Giulietti, A., Overbergh, L., Valckx, D. et al. (2001). An overview of real-time quantitative PCR: applications to quantify cytokine gene expression. *Methods* 25: 386–401.

41 SEQC/MAQC-III Consortium (2014). A comprehensive assessment of RNA-seq accuracy, reproducibility and information content by the Sequencing Quality Control Consortium. *Nat. Biotechnol.* 32: 903–914.

42 Livak, K.J. and Schmittgen, T.D. (2001). Analysis of relative gene expression data using real-time quantitative PCR and the $2{-}\Delta\Delta CT$ method. *Methods* 25: 402–408.

43 Wong, M.L. and Medrano, J.F. (2005). Real-time PCR for mRNA quantitation. *Biotechniques* 39: 75–85.

44 Patel, V.M., Balakrishnan, K., Douglas, M. et al. (2017). Duvelisib treatment is associated with altered expression of apoptotic regulators that helps in sensitization of chronic lymphocytic leukemia cells to venetoclax (ABT-199). *Leukemia* 31: 1872–1881.

45 Bomben, R., Ferrero, S., D'Agaro, T. et al. (2018). A B-cell receptor-related gene signature predicts survival in mantle cell lymphoma: results from the Fondazione Italiana Linfomi MCL-0208 trial. *Haematologica* 103: 849–856.

46 Rasmussen, R. (2001). Quantification on the light cycler. In: *Rapid Cycle Real-Time PCR*, 21–34. Berlin, Heidelberg: Springer Berlin Heidelberg.

47 Dwivedi, N., Mondal, S., Smitha, P.K. et al. (2020). Relative quantification of BCL2 mRNA for diagnostic usage needs stable uncontrolled genes as reference. *PLoS One* 15: e0236338.

48 Ma, H., Bell, K.N., and Loker, R.N. (2021). qPCR and qRT-PCR analysis: regulatory points to consider when conducting biodistribution and vector shedding studies. *Mol. Ther. Methods Clin. Dev.* 20: 152–168.

49 Brown, D., Altermatt, M., Dobreva, T. et al. (2021). Deep parallel characterization of AAV tropism and AAV-mediated transcriptional changes via single-cell RNA sequencing. *Front. Immunol.* 12: https://doi.org/10.3389/fimmu.2021.730825.

50 Liu, X., Zhao, J., Xue, L. et al. (2022). A comparison of transcriptome analysis methods with reference genome. *BMC Genomics* 23: 232.

51 Speranza, E., Altamura, L.A., Kulcsar, K. et al. (2017). Comparison of transcriptomic platforms for analysis of whole blood from ebola-infected cynomolgus macaques. *Sci. Rep.* 7: 14756.

52 Ayers, J.I., Fromholt, S., Sinyavskaya, O. et al. (2015). Widespread and efficient transduction of spinal cord and brain following neonatal AAV injection and potential disease modifying effect in ALS mice. *Mol. Ther.* 23: 53–62.

53 Corchete, L.A., Rojas, E.A., Alonso-López, D. et al. (2020). Systematic comparison and assessment of RNA-seq procedures for gene expression quantitative analysis. *Sci. Rep.* 10: 19737.

54 Chen, J.-X., He, S., Wang, Y.-J. et al. (2021). Comprehensive analysis of mRNA expression profiling and identification of potential diagnostic biomarkers in coronary artery disease. *ACS Omega* 6: 24016–24026.

55 Bustin, S.A., Benes, V., Garson, J.A. et al. (2009). The MIQE guidelines: minimum information for publication of quantitative real-time PCR experiments. *Clin. Chem.* 55: 611–622.

56 Udvardi, M.K., Czechowski, T., and Scheible, W.-R. (2008). Eleven golden rules of quantitative RT-PCR. *Plant Cell* 20: 1736–1737.

57 FDA, CDER (2018). *Bioanalytical Method Validation Guidance for Industry Biopharmaceutics Bioanalytical Method Validation Guidance for Industry Biopharmaceutics Contains Nonbinding Recommendations*, https://www.fda.gov/files/drugs/published/Bioanalytical-Method-Validation-Guidance-for-Industry.pdf.

58 *Committee for Medicinal Products for Human Use ICH guideline M10 on bioanalytical method validation and study sample analysis Step5*, www.ema .europa.eu/contact (2022).

59 CBER (2006). *Guidance for Industry: Gene Therapy Clinical Trials – Observing Subjects for Delayed Adverse Events*, http://www.fda.gov/cber/guidelines.htm (accessed 21 November 2023).

60 FDA, CBER (2020). *Long Term Follow-Up After Administration of Human Gene Therapy Products; Guidance for Industry*, https://www.fda.gov/regulatory-information/search-fda-guidance-documents/long-term-follow-after-administration-human-gene-therapy-products (accessed 21 November 2023).

61 Medicines Agency E (2014). *Guideline on the quality, non-clinical and clinical aspects of gene therapy medicinal products*, https://www.ema.europa.eu/en/documents/scientific-guideline/guideline-quality-non-clinical-clinical-aspects-gene-therapy-medicinal-products_en.pdf (accessed 21 November 2023).

62 *Expectations for Biodistribution (BD) Assessments for Gene Therapy (GT) Products*, https://admin.iprp.global/sites/default/files/2018-09/IPRP_GTWG_ReflectionPaper_BD_Final_2018_0713.pdf (accessed 21 November 2023).

63 Bedbrook, C.N., Deverman, B.E., and Gradinaru, V. (2018). Viral strategies for targeting the central and peripheral nervous systems. *Annu. Rev. Neurosci.* 41: 323–348.

64 Dupont, J.-B., Guo, J., Renaud-Gabardos, E. et al. (2020). AAV-mediated gene transfer restores a normal muscle transcriptome in a canine model of X-linked myotubular myopathy. *Mol. Ther.* 28: 382–393.

65 Gorovits, B., Marshall, J.-C., Smith, J. et al. (2019). Bioanalysis of adeno-associated virus gene therapy therapeutics: regulatory expectations. *Bioanalysis* 11: 2011–2024.

66 Marco-Puche, G., Lois, S., Benítez, J. et al. RNA-Seq perspectives to improve clinical diagnosis. *Front. Genet.* 10: https://doi.org/10.3389/fgene.2019.01152.

67 Branched DNA technology, https://www.mybiosource.com/learn/branched-dna-bdna-technology/.

68 Wang, F., Flanagan, J., Su, N. et al. (2012). RNAscope: a novel in situ RNA analysis platform for formalin-fixed, paraffin-embedded tissues. *J. Mol. Diagn.* 14: 22–29.

69 *RNASCOPE ® Manual Solutions Biomarker Analysis By Rna In Situ Hybridization Get Quantitative Molecular Detection with Morphological Context in a Single Assay*. (accessed 21 November 2023).

70 https://www.thermofisher.com/order/catalog/product/4316034?SID=srch-hj-4316034.

71 https://www.idtdna.com/pages/products/qpcr-and-pcr/gene-expression/primetime-qpcr-probes.

11

Detection of Cellular Immunity to Viral Capsids and Transgene Proteins

Maurus de la Rosa[1] *and Magdalena Tary-Lehmann*[2]

[1] *Sangamo Therapeutic Allée de la Nertière, Valbonne, France*
[2] *CTL-Contract Laboratory, Cellular Technology Limited, Shaker Heights, OH, USA*

11.1 Introduction

Gene delivery with recombinant adeno-associated virus (rAAV) vectors is one of the most promising gene therapy approaches, but immunity against the vector capsid or the transgene poses a major challenge to successful long-term efficacy in humans. The immunity to rAAV vectors can be divided into humoral antibody-mediated immunity and cellular T-cell-mediated immune responses.

11.1.1 Humoral and Cellular Immune Responses to Gene Therapy

A preexisting natural neutralizing antibody (NAb) response against the AAV capsid is a crucial hurdle for successful gene therapy treatments [1–4]. Preexisting NAb responses against the capsid can efficiently neutralize the application of gene therapy drug candidates [5]. Antibody responses against AAV serotypes are mostly switched, indicating the presence of cluster of differentiation CD4 T-cells [2]. This finding suggests a full antiviral immune response against a natural serotype, including cytotoxic CD8 T-cell responses, that could kill transduced cells. Patients with preexisting anti-AAV antibodies are often excluded from clinical trials to avoid any impact on safety and efficacy [1, 2, 6]. Additionally, a NAb response against the rAAV capsid post treatment in an initially rAAV antibody-negative patient may block re-dosing of the patient with the same vector.

Drug Development for Gene Therapy: Translational Biomarkers, Bioanalysis, and Companion Diagnostics, First Edition. Edited by Yanmei Lu and Boris Gorovits.

Antibody responses against the transgene of an AAV vector have the potential to limit drug efficacy [6–9]. In hemophilic patients, a persistent anti-transgene response could even limit protein replacement therapy. Hence, antibody responses to the AAV capsid and the transgene are carefully monitored in all rAAV clinical trials.

Although clinical trials have advanced the knowledge of interactions between rAAV vectors and the immune system, the role of T-cell immunity against viral capsids is still not fully understood. For AAV liver-tropic gene therapies, it is considered that rAAV capsid-specific cytotoxic CD8 T cells kill rAAV-transduced cells and are responsible for liver enzyme elevations and the loss of transgene expression [6]. This suggests a full anti-capsid T-cell immune response, where CD8 T cells might be supported by rAAV-specific CD4 helper T cells to exert an efficient cytotoxic immune response. However, a T-cell-mediated response in enzyme-linked immunosorbent spot (ELISPOT) analyses of peripheral blood mononuclear cells (PBMCs) could only be detected in some cases [1, 6, 10]. A clear correlation between the T-cell-mediated cellular immune response, an increase in liver enzymes, and a drop in transgene expression could not be observed for AAV liver-tropic gene therapies. However, for other gene therapy approaches like muscle-targeting gene therapies, T-cell immunity is considered a major driver of immunogenicity events and loss of transduced cells in patients [6, 10]. The lack of reliable predictive animal models also underlines the need for careful monitoring of T-cell responses in patients.

The attempt to understand these points by analyzing published clinical trial data is challenging due to different patient populations, sampling regimes, vector designs, and application routes. Moreover, the methods used to assess anti-vector immune responses in clinical trials are not standardized regarding sensitivity, methodology, cut-offs, sampling, sample quality, and reagents, adding additional complexity in evaluating clinical results from different trials. In fact, these parameters most likely contributed to the varying data sets described by different groups.

This chapter will focus on methods of detecting and characterizing T-cell-mediated cellular immune responses in clinical trials. We share our view on assay parameters that are considered relevant for high-quality data and hence an improved understanding of a crucial part of the immune response impacting efficacy and safety.

Noteworthy, we understand T-cell analytics as only one part of the clinical analytics required to improve the understanding of immune responses toward gene therapeutic vectors.

11.1.2 Selected Clinical Observations Showing the Lack of Understanding About T-Cell-Mediated Immune Responses and the Need for Sensitive T-Cell Analytics

While some clinical trials of rAAV gene therapy succeeded, even resulting in the approval of the first gene transfer therapies, others were less successful [1]. Mixed results have been obtained so far in gene therapy trials for hemophilia, with some trials achieving long-term efficacy, while others observed loss of transgene expression, hypothesized to be due to immune-mediated loss of transduced cells, often in conjunction with liver transaminase increases [1, 11]. However, although a correlation between the loss of transgene expression, liver transaminase increases, and cellular anti-AAV T-cell responses could be observed in some patients, this is not the case in other patients. A clear correlation could not be shown yet.

In the first trial of rAAV-mediated gene transfer of human FIX in hemophilia B patients with successful proof of concept, transgene expression was limited to approximately eight weeks, followed by a gradual decline in FIX expression that was accompanied by transient, asymptomatic liver transaminase increases [9]. It was hypothesized that both were caused by the T-cell-mediated elimination of transduced hepatocytes, elicited by capsid antigens. Subsequently, it was demonstrated that AAV-specific memory CD8 T cells are present in the human population and can be reactivated following rAAV-mediated gene transfer; prior infection with wild-type AAV with generation of capsid-specific memory T cells was hypothesized to be the main cause of the observed immune response [8, 12, 13]. In the first trial with long-term success in hemophilia B, patients with preexisting natural AAV-NAbs were excluded [14, 15]. Some patients who received high doses developed transient increases in liver transaminases that were successfully managed with glucocorticoids to suppress anti-rAVV T-cell responses and did not cause substantial loss of transgene expression.

In other trials, glucocorticoid treatment alleviated liver transaminase increases but did not prevent the loss of transgene expression. In a trial with six hemophilia B patients, liver transaminase increases followed by transgene loss were observed in all but one patient, and transgene loss could not be prevented by glucocorticoid treatment [16]. Loss of transgene expression was also observed in seven of eight hemophilia B patients in another trial [10]. In the patients with transgene loss in the low-dose cohort, there were neither concomitant transaminase increases nor signs of T-cell or antibody responses to the vector or transgene. Concomitant liver transaminase increases were only observed in the high-dose cohort, and capsid-directed T-cell activation was only observed in two of these patients by interferon (IFN)-gamma ELISPOT assay. While the transaminase increases and T-cell activation could be managed by glucocorticoid treatment, this did not stop the loss of transgene expression.

Gene therapy with rAAV vectors has also been explored for other indications, including delivery to the muscle, eye, or central nervous system [17]. A range of successful clinical trials have been reported for neuromuscular diseases of genetic origin, where liver transaminase increases and capsid-directed T-cell responses were often observed, but usually not associated with the loss of transgene [1]. However, in two trials in Duchenne muscular dystrophy, patients experienced rapid activation of the immune system with T-cell and antibody responses, and individual patients showed symptoms of complement activation and acute kidney injury [18, 19].

Cellular immune responses to the transgene have only been observed in a few clinical trials of rAAV-mediated gene transfer so far, mostly in the setting of intramuscular gene delivery [1], suggesting that the route of administration could increase the immunogenicity of the vector. Cytotoxic T-cell responses against the transgene have been reported in clinical trials with intramuscular delivery in Duchenne muscular dystrophy [20] and alpha-1 antitrypsin deficiency [21] and with intracerebral delivery in mucopolysaccharidosis Type IIIB [22].

Noteworthy, all patients treated with rAAV vectors develop antibodies against the viral capsids, indicating high probability of immune response against viral vectors [6, 10]. Moreover, these immune responses are switched antibody responses, like the naturally occurring anti-AAV antibodies, suggesting the involvement of CD4 T helper cells [2]. A preclinical model showed that blocking the cellular immune responses, and hence CD4 T helper cells, prevented the formation of anti-rAAV antibodies [7]. However, CD4 responses were rarely detected in clinical trials despite this. The reason for this might be that AAV-specific T cells are only partially circulating and rather tissue-resident, or are rare in numbers, despite the high rAVV doses.

In conclusion, despite the lack of understanding of the involvement of the T-cell-mediated cellular immune response, it is recommended to apply highly robust and sensitive methods designed to detect T-cell-mediated responses.

11.2 Methods for the Detection of Cellular Immune Responses

11.2.1 Methods to Detect T-Cell Responses in Clinical Trials

The sensitive detection and characterization of rAAV-specific T-cell populations is critical for understanding cellular immune responses during the development of gene therapeutic drugs. Evaluation of a combination of orthogonal T-cell assays or other assays assessing the immune status is therefore recommended, using suitable methods that are appropriately robust and sensitive. An overview of methods for detection of cellular immunity to viral capsids and transgene proteins is provided in this section.

11.2.1.1 Enzyme-Linked Immunosorbent Spot Assay

The ELISPOT (Figure 11.1) assay is a highly sensitive method that measures the frequency of antigen-specific T cells that secrete cytokines, such as IFN-gamma, by antigen-specific T cells [23–25]. The assay is based on the principle of capturing the cytokines secreted by T cells on a membrane coated with specific antibodies.

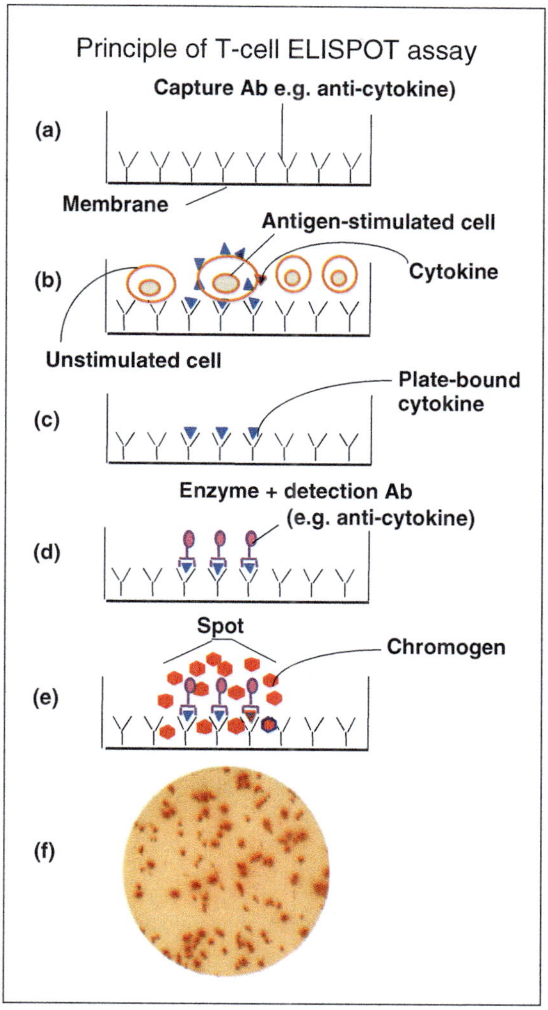

Figure 11.1 Schematic diagram of ELISPOT assay. Ninety-six well plates are coated with capture antibodies (a). PBMC are added to the wells with or without antigenic stimulus. As the individual T cells respond to the antigen (Ag) and secrete cytokines, the cytokines are bound by the antibody directly at the site of secretion (b). Cells are washed off (c), detection antibodies are added (d), and the spots are developed with chromogen (e). An example individual well from a 96-well ELISPOT plate is shown. Each spot represents the captured cytokine produced by a single T cell to the given recall antigen (f).

The captured cytokines are then detected using enzyme-linked antibodies, and the spots generated are counted to quantify the number of cytokine-secreting T cells in the presence of specific antigen. As antigens frequently peptide pools or other materials such as proteins, vectors, and drugs are used. This assay can be used to detect both CD4 and CD8 T-cell responses and is commonly used in gene therapy trials to evaluate the immune response to the therapy. Additionally, there are different ELISPOT variations. In the related FluoroSpot, the detection antibody is conjugated with a fluorescent dye [26–29]. The advantage of FluoroSpot over ELISPOT is that it allows for multiplexing, meaning multiple cytokines of one stimulated T cell can be detected simultaneously.

11.2.1.2 Intracellular Cytokine Staining

Intracellular cytokine staining (ICS) is a flow cytometry-based assay that measures the production of cytokines by T cells in response to stimulation with specific antigens [30–33]. The assay involves the stimulation of T cells with the antigen of interest and the addition of Brefeldin A, which inhibits the secretion of various cytokines on a single-cell basis. The cells are then stained with fluorescently labeled antibodies to identify the specific T-cell subsets and the cytokines produced. This assay is sensitive and allows to measure polyfunctional T cells, making it an ideal method for assessing the T-cell response to gene therapies. However, the ICS assay is considered less sensitive than the ELISPOT assay [30–32, 34].

11.2.1.3 Tetramer Staining

Tetramer staining is a flow cytometry-based assay that uses fluorescently labeled major histocompatibility complex (MHC) class I tetramers loaded with specific antigenic peptides to detect antigen-specific T cells [35]. The tetramers bind to the T-cell receptors (TCRs) of antigen-specific T cells and can be detected using flow cytometry. This method is specific and considered to have similar sensitivity as the ICS assay and can be used to detect low-frequency T cells that are able to bind the peptide of interest [36].

11.2.1.4 Proliferation Assays

This assay reveals the capability of T cells to proliferate upon ex vivo antigen stimulation. The assay involves the labeling of T cells with a fluorescent dye, such as carboxyfluorescein succinimidyl ester (CFSE), which is then diluted with each cell division [37]. Alternatively, proliferation can be assessed by 3H-thymidine incorporation into the DNA of the cell. The cells are then stimulated with the antigen of interest, and the dilution of the dye or 3H-thymidine incorporation is measured using flow cytometry or with a scintillation counter, respectively. It has been demonstrated that proliferation is correlated with interleukin-2 production [38].

11.2.1.5 Cytokine Bead Array

The cytokine bead array is a multiplexed assay that measures the secretion of multiple cytokines from liquids (e.g. serum, plasma, and supernatants) simultaneously [39]. For the assay, T cells are stimulated with specific antigens, and the supernatant is collected to measure cytokine secretion using fluorescently labeled antibodies specific to each cytokine. This method can be used to detect the secretion of multiple cytokines from T cells, providing a more comprehensive analysis of the T-cell response to gene therapies. However, it must be considered that cytokines can also be produced by other potentially co-cultured cells. This has to be taken into account additionally, in context of the frequency of antidrug-specific T-cell responses, to avoid data misinterpretation.

11.2.1.6 Gene Expression Profiling

Gene expression profiling is a molecular technique that can be used to identify the genes that are differentially expressed in T cells in response to gene therapy [40, 41]. The technique involves the isolation of T cells from blood or tissue samples and the measurement of gene expression using microarrays or RNA sequencing. The gene expression profiles can then be analyzed to identify the pathways and processes that are activated in T cells in response to the therapy. This method provides a comprehensive analysis of the T-cell response to gene therapy at the molecular level.

In combination with single-cell analysis, gene expression profiling can be a very powerful tool. However, validation of such an assay is considered to require the validation of complex processes [42].

11.2.1.7 Multiplexed Epitope Mapping

Multiplexed epitope mapping is a high-throughput method that can be used to identify the T-cell epitopes that are recognized in response to gene therapy [9, 43]. The technique involves the synthesis of overlapping peptide matrices that span the entire sequence of the antigen of interest. The peptides or matrix peptide pools are then screened for their ability to stimulate T-cell responses using ELISPOT or ICS assays. The T-cell responses to each peptide are then mapped to identify the specific epitopes that are recognized by T cells in response to the therapy.

11.2.1.8 Conclusion

In conclusion, the methods listed above are available to detect T-cell responses in gene therapy trials. There are additional options like TCR sequencing and IFN-gamma release assays or even in vivo assays using humanized mouse models. Each method has its advantages and limitations, and the choice of method depends on the specific needs of the study. A combination of these methods and other immune monitoring assays can provide a comprehensive analysis of the T-cell response to gene therapy.

11.2.2 Technical Challenges of Detecting Cellular Immune Responses

While B-cell immunity can be monitored by robust, reproducible, and validated assays for antibody detection, monitoring of T-cell immunity has been proven to be more challenging [44]. Interlaboratory standardization is necessary to allow immune monitoring in multicenter clinical trials and obtain reproducible results. It has been demonstrated that the lack of reproducibility of methods such as ICS, ELISPOT, and tetramer detection originates from a lack of assay standardization rather than from inherent difficulties in working with live cells and complex assay systems. However, sample and material quality contribute significantly to the quality of results.

Although in some publications the ELISPOT has been reported to have the lowest precision after tetramer staining and ICS, the ELISPOT is still considered a robust, sensitive, and powerful technique to measure immune activation when performed in a standardized manner using an optimized protocol [45]. Optimization of parameters to improve ELISPOT assay performance is described in Section 11.3.2. The authors highlight that optimization of parameters in Section 11.3.2 is an example that can be applied to other methods too, with certain limitations.

In general, other methods can be chosen to assess anti-AAV therapy T-cell responses, depending on the scope of the study and the frequency of specific T-cell responses. In fact, when standardized assays are used in the same samples, ELISPOT results have been shown to correlate with the results of other methods, such as cytokine flow cytometry [46].

11.3 Validation of Cellular Assays Using PBMC (Example ELISPOT)

11.3.1 Validation Strategies

Cell-mediated, antigen-specific qualification and validation of the recall antigen-specific ELISPOT method currently has no specific US Food and Drug Administration (FDA) guidance available. The concept of the FDA guidance on bioanalytical method validation [47] can be adapted to some extent to the ELISPOT assays when applicable. To start the qualification/validation of the given ELISPOT assay, the laboratory will need to have standard operating procedures in place, specifying how the assay is performed. The quality of the assay performance in the laboratory needs to be maintained by Quality Control (QC) and Quality Assurance (QA) for the duration of the study. A separate QC unit is not required in a Good Laboratory Practice (GLP)-regulated environment, but to QC the work performed is important for data integrity. A separate QA unit is required for inspections in regulated studies (GLP, GCP, etc.) and reporting to the Testing Facility Management. Once this is in place, qualification/validation of the test method is performed before testing the samples.

One challenge in ELISPOT assay qualification/validation is that recall antigen-specific PBMCs are evaluated and, ideally, other samples, e.g. antigen-positive reference samples, responding to the same antigens are required. Qualification/

validation should be planned for each antigen and cytokine readout to be evaluated. Validation approaches for cytokine and recall antigen-specific ELISPOT assays have been discussed and published in the scientific community [34] and recommendations provided by the Global CRO Council for Bioanalysis (GCC) [48].

In general, after the given recall antigen-specific ELISPOT assay has been optimized, the validation study is divided into three segments: (1) the validation plan; (2) the validation test data (including Phase 1 of the assay validation, the qualification, generating the acceptance criteria, and Phase 2 of the assay validation, confirming the performance of the selected acceptance criteria); and (3) the validation report.

The validation of the ELISPOT assay should be performed in an environment that is functionally the same as planned for analysis of the human clinical study samples (i.e. regulated environment, same equipment, same software, and same location). The analyte in the ELISPOT assay is defined as the number of spots secreted by the antigen-specific cell and captured on the ELISPOT plate for the given released product measured. Please note that analytical assay qualifications/validations usually fall under GLP regulations. The validated assay is then used for the clinical study sample testing. The testing of the clinical samples, depending on the intended use of the results obtained, may be performed under GCP or CLIA regulations.

11.3.1.1 Precision

The precision, determined during assay validation, is the variability in the data from replicate determinations of the same homogeneous sample with the same collection date. Precision data represent the scatter of data and can be established without having a "gold standard." For example, for a qualification experiment, a total of three different technicians would be performing the ELISPOT assay with the same samples and antigens on three different days, each running the experiment on different plates. The tested precision parameters are intra-assay precision, also called repeatability (well-to-well), and intermediate precision (day-to-day, analyst-to-analyst).

Reproducibility (laboratory-to-laboratory) is usually excluded from the validation plan, as no other laboratory is involved. However, laboratory-to-laboratory precision data have been evaluated and published [44]. The data clearly showed that reproducible and precise data were obtained by ELISPOT-inexperienced investigators when following standardized procedures and using standardized samples and materials. All laboratories that participated could detect the recall responses within three standard deviations (SDs) of the used reference values.

11.3.1.2 Specificity

Specificity in the ELISPOT assay is the ability to detect the number of PBMCs that are releasing, e.g. IFN-gamma that is directly bound to the plate, as determined by the antibodies used to capture and detect IFN-gamma. The cell population tested in ELISPOT assays is specific toward a certain antigen. The assay determines whether the sample includes cells with specificity to the test antigen. The secretion of the cytokine of interest is only induced and captured on the ELISPOT plate if the cells are specific to the tested antigen. Cells not activated by the test antigen will not secrete the tested cytokine. The ability of the ELISPOT assay to determine the antigen-specific and mitogen-specific lymphocyte responses depends on the antigens employed. The ability of the assay to identify IFN-gamma from antigen-stimulated cells versus spontaneously secreted cytokine is controlled by including cells exposed to medium only for each sample, referred to as medium control.

Overall, evaluation of ELISPOT assays can be categorized into empirical approaches and statistical approaches, and different groups utilize different approaches [49–54].

A formula for the calculation of antigen-specific positive response based on the results of each individual sample has been published in 2008 and has been performing well [34]. An antigen-specific response is defined as positive when both of the following conditions are met: (1) antigen-specific value (ASV) is greater than 3. ASV is calculated as X − (SD of X) − (Y + [2SD of Y]), where X is the average spot number per well induced by a given antigen (cells and antigen), Y is the average spot number in the medium control wells (cells and no antigen), and SD is standard deviation of the replicates. (2) Positive response determined as Y + O + 2SD of Y is greater than 10, where O is the average spot number in the negative control wells with only medium (no cells and no antigen). All negative values generated by the formula are reported as 0 spot values, as negative values are biologically irrelevant.

For all test conditions, mean spot-count-per-well data are generated. For mathematical reasons, ELISPOT data (such as any counts covering natural numbers) suffer from increasing variability the closer the mean value is to 0, as no linear correlation is maintained. It has been demonstrated that the data precision of mean spot counts decreased dramatically the closer the mean values were to 0. On the other hand, the precision of high mean spot counts can reach a constant and relatively low level [34, 55]. Spot counts below 30 spots per well should not be evaluated by percentage coefficient of variation (CV) but only by SD [55].

11.3.1.3 Limit of Detection and Range

The limit of detection (LOD) in the ELISPOT assay is the lowest number of spots that can be detected and counted in one well for one sample with no antigen stimulation. The lower LOD (LLOD) is the lowest number of spots that can be detected in wells with cells but without antigen and is statistically distinguishable from background or negative control (O), where no cells are plated in a given well. The upper LOD (ULOD), also known as upper limit of quantitation, is the highest number of spots that can be counted in that specific assay system, not reaching "'too numerous to count" (TNTC) by the used ELISPOT image analyzer.

The following acceptance values of the spot number in medium wells (LLOD) and in mitogen/positive control wells (ULOD) could be used in the recall antigen-specific regulated ELISPOT study:

$$\text{ULOD is less than TNTC} = \text{number defined by given assay system and used image analyzer (spots/well).}$$

The range in a human IFN-gamma ELISPOT assay will be defined by the LOD value and the ULOD below TNTC.

11.3.1.4 Common Exceptions for ELISPOT Validation: Accuracy, Linearity, and Reproducibility

Exceptions from the bioanalytical method validation guidance are commonly made for ELISPOT validation for accuracy, linearity, and reproducibility. Accuracy is defined as the closeness of agreement between the value determined by the assay and the value that is accepted as a true value or an accepted reference value. As no true reference values are available for this assay, accuracy can be determined by performing measurements employing two different test methods. The same test samples may be used for the IFN-gamma ELISPOT assay and an orthogonal method, IFN-gamma ICS. Both IFN-gamma measurements can provide the percentage of responding cell population to the same antigen. This percentage or frequency measurement allows comparison of values generated by the two systems. The test sample used for these experiments, however, must have high frequencies as measured in ELISPOT (e.g. more than 400 spots to a given antigen in 400,000 cells per well) to qualify for testing by ICS and for accuracy validation. Accuracy can be determined, but it is not useful for ELISPOT assays in a clinical setting as not enough cells are available from the clinical samples to evaluate accuracy by the proposed methods in a routine manner, and the response frequency to the individual antigen is not known. Therefore, accuracy is usually not evaluated. Moreover, determination of accuracy is not necessary because ELISPOT is not a quantitative assay. It does not quantify the amount of IFN-gamma produced but reports the results as number of positive or no spots counted that produced IFN-gamma if stimulated by the recall antigen. It is also expected that different tested individuals react differently to the same antigen. However, it has been demonstrated that the ELISPOT assay can be performed accurately, using the above described assay systems [30, 31, 56–58].

Linearity is the ability of the assay to return values that are directly proportional to the concentration of analyte in the sample. Mathematical data transformations to promote linearity may be allowed if there is scientific evidence that the transformation is appropriate for the method. In general, linearity of the ELISPOT method has been explored in multiple scientific publications, providing evidence for the single-cell resolution of the ELISPOT assay [30, 31, 56–58]. Linearity can be evaluated, however, is not suited for the clinical sample testing as it is cumbersome, expensive, and requires a lot of cell material and antigens, which is unpractical or even impossible to obtain. Moreover, determination of linearity is not required, considering that negative and positive controls are used in the ELISPOT assay run for each clinical sample tested. The medium control commonly employs culture medium with the PBMC samples without antigen stimulation, and the positive control employs a mitogen such as phytohemagglutinin for each sample tested. These control samples are tested in parallel and as a reference for comparison to the antigen recall (cell and antigen) sample.

11.3.2 Parameters Affecting ELISPOT Assay Performance

11.3.2.1 PBMC Sample Handling: Temperature, Resting, and Serum

ELISPOT assays use functional PBMCs, either freshly isolated from blood or cryopreserved. As the PBMCs are the sample and the most critical component of the ELISPOT assay, handling of the whole blood from which the PBMCs are derived needs to be standardized in a way that maintains functionality from the time of blood draw throughout handling, shipment and separation, and cryopreservation. Cryopreservation of PBMCs makes clinical trial measurements of cell-mediated recall antigen-specific immune responses feasible, allows standardization of the ELISPOT assay and batch testing of samples from multiple timepoints from a given patient, and reduces the cost of the assays. For this reason, it has found wide use in immune monitoring studies [9, 59–63]. Standardization of cryopreservation methods of PBMCs to maintain functionality has been achieved, as confirmed by recall antigen-specific ELISPOT assays for T cells, B cells, natural killer cells, and others, when whole blood is handled appropriately and kept at room temperature starting from phlebotomy, packaging, shipment to the whole blood processing laboratory, separation of PBMCs, and cryopreservation [55]. The cryopreserved PBMCs were found to perform equally well to freshly non-frozen, separated, and tested PBMCs in the diverse recall antigen-specific ELISPOT assays for CD4 and CD8 T cells for Th1 and Th2 readouts [55].

It has been established that keeping the whole blood at room temperature during the entire sample handling process is critical, as even exposure to 4 °C for 20 minutes reduced the functionality of PBMC samples by a minimum of 30% [55] (also illustrated by Figure 11.2). This temperature sensitivity of PBMCs was also observed when cold freezing solutions (e.g. dimethyl sulfoxide solutions) were used for cryopreservation, as this also reduced the functionality of the thawed PBMCs [55]. The cell recovery and functionality are critically temperature-dependent (see Figure 11.2) rather than time-dependent [55] (see Figure 11.3).

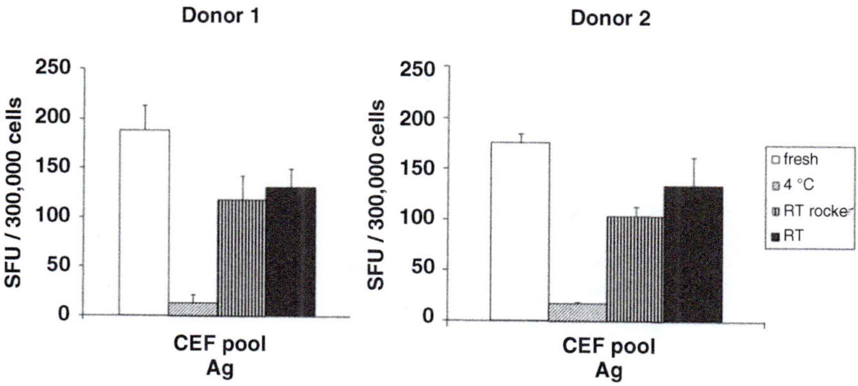

Figure 11.2 Storage temperature of whole blood influences the functionality of the cryopreserved and thawed PBMCs in the recall antigen-specific IFN-gamma ELISPOT assay.

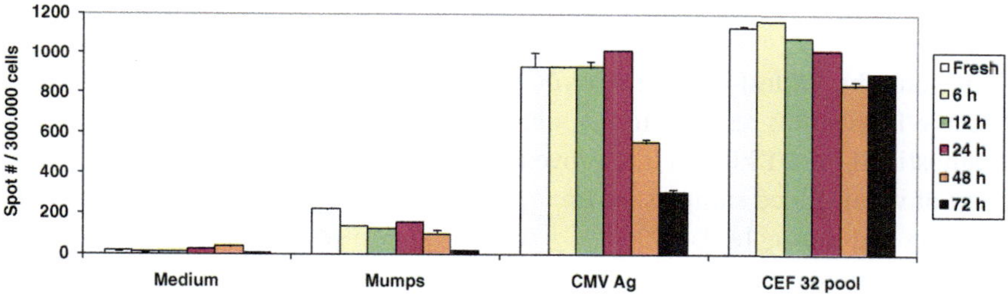

Figure 11.3 Cryopreserved PBMCs generated from whole blood at different times after blood draw are tested in a recall antigen-specific IFN-gamma ELISPOT assay.

In addition to T cells, such observations have also been reported for antigen-specific B-cell readouts from cryopreserved PBMCs [64]. In addition to the logistics, separation, and freezing processes, optimal thawing procedures of the cryopreserved PBMC should be used, making the handling of PBMCs complex and requiring trained personnel to do so, in order not to jeopardize the results of the ELISPOT assay and with it the clinical trial it is supporting [65].

Figure 11.3 demonstrates that whole blood, handled appropriately (using serum-free solutions) and shipped at room temperature before processing to cryopreserved PBMCs, can maintain functionality compared with freshly processed blood. There was no statistical difference in recall antigen-specific IFN-gamma ELISPOT assays of cryopreserved PBMCs from whole blood kept at room temperature when PBMCs were separated immediately or at 6, 12, and 24 hours, confirming previously published results [55].

If whole blood samples are handled and processed correctly, fresh and cryopreserved PBMCs perform equally well [55], as shown in Figure 11.4 for the recall antigen cytomegalovirus (CMV) peptide pool.

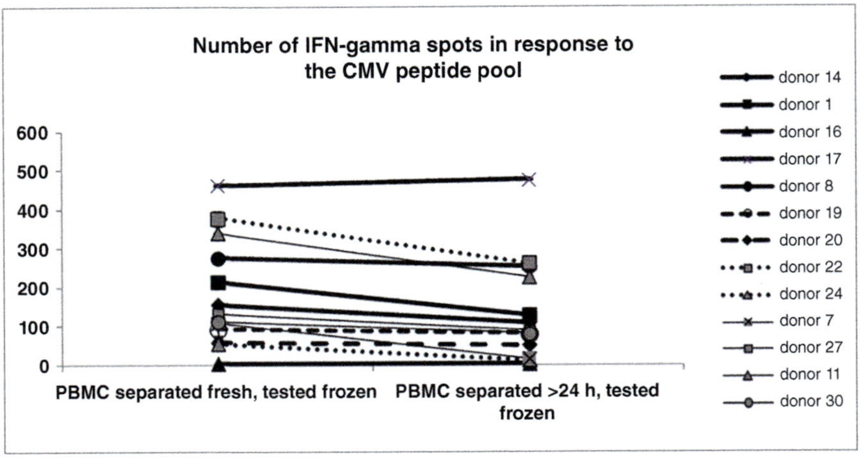

Figure 11.4 Fresh and cryopreserved PBMC perform equally well in recall antigen-specific ELISPOT assay (CMV peptide pool).

Of note, the *TB* test is an in vitro diagnostic ELISPOT test for the detection of effector T cells that respond to stimulation by Mycobacterium *tuberculosis*. This diagnostic ELISPOT kit that is approved by the FDA (T-SPOT TB test) allows shipment of whole blood samples for up to 54 hours post venipuncture before PBMC isolation and testing [66].

For the standardization of separation and cryopreservation of PBMCs from whole blood for use in recall, antigen-specific assays such as ELISPOT and serum-free solutions are recommended for washing, cryopreservation, thawing, and testing. Even if serum-containing solutions have been screened and qualified for cell culture use, an approximately 30% effect on the human sample population can still be expected due to toxicity or mitogenic effects of serum on the PBMC samples used in the recall antigen-specific ELISPOT assay, which can affect the results. It has been established that, while reproducible results among laboratories can be obtained using serum-free solutions, different qualified sera used in the different laboratories can introduce variability in the ELISPOT assay results [44].

Some laboratories prefer to let the thawed cryopreserved PBMCs rest, which was established for $CD8^+$ T cells that do not require APCs, using the CEF peptide pool (derived from CMV, Epstein–Barr virus, and flu virus epitopes), presented by 11 Class 1 HLA-A and HLA-B alleles [67–70].

However, it has been shown that resting of thawed PBMCs is not generally beneficial in comparison to direct ex vivo ELISPOT assays performed [71]. It doubles the required cell numbers for the assay, increases complexity of work, and with it the cost. It might lead to loss of effector cells, as these are sticky and have thin membranes. The activation of the cells increases the cell size and thins the membrane, making them more vulnerable when handled. Accordingly, effector T cells or plasma B cells might be selectively lost over memory T and B cells, which have a thicker membrane and are smaller, and thus more resistant to handling. In addition, if antigen processing by APCs is required for more complex recall antigens (proteins, viruses, virus particles, etc.) then the results of the ELISPOT testing might be more impacted by resting. The scientific community is split on this direct ex vivo and resting topic.

Investigators should consider all these factors and test if resting would be beneficial or not for their given study and the given recall antigens to be used in the recall antigen-specific ELISPOT assay.

11.3.2.2 Antigen Concentration and Number of Replicates

Besides the standardization of the sample material, that is, the PBMCs, the antigen intended to be used needs to be optimized to qualify and validate the recall antigen-specific ELISPOT assay. The optimal antigen concentration will need to be determined, as the concentration of the antigen will influence the obtained results [57]. The antigen concentration will influence the number of spots recalled from the PBMCs (i.e. whether all cells that are present or only a fraction of the cells respond), as well as the quality of the spots measured, leading to distinct spots with reduced signal-to-noise ratio.

The number of replicates required for the ELISPOT assay depends on the assay system, the antigen, the affinity with which the T cells are assumed to respond to the antigen, and the frequency of the specific T-cell population within the PBMC sample [54].

The FDA-approved T-SPOT kit uses single wells for their recall antigen in the IFN-gamma ELISPOT assay [66]. The usage of triplicates is more common in the scientific community, and sometimes even six-well replicates have been used if the frequency of the recall antigen-specific T cells was very low. The decision on the required number of replicates for testing of clinical samples should be made during the optimization phase of the recall antigen-specific ELISPOT assay and tested with the decision made after Phase 1 of the validation/qualification have been completed and acceptance criteria have been generated (see Section 3.1) since this provides the precision data.

In summary, recall antigen-specific ELISPOT assays are very robust and can be qualified and validated to support clinical trials in all phases of the drug lifecycle. The most crucial component for the ELISPOT assay is the quality of the PBMCs provided, which should be carefully considered.

References

1 Ronzitti, G., Gross, D.-A., and Mingozzi, F. (2020). Human immune responses to adeno-associated virus (AAV) vectors. *Front. Immunol.* 11: 670.

2 Kruzik, A., Fetahagic, D., Hartlieb, B. et al. (2019). Prevalence of anti-adeno-associated virus immune responses in international cohorts of healthy donors. *Mol. Ther. Methods Clin. Dev.* 14: 126–133.

3 Kruzik, A., Koppensteiner, H., Fetahagic, D. et al. (2019). Detection of biologically relevant low-titer neutralizing antibodies against adeno-associated virus require sensitive in vitro assays. *Hum. Gene Ther. Methods* 30 (2): 35–43.

4 Wang, L., Calcedo, R., Bell, P. et al. (2011). Impact of pre-existing immunity on gene transfer to nonhuman primate liver with adeno-associated virus 8 vectors. *Hum. Gene Ther.* 22 (11): 1389–1401.

5 Falese, L., Sandza, K., Yates, B. et al. (2017). Strategy to detect pre-existing immunity to AAV gene therapy. *Gene Ther.* 24 (12): 768–778.

6 Mingozzi, F. and High, K.A. (2013). Immune responses to AAV vectors: overcoming barriers to successful gene therapy. *Blood* 122 (1): 23–36.

7 Frentsch, M., Japp, A.S., Dingeldey, M. et al. (2020). Blockade of the costimulatory CD28-B7 family signal axis enables repeated application of AAV8 gene vectors. *J. Thromb. Haemost.* 18 (5): 1075–1080.

8 Mingozzi, F. and High, K.A. (2011). Immune responses to AAV in clinical trials. *Curr. Gene Ther.* 11 (4): 321–330.

9 Manno, C.S., Pierce, G.F., Arruda, V.R. et al. (2006). Successful transduction of liver in hemophilia by AAV-Factor IX and limitations imposed by the host immune response. *Nat. Med.* 12 (3): 342–347.

10 Konkle, B.A., Walsh, C.E., Escobar, M.A. et al. (2021). BAX 335 hemophilia B gene therapy clinical trial results: potential impact of CpG sequences on gene expression. *Blood* 137 (6): 763–774.

11 Hamilton, B.A. and Wright, J.F. (2021). Challenges posed by immune responses to AAV vectors: addressing root causes. *Front. Immunol.* 12: 675897.

12 Mingozzi, F., Maus, M.V., Hui, D.J. et al. (2007). CD8(+) T-cell responses to adeno-associated virus capsid in humans. *Nat. Med.* 13 (4): 419–422.

13 Mingozzi, F. and High, K.A. (2007). Immune responses to AAV in clinical trials. *Curr. Gene Ther.* 7 (5): 316–324.

14 Nathwani, A.C., Tuddenham, E.G., Rangarajan, S. et al. (2011). Adenovirus-associated virus vector-mediated gene transfer in hemophilia B. *New Engl. J. Med.* 365 (25): 2357–2365.

15 Nathwani, A.C., Reiss, U.M., Tuddenham, E.G. et al. (2014). Long-term safety and efficacy of factor IX gene therapy in hemophilia B. *New Engl. J. Med.* 371 (21): 1994–2004.

16 Pipe, S., Stine, K., Rajasekhar, A. et al. (2017). 101HEMB01 is a phase 1/2 open-label, single ascending dose-finding trial of DTX101 (AAVrh10FIX) in patients with moderate/severe hemophilia B that demonstrated meaningful but transient expression of human factor IX (hFIX). *Blood* 130 (Supplement 1): 3331.

17 Ertl, H.C.J. (2021). T cell-mediated immune responses to AAV and AAV vectors *Front. Immunol.* 12: 666666.

18 Solid Biosciences Provides SGT-001 Program Update [press release]. https://www.solidbio.com/about/media/press-releases [internet]: Solid Biosciences Inc.2019.

19 Pfizer Presents Initial Clinical Data on Phase 1b Gene Therapy Study for Duchenne Muscular Dystrophy (DMD) [press release]. https://www.pfizer.com/news/press-release [internet]: Pfizer Inc 2019.

20 Mendell, J.R., Campbell, K., Rodino-Klapac, L. et al. (2010). Dystrophin immunity in Duchenne's muscular dystrophy. *New Engl. J. Med.* 363 (15): 1429–1437.

21 Calcedo, R., Somanathan, S., Qin, Q. et al. (2017). Class I-restricted T-cell responses to a polymorphic peptide in a gene therapy clinical trial for α-1-antitrypsin deficiency. *Proc. Natl. Acad. Sci.* 114 (7): 1655–1659.

22 Tardieu, M., Zérah, M., Gougeon, M.-L. et al. (2017). Intracerebral gene therapy in children with mucopolysaccharidosis type IIIB syndrome: an uncontrolled phase 1/2 clinical trial. *Lancet Neurol.* 16 (9): 712–720.

23 Taguchi, T., McGhee, J.R., Coffman, R.L. et al. (1990). Analysis of Th1 and Th2 cells in murine gut-associated tissues. Frequencies of CD4+ and CD8+ T cells that secrete IFN-gamma and IL-5. *J. Immunol.* 145 (1): 68–77.

24 Forsthuber, T., Yip, H.C., and Lehmann, P.V. (1996). Induction of TH1 and TH2 immunity in neonatal mice. *Science.* 271 (5256): 1728–1730.

25 Slota, M., Lim, J.B., Dang, Y., and Disis, M.L. (2011). ELISpot for measuring human immune responses to vaccines. *Expert Rev. Vaccines* 10 (3): 299–306.

26 Ahlborg, N. and Axelsson, B. (2012). Dual- and triple-color fluorospot. *Methods Mol. Biol.* 792: 77–85.

27 Rebhahn, J.A., Bishop, C., Divekar, A.A. et al. (2008). Automated analysis of two- and three-color fluorescent Elispot (Fluorospot) assays for cytokine secretion. *Comput. Methods Programs Biomed.* 92 (1): 54–65.

28 Chauvat, A., Benhamouda, N., Gey, A. et al. (2014). Clinical validation of IFNγ/IL-10 and IFNγ/IL-2 FluoroSpot assays for the detection of Tr1 T cells and influenza vaccine monitoring in humans. *Hum. Vaccines Immunother.* 10 (1): 104–113.

29 Gazagne, A., Claret, E., Wijdenes, J. et al. (2003). A Fluorospot assay to detect single T lymphocytes simultaneously producing multiple cytokines. *J. Immunol. Methods* 283 (1): 91–98.

30 Heeger, P.S., Greenspan, N.S., Kuhlenschmidt, S. et al. (1999). Pretransplant frequency of donor-specific, IFN-gamma-producing lymphocytes is a manifestation of immunologic memory and correlates with the risk of posttransplant rejection episodes. *J. Immunol.* 163 (4): 2267–2275.

31 Helms, T., Boehm, B.O., Asaad, R.J. et al. (2000). Direct visualization of cytokine-producing recall antigen-specific CD4 memory T cells in healthy individuals and HIV patients. *J. Immunol.* 164 (7): 3723–3732.

32 Tassignon, J., Burny, W., Dahmani, S. et al. (2005). Monitoring of cellular responses after vaccination against tetanus toxoid: comparison of the measurement of IFN-γ production by ELISA, ELISPOT, flow cytometry and real-time PCR. *J. Immunol. Methods* 305 (2): 188–198.

33 Maecker, H.T., Maino, V.C., and Picker, L.J. (2000). Immunofluorescence analysis of T-cell responses in health and disease. *J. Clin.Immunol.* 20 (6): 391–399.

34 Tary-Lehmann, M., Hamm, C.D., and Lehmann, P.V. (2008). Validating reference samples for comparison in a regulated ELISPOT assay. In: *Validation of Cell-Based Assays in the GLP Setting* (ed. U. Prabhakar and M. Kelley), 127–146. Chichester: John Wiley & Sons.

35 Sims, S., Willberg, C., and Klenerman, P. (2010). MHC–peptide tetramers for the analysis of antigen-specific T cells. *Expert Rev. Vaccines* 9 (7): 765–774.

36 Berner, B.R., Tary-Lehmann, M., Yonkers, N.L. et al. (2005). Phenotypic and functional analysis of EBV-specific memory CD8 cells in SLE. *Cell. Immunol.* 235 (1): 29–38.

37 Mannering, S.I., Morris, J.S., Jensen, K.P. et al. (2003). A sensitive method for detecting proliferation of rare autoantigen-specific human T cells. *J. Immunol. Methods* 283 (1-2): 173–183.

38 Anthony, D.D., Milkovich, K.A., Zhang, W. et al. (2012). Dissecting the T cell response: proliferation assays vs. Cytokine Signatures by ELISPOT. *Cells* 1 (2): 127–140.

39 Morgan, E., Varro, R., Sepulveda, H. et al. (2004). Cytometric bead array: a multiplexed assay platform with applications in various areas of biology. *Clin. Immunol.* 110 (3): 252–266.

40 Zhao, S., Fung-Leung, W.-P., Bittner, A. et al. (2014). Comparison of RNA-Seq and microarray in transcriptome profiling of activated T cells. *PLoS One* 9 (1): e78644.

41 Piersanti, S., Martina, Y., Cherubini, G. et al. (2004). Use of DNA microarrays to monitor host response to virus and virus-derived gene therapy vectors. *Am. J. Pharmacogenomics* 4 (6): 345–356.

42 Boyle, T.A., Mondal, A.K., Saeed-Vafa, D. et al. (2021). Guideline-adherent clinical validation of a comprehensive 170-gene DNA/RNA panel for determination of small variants, copy number variations, splice variants, and fusions on a next-generation sequencing platform in the CLIA setting. *Front. Genet.* 12: 503830.

43 Gross, D.M., Forsthuber, T., Tary-Lehmann, M. et al. (1998). Identification of LFA-1 as a candidate autoantigen in treatment-resistant lyme arthritis. *Science* 281 (5377): 703–706.

44 Zhang, W., Caspell, R., Karulin, A. et al. (2009). ELISPOT assays provide reproducible results among different laboratories for T-cell immune monitoring—even in hands of ELISPOT-inexperienced investigators. *J. Immunotoxicol.* 6 (4): 227–234.

45 Maecker, H.T., Hassler, J., Payne, J.K. et al. (2008). Precision and linearity targets for validation of an IFNgamma ELISPOT, cytokine flow cytometry, and tetramer assay using CMV peptides. *BMC Immunol.* 9: 9.

46 Maecker, H.T., Moon, J., Bhatia, S. et al. (2005). Impact of cryopreservation on tetramer, cytokine flow cytometry, and ELISPOT. *BMC Immunol.* 6: 17.

47 Food and Drug Administration Center for Drug Evaluation and Research. Bioanalytical method validation guidance for industry. US Department of Health and Human Services Food and Drug Administration Center for Drug Evaluation and Research and Center for Veterinary Medicine; 2018.

48 Islam, R., Vance, J., Poirier, M. et al. (2022). Recommendations on ELISpot assay validation by the GCC. *Bioanalysis* 14 (4): 187–193.

49 Cox, J.H., Ferrari, G., Kalams, S.A. et al. (2005). Results of an ELISPOT proficiency panel conducted in 11 laboratories participating in international human immunodeficiency virus type 1 vaccine trials. *AIDS Res. Human Retroviruses* 21 (1): 68–81.

50 Gebauer, B.S., Hricik, D.E., Atallah, A. et al. (2002). Evolution of the enzyme-linked immunosorbent spot assay for post-transplant alloreactivity as a potentially useful immune monitoring tool. *Am. J. Transplant.* 2 (9): 857–866.

51 Janetzki, S., Cox, J.H., Oden, N., and Ferrari, G. (2005). Standardization and validation issues of the ELISPOT assay. *Methods Mol. Biol.* 302: 51–86.

52 Smith, J.G., Liu, X., Kaufhold, R.M. et al. (2001). Development and validation of a gamma interferon ELISPOT assay for quantitation of cellular immune responses to varicella-zoster virus. *Clin. Diagn. Lab. Immunol.* 8 (5): 871–879.

53 Moodie, Z., Price, L., Gouttefangeas, C. et al. (2010). Response definition criteria for ELISPOT assays revisited. *Cancer Immunol., Immunother.* 59 (10): 1489–1501.

54 Dittrich, M. and Lehmann, P.V. (2012). Statistical analysis of ELISPOT assays. In: *Handbook of ELISPOT: Methods and Protocols* (ed. A.E. Kalyuzhny), 173–183. Totowa, NJ: Humana Press.

55 Kreher, C.R., Dittrich, M.T., Guerkov, R. et al. (2003). CD4+ and CD8+ cells in cryopreserved human PBMC maintain full functionality in cytokine ELISPOT assays. *J. Immunol. Methods* 278 (1): 79–93.

56 Karulin, A.Y., Hesse, M.D., Tary-Lehmann, M., and Lehmann, P.V. (2000). Single-cytokine-producing CD4 memory cells predominate in type 1 and type 2 immunity. *J. Immunol.* 164 (4): 1862–1872.

57 Hesse, M.D., Karulin, A.Y., Boehm, B.O. et al. (2001). A T cell clone's avidity is a function of its activation state. *J. Immunol.* 167 (3): 1353–1361.

58 Schlingmann, T.R., Shive, C.L., Targoni, O.S. et al. (2009). Increased per cell IFN-γ productivity indicates recent in vivo activation of T cells. *Cell. Immunol.* 258 (2): 131–137.

59 Frey, S.E., Winokur, P.L., Salata, R.A. et al. (2013). Safety and immunogenicity of IMVAMUNE® smallpox vaccine using different strategies for a post event scenario. *Vaccine* 31 (29): 3025–3033.

60 Osorio, J.E., Brewoo, J.N., Silengo, S.J. et al. (2011). Efficacy of a tetravalent chimeric dengue vaccine (DENVax) in *Cynomolgus macaques*. *Am. J. Trop. Med. Hyg.* 84 (6): 978–987.

61 Jaski, B.E., Jessup, M.L., Mancini, D.M. et al. (2009). Calcium upregulation by percutaneous administration of gene therapy in cardiac disease (CUPID Trial), a first-in-human phase 1/2 clinical trial. *J. Card. Fail.* 15 (3): 171–181.

62 Brudno, J.N., Lam, N., Vanasse, D. et al. (2020). Safety and feasibility of anti-CD19 CAR T cells with fully humanbinding domains in patients with B-cell lymphoma. *Nat. Med.* 26 (2): 270–280.

63 Eiden, J., Gordon, G., Fierro, C. et al. (2021). Safety and immunogenicity of M2-deficient, single replication, live influenza vaccine (M2SR) in adults. *Vaccines* 9 (12): 1388.

64 Fecher, P., Caspell, R., Naeem, V. et al. (2018). B cells and B cell blasts withstand cryopreservation while retaining their functionality for producing antibody. *Cells* 7 (6): 50.

65 Ramachandran, H., Laux, J., Moldovan, I. et al. (2012). Optimal thawing of cryopreserved peripheral blood mononuclear cells for use in high-throughput human immune monitoring studies. *Cells* 1 (3): 313–324.

66 Immunotec O. T-Spot TB Package Insert For In Vitro Diagnostic Use. V3. Available at: https://www.tspot.com/uk/resources/. Accessed 09 May 2023.

67 Janetzki, S., Panageas, K.S., Ben-Porat, L. et al. (2008). Results and harmonization guidelines from two large-scale international Elispot proficiency panels conducted by the Cancer Vaccine Consortium (CVC/SVI). *Cancer Immunol., Immunother.* 57 (3): 303–315.

68 Kutscher, S., Dembek, C.J., Deckert, S. et al. (2013). Overnight resting of PBMC changes functional signatures of antigen specific T-cell responses: impact for immune monitoring within clinical trials. *PLoS One* 8 (10): e76215.

69 Currier, J.R., Kuta, E.G., Turk, E. et al. (2002). A panel of MHC class I restricted viral peptides for use as a quality control for vaccine trial ELISPOT assays. *J. Immunol. Methods* 260 (1): 157–172.

70 Mwau, M., McMichael, A.J., and Hanke, T. (2002). Design and validation of an enzyme-linked immunospot assay for use in clinical trials of candidate HIV vaccines. *AIDS Res. Human Retroviruses* 18 (9): 611–618.

71 Kuerten, S., Batoulis, H., Recks, M.S. et al. (2012). Resting of cryopreserved PBMC does not generally benefit the performance of antigen-specific T cell ELISPOT assays. *Cells* 1 (3): 409–427.

12

Detection of Humoral Response to Transgene Protein and Gene Editing Reagents

George Buchlis[1] and Boris Gorovits[2]

[1] *Department of Medicine, University of Pennsylvania, Philadelphia, PA, USA*
[2] *Gorovits BioSolutions, LLC, Cambridge, MA, USA*

12.1 Pre- and Post-dose Humoral Immunity to Transgene-expressed Proteins

Much emphasis on the development of gene therapies has been placed on measuring, avoiding, and/or limiting immune responses to the viral capsid. While anti-capsid immune responses can indeed render the therapeutic ineffective, anti-transgene immune responses can prevent patients from benefiting from either gene therapies or existing protein therapeutics. Humoral responses to secreted transgene proteins can neutralize an enzyme's activity, block its uptake, and reduce its availability. In addition, antibody responses to non-secreted proteins can potentially lead to the destruction of transduced cells through complement-dependent or antibody-dependent cellular cytotoxicity (ADCC).

12.1.1 Risk-based Analysis of Response Probability and Impact

12.1.1.1 Route of Administration

One of the greatest determinants of humoral responses to transgene proteins is the route of administration of the gene therapy. The most well-documented route of administration resulting in anti-transgene antibodies is intramuscular injection. Antibodies to transgene expressed in muscle fibers have been observed in multiple preclinical animal models, including mice [1–7], rabbit [8], and non-human

Drug Development for Gene Therapy: Translational Biomarkers, Bioanalysis, and Companion Diagnostics, First Edition. Edited by Yanmei Lu and Boris Gorovits.

primates (NHP) [9, 10]. However, there are ways to mitigate transgene-elicited antibody responses to muscle gene transfer. One approach is to administer the vector intravenously but in an isolated region of the anatomy. In a method termed locoregional intravenous administration or regional administration, transvenular extravasation of the vector is achieved under high pressure in a cannulated limb vein isolated from the systemic circulation via tourniquet [9, 11, 12]. Similarly, in a technique termed afferent transvenular retrograde extravasation (ATVRX), a similar isolated vector delivery achieves substantial muscle transduction without accompanying transgene antibody responses [13, 14].

One method to address the antigenicity of muscle-directed gene transfer is the use of immunomodulation or immunosuppression. These approaches in preclinical studies have involved the use of cyclophosphamide [15], anti-CD4 antibody with cyclosporin [16], cyclosporin with mycophenolate mofetil (MMF) and anti-thymocyte globulin (ATG) [17], cyclosporin with rituximab [18], MMF with sirolimus [19], and rapamycin with ibrutinib [20]. Another interesting approach is downregulating expression in antigen-presenting cells via miRNA inclusion in the transgene cassette [21]. Restricting expression to myofibers by employing a muscle-specific promoter has shown efficacy in human gene therapy trials of limb-girdle muscular dystrophy type 2D [22, 23].

Intravenous administration of a gene therapy vector is commonly employed for transduction of well-vascularized organ systems or to access broad areas of the anatomy. For many transgene proteins that are expressed in the liver, this route of gene transfer does not provoke deleterious antibody responses. This phenomenon has been observed in mice [24–36], dogs [37, 38], and humans [39, 40]. Of course, there are instances where antibodies do form against the therapeutic protein in non-disease models of IV gene therapy. This can often be due to the immunogenic nature of the transgene protein [41–44], or the propensity of NHP to mount humoral immune responses to foreign transgene products [41, 45].

Monogenic diseases of the nervous system are a frequent target for gene therapy research approaches. However, anti-transgene antibody responses have been observed in preclinical animal models against nonself proteins when the gene therapy vector is infused into brain parenchyma [45, 46]. This nonself antibody response can be reduced with administration of rapamycin [47].

Intraocular gene transfer of a subretinal-administered Adeno-associated virus (AAV) expressing RPE65 became the first FDA-approved gene therapy in the United States. In this limited setting, anti-transgene immune responses are rarely observed, even while detectable yet non-consequential anti-capsid humoral responses are measured in the circulation. This has been observed in small animals [48], large animals [49, 50], and human patients [51, 52]. In fact, ocular readministration of the gene therapy has proven efficacious in human subjects [53].

12.1.1.2 Biodistribution of Vector, Vector Serotype, Dose, and Expression Level

While the administration route often dictates the location (biodistribution) of tissue transduction, intrinsic properties of the gene therapy vector itself, serotype, dose, promoter specificity, and promoter strength can have a strong impact on the formation of antibodies against the transgene product.

Importantly, the transgene genome itself can play a role in enhanced antibody production. Unmethylated CpG sequences in the transgene cassette can stimulate TLR9 and MyD88-dependent type I interferon production shown to be essential in anti-transgene responses [5, 7]. Along with the sequence of the gene therapy transgene, the structure of the packaged DNA can also impact responses to the expressed protein. Self-complementary AAV vectors were developed to package the genome as a double-stranded structure, overcoming the rate-limiting step of second-strand synthesis thought to result in less efficient protein expression from the traditionally single-stranded AAV genome [54, 55]. However, this has been shown to elicit stronger immune responses against the transgene product [56, 57]. The promoter that drives gene expression of the transgene also factors into humoral responses against the translated protein. Studies that have restricted expression to the liver via a hepatocyte-specific promoter have generally shown a lack of anti-transgene antibody responses [25, 26, 39, 40, 58].

The AAV capsid serotype can also influence whether responses are mounted against the transgene as well [2]. This will vary in different species based on tissue tropism variation. While serotype, biodistribution, and route of administration are all intimately related and responsible for anti-transgene responses, the dose of vector influences humoral immunity to the protein. In general terms, the higher the dose, the higher the expression, and greater the potential for eliciting anti-transgene antibody responses [59]. This was shown in mice to be correlated to the amount of expression on a per-cell basis and not simply overall expression in the target organ [60].

12.1.1.3 Patient Immune Status: Age, Prior Exposure, No Endogenous Production, Immunosuppression, and Autoimmunity

After understanding how the gene therapy vector can influence transgene antibody formation, it's important to consider the immune status and antigenic exposure of the recipient. An interesting phenomenon occurs when administering gene therapy very early in life: in utero, perinatally, or neonatally. This early therapeutic administration of a vector leads to immunological tolerance to the transgene product. This effect has been observed in mice [61–65], dogs [66], sheep [67], and nonhuman primates [66, 68].

Another factor influencing antibody responses to the transgene is the immune state of the patient (i.e. immunosuppressed, autoimmune condition). Mice strains prone to autoimmunity developed anti-transgene antibodies after gene transfer [69].

As we previously described, immunosuppression can reduce or prevent antibody responses to the transgene product. Thus, it is important to consider if the recipient has any altered immune state due to underlying immunodeficiencies, autoimmunity, or treatment involving immunosuppression. Another component of immunity is prior exposure to an antigen. Many genetic diseases present as a spectrum of mutations, with varying degrees of impact on the translatability of the protein in question. Depending on the mutation, the protein may be fully produced but nonfunctional due to point mutations, may be truncated, or may not be translated at all. In those subjects with no portion of the protein translated, termed cross-reactive immunogenic material (CRIM) negative patients, the risk for anti-transgene antibody formation is highest [3, 15, 70]. These subjects have typically been excluded in some of the earliest gene therapy trials on account of this risk [39, 40]. In addition, some subjects have formed antibodies to recombinant protein therapies used to treat their disease prior to receiving gene therapy. These subjects have also been excluded from many early gene therapy clinical trials, though the reversal of antibodies against clotting factors in preclinical animal models has provided hope that gene transfer to the liver with or without immunomodulation can eventually tolerize the recipient to these proteins [31, 38, 71]. Indeed, several current trials are stratifying patients into transgene antibody positive and negative arms, to assess the impact of these preexisting antibodies on therapeutic efficacy (studies NCT04046224, NCT04684940, and NCT03734588).

12.1.1.4 Response Induction vs. Response Boosting

Antibodies in gene therapy can be consequential to therapeutic efficacy, whether they are primary or memory responses. With a secreted transgene product, antibodies in circulation can have a direct impact in blocking the protein and/or clearing it from circulation. For extracellularly expressed proteins, antibody binding can potentially lead to antibody-dependent or complement-dependent cytotoxicity (CDC) [72, 73]. For intracellular proteins, an antibody response may appear if the antigen is somehow transported to the cell surface, but it is likely to be of low frequency and of little consequence. A recall response in a recipient that has already formed a primary humoral response to the protein will result in antibodies that are of higher affinity and at a higher concentration in the circulation [74, 75].

12.2 Relevance of Analytical Protocols Applied in Determining Immune Response to Protein Therapeutics to the Detection of Anti-Transgene Protein Responses

When comparing analytical methods used for protein therapeutics to methods that have been developed for immune response characterization of a gene therapy transgene product, there is a great deal of overlap. While the next section describes these assays in more detail, there are some unique considerations that make a transgene product different from a protein therapy. As mentioned in the previous paragraph, transgene products can either be secreted, expressed extracellularly, or expressed intracellularly. In addition to the unique immunogenicity risks posed, the assays to measure these antibody responses need modification depending on the expression pattern. For secreted proteins, the anti-drug antibody assays are identical to that of an exogenously administered recombinant protein [76]. For an extracellularly expressed protein, the same methods can be used while using only the extracellular portion of the protein in the assay, instead of the entire protein. For intracellularly expressed proteins, the full protein suffices as the capture and detection reagent for antibody measurements, bearing in mind that there may be little relevance to a response against a nonexposed antigen. In addition, some parameters in the development of the assay, such as drug tolerance, may be harder to overcome with a consistently produced protein without any trough serum samples. While not precluding successful assay development, this can complicate the method validation process [77].

12.3 Analysis of Immune Response by Binding and Functional Antibody Assay Protocols

Measuring binding antibodies to the transgene product involves immunoassay methods that can employ a variety of readouts, including traditional optical density measurements and chemiluminescence. Perhaps, the most common format is a standard sandwich ELISA or electrochemiluminescence assay, where a capture antibody is attached to the assay plate, the transgene protein is incubated on the plate, the sample with potential anti-transgene antibodies follows, and then a secondary anti-Ig detection antibody with enzymatic or chemiluminescent activity is added, developed, and read on an appropriate plate reading device (Figure 12.1a) [78, 79]. Alternatively, a bridging assay is employed using the transgene protein as a capture and detection reagent, with a caveat that the antibody isotype "bridging" the capture and detection reagents is not determined (Figure 12.1b) [79]. This can be modified as an indirect binding assay that can have the secondary antibody be an anti-isotype antibody if Ig subtyping is of interest. These methods can be done in a stepwise, layer-by-layer incubation format, or they can be performed in what is known as a homogenous or co-incubation assay. The homogenous method involves incubating the capture antibody, analyte, and detection antibodies prior to plate incubation and chemiluminescent signal readout [79].

Neutralizing antibodies are a component of the total antibody response that inhibit or "neutralize" the cellular uptake and/or biological activity of the

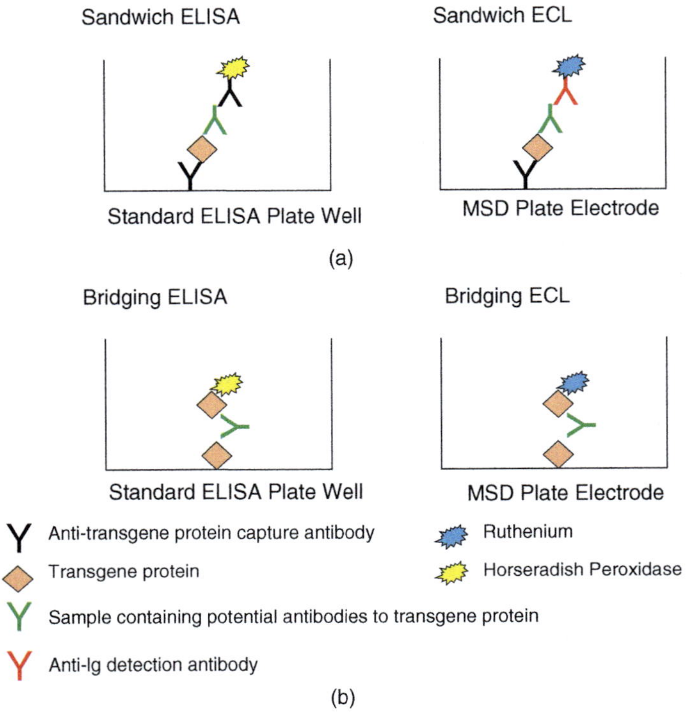

Figure 12.1 ELISA and ECL assay formats for the detection of total antibodies to transgene proteins. (a) In both ELISA and ECL sandwich assay formats, the plate wells are coated with anti-transgene capture antibody. Then transgene protein is incubated on the plate, the sample with potential anti-transgene antibodies is added, and then a secondary anti-Ig detection antibody with enzymatic (left panel) or chemiluminescent (right panel) activity leads to detection or not of anti-transgene antibodies. (b) In both ELISA and ECL bridging assay formats, the wells are coated with transgene protein. Then, in the presence of sample antibodies, labeled transgene will be bridged to the coated transgene by way of the anti-transgene antibodies present, thus producing a signal.

protein [80]. For a secreted transgene protein, a non-cell-based neutralizing assay can be developed using a labeled competitive ligand. The labeled ligand will compete for the same receptor as the transgene protein, and in the presence of neutralizing antibodies, the labeled ligand will outcompete the transgene protein that is being blocked from binding its receptor (Figure 12.2) [81]. Alternatively, as is the case with enzymes or clotting factors that involve proteolytic cleavage with a biological readout, activity assays can be used to determine the effect of neutralizing antibodies on protein function. One common method in hematology is the Bethesda assay, which measures the amount of inhibitor that will neutralize 50% of factor VIII activity in an activated partial thromboplastin time (APTT) clotting assay [38].

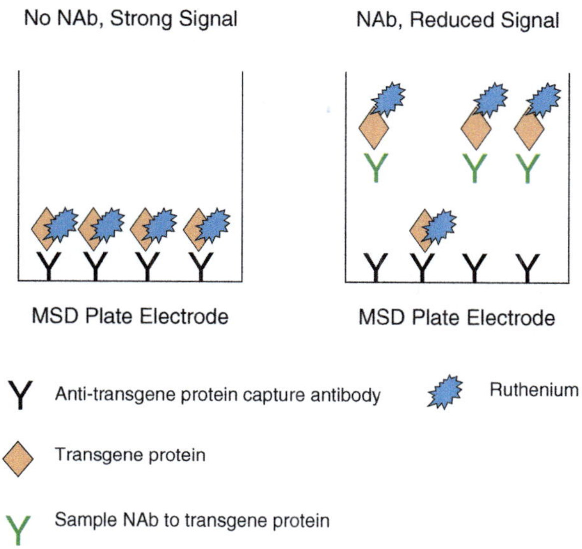

Figure 12.2 ECL assay for the detection of neutralizing antibodies to transgene proteins. electrochemical assay plates from MSD are pre-coated with an anti-transgene antibody. Subsequently, ruthenium-labeled transgene protein is added in the presence of test subject serum and incubated over the coated well. In the absence of neutralizing antibodies in the subject serum, the labeled transgene will bind to the antibody on the coated well and upon electrical stimulation, will luminesce. Any reduction in signal will indicate the presence of neutralizing antibodies to the transgene protein blocking its binding to the coated well.

12.4 Comparative Analysis of the Immune Response Evaluation for Transgene Proteins that are Expressed Extracellularly vs. Intracellularly

Extracellularly expressed proteins are unique in that when bound by antibody, they theoretically expose the transduced cell to multiple mechanisms of cytotoxicity, with deleterious consequences. These functional responses, which essentially "neutralize" the extracellular protein by binding and effecting cell killing, can be measured in ADCC or CDC assays. In ADCC, FcgRIIIa (CD16a) on NK cells binds the Fc portion of the antibody bound to the extracellular protein of interest. When enough signal is received, the NK cells will release cytolytic/cytotoxic granules, leading to cell death [82]. Various commercial sources exist with reporter cell lines that can be activated upon FcgRIIIa signaling, with a signal-dependent output such as luciferase production. Another important antibody-dependent process is complement fixation on the cell surface. Complement factor C1q binds to antibody/antigen complexes on the cell surface, initiating a proteolytic cascade that results in forming a membrane pore complex that lyses the cell [73]. Assays involving CDC are more direct lytic assays, where incubation with cells expressing antigen leads to complement fixation, activation, and cell killing. These can employ various viability dyes or markers of cell death [83].

While it is possible to provoke antibody responses to intracellular antigens, they are unlikely to play a role in antigen clearance or loss of therapeutic efficacy. This is commonly observed in autoimmune conditions, where these antibodies serve as diagnostic hallmarks, but are not causal of disease [84]. Yet, intracellular proteins are still capable of eliciting cellular immune responses to antigenic peptides presented to T cells on MHC Class I or MHC Class II, which can have negative impacts on therapeutic efficacy.

12.5 Humoral Immune Response to Gene Editing Reagents

12.5.1 Diversity of Gene Editing Systems

Recent advances in gene editing systems have generated significant interest, with the promise to provide a precise and targeted tool for conducting genetic changes in eukaryotic cells ex vivo and in vivo. One of the most broadly applied technologies is based on clustered regularly interspaced short palindromic repeats (CRISPR) associated nuclease 9 (Cas9). CRISPR-Cas9 is an RNA-guided genome editing tool that includes a nuclease, Cas9, and a single guide RNA. Other Cas proteins have been identified, for example, Cas12 [85–87]. The specificity of the CRISPR system is based on the ability of a single guide RNA to recognize target DNA and guide Cas9 nuclease to a targeted site, where a site-specific double-strand break is generated. The latter is subsequently repaired, for example, based on a donor template [88]. The CRISPR/Cas9 system was originally identified as a bacterial immune protection mechanism [89] and holds significant potential for the treatment of a number of human diseases through precise correction of disease-causing mutations, termination of aberrant protein expression, or insertion of therapeutically critical transgene. Other methodologies for generating site-specific double-stranded breaks include protein-based DNA recognition systems with designed DNA-binding properties, for example, meganucleases, zinc finger nucleases (ZFNs), and transcription activator-like effector nucleases (TALENs) [90, 91].

Application of meganucleases, ZFNs, and TALENs has been challenged by the complexity of the systems, lack of robust technologies to construct effective and specific structure, and complexity of molecular cloning technologies required [92–94].

The advantage of CRISPR technology over ZFN and TALEN methods which are based on the specificity of protein-DNA interaction, is that the CRISPR technology employs direct base-pairing rules in guide RNA molecule complementary binding to the targeted DNA sequence, to precisely target nuclease activity.

Gene editing technologies, including ZFN, TALEN, and CRISPR/Cas9, have already been applied to generate autologous or allogeneic chimeric antigen receptor (CAR) T lymphocytes to fight various B-cell malignancies [95, 96]. The TALEN technology has been successfully applied to generate gene-edited T cells designed for treatment of refractory relapsed B cell acute lymphoblastic leukemia (B-ALL) [95]. In that study, TALEN technology was used to disrupt the CD52 gene, a target antigen for lymphodepleting agent alemtuzumab, and αβ T-cell receptor (TCR αβ) on the T-cell surface. ZFN technology was applied as the gene editing approach in an ex vivo cell therapy to deactivate the CCR5 gene, believed to be responsible for progression of human immunodeficiency virus infection [97]. The CRISPR-Cas gene editing technology is generally viewed as more practical and with higher specificity for introduced modifications [89, 98].

The CRISPR/Cas systems can be categorized into two classes and several subclasses or types, depending on the number of protein complexes involved in the cleavage process and the location of the sequence recognized by the system. For example, the Cas12 and Cas9 nucleases recognize protospacer flanking sequence located directly before or downstream of the protospacer motif sequences (PAM) [85–87]. Application of innovative versions of Cas enzyme has great promise in improving the specificity and efficiency of gene editing techniques [99].

12.5.2 Immunological Potential of CRISPR-Cas System

At this time, CRISPR/Cas9 is the most characterized and applied gene editing system. It can be delivered into a cell as a plasmid packaged in a viral vector. Examples include in vivo delivery using AAV-based vector, encoding Cas nuclease and gRNA, or as a ribonucleoprotein complex of Cas9 nuclease bound to gRNA [100–102].

Two broadly used Cas9 proteins are adopted from sequences found in bacterial strains, namely Staphylococcus aureus (S. aureus) and Streptococcus pyogenes (S. pyogenes). Both S. aureus and S. pyogenes are common pathogens found in humans. As expected, immunity against S. aureus and S. pyogenes bacteria is widespread and readily detectable [103–105].

Although it was originally expected that the intracellular nature of Cas9 protein limited observed human immune responses, antibody-based response to S. aureus and S. pyogenes-originating Cas9 (SaCas9 and SpCas9) was readily detected in a majority (78% and 58%) of blood samples collected from healthy adults [106]. In addition to humoral immunity, cellular-based immune response was also detected against both SaCas9 and SpCas9 in a large fraction of tested samples (78% and 58%) [106].

The presence of humoral and T-cell response in the normal human population presents a potential concern as anti-Cas9 immunity may lead to destruction of Cas9-expressing cells with a negative impact on treatment.

An ex vivo application of gene editing based on CRISPR/Cas technology offers a unique opportunity to clear any residual Cas protein expressed during the editing procedure. Cells may be administered after such a washout period to avoid Cas protein expression in patients [107, 108].

Several risk factors should be considered for an in vivo administration of Cas protein-expressing vector, particularly when treatment of patients with fully competent immune systems is proposed. Based on information from AAV-based delivery, these factors may include the nature of gene editing vector, route of administration and/or targeted tissue, dose, and nature of the targeted gene [109].

The presence of anti-Cas9-specific T-cells was detected after administration of AAV vector encoding SpCas9 protein. Such T-cell response was deemed to be naïve and immature and did not result in T-cell-based cell killing [110]. Cytotoxic anti-Cas9 immune response was detected in an experiment where Cas9 DNA was physically delivered in mice via electroporation [110]. Mouse models utilizing AAV-based delivery of CRISPR-Cas9 system showed the development of a robust anti-Cas9 antibody and cellular immune response, with limited impact on the persistence of gene-edited cells [110–113]. Anti-Cas9-specific T cells were shown to effectively recognize and destroy cells pulsed with Cas9 peptide epitopes [111].

Administration of Cas9-expressing vector to animals with preexisting anti-Cas9 immunity appears to negatively impact treatment efficacy [111]. A connection between anti-SaCas9 protein T-cell immune response and a significant decline of gene-edited cells was demonstrated by Li et al. [114]. In this study, one group of animals was pretreated with SaCas9 protein leading to the generation of robust anti-Cas immunity and a strong memory T-cell response. Anti-Cas immune response did not immediately block the genome editing ability of the AAV-CRISPR vector in the target organ (liver) during the initial treatment. Subsequent administration of AAV-Cas9 constructs to animals with existing anti-Cas9 immunity induced a Cas9-specific CD8+ T-cell response and killing of cells infected with the Cas9 construct. A significant loss of genome-edited cells was reported as soon as 12 weeks after injection. The proportion of CD8+ T cells in liver of mice pre-immunized with SaCas9 increased starting at 1 week after injection and remained elevated through week 4. Similarly, elevated levels of alanine transaminase activity were observed at week 2, returning to normal levels by week 12. Evidence of active liver regeneration processes was reported, indicated by the level of Ki-67 mRNA in liver sections. Loss of either transgene protein or the evidence of gene editing was reported by week 12 after treatment. Consistent presence of gene-edited cells was observed in the control group of animals that were pre-immunized with ovalbumin, and which did not express preexisting anti-Cas9 T cells at the point of AAV-Cas9 construct administration.

In a separate investigation, an AAV-based delivery approach was evaluated for SpCas9 structure-guided split-intein fusion proteins [110]. The Cas9 N-terminal lobe was fused with the Rhodothermus marinus N-split intein (Cas9N) and the C-terminal lobe with C-split intein (Cas9C). Structure-guided split-intein reconstitution produced fully active Cas9 in transfected cells. The Cas9C and Cas9N constructs were delivered via an AAV9 vector. The resulting viral construct was injected interperitoneally into neonatal mice, and genome editing rates demonstrated detectable functional activity of the construct. Administration of AAV9-Cas9 resulted in immune system activation which included enlargement of lymph nodes and elevation of CD45+ leukocyte count, including the T-cell population. In addition to activation of cellular immune responses, Cas9 encoding construct administration resulted in the generation of Cas9-specific antibodies with variable epitope specificity and titer levels. Despite observed evidence of activation of cellular immunity, lack of significant muscle cell damage was reported at 2 weeks after AAV-Cas9 administration [110].

A notable difference in anti-SaCas9 immune response between adult and neonate animal populations was reported by Nelson et al.[112] The humoral anti-SaCas9 response was detected in the vast majority of adult mice treated with dual AAV-based Cas9 and guide RNA delivery system by intramuscular injection, thus resulting in a substantial decrease in degree of total gene modification over 6 month period. Conversely, anti-SaCas9 antibodies were not elicited in neonatal mice treated systemically (intravenous facial vein injection, FVI) at postnatal Day 2, with a relatively low but noticeable increase in total gene modifications over 1 year. The same trend was observed for anti-Cas9 T-cell-based cellular immunity. It was concluded that anti-Cas9 immune responses in mice could be avoided if the delivery system is administered very early in life, that is, neonatal, when the immune system is undeveloped. It remains to be demonstrated whether this approach is translatable and relevant to newborn humans.

In summary, anti-Cas protein humoral and cellular immunity was reported in the normal human population. Animal models suggest that preexisting anti-Cas immunity can greatly impact efficacy and durability of CRISPR-based gene editing. It must be understood that conclusions made based on nonclinical models of immune responses may not directly predict clinical outcomes.

12.5.3 Detection of Anti-Cas9 Protein Immunity in Animal and Human Matrix

A significant proportion of the human population has been reported as positive for preexisting anti-Cas9 immunity, including both humoral and cellular responses [106, 115–117].

The reported prevalence of antibody and T-cell-based immunity against SpCas9 and SaCas9 proteins varies considerably, likely driven by the number of tested samples, type of evaluated population, and parameters of the detection methods used in each study. These assay parameters include the methodology used to determine the cut point (to identify positive/negative sample), assay sensitivity, and potential cross-reactivity with other components of the matrix.

Several examples of protocols used to detect anti-Cas immunity are presented below. Ferdosi et al. applied an ELISA protocol to detect anti-SpCas9 IgG response in a direct binging-based method while using 99th %-tile of anti-human hemoglobin response to determine the assay cutoff value [115]. In that method, recombinant SpCas9 protein and anti-human IgG horseradish peroxidase were used as capture and detection reagents, respectively. The assay cut-point was defined as the 99th %-tile value of the signal generated by the assay negative control, human hemoglobin. Although 57.3% (82 of 143) of the tested healthy control sera samples were reported positive for the presence of anti-S. pyogenes lysate antibody, only 5% of individual samples were reported as positive against Cas9-specific immunoglobulins.

A high prevalence of anti-Cas9 immunity was reported by Charlesworth et al. [106]. Serum samples derived from human cord blood were evaluated to detect the presence of antibodies against SaCas9 and SpCas9 proteins utilizing an immunoblot protocol. Among 22 samples assessed in the assay, up to 86% and 73% were deemed positive for SaCas9 and SpCas9 proteins, respectively. The limited number of tested samples and semi-quantitative nature of the immunoblot protocol may have resulted in the relatively high reported prevalence of anti-Cas9 immunity. The same study reported the results of analysis of 125 individual adult blood donor samples in an ELISA method for the presence of anti-Cas9 antibodies. Assay response against human albumin was used as the assay negative control value. By applying the cut-point value that was set at the mean of the assay negative control signal plus three standard deviations from the mean, 78% and 58% of tested samples were positive for the presence of anti-SaCas9 and anti-SpCas9 protein antibodies, respectively [106].

Relatively low prevalence of preexisting anti-Cas9 humoral immunity was reported by Simhadri et al. [116] with 10% and 2.5% of tested individual human serum samples reported positive for antibodies against SaCas9 and SpCas9 proteins, respectively. A total of 200 samples were analyzed in a direct ELISA method. Assay validation followed industry- and regulatory-accepted approaches broadly applied to methods used in evaluating antibody responses to protein-based biotherapeutics [78]. In this assay, labeled protein G was used as the detection reagent, suggesting that only the IgG isotype of immunoglobulins could be detected, potentially limiting the overall assessment of anti-Cas9 protein antibody response. Assay validation included assessment of method precision and matrix interference, selection and evaluation of the assay positive control performance, and determination of the assay cut-point value. It is likely that prior exposure of human donors to S. pyogenes and S. aureus bacteria resulted in a high degree of reactivity in naïve samples. This led to relatively high cut-point values in both anti-SaCas9 and SpCas9 antibody detection methods when using a traditional approach for data analysis (original cut-point values, 0.874 and 1.012 OD values for SpCas9 and SaCas9 methods, respectively). To mitigate this issue, samples were supplemented with an excess of Cas9 protein, effectively inhibiting anti-Cas9 antibodies, and then tested in the underlying method. The addition of Cas9 led to a significant reduction of the assay signal, up to 74.7% and 87.8% for SaCas9 and SpCas9 methods, respectively. Data obtained in analysis of Cas9-supplemented samples were statistically analyzed to obtain assay cut-points following standard 5% false-positive rate reporting. Such an approach resulted in reduced (alternative) cut-points, down to 0.615 and 0.513 OD values, for SpCas9 and SaCas9 methods, respectively. Finally, a confirmatory assay cut-point was established by statistical evaluation of data produced by analysis of individual samples supplemented with excess amount of Cas9 protein [78]. The confirmatory cut-points reported in the study were 73.1% and 71.6% for SpCas9 and SaCas9 methods, respectively.

Individual serum samples were tested in a tiered-based approach that included initial screening followed by confirmatory and titer tests. A lower value of the alternative vs. original screening assay cut-point resulted in a notably higher rate of samples that screened antibody-positive in both anti-SpCas9 and SaCas9 assays. Yet, the confirmatory analysis yielded a significantly reduced rate of screening and confirmed positive responses (5% and 10% for SaCas9 and 1.5% and 2.5% for SpCas9). Lower rates of anti-Cas9 responses reported in this study vs. other anti-Cas9 protein antibody detecting protocols were attributed to several factors, including the solution-based and more quantitative nature of the ELISA and higher number of samples used during the method validation phase. The lower prevalence of anti-Cas9 samples reported in this study may also be due to a significant and potentially too high confirmatory cut-point value exceeding 70% for both methods. Such high confirmatory cut-point values suggest the presence of anti-Cas9-specific antibodies in the samples used during assay validation. It has been previously stated that ideally, treatment naïve and, therefore, expected antibody-negative samples should be used in assay cut-point assessment. Clearly, the high prevalence of anti-bacterial immunity for both S. aureus and S. pyogenes microorganisms precludes easy selection of such negative samples resulting in high range of screening test responses and complexities in data analysis. Such protocols have been considered for other biotherapeutic modalities, including AAVs [79] and antibody drug conjugates [118]. The use of high confirmatory cut-points may result in misrepresentation of the overall positivity rates. Alternative approaches to sampling and data analysis need to be identified.

Considerable difference in reported prevalence of anti-Cas immunity underlines clear need for a harmonized approach in protocols used to detect humoral and cellular responses. Well-defined standards have been established in support of protein-based biotherapeutics [77, 78]. Similar methodologies can be applied when developing anti-Cas responses allowing for a robust and aligned representation of immunity across various study populations.

12.5.4 Strategies Proposed to Mitigate Anti-Cas9 Immunity

Several strategies have been proposed and investigated to mitigate Cas9 immunity. These include epitope masking, altering antigen presentation, use of orthologs, including from nonpathogenic bacteria, administration of the treatment to immune-privileged organs, and induction of immune tolerance [119]. Similar solutions have been successfully applied for protein-based and other biotherapeutic modalities with the clear need for continued evaluation in the clinic [120].

Several HLA-A*02:01-restricted T-cell epitopes were identified on SpCas9 protein using an in silico sequence analysis model with two peptide sequences confirmed for their ability to activate healthy donor PBMC samples in vitro. Mutated variants of Cas9 protein were generated aiming to disrupt HLA binding of the two identified epitopes. Based on in vitro data, it was suggested that the introduction of mutations in highly immunogenic epitopes can successfully reduce the overall immunogenicity of Cas9 while maintaining the gene editing ability of the enzyme. Additional in silico and T-cell activation tests of the SpCas9 sequence identified other potential HLA class I and II interacting epitopes, suggesting further modifications to the sequence may be required for complete de-immunization of the protein.

References

1 Fisher KJ, Jooss K, Alston J, Yang Y, Haecker SE, High K, et al. Recombinant adeno-associated virus for muscle directed gene therapy. *Nat. Med.* 1997;3(3):306-12. doi: https://doi.org/10.1038/nm0397-306.

2 Wang L, Louboutin JP, Bell P, Greig JA, Li Y, Wu D, et al. Muscle-directed gene therapy for hemophilia B with more efficient and less immunogenic AAV vectors. *J. Thromb. Haemost.* 2011;9(10):2009-19. doi: https://doi.org/10.1111/j.1538-7836.2011.04491.x.

3 Yuasa K, Sakamoto M, Miyagoe-Suzuki Y, Tanouchi A, Yamamoto H, Li J, et al. Adeno-associated virus vector-mediated gene transfer into dystrophin-deficient skeletal muscles evokes enhanced immune response against the transgene product. *Gene Ther.* 2002;9(23):1576-88. doi: https://doi.org/10.1038/sj.gt.3301829.

4 Wang L, Cao O, Swalm B, Dobrzynski E, Mingozzi F, Herzog RW. Major role of local immune responses in antibody formation to factor IX in AAV gene transfer. *Gene Ther.* 2005;12(19):1453-64. doi: https://doi.org/10.1038/sj.gt.3302539.

5 Herzog RW, Cooper M, Perrin GQ, Biswas M, Martino AT, Morel L, et al. Regulatory T cells and TLR9 activation shape antibody formation to a secreted transgene product in AAV muscle gene transfer. *Cell. Immunol.* 2019;342:103682. doi: https://doi.org/10.1016/j.cellimm.2017.07.012.

6 Herzog RW, Hagstrom JN, Kung SH, Tai SJ, Wilson JM, Fisher KJ, et al. Stable gene transfer and expression of human blood coagulation factor IX after intramuscular injection of recombinant adeno-associated virus. *PNAS* 1997;94(11):5804-9. doi: https://doi.org/10.1073/pnas.94.11.5804.

7 Zhu J, Huang X, Yang Y. The TLR9-MyD88 pathway is critical for adaptive immune responses to adeno-associated virus gene therapy vectors in mice. *J. Clin. Invest.* 2009;119(8):2388-98. doi: https://doi.org/10.1172/jci37607.

8 Favaro P, Finn JD, Siner JI, Wright JF, High KA, Arruda VR. Safety of liver gene transfer following peripheral intravascular delivery of adeno-associated virus (AAV)-5 and AAV-6 in a large animal model. *Hum. Gene Ther.* 2011;22(7):843-52. doi: https://doi.org/10.1089/hum.2010.155.

9 Gao G, Lebherz C, Weiner DJ, Grant R, Calcedo R, McCullough B, et al. Erythropoietin gene therapy leads to autoimmune anemia in macaques. *Blood* 2004;103(9):3300-2. doi: https://doi.org/10.1182/blood-2003-11-3852.

10 Toromanoff A, Adjali O, Larcher T, Hill M, Guigand L, Chenuaud P, et al. Lack of immunotoxicity after regional intravenous (RI) delivery of rAAV to nonhuman primate skeletal muscle. *Molecular Therapy* 2010;18(1):151-60. doi: https://doi.org/10.1038/mt.2009.251.

11 Guilbaud M, Devaux M, Couzinié C, Le Duff J, Toromanoff A, Vandamme C, et al. Five years of successful inducible transgene expression following locoregional adeno-associated virus delivery in nonhuman primates with no detectable immunity. *Hum. Gene Ther.* 2019;30(7):802-13. doi: https://doi.org/10.1089/hum.2018.234.

12 Toromanoff A, Chérel Y, Guilbaud M, Penaud-Budloo M, Snyder RO, Haskins ME, et al. Safety and efficacy of regional intravenous (RI) versus intramuscular (IM) delivery of rAAV1 and rAAV8 to nonhuman primate skeletal muscle. *Molecular Therapy* 2008;16(7):1291-9. doi: https://doi.org/10.1038/mt.2008.87.

13 Haurigot V, Mingozzi F, Buchlis G, Hui DJ, Chen Y, Basner-Tschakarjan E, et al. Safety of AAV factor IX peripheral transvenular gene delivery to muscle in hemophilia B dogs. *Molecular Therapy* 2010;18(7):1318-29. doi: https://doi.org/10.1038/mt.2010.73.

14 Arruda VR, Stedman HH, Haurigot V, Buchlis G, Baila S, Favaro P, et al. Peripheral transvenular delivery of adeno-associated viral vectors to skeletal muscle as a novel therapy for hemophilia B. *Blood* 2010;115(23):4678-88. doi: https://doi.org/10.1182/blood-2009-12-261156.

15 Herzog RW, Mount JD, Arruda VR, High KA, Lothrop CD, Jr. Muscle-directed gene transfer and transient immune suppression result in sustained partial correction of canine hemophilia B caused by a null mutation. *Molecular Therapy* 2001;4(3):192-200. doi: https://doi.org/10.1006/mthe.2001.0442.

16 McIntosh JH, Cochrane M, Cobbold S, Waldmann H, Nathwani SA, Davidoff AM, et al. Successful attenuation of humoral immunity to viral capsid and transgenic protein following AAV-mediated gene transfer with a non-depleting CD4 antibody and cyclosporine. *Gene Ther.* 2012;19(1):78-85. doi: https://doi.org/10.1038/gt.2011.64.

17 Wang Z, Storb R, Halbert CL, Banks GB, Butts TM, Finn EE, et al. Successful regional delivery and long-term expression of a dystrophin gene in canine muscular dystrophy: a preclinical model for human therapies. *Molecular Therapy* 2012;20(8):1501-7. doi: https://doi.org/10.1038/mt.2012.111.

18 Mingozzi F, Chen Y, Murphy SL, Edmonson SC, Tai A, Price SD, et al. Pharmacological modulation of humoral immunity in a nonhuman primate model of AAV gene transfer for hemophilia B. *Molecular Therapy* 2012;20(7):1410-6. doi: https://doi.org/10.1038/mt.2012.84.

19 Mingozzi F, Hasbrouck NC, Basner-Tschakarjan E, Edmonson SA, Hui DJ, Sabatino DE, et al. Modulation of tolerance to the transgene product in a nonhuman primate model of AAV-mediated gene transfer to liver. *Blood* 2007;110(7):2334-41. doi: https://doi.org/10.1182/blood-2007-03-080093.

20 Xiang Z, Kuranda K, Quinn W, Chekaoui A, Ambrose R, Hasanpourghai M, et al. The effect of rapamycin and ibrutinib on antibody responses to adeno-associated virus vector-mediated gene transfer. *Hum. Gene Ther.* 2022;33(11-12):614-24. doi: https://doi.org/10.1089/hum.2021.258.

21 Majowicz A, Maczuga P, Kwikkers KL, van der Marel S, van Logtenstein R, Petry H, et al. Mir-142-3p target sequences reduce transgene-directed immunogenicity following intramuscular adeno-associated virus 1 vector-mediated gene delivery. *J. Gene Med.* 2013;15(6-7):219-32. doi: https://doi.org/10.1002/jgm.2712.

22 Mendell JR, Rodino-Klapac LR, Rosales-Quintero X, Kota J, Coley BD, Galloway G, et al. Limb-girdle muscular dystrophy type 2D gene therapy restores alpha-sarcoglycan and associated proteins. *Ann. Neurol.* 2009;66(3):290-7. doi: https://doi.org/10.1002/ana.21732.

23 Mendell JR, Rodino-Klapac LR, Rosales XQ, Coley BD, Galloway G, Lewis S, et al. Sustained alpha-sarcoglycan gene expression after gene transfer in limb-girdle muscular dystrophy, type 2D. *Ann. Neurol.* 2010;68(5):629-38. doi: https://doi.org/10.1002/ana.22251.

24 Mingozzi F, Liu YL, Dobrzynski E, Kaufhold A, Liu JH, Wang Y, et al. Induction of immune tolerance to coagulation factor IX antigen by in vivo hepatic gene transfer. *J. Clin. Invest.* 2003;111(9):1347-56. doi: https://doi.org/10.1172/jci16887.

25 Sun B, Li S, Bird A, Yi H, Kemper A, Thurberg BL, et al. Antibody formation and mannose-6-phosphate receptor expression impact the efficacy of muscle-specific transgene expression in murine Pompe disease. *J. Gene Med.* 2010;12(11):881-91. doi: https://doi.org/10.1002/jgm.1511.

26 Cao O, Dobrzynski E, Wang L, Nayak S, Mingle B, Terhorst C, et al. Induction and role of regulatory CD4+CD25+ T cells in tolerance to the transgene product following hepatic in vivo gene transfer. *Blood* 2007;110(4):1132-40. doi: https://doi.org/10.1182/blood-2007-02-073304.

27 Ishiwata A, Mimuro J, Mizukami H, Kashiwakura Y, Takano K, Ohmori T, et al. Liver-restricted expression of the canine factor VIII gene facilitates prevention of inhibitor formation in factor VIII-deficient mice. *J. Gene Med.* 2009;11(11):1020-9. doi: https://doi.org/10.1002/jgm.1391.

28 Hoffman BE, Martino AT, Sack BK, Cao O, Liao G, Terhorst C, et al. Nonredundant roles of IL-10 and TGF-β in suppression of immune responses to hepatic AAV-factor IX gene transfer. *Molecular Therapy* 2011;19(7):1263-72. doi: https://doi.org/10.1038/mt.2011.33.

29 Zhang P, Sun B, Osada T, Rodriguiz R, Yang XY, Luo X, et al. Immunodominant liver-specific expression suppresses transgene-directed immune responses in murine pompe disease. *Hum. Gene Ther.* 2012;23(5):460-72. doi: https://doi.org/10.1089/hum.2011.063.

30 Ziegler RJ, Lonning SM, Armentano D, Li C, Souza DW, Cherry M, et al. AAV2 vector harboring a liver-restricted promoter facilitates sustained expression of therapeutic levels of alpha-galactosidase A and the induction of immune tolerance in Fabry mice. *Mol. Ther.* 2004;9(2):231-40. doi: https://doi.org/10.1016/j.ymthe.2003.11.015.

31 Nayak S, Sarkar D, Perrin GQ, Moghimi B, Hoffman BE, Zhou S, et al. Prevention and reversal of antibody responses against factor IX in gene therapy for hemophilia B. *Front. Microbiol.* 2011;2:244. doi: https://doi.org/10.3389/fmicb.2011.00244.

32 Franco LM, Sun B, Yang X, Bird A, Zhang H, Schneider A, et al. Evasion of immune responses to introduced human acid alpha-glucosidase by liver-restricted expression in glycogen storage disease type II. *Molecular Therapy* 2005;12(5):876-84. doi: https://doi.org/10.1016/j.ymthe.2005.04.024.

33 Chan CC, Lai CW, Wu CJ, Chen LC, Tao MH, Kuo ML. Liver-specific allergen gene transfer by adeno-associated virus suppresses allergic airway inflammation in mice. *Hum. Gene Ther.* 2016;27(8):631-42. doi: https://doi.org/10.1089/hum.2015.161.

34 Dobrzynski E, Mingozzi F, Liu YL, Bendo E, Cao O, Wang L, et al. Induction of antigen-specific CD4+ T-cell anergy and deletion by in vivo viral gene transfer. *Blood* 2004;104(4):969-77. doi: https://doi.org/10.1182/blood-2004-03-0847.

35 Martino AT, Nayak S, Hoffman BE, Cooper M, Liao G, Markusic DM, et al. Tolerance induction to cytoplasmic beta-galactosidase by hepatic AAV gene transfer: implications for antigen presentation and immunotoxicity. *PLoS One* 2009;4(8):e6376. doi: https://doi.org/10.1371/journal.pone.0006376.

36 Follenzi A, Battaglia M, Lombardo A, Annoni A, Roncarolo MG, Naldini L. Targeting lentiviral vector expression to hepatocytes limits transgene-specific immune response and establishes long-term expression of human antihemophilic factor IX in mice. *Blood* 2004;103(10):3700-9. doi: https://doi.org/10.1182/blood-2003-09-3217.

37 Niemeyer GP, Herzog RW, Mount J, Arruda VR, Tillson DM, Hathcock J, et al. Long-term correction of inhibitor-prone hemophilia B dogs treated with liver-directed AAV2-mediated factor IX gene therapy. *Blood* 2009;113(4):797-806. doi: https://doi.org/10.1182/blood-2008-10-181479.

38 Finn JD, Ozelo MC, Sabatino DE, Franck HW, Merricks EP, Crudele JM, et al. Eradication of neutralizing antibodies to factor VIII in canine hemophilia A after liver gene therapy. *Blood* 2010;116(26):5842-8. doi: https://doi.org/10.1182/blood-2010-06-288001.

39 Nathwani AC, Tuddenham EG, Rangarajan S, Rosales C, McIntosh J, Linch DC, et al. Adenovirus-associated virus vector-mediated gene transfer in hemophilia B. *N. Engl. J. Med.* 2011;365(25):2357-65. doi: https://doi.org/10.1056/NEJMoa1108046.

40 Manno CS, Pierce GF, Arruda VR, Glader B, Ragni M, Rasko JJ, et al. Successful transduction of liver in hemophilia by AAV-Factor IX and limitations imposed by the host immune response. *Nat. Med.* 2006;12(3):342-7. doi: https://doi.org/10.1038/nm1358.

41 Patel SR, Lundgren TS, Spencer HT, Doering CB. The immune response to the fVIII gene therapy in preclinical models. *Front. Immunol.* 2020;11:494. doi: https://doi.org/10.3389/fimmu.2020.00494.

42 Salabarria SM, Nair J, Clement N, Smith BK, Raben N, Fuller DD, et al. Advancements in AAV-mediated gene therapy for pompe disease. *J. Neuromuscul. Dis.* 2020;7(1):15-31. doi: https://doi.org/10.3233/jnd-190426.

43 Koeberl DD, Kishnani PS. Immunomodulatory gene therapy in lysosomal storage disorders. *Curr. Gene Ther.* 2009;9(6):503-10. doi: https://doi.org/10.2174/156652309790031094.

44 Samelson-Jones BJ, Finn JD, Favaro P, Wright JF, Arruda VR. Timing of intensive immunosuppression impacts risk of transgene antibodies after AAV gene therapy in nonhuman primates. *Mol. Ther. Methods Clin. Dev.* 2020;17:1129-38. doi: https://doi.org/10.1016/j.omtm.2020.05.001.

45 Murrey DA, Naughton BJ, Duncan FJ, Meadows AS, Ware TA, Campbell KJ, et al. Feasibility and safety of systemic rAAV9-hNAGLU delivery for treating mucopolysaccharidosis IIIB: toxicology, biodistribution, and immunological assessments in primates. *Hum. Gene Ther. Clin. Dev.* 2014;25(2):72-84. doi: https://doi.org/10.1089/humc.2013.208.

46 Ciesielska A, Hadaczek P, Mittermeyer G, Zhou S, Wright JF, Bankiewicz KS, et al. Cerebral infusion of AAV9 vector-encoding non-self proteins can elicit cell-mediated immune responses. *Molecular Therapy* 2013;21(1):158-66. doi: https://doi.org/10.1038/mt.2012.167.

47 Ramsingh AI, Gray SJ, Reilly A, Koday M, Bratt D, Koday MT, et al. Sustained AAV9-mediated expression of a non-self protein in the CNS of non-human primates after immunomodulation. *PLoS One* 2018;13(6):e0198154. doi: https://doi.org/10.1371/journal.pone.0198154.

48 Anand V, Duffy B, Yang Z, Dejneka NS, Maguire AM, Bennett J. A deviant immune response to viral proteins and transgene product is generated on subretinal administration of adenovirus and adeno-associated virus. *Molecular Therapy* 2002;5(2):125-32. doi: https://doi.org/10.1006/mthe.2002.0525.

49 Acland GM, Aguirre GD, Ray J, Zhang Q, Aleman TS, Cideciyan AV, et al. Gene therapy restores vision in a canine model of childhood blindness. *Nat. Genet.* 2001;28(1):92-5. doi: https://doi.org/10.1038/ng0501-92.

50 Amado D, Mingozzi F, Hui D, Bennicelli JL, Wei Z, Chen Y, et al. Safety and efficacy of subretinal readministration of a viral vector in large animals to treat congenital blindness. *Sci. Transl. Med.* 2010;2(21):21ra16. doi: https://doi.org/10.1126/scitranslmed.3000659.

51 Maguire AM, High KA, Auricchio A, Wright JF, Pierce EA, Testa F, et al. Age-dependent effects of RPE65 gene therapy for Leber's congenital amaurosis: a phase 1 dose-escalation trial. *Lancet* 2009;374(9701):1597-605. doi: https://doi.org/10.1016/s0140-6736(09)61836-5.

52 Maguire AM, Simonelli F, Pierce EA, Pugh EN, Jr., Mingozzi F, Bennicelli J, et al. Safety and efficacy of gene transfer for Leber's congenital amaurosis. *N. Engl. J. Med.* 2008;358(21):2240-8. doi: https://doi.org/10.1056/NEJMoa0802315.

53 Bennett J, Ashtari M, Wellman J, Marshall KA, Cyckowski LL, Chung DC, et al. AAV2 gene therapy readministration in three adults with congenital blindness. *Sci. Transl. Med.* 2012;4(120):120ra15. doi: https://doi.org/10.1126/scitranslmed.3002865.

54 McCarty DM, Fu H, Monahan PE, Toulson CE, Naik P, Samulski RJ. Adeno-associated virus terminal repeat (TR) mutant generates self-complementary vectors to overcome the rate-limiting step to transduction in vivo. *Gene Ther.* 2003;10(26):2112-8. doi: https://doi.org/10.1038/sj.gt.3302134.

55 McCarty DM, Monahan PE, Samulski RJ. Self-complementary recombinant adeno-associated virus (scAAV) vectors promote efficient transduction independently of DNA synthesis. *Gene Ther.* 2001;8(16):1248-54. doi: https://doi.org/10.1038/sj.gt.3301514.

56 Wu T, Töpfer K, Lin SW, Li H, Bian A, Zhou XY, et al. Self-complementary AAVs induce more potent transgene product-specific immune responses compared to a single-stranded genome. *Molecular Therapy* 2012;20(3):572-9. doi: https://doi.org/10.1038/mt.2011.280.

57 Rogers GL, Martino AT, Zolotukhin I, Ertl HC, Herzog RW. Role of the vector genome and underlying factor IX mutation in immune responses to AAV gene therapy for hemophilia B. *J. Transl. Med.* 2014;12:25. doi: https://doi.org/10.1186/1479-5876-12-25.

58 Feng Y, Jacobs F, Van Craeyveld E, Lievens J, Snoeys J, Van Linthout S, et al. The impact of antigen expression in antigen-presenting cells on humoral immune responses against the transgene product. *Gene Ther.* 2010;17(2):288-93. doi: https://doi.org/10.1038/gt.2009.125.

59 Herzog RW, Fields PA, Arruda VR, Brubaker JO, Armstrong E, McClintock D, et al. Influence of vector dose on factor IX-specific T and B cell responses in muscle-directed gene therapy. *Hum. Gene Ther.* 2002;13(11):1281-91. doi: https://doi.org/10.1089/104303402760128513.

60 Yin Y, Takahashi Y, Hamana A, Nishikawa M, Takakura Y. Effects of transgene expression level per cell in mice livers on induction of transgene-specific immune responses after hydrodynamic gene transfer. *Gene Ther.* 2016;23(7):565-71. doi: https://doi.org/10.1038/gt.2016.26.

61 Matsui H, Shibata M, Brown B, Labelle A, Hegadorn C, Andrews C, et al. A murine model for induction of long-term immunologic tolerance to factor VIII does not require persistent detectable levels of plasma factor VIII and involves contributions from Foxp3+ T regulatory cells. *Blood* 2009;114(3):677-85. doi: https://doi.org/10.1182/blood-2009-03-202267.

62 Sabatino DE, Mackenzie TC, Peranteau W, Edmonson S, Campagnoli C, Liu YL, et al. Persistent expression of hF.IX After tolerance induction by in utero or neonatal administration of AAV-1-F.IX in hemophilia B mice. *Molecular Therapy* 2007;15(9):1677-85. doi: https://doi.org/10.1038/sj.mt.6300219.

63 Waddington SN, Buckley SM, Nivsarkar M, Jezzard S, Schneider H, Dahse T, et al. In utero gene transfer of human factor IX to fetal mice can induce postnatal tolerance of the exogenous clotting factor. *Blood* 2003;101(4):1359-66. doi: https://doi.org/10.1182/blood-2002-03-0779.

64 Carlon MS, Vidović D, Dooley J, da Cunha MM, Maris M, Lampi Y, et al. Immunological ignorance allows long-term gene expression after perinatal recombinant adeno-associated virus-mediated gene transfer to murine airways. *Hum. Gene Ther.* 2014;25(6):517-28. doi: https://doi.org/10.1089/hum.2013.196.

65 Nivsarkar MS, Buckley SM, Parker AL, Perocheau D, McKay TR, Rahim AA, et al. Evidence for contribution of CD4+ CD25+ regulatory T cells in maintaining immune tolerance to human factor IX following perinatal adenovirus vector delivery. *J. Immunol. Res.* 2015;2015:397879. doi: https://doi.org/10.1155/2015/397879.

66 Hinderer C, Bell P, Louboutin JP, Zhu Y, Yu H, Lin G, et al. Neonatal systemic AAV induces tolerance to CNS gene therapy in MPS I dogs and nonhuman

primates. *Molecular Therapy* 2015;23(8):1298-307. doi: https://doi.org/10.1038/mt.2015.99.

67 Davey MG, Riley JS, Andrews A, Tyminski A, Limberis M, Pogoriler JE, et al. Induction of immune tolerance to foreign protein via adeno-associated viral vector gene transfer in mid-gestation fetal sheep. *PLoS One* 2017;12(1):e0171132. doi: https://doi.org/10.1371/journal.pone.0171132.

68 Mattar CNZ, Gil-Farina I, Rosales C, Johana N, Tan YYW, McIntosh J, et al. In utero transfer of adeno-associated viral vectors produces long-term factor IX levels in a cynomolgus macaque model. *Molecular Therapy* 2017;25(8):1843-53. doi: https://doi.org/10.1016/j.ymthe.2017.04.003.

69 Zhang YC, Powers M, Wasserfall C, Brusko T, Song S, Flotte T, et al. Immunity to adeno-associated virus serotype 2 delivered transgenes imparted by genetic predisposition to autoimmunity. *Gene Ther.* 2004;11(3):233-40. doi: https://doi.org/10.1038/sj.gt.3302144.

70 Fields PA, Arruda VR, Armstrong E, Chu K, Mingozzi F, Hagstrom JN, et al. Risk and prevention of anti-factor IX formation in AAV-mediated gene transfer in the context of a large deletion of F9. *Molecular Therapy* 2001;4(3):201-10. doi: https://doi.org/10.1006/mthe.2001.0441.

71 Biswas M, Palaschak B, Kumar SRP, Rana J, Markusic DM. B cell depletion eliminates FVIII memory B cells and enhances AAV8-coF8 immune tolerance induction when combined with rapamycin. *Front. Immunol.* 2020;11:1293. doi: https://doi.org/10.3389/fimmu.2020.01293.

72 Wang SY, Weiner G. Complement and cellular cytotoxicity in antibody therapy of cancer. *Expert Opin. Biol. Ther.* 2008;8(6):759-68. doi: https://doi.org/10.1517/14712598.8.6.759.

73 Parrillo JE, Fauci AS. Comparison of the effector cells in human spontaneous cellular cytotoxicity and antibody-dependent cellular cytotoxicity: differential sensitivity of effector cells to in vivo and in vitro corticosteroids. *Scand. J. Immunol.* 1978;8(2):99-107. doi: https://doi.org/10.1111/j.1365-3083.1978.tb00501.x.

74 Van Beek M, Nussenzweig MC, Chakraborty AK. Two complementary features of humoral immune memory confer protection against the same or variant antigens. *PNAS* 2022;119(37):e2205598119. doi: https://doi.org/10.1073/pnas.2205598119.

75 Steiner LA, Eisen HN. The relative affinity of antibodies synthesized in the secondary response. *J. Exp. Med.* 1967;126(6):1185-205. doi: https://doi.org/10.1084/jem.126.6.1185.

76 Yang TY, Braun M, Lembke W, McBlane F, Kamerud J, DeWall S, et al. Immunogenicity assessment of AAV-based gene therapies: an IQ consortium industry white paper. *Mol. Ther. Methods Clin. Dev.* 2022;26:471-94. doi: https://doi.org/10.1016/j.omtm.2022.07.018.

77 FDA (2019). Immunogenicity testing of therapeutic protein products—developing and validating assays for anti-drug antibody detection. Guidance for Industry. https://www.fda.gov/media/119788/download (accessed July 2019).

78 Shankar G, Devanarayan V, Amaravadi L, Barrett YC, Bowsher R, Finco-Kent D, et al. Recommendations for the validation of immunoassays used for detection of host antibodies against biotechnology products. *J. Pharm. Biomed. Anal.* 2008;48(5):1267-81. doi: doi: https://doi.org/10.1016/j.jpba.2008.09.020.

79 Gorovits B, Azadeh M, Buchlis G, Harrison T, Havert M, Jawa V, et al. Evaluation of the humoral response to adeno-associated virus-based gene therapy modalities using total antibody assays. *Aaps J.* 2021;23(6):108. doi: https://doi.org/10.1208/s12248-021-00628-3.

80 Gorovits B, Fiscella M, Havert M, Koren E, Long B, Milton M, et al. Recommendations for the development of cell-based anti-viral vector neutralizing antibody assays. *Aaps J.* 2020;22(2):24. doi: https://doi.org/10.1208/s12248-019-0403-1.

81 Zhang XT, Chen H, Shao W, Lin ZJ, Melhem M, Lu S. A competitive ligand-binding assay for the detection of neutralizing antibodies against dostarlimab (TSR-042). *AAPS Open* 2021;7(1):8. doi: https://doi.org/10.1186/s41120-021-00039-w.

82 Quast I, Peschke B, Lünemann JD. Regulation of antibody effector functions through IgG Fc N-glycosylation. *Cell. Mol. Life Sci.* 2017;74(5):837-47. doi: https://doi.org/10.1007/s00018-016-2366-z.

83 Gadjeva M. The complement system. Overview. *Methods Mol Biol.* 2014;1100:1-9. doi: https://doi.org/10.1007/978-1-62703-724-2_1.

84 Burbelo PD, Iadarola MJ, Keller JM, Warner BM. Autoantibodies targeting intracellular and extracellular proteins in autoimmunity. *Front. Immunol.* 2021;12:548469. doi: https://doi.org/10.3389/fimmu.2021.548469.

85 Yang Z, Edwards H, Xu P. CRISPR-Cas12a/Cpf1-assisted precise, efficient and multiplexed genome-editing in Yarrowia lipolytica. *Metab. Eng. Commun.* 2020;10:e00112. doi: https://doi.org/10.1016/j.mec.2019.e00112.

86 Leenay RT, Beisel CL. Deciphering, communicating, and engineering the CRISPR PAM. *J. Mol. Biol.* 2017;429(2):177-91. doi: https://doi.org/10.1016/j.jmb.2016.11.024.

87 Gleditzsch D, Pausch P, Müller-Esparza H, Özcan A, Guo X, Bange G, et al. PAM identification by CRISPR-Cas effector complexes: diversified mechanisms and structures. *RNA Biol.* 2019;16(4):504-17. doi: https://doi.org/10.1080/15476286.2018.1504546.

88 Wang H, La Russa M, Qi LS. CRISPR/Cas9 in genome editing and beyond. *Annu. Rev. Biochem.* 2016;85:227-64. doi: https://doi.org/10.1146/annurev-biochem-060815-014607.

89 Jinek M, Chylinski K, Fonfara I, Hauer M, Doudna JA, Charpentier E. A programmable dual-RNA-guided DNA endonuclease in adaptive bacterial immunity. *Science* 2012;337(6096):816-21. doi: https://doi.org/10.1126/science.1225829.

90 Bogdanove AJ, Voytas DF. TAL effectors: customizable proteins for DNA targeting. *Science* 2011;333(6051):1843-6. doi: https://doi.org/10.1126/science.1204094.

91 Urnov FD, Miller JC, Lee YL, Beausejour CM, Rock JM, Augustus S, et al. Highly efficient endogenous human gene correction using designed zinc-finger nucleases. *Nature* 2005;435(7042):646-51. doi: https://doi.org/10.1038/nature03556.

92 Joung JK, Sander JD. TALENs: a widely applicable technology for targeted genome editing. *Nat. Rev. Mol. Cell Biol.* 2013;14(1):49-55. doi: https://doi.org/10.1038/nrm3486.

93 Urnov FD, Rebar EJ, Holmes MC, Zhang HS, Gregory PD. Genome editing with engineered zinc finger nucleases. *Nat. Rev. Genet.* 2010;11(9):636-46. doi: https://doi.org/10.1038/nrg2842.

94 Sander JD, Dahlborg EJ, Goodwin MJ, Cade L, Zhang F, Cifuentes D, et al. Selection-free zinc-finger-nuclease engineering by context-dependent assembly (CoDA). *Nat. Methods* 2011;8(1):67-9. doi: https://doi.org/10.1038/nmeth.1542.

95 Qasim W, Zhan H, Samarasinghe S, Adams S, Amrolia P, Stafford S, et al. Molecular remission of infant B-ALL after infusion of universal TALEN gene-edited CAR T cells. *Sci. Transl. Med.* 2017;9(374). doi: https://doi.org/10.1126/scitranslmed.aaj2013.

96 Stadtmauer EA, Fraietta JA, Davis MM, Cohen AD, Weber KL, Lancaster E, et al. CRISPR-engineered T cells in patients with refractory cancer. *Science.* 2020;367(6481). doi: https://doi.org/10.1126/science.aba7365.

97 Tebas P, Stein D, Tang WW, Frank I, Wang SQ, Lee G, et al. Gene editing of CCR5 in autologous CD4 T cells of persons infected with HIV. *N. Engl. J. Med.* 2014;370(10):901-10. doi: https://doi.org/10.1056/NEJMoa1300662.

98 Doudna JA. The promise and challenge of therapeutic genome editing. *Nature* 2020;578(7794):229-36. doi: https://doi.org/10.1038/s41586-020-1978-5.

99 Marzec M, Hensel G. More precise, more universal and more specific - the next generation of RNA-guided endonucleases for genome editing. *Febs J.* 2019;286(23):4657-60. doi: https://doi.org/10.1111/febs.15079.

100 Gao Q, Dong X, Xu Q, Zhu L, Wang F, Hou Y, et al. Therapeutic potential of CRISPR/Cas9 gene editing in engineered T-cell therapy. *Cancer Med.* 2019;8(9):4254-64. doi: https://doi.org/10.1002/cam4.2257.

101 Glass Z, Lee M, Li Y, Xu Q. Engineering the delivery system for CRISPR-based genome editing. *Trends Biotechnol.* 2018;36(2):173-85. doi: https://doi.org/10.1016/j.tibtech.2017.11.006.

102 Kholosy WM, Visscher M, Ogink K, Buttstedt H, Griffin K, Beier A, et al. Simple, fast and efficient iTOP-mediated delivery of CRISPR/Cas9 RNP in difficult-to-transduce human cells including primary T cells. *J. Biotechnol.* 2021;338:71-80. doi: https://doi.org/10.1016/j.jbiotec.2021.07.006.

103 Mortensen R, Nissen TN, Blauenfeldt T, Christensen JP, Andersen P, Dietrich J. Adaptive Immunity against *Streptococcus pyogenes* in adults involves increased IFN-γ and IgG3 responses compared with children. *J. Immunol.* 2015;195(4):1657-64. doi: https://doi.org/10.4049/jimmunol.1500804.

104 Kolata JB, Kühbandner I, Link C, Normann N, Vu CH, Steil L, et al. The fall of a dogma? unexpected high T-cell memory response to *Staphylococcus aureus* in humans. *J. Infect. Dis.* 2015;212(5):830-8. doi: https://doi.org/10.1093/infdis/jiv128.

105 Colque-Navarro P, Jacobsson G, Andersson R, Flock JI, Möllby R. Levels of antibody against 11 *Staphylococcus aureus* antigens in a healthy population. *Clin. Vac. Immunol.* 2010;17(7):1117-23. doi: https://doi.org/10.1128/cvi.00506-09.

106 Charlesworth CT, Deshpande PS, Dever DP, Camarena J, Lemgart VT, Cromer MK, et al. Identification of preexisting adaptive immunity to Cas9 proteins in humans. *Nat. Med.* 2019;25(2):249-54. doi: https://doi.org/10.1038/s41591-018-0326-x.

107 Guo Y, Tong C, Su L, Zhang W, Jia H, Liu Y, et al. CRISPR/Cas9 genome-edited universal CAR T cells in patients with relapsed/refractory lymphoma. *Blood Adv.* 2022;6(8):2695-9. doi: https://doi.org/10.1182/bloodadvances.2021006232.

108 Hu Y, Zhou Y, Zhang M, Ge W, Li Y, Yang L, et al. CRISPR/Cas9-engineered universal CD19/CD22 dual-targeted CAR-T cell therapy for relapsed/refractory B-cell acute lymphoblastic leukemia. *Clin. Cancer Res.* 2021;27(10):2764-72. doi: https://doi.org/10.1158/1078-0432.Ccr-20-3863.

109 Mays LE, Wilson JM. The complex and evolving story of T cell activation to AAV vector-encoded transgene products. *Molecular Therapy* 2011;19(1):16-27. doi: https://doi.org/10.1038/mt.2010.250.

110 Chew WL, Tabebordbar M, Cheng JK, Mali P, Wu EY, Ng AH, et al. A multifunctional AAV-CRISPR-Cas9 and its host response. *Nat. Methods* 2016;13(10):868-74. doi: https://doi.org/10.1038/nmeth.3993.

111 Moreno AM, Palmer N, Alemán F, Chen G, Pla A, Jiang N, et al. Immune-orthogonal orthologues of AAV capsids and of Cas9 circumvent the immune response to the administration of gene therapy. *Nat. Biomed. Eng.* 2019;3(10):806-16. doi: https://doi.org/10.1038/s41551-019-0431-2.

112 Nelson CE, Wu Y, Gemberling MP, Oliver ML, Waller MA, Bohning JD, et al. Long-term evaluation of AAV-CRISPR genome editing for Duchenne muscular dystrophy. *Nat. Med.* 2019;25(3):427-32. doi: https://doi.org/10.1038/s41591-019-0344-3.

113 Thakore PI, Kwon JB, Nelson CE, Rouse DC, Gemberling MP, Oliver ML, et al. RNA-guided transcriptional silencing in vivo with *S. aureus* CRISPR-Cas9 repressors. *Nat. Commun.* 2018;9(1):1674. doi: https://doi.org/10.1038/s41467-018-04048-4.

114 Li A, Tanner MR, Lee CM, Hurley AE, De Giorgi M, Jarrett KE, et al. AAV-CRISPR gene editing is negated by pre-existing immunity to Cas9. *Molecular Therapy* 2020;28(6):1432-41. doi: https://doi.org/10.1016/j.ymthe.2020.04.017.

115 Ferdosi SR, Ewaisha R, Moghadam F, Krishna S, Park JG, Ebrahimkhani MR, et al. Multifunctional CRISPR-Cas9 with engineered immunosilenced human T cell epitopes. *Nat. Commun.* 2019;10(1):1842. doi: https://doi.org/10.1038/s41467-019-09693-x.

116 Simhadri VL, McGill J, McMahon S, Wang J, Jiang H, Sauna ZE. Prevalence of pre-existing antibodies to CRISPR-associated nuclease Cas9 in the USA population. *Mol. Ther. Methods Clin. Dev.* 2018;10:105-12. doi: https://doi.org/10.1016/j.omtm.2018.06.006.

117 Wagner DL, Amini L, Wendering DJ, Burkhardt LM, Akyüz L, Reinke P, et al. High prevalence of *Streptococcus pyogenes* Cas9-reactive T cells within the adult human population. *Nat. Med.* 2019;25(2):242-8. doi: https://doi.org/10.1038/s41591-018-0204-6.

118 Kumar SC, DelCarpini JA, Qu Q, Kane M, Gorovits B. Mitigation of pre-existing antibodies to a biotherapeutic in non-clinical species when establishing anti-drug antibody assay cutpoint. *AAPS J.* 2017;19(1):313-9. doi: https://doi.org/10.1208/s12248-016-0011-2.

119 Mehta A, Merkel OM. Immunogenicity of Cas9 protein. *J. Pharm. Sci.* 2020;109(1):62-7. doi: https://doi.org/10.1016/j.xphs.2019.10.003.

120 Safdari Y, Farajnia S, Asgharzadeh M, Khalili M. Antibody humanization methods - a review and update. *Biotechnol. Genet. Eng. Rev.* 2013;29:175-86. doi: https://doi.org/10.1080/02648725.2013.801235.

13

rAAV Integration: Detection and Risk Assessment

Jing Yuan[1], *Irene Gil-Farina*[2], *Raffaele Fronza*[2], *and Laurence O. Whiteley*[3]

[1] *Department of Toxicology, Kymera Therapeutics, Watertown, MA, USA*
[2] *ProtaGene CGT GmbH, Heidelberg, Germany*
[3] *Pfizer Inc. Drug Safety Research and Development, Cambridge MA, USA*

13.1 Introduction

Several attributes of recombinant adeno associated virus (rAAV) vectors contribute to the popularity of their use in in vivo gene therapy (GTx). First, multiple serotypes with different tissue tropism, second, widespread infection of wild-type AAV (wtAAV) in mammalian populations without evidence of pathogenicity and, third, they are able to express a transgene without integrating into the host cell genome (e.g. episomal). While considered to be a nonintegrating vector, data, primarily from neonatal mice, indicates that AAV can integrate and result in genotoxic effects, leading to the development of hepatocellular carcinoma (HCC) [1, 2]. The human relevance of recombinant AAV vector-induced liver tumors in neonatal mice is controversial [3]. The uncertainty of human relevance of these findings in mice and the increased interest in developing AAV GTx as a therapeutic modality has resulted in intense interest in the scientific and regulatory community as evidenced by several recent public forum that have discussed this topic: September 2021 FDA Cellular, Tissue, and Gene Therapies Advisory Committee (CTGTAC) meeting [4], the 2021 ASGCT Policy Summit, and the November 2021 ASGCT-FDA Liaison Meeting [5].

This chapter will: (1) review the biology of AAV integration and published literature related to AAV-associated carcinogenesis, (2) discuss study design considerations for assessing AAV integration in nonclinical safety studies, (3) outline

Drug Development for Gene Therapy: Translational Biomarkers, Bioanalysis, and Companion Diagnostics, First Edition. Edited by Yanmei Lu and Boris Gorovits.

methods for detecting and quantitating AAV integration, (4) discuss the regulatory landscape related to AAV integration, and (5) provide perspective on human risk assessment considerations.

13.1.1 Biology of AAV Vectors as it Relates to Mechanisms of AAV Integration

wtAAV has a single-stranded DNA genome that is flanked by inverted terminal repeat sequences (ITRs). The genome encodes two genes: the *rep* gene that is required for replication and packaging of the DNA and *cap* gene that encodes the proteins that assemble in the viral capsid. Following entry into the cell, wtAAV can be maintained as an episome in the host genome or integrated into host genome, AAVS1 locus on chromosome 19 was identified as a preferential integration region in humans. The integration is mediated by viral Rep protein. In rAAV, the *rep* and *cap* genes are removed and replaced with the transgene expression cassette; thus, the only remaining viral genome sequences are the ITRs that are essential for packaging of the vector genome and second strand synthesis. rAAV vector genome processing toward stable transduction relies on host cellular machinery. Following AAV vector transduction, single-stranded (ss) rAAV vector genomes are used as template to form double-stranded (ds) linear rAAV monomers as the intermediates which then transform to stable double-stranded circular monomers and concatemers DNA [6]. Unlike wtAAV, rAAV DNA lacks the Rep-mediated active integration and primarily remains in a circular concatemeric episomal form after transduction. However, AAV vectors passively integrate at low frequency into the target cell genome. rAAV integration events may happen both through nonhomologous end-joining (NHEJ) DSB repair pathway and homologs recombination. Unlike retroviral vectors, which induce double-stranded breaks to facilitate integration, AAV uses the spontaneous double-strand breaks that occur during the normal cell cycle to integrate into host cell DNA.

13.1.2 Literature Review of AAV Studies in Relation to Neoplasia Development

Integration of rAAV into host cell genome has recently been reviewed and the nonclinical studies are summarized in Tables 13.1 and 13.2 [21]. Evidence of rAAV vector integration into the host genome with subsequent tumorigenesis first came from rAAV-treated neonatal mice. Mucopolysaccharidoses (MPS)VII is a lysosomal storage disease caused by a deficiency of beta-glucuronidase (GUSB). When MPSVII mice were intravenously injected with an AAV2 vector containing a cytomegalovirus early enhancer element and chicken beta-actin promoter (CAG) and a human GUSB cDNA at the newborn stage (postnatal day 2),

Table 13.1 List of studies with rAAV administration in neonatal and adult mice.

Publication	AAV serotype	Genome Configuration	Enhancer/ promoter (*promoters only)	Transgene	Route of delivery	Dose	Time of treatment	Loci with integration(s) associated with HCC
Donsante [7] Donsante [1]	2	SS	CMV/CBA	*GUSB*	Superficial temporal vein	1×10^{14} vg/kg	Neonatal	*Rian*
Bell [8]	1,2,5,7,8,9, others	SS	Alb, CBA, SV40, TBG, Z12*	*A1AT, EGFP, Epo, FIX, LacZ, LDLR, TF1*	Portal vein Tail vein	3×10^{9}–1×10^{12} vg/mouse	6–8 weeks	*None*
Bell [9] Zhong [10]	2,7,8,9	SS	AMPB/TBG	*Oct, LacZ*	Portal vein	3×10^{9}–1×10^{12} vg/mouse	9–11 weeks	*Tax1bp1, Rian*
Reiss [11]	1/2	SS	CMV/CBA	*MOCS1*	Intrahepatic	1–3×10^{9} vg/pup	Neonatal	*None*
Embury [12]	2	SS	CBA* (WPRE)	*mPah*	Portal vein	3.7×10^{9}–7.3×10^{13} vg/mouse	10–14 weeks	*None*
Li [13]	2	SS	APOE/hAAT	*FIX*	Portal vein	5×10^{12}–1×10^{14} vg/kg	4 weeks	*None*
Wang [14]	2	SS	CMV/CBA	*Homology Arms to Rian*	Temporal vein	3×10^{9} vg/pup	Neonatal	*Rian*

(Continued)

Table 13.1 (Continued)

Publication	AAV serotype	Genome Configuration	Enhancer/ promoter (*promoters only)	Transgene	Route of delivery	Dose	Time of treatment	Loci with integration(s) associated with HCC
Rosas [15]	Rh74	SC	CMV, CBA*	*GFP, null*	Tail vein	5×10^{10} vg/mouse	8–12 weeks	*Hras1, Sos1, Fgf3*
Kao [16]	8	SS	APOE/hAAT	*FIX-WT, FIX-Triple, FIX-Padua*	Tail vein	1.6×10^{12}–2×10^{10} vg/mouse	Adult	*None*
Chandler [2]	2,8,9	SS	AMPB/TBG CMV/CBA APOE/hAAT	*hMUT, mMut, GFP*	Intrahepatic	7×10^{12}–1×10^{14} vg/kg	Neonatal	*Rian*
Walia [17]	9	SS	CMV/CMV	*HexB, LacZ*	Superficial temporal vein	2.5×10^{14} vg/kg	Neonatal	*Rian*
Kang [18]	9	SS	TBG	*G6PC, ZFN*	Tail vein	4.8×10^{12}–1.3×10^{13} vg/kg	Adult	*Rian*
Dalwadi [19]	DJ	SS	CMV/CAG	*Homology Arms to Rian, tdTomato*	Superficial temporal vein	3×10^{10} vg/pup 8×10^{11} vg/adult	Neonatal Adults	*Rian*
Li [20]	9	SS	CMV/CBA	*GALC*	Intracranial or Intrathecal	2.7×10^{10} vg/pup	Neonatal	*Rian, FoxP1, Tusc3, Magi1, Kcnh1, Akap13*

HCC, hepatocellular carcinoma; rAAV, recombinant adeno-associated virus; CMV, cytomegalovirus; CBA, chicken beta-actin; Alb, albumin; TBG, thyroxin-binding globulin; WPRE, woodchuck hepatitis virus post-transcriptional regulatory element; TBG, thyroxin-binding globulin; APOE, apolipoprotein E; hAAT, human alpha-1 antitrypsin; CAG, cytomegalovirus enhancer/chicken beta-actin promoter.

Table 13.2 List of studies with rAAV administration in large animal species.

Publication	Species	Disease model	AAV vector; route	AAV dose (vg/kg)	Time of treatment	Duration of F/U
Nowrouzi et al. [22]	NHP	WT $n=18$	AAV-RSV-LEA29Y AAV1 or AAV8 IV or IM	5×10^{12}	Adult	1.2–2.8 yr
Gil Farina et al. [3]	NHP	WT $n=6$	AAV5-hPBGD (IV)	1×10^{13} 5×10^{13}	Adult	1 mo
Mattar et al. [23]	NHP	WT $n=6$	AAV-hFIX AAV8 or AAV5 (umbilical vein)	1.4–1.9×10^{13}	In utero	11–71 mo
Spronck et al. 2020 [24]	NHP	WT $n=12$	AAV5-hFIX (IV)	5×10^{11} 5×10^{12} 2×10^{13} 9×10^{13}	Adult	6 mo
Sullivan et al. [25]	NHP	WT $n=12$	AAV5-hFVIII (IV)	2×10^{13} 6×10^{13}	Adult	13 and 26 wk (6 mo)
Niemeyer et al. 2009 [26]	Dog	Hemophilia B $n=4$	AAV2-CMV-cFIX (IM) AAV2-(Apoe)4/hAAT-cFIX (IV)	1.1×10^{12} to 3.4×10^{12}	Juvenile/Adult 5.5–12 mo	8 yr
Nguyen, Everett et al. [27]	Dog	Hemophilia A $n=9$	AAV-TBG-cFVIII AAV-hAAT-cFVIII AAV8 or AAV9 (PV or IV)	1×10^{13} 2×10^{13} 4×10^{13}	Juvenile/Adult 5 mo–4 yr	2–10 yr
Batty et al. [28], Batty et al. [29]	Dog	Hemophilia A $n=8$	AAV-TTR-cFVIII AAV2, 6, 8 PV	6×10^{12} to 2.7×10^{13}	Juvenile/Adult 6 mo–2 yr	8–12 yr

Source: Sabatino et al. [21]/Elsevier/Licensed under CC BY 4.0.

long-term transgene expression, GUSB activity, and phenotypic correction were achieved. A number of long-lived rAAV-treated MPSVII mice developed HCC and angiosarcomas over a year after treatment [30]. A follow-up study in newborn

MPSVII with intravenous injection of the same AAV2 vector confirmed HCC development in in both normal and MPSVII mice 13 months after dosing. In addition, this study detected rAAV integrations in the tumor tissue. All of the integrants were rearranged such that the GUSB cDNA was deleted [1]. This set of neonatal mice studies provided the main evidence on integration and tumorigenesis risk of AAV therapy.

Several additional studies reported liver tumorigenesis following rAAV administration in mice. However, the relevance of AAV integration as the primary event is questionable, especially in adult mice. The confounding factors, such as transgene expression [9, 12], genetic background in certain mouse strains [13], or pretreatment in study mice that predispose to tumor risk maybe the primary cause of tumor formation in many cases. The tumorigenesis findings related to AAV integration are considered age-specific to neonatal mice (PND1-2) [1, 30] and, in some cases, mouse strain specific to those sensitive to tumor formation. This argument has been supported by several studies in juvenile and adult animals where administration of AAV either failed to induce tumorigenesis in rodents [8, 13, 31], or the frequency of liver tumors in AAV-treated animals was comparable with an average frequency of spontaneous liver tumors in C57BL/6 mice (0–10%). A summary of AAV integration studies in rodents is listed in Table 13.1. Neonatal-specific HCC in AAV-treated mice can be explained by the fact that the neonatal liver contains proliferating hepatocytes, which may lead to a high frequency of rAAV integrations at the sites of DNA strand break, thus has an intrinsic susceptibility to tumor formation. This hypothesis is supported by the findings that AAV administration in adult mice can lead to HCC in the context of chronic liver disease, a state that induces hepatocyte proliferation [19].

HCC related to AAV administration has not been identified in large animals nor human. AAV integration and clonal expansion have been observed in hemophilia dogs following treatment with a variety of AAV serotypes. No tumor formation has been observed following more than 10 years observation [18, 26, 27, 32, 33]. In a long-term study of AAV gene therapy with hemophilia A dogs [34], two dogs treated with AAV vectors expressing canine factor VIII (AAV-cFVIII) and followed for up to 10 years had vector integration in host genome and clonal expansion. Integration events were enriched in or near genes involved in cell growth. Most of the integration recovered lacked transgene sequence and showed vector deletion and rearrangement, a phenomenon observed with many AAV integration events. None of the dogs with clonal expansions showed evidence of tumors or altered liver function revealed by liver enzyme levels and serum alpha-fetoprotein (AFP), a clinical biomarker for HCC. However, in another long-term study with hemophilia A dogs treated with AAV-cFVIII, while the genome integration was confirmed as rare events, no clonal expansion cells and tumors were detected after 8–12 years follow-up [28]. The discrepancy may be related to different methods and genome annotation is used for integration analysis. This highlights the complexity of data analysis and methodological challenges in identifying the rare integration events and clonal expansion.

Additional integration studies have been performed in large animals, including nonhuman primates, and animals have been followed up for extended years after various AAV treatments (Table 13.2). The integration is confirmed to be rare event and integration sites (IS) are random across the host genome with no clustering of integration. One nonhuman primate (NHP) study delivered a self-complementary (sc) AAV5 and AAV8 with a LP1-driven-human factor IX (hFIX) transgene (scAAV-LP1-hFIXco) to late-gestation fetuses or adult animals through a single intravenous injection [23]. Sustained clinically relevant levels of hFIX with liver-specific expression were observed without any clinical concerns four to six years after AAV administration. However long-term genotoxicity evaluation in NHP remained to be determined.

AAV integration data in patients is generally lacking as patient liver biopsies are not readily accessible. However, several different AAV serotypes (AAV1, AAV2, AAV5, AAV8, and AAV9) have been or are currently being used in clinical trials (including children) and no increase of cancer of any type has been reported [31, 35–37]. In a study investigating the potential of AAV therapy-induced tumorigenesis in humans, patients participating in an AAV2/5 gene therapy trial for acute intermitted porphyria provided liver biopsies for analysis [3]. In a context of low transduction levels, the study confirmed that AAV integration is both low in frequency and random in nature, with no clustered IS near genes that had been previously implicated in the mouse studies. Another study followed a small number of hemophilia B patients up to 15 years after liver-directed AAV2-FIX gene transfer and found no evidence of tumor formation as assessed by liver transaminase values, serum α-fetoprotein, and liver ultrasound [38]. Finally, FDA-approved AAV therapies, such as Luxturna and Zolgensma, have not reported preneoplastic proliferative lesions after multiple years of posttreatment follow-up.

In general, it is believed that the strong promoters, such CAG or TBG promoters, likely drive the read through after genome integration and trigger the nearby proto-oncogene expression that leads to HCC development [2]. This hypothesis has been challenged by a few liver targeting AAV therapies with strong promoters, such as Zolgensma for spinal muscular atrophy-type 1 (SMA1), where the vectors contain a strong CBA promoter and have been administrated at a high dose of 2×10^{14} vg/kg to infants less than two years of age with rapid liver growth. Tumor or neoplastic changes have not been identified in any of the hemophilia dogs that were dosed as juveniles (1.5–10 months of age) and with the vectors containing the CBA promoter, suggesting that the combination of a strong promoter and liver growth/cell proliferation do not necessarily contribute to enhanced neoplastic risk in nonrodents when treated with an AAV gene therapy vector.

Similar to AAV GTx vectors, there is little evidence that natural infection with wtAAV in humans is associated with neoplasia. Infection of wtAAV, mainly AAV2, is frequent in the human population. However, there is no clear evidence of wtAAV integration leading to neoplasia in humans. Insertion analyses of the liver tumors identified AAV host genome integration in less than 10% of the subjects with HCC [39–41]. AAV integrants were found in both tumors and non-tumor tissue, and in some cases, the integration rates were higher in the non-tumor tissues than in the tumors [39]. AAV was also detected with a similar frequency in malignant and benign tumors. Thus, the concurrence of AAV infection and integration and HCC does not support a pathogenic role for wtAAV infection in the HCC patients. While some of the IS were identified to be in the genes related to cancer development, those genes are also the recurrent targets by tumorigenic hepatitis B virus (HBV) integration [42], hypothesizing that HBV infection could be the main driver for malignant tumor, and wtAAV integration could be bystanders in HCC development. Consistent with the notion, non-clonal AAV2 insertions distributed throughout the genome in non-tumor samples, and AAV integration were only enriched in cancer genes in the malignant tumor cell.

13.2 Review of Regulatory Guidance and Discussion Points that Are Raised on AAV Carcinogenesis

Four classes of viruses have been used as GTx vectors that are considered episomal vectors (AAV, adenovirus, poxvirus, and herpes simplex) because they do not need to integrate into host cell genomes to produce their transgene [43]. This is in contrast to retroviral vectors (gamma retroviruses and lentiviruses) that need to integrate into the host cell genome to express their transgene. Amongst all these vectors used as GTx vectors, only vectors derived from gamma retroviruses have been definitively linked to insertional mutagenesis, leading to neoplasia in humans [44].

Because AAV does integrate at low frequency, the risks of AAV vector genotoxicity and carcinogenicity from insertional mutagenesis remain an ongoing concern by health authorities even though FDA and EMA guidance documents consider AAV vectors as non-integrating [43, 45–48]. However, both agencies discuss research publications by Donsante and coworkers that described the induction of HCC, associated with AAV vector DNA integration, in mice that were treated with AAV as neonates and imply that this should be considered in safety evaluations [1, 7].

Based on review of regulatory document of approved AAV gene therapies (Glybera, Luxturna, and Zolgensma), both EMA and FDA have acknowledged that AAV integration is a low risk [49–53]. The summary basis of approval for Luxturna indicates that the manufacturer of this GTx justified not conducting genotoxicity studies on the basis that the current scientific literature reported a low integration frequency of AAV vectors and delivery to a post mitotic cell population [54]. For Glybera, after extensive evaluation of integration in animals, the EMA concluded: "Data do not substantiate a concern for tumorigenicity. Theoretically, the product could integrate and cause a tumor, however no further animal testing or experiments can usefully address these concerns" [49]. Given the remaining uncertainty on the relevance of AAV integration for human risk, the assessment of integration events in nonclinical and clinical studies will be a discussion point with regulatory authorities during the development and post-approval monitoring of AAV GTx.

13.2.1 Factors to Consider in the Design of Nonclinical Studies Evaluating AAV Integration

Drug product that is evaluated in pivotal nonclinical studies to support administration to humans should be comparable to that administered to subjects in clinical trials. One of the parameters assessed for comparability that is relevant to characterizing AAV integration is the DNA content and form of the administered vector. Where studied, integrated rAAV DNA is often found to be rearranged, and this is also true of rAAV DNA in vector particles prior to administration [55, 56]. While sequencing technologies for characterizing different forms of DNA that are packaged in AAV vector particles are in their infancy, the ability to characterize these forms may become an important criterion in assessing comparability of AAV preparations [55–57]. Rearranged rAAV genomes and production plasmid sequences have been found to be integrated in the genome of human hepatocytes transduced ex vivo or in vivo [19]. Consequently, the molecular characterization of AAV vector preparations may be important in understanding the effects on genome integration and functional implications of an AAV's integration profile.

The need to assess DNA integration may vary between regulatory agencies in different geographies and may be requested by regulatory authorities at the latter stages of development (author's experience). Consequently, this topic should be a discussion point with regulatory agencies. Considerations should be given to collecting appropriate samples from toxicology studies and/or retaining DNA extracted from tissues for evaluating vector biodistribution, so that DNA integration can be assessed, at the request of a regulatory authority, as a program develops over time or if an observation of concern arises in a preclinical study [45, 58]. The proactive collection of tissues will avoid the need to repeat studies to address this issue, thus minimizing animal use.

13.2.2 Methods for rAAV Integration Analysis

Integration site analyses are based on the identification of host cell genomic sequences flanking the integrated vector genome in order to determine the position of the insertion event. The approaches for integration site screening traditionally consisted of the enrichment for vector-genome junctions present within a sample to enable the subsequent identification process. In this context, rAAV vectors with a predominance of vector genomes persisting episomally, that will be co-purified during the junction retrieval process, together with rare integration frequencies due to the lack of integrase activity, require highly sensitive approaches to detect infrequent integration events.

rAAV integration has been predominantly evaluated by ligation-mediated (LM-) PCR [59] or linear amplification-mediated (LAM-)PCR [60] approaches using restriction endonucleases. In essence, LAM-PCR consists of an initial linear amplification performed with a biotinylated primer located at the vector end, followed by purification utilizing streptavidin magnetic beads. Subsequent steps are performed on the DNA immobilized on the bead surface in order to cut the flanking genomic sequence with a restriction endonuclease enabling the ligation of a linker cassette of known sequence. This provides the binding site for a reverse primer that together with a forward oligonucleotide binding to the vector is used in the exponential amplification steps. LAM-PCR represented the state-of-the-art for vector integration site retrieval for almost two decades due to its superior sensitivity. Nonetheless, LM-PCR approaches have been widely used and their general principle was the digestion of the genomic DNA with a restriction endonuclease followed by a linker cassette ligation and exponential amplification. Initially, both approaches were followed by Sanger sequencing of the amplicons, sometimes including an interim cloning step [15]. However, all methods are now combined with next-generation sequencing, allowing for a time- and cost-effective high-throughput screening of vector insertion sites [13].

The limitations associated with the usage of restriction enzymes were recognized early, since integration site retrieval was limited to those events having a suitable restriction site nearby, and initially approached by the usage of multiple restriction endonucleases [61, 62]. This, in addition to biases introduced by deep sequencing of amplicons of divergent sizes, has encouraged the usage of approaches utilizing random DNA shearing [63, 64]. Accordingly, a new generation of LM-PCR-based methods utilizing DNA sonication has emerged and been successfully used to retrieve AAV vector insertion sites [27].

In addition to the determination of the integration site position, the quantification of the number of integration events constitutes a key part of integration site analyses. Different strategies are being utilized in order to correct insertion site frequencies for PCR-associated as well as other technical biases [65].

Although these mainly occur at the data processing stage as it will be detailed later, laboratory protocols contemplating the addition of molecular barcodes allow for broader dynamic ranges by overcoming eventual saturation phenomena due to the limited availability of diverse shearing sites in scenarios of clonal dominance. These molecular barcodes consist of unique molecular identifiers introduced within the linker cassettes ligated to the DNA fragments prior to the exponential amplification steps [66, 67].

PCR-based approaches have allowed the identification of AAV vector IS and clonal expansions. However, their dependence on the presence of the selected primer binding site limits the ability to detect all components of the vector that may integrate into host cell DNA. To address integrated fragments undergoing internal breakage and rearrangements, multiplexed LAM-PCR approaches were developed also allowing for the retrieval of insertion events arising from internal vector regions [3]. Improved protocols for multiplex PCR-based techniques are still needed to overcome the LAM-PCR limitations detailed above. Recently, target enrichment (TE) approaches offer a promising alternative [68]. Sequence adaptors are ligated to fragmented genomic DNA, and the enrichment is performed by hybridization with a set of RNA or DNA baits complementary to the entire vector sequence. This approach pulls down any DNA molecule-bearing vector fragments, including both episomal and integrated genomes. Considering the lower number of amplification cycles and the relatively small size of the targeted region (the AAV genome), TE approaches are considered to present a lower sensitivity when compared to PCR-based methods, although systematic comparisons are still needed. Nonetheless, these approaches allow to capture and sequence internal vector fragments, thus providing complete information on persisting vector genomes. Indeed, a study coupling a capture approach with long-read sequencing has showed the presence of integrated concatemeric sequences and a high degree of rearrangements within the vector genomes revealing the presence of vector structures that may have been missed by PCR-based technologies [19].

Besides the vector-targeted approaches, whole genome sequencing (WGS) has also been utilized to retrieve vector IS [69]. The procedure is analog to the one used in classical WGS studies and the difference relies on the sequencing data analyses. Despite its unbiased nature, this approach presents a low sensitivity, given by the absence of enrichment steps, thus limiting its usage in samples with rare and low-frequency integration events. Despite the advances in sequencing technologies, performing integration site analyses through WGS still remains time and cost-intensive. Nevertheless, it still constitutes a valid orthogonal method when eventual clonal expansions are observed.

A consensus on the precise range of AAV integration frequency detected by different approaches is still missing. The diversity of approaches utilized and the absence of reference materials, impose limitations to systematic comparisons. For integrating vectors (e.g. lentivirus vectors), reference materials have been generated by cloning cells with a limited and known number of insertion events. However, such process would be challenging for AAV vectors considering their passive integration nature and the high diversity of vector designs, with only the ITRs being almost universal throughout different vector designs. Therefore, the definition of method performance parameters and novel strategies enabling inter-study comparison are still required.

13.2.3 AAV Data Analysis Methods

In the past decade, all the approaches evaluating rAAV genome interaction with the host DNA employed next-generation sequencing technologies. Nowadays, standardized procedures automatically and cost-effectively reveal thousands of IS in a few days. Figure 13.1 presents a simplified flowchart that generalizes the stages required to generate the common safety artifacts concerning the scientific and regulatory GTx community.

For rAAV, safety studies incorporating an assessment of DNA integration, sequencing technologies, and bioinformatics have become an important consideration. Low-throughput technologies are frequently used in studies with low numbers of samples to permit manual analysis [15, 26, 70]. The analysis does not necessitate specialized bioinformatics and can be performed using online alignment tools [71, 72] and visually inspecting the results. This approach was frequently used in older studies with low numbers of samples [15, 26, 70].

Advancements in the wet lab, sequencing, and data management technologies require developing progressively enhanced data analysis procedures. For example, the different chemistries employed for library preparation require distinct data processing strategies. This section combines the library preparation methods into three main categories: PCR, TE or viral capture, and WGS (Figure 13.2). All three, are described in Section 13.2.2, yield datasets requiring specialized data analysis workflows to enable integration site retrieval and quantification.

WGS and whole transcriptome sequencing (WTS) libraries are utilized for analysis like germinal/somatic variants detection and copy number variation, where the IS analysis is typically an accessory safety assessment primarily used in large diagnostic projects [69]. This data type does not require specific preprocessing steps, such as trimming-specific sequences or combining reads, and the alignment stage directly processes the reads [73, 74]. Some tools, such as VirusFinder [73], Virus-clip [74], and VirusBreakend [69], are explicitly developed for WGS data. They are not AAV-specialized tools and do not make any assumptions regarding the reads structure or the integration process. To achieve the high sensitivity required for unspecialized methods, the tools perform a

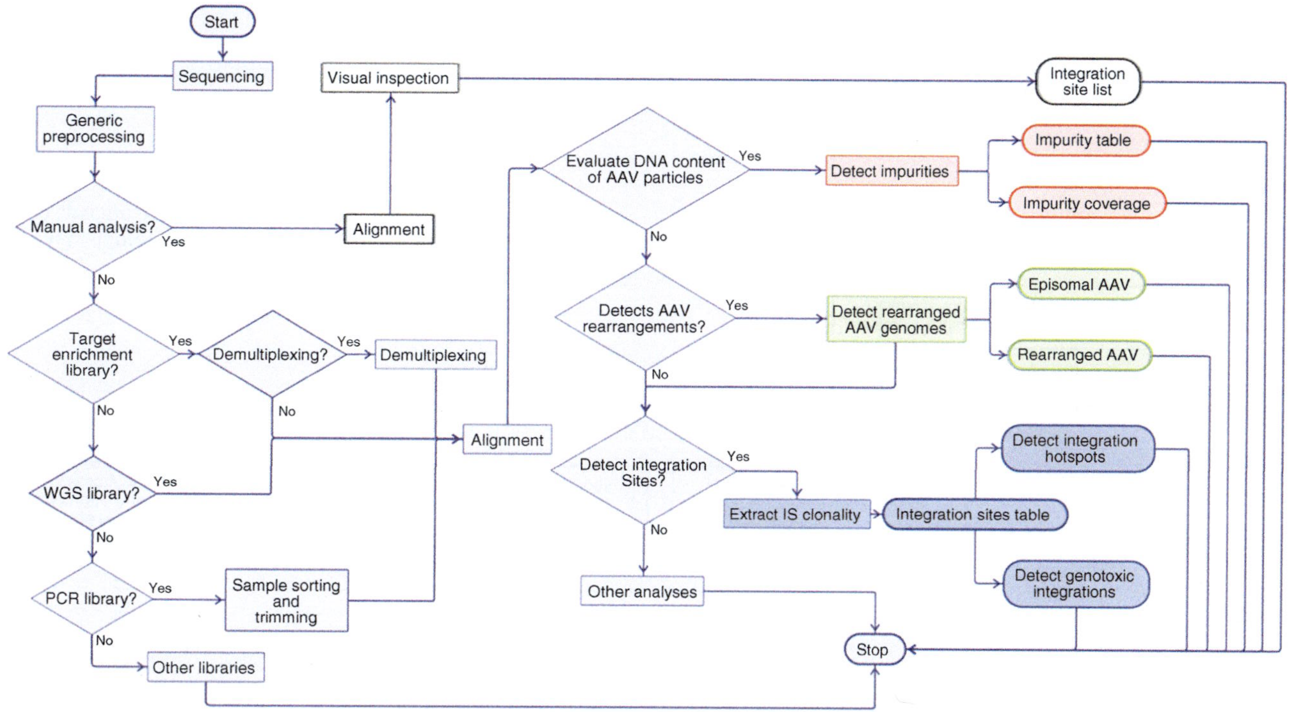

Figure 13.1 Common analysis steps in AAV GTx safety analysis. In light blue are the steps that are common to all the analyses. The three main categories are: impurities detection (red), AAV genome rearrangement (green), and integration site analysis (blue). White boxes refer to the analyses that detect integrations inspecting the reads manually. "Other libraries" represents emerging protocols and technologies as long-range reads. "Other Analyses" represent analyses that are not discussed in this chapter as the nuclease on/off-target activity detection.

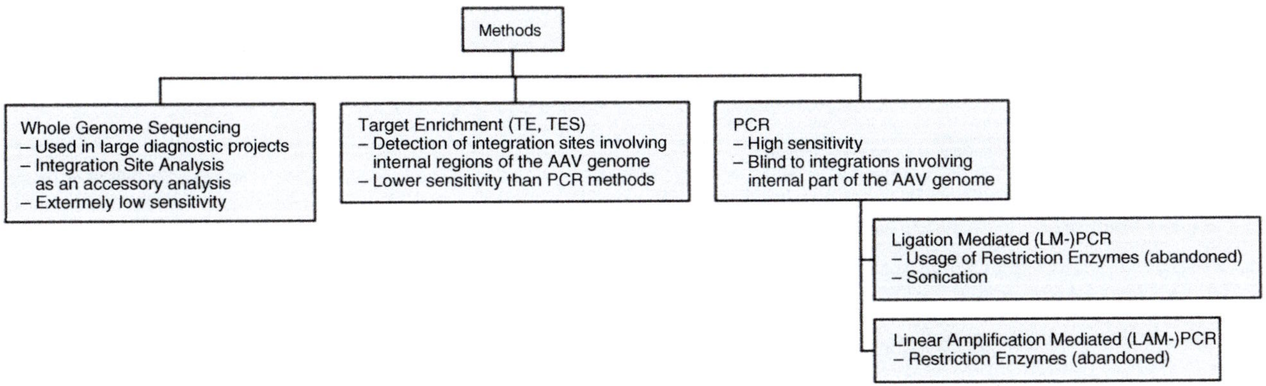

Figure 13.2 Library preparation methods for vector integration site analysis yielding sequencing datasets requiring specific data analysis workflows.

preselection step where the reads are aligned on the vector genome. In this way, any vector signal coming from the data can be further elaborated for extracting IS.

TE is a method developed to enrich the concentration of fragments that contain a preselected portion of the vector. Like WGS libraries, the read structure is simple, and the strategies used for WGS analysis are still usable. Given that the sample is enriched in the AAV genome fragments, some pipelines [19, 20, 39, 75] analyze TE data by implementing a strategy that uses this information aligning the reads directly with a hybrid genome that combines the organism assembly and the vector sequence as an additional chromosome. TE was used to study AAV integration [20] and to characterize HCC development after AAV infection [39]. rAAV integrations and the genomic rearrangements in transduced human hepatocytes, expanded in a mouse model, were recently discovered by employing TE in combination with PacBio long-read sequencing [19].

The PCR methods are probably the most widely used to assess rAAV integration as they are considered more sensitive than WGS and TE [3, 13, 18, 27, 76–78]. Medium-throughput sequencers, such as MiSeq, are sufficient for analyzing several samples in one run. Mixing the samples in a single library requires advanced barcoding strategies for the sample fragments. As an example, a unique combination of two barcodes (short sequences) are ligated at the 5′ and 3′of the sample fragment. This combination is recognized on the forward and reverse read by the sorting tool and is used to group the reads in the corresponding sample. An essential preprocessing stage requires assigning (sorting) the reads to the original sample. The AAV genome sequence used as a template for the PCR primer is employed as an anchor point for preselecting the vector positive reads. Typically, this sequence is removed in the trimming phase.

13.2.3.1 AAV Primary Analysis

An alignment tool, independent from the algorithm implemented for finding hits, returns the coordinates of the alignment along with the information helpful in evaluating the match. The most straightforward score is the identity and the percentage of identical bases in the alignment. It is customary to consider valid a match when the identity is more significant than a predetermined threshold (usually 95%) [3, 13, 19, 27, 76, 77]. Moreover, under the simple hypothesis that all the bases are equiprobable, we expect that any random fragment of 16 bases should appear about once into a random string of the dimension of the human genome. For this reason in many approaches, the aligned region must be longer than a minimum [75] to reduce the risk of false-positive hits. Some alignment tools present a supplementary score that combines different quality metrics, simplifying the match quality estimation. For example, BLAST returns a score that estimates the number of times an alignment with the same characteristics is expected by chance (E-value). The information about the alignment is stored in large data files and is the primary source for answering one or more questions addressed by the study. A *de facto* standard is the Sequence Alignment Map format (SAM) and most of the modern alignment tools return the alignment in this or in the corresponding binary format (BAM).

13.2.3.2 Impurity Analysis

One area of emerging interest is the evaluation of rearranged vector DNA and DNA-process-related contaminates that are packaged in the AAV viral particle. The bioinformatics methods are in active development. However, the fundamental analysis can be reduced to detect potential contaminants or misarranged AAV encapsidated during the virus production. From the alignment file, where all the potential DNA contaminants, derived from the production process, are used as the target genome, the number of reads and the position of the contaminants are returned and presented as count table and coverage figures [57, 79, 80].

13.2.3.3 AAV Genome Rearrangements

In postmitotic cells, rAAV mainly persists as an extrachromosomal element able to stably express the transgene for a long period of time [6, 81–83]. The inter and intra-molecular recombination of the ITRs produces three primary types of circles: 5′ITR-3′ITR (head to tail), 5′ITR-5′ITR (head to head), and 3′ITR-3′ITR (tail to tail). While in dividing cells, the mechanism that ensures the persistence of the rAAV DNA seems to be the integration into the host DNA. Assessing the amount of episomal circular/concatemeric forms of the vector and distinguishing the elements from the integrated vector genome is a challenge for the analysis tools. Searching for reads that map discontinuously on the AAV genome is the primary strategy for determining the frequency of the episomal forms in the samples [22, 77] assuming that rearrangement among vector is an indication of an extrachromosomal AVV genome. It has been shown that rearranged vectors can integrate into the host genome [3, 19, 84], and that numerous rearrangement events can lead to complicated structures of concatenated AAV fragments. For this reason, it is not possible to univocally distinguish between integrated/nonintegrated AAV DNA without more specialized methods that make use of long-read sequencing [19, 85] and new bioinformatics tools [3, 19].

13.2.3.4 Integration Site Analysis

The most common question answered by advanced bioinformatics methods is the location of IS. Whilst only a few reads are inspected visually in the manual analysis, in all the modern analysis tools, a multitude of criteria are introduced to select, evaluate, and report integration events. Here, we will discuss the most relevant standard analysis.

It is not uncommon that a single DNA fragment aligns with multiple genomic regions, and these analyses must be handled with caution to minimize signal loss.

For example, integrations into a repetitive area of the host genome cannot be identified uniquely and represents potential unresolved IS. Different strategies can be adopted to reduce the impact of unresolved IS. A straightforward but not always implementable experimental solution is to increase the reads' length to decrease the number of unresolved IS. Alternatively, a computational strategy that increases the reads' dimension *in silico* can be implemented as established in gamma-TRIS aligner [86], a tool for detecting retroviral integrations. More straightforward but less efficient approaches discard the reads that show alignments with the same score on multiple locations [13] or when the score between the best match and the second best is more significant than a threshold [75]. Additionally, the mapping of IS into low complexity and repetitive regions depends on the alignment tool sensitivity. For example, by default, BLAST masks low complexity regions from the search [87], making it practically blind in a not negligible portion of the human genome, which results in a loss of sensitivity [3, 13, 15, 26, 77].

Besides host genome mapping, identification of the junction, i.e. the region where the AAV genome and the reference are fused can be challenging. Ideally, when the AAV sequence ends, the genomic begins so that the junction region is clearly defined. While preparing the sequencing libraries, different operation on the samples introduces noise in the form of (1) minor variations, (2) biased amplification rate, and (3) *in vitro* recombination of the DNA fragments. The last one, producing biological artifacts [88], affects the integration site detection. For this reason, the analysis tool should control the formation of the chimeric reads by requiring that the vector-genome junction sequences are well-formed.

In WGS and TE methods, the reads aligned to the vector are parsed to validate the alignment structure. The Concise Idiosyncratic Gapped Alignment Report (CIGAR) string in the Sequence Alignment/Map (SAM) format is used to authenticate the alignment structure [19, 75]. A less stringent requirement sometimes adopted is the presence in the read pairs of at least one read aligned to the AAV genome [20, 39]. Analysis tools for PCR methods use filtering strategies that may involve a) a limited number of unaligned bases between the vector and the genomic regions [27, 75, 77] and the absence of more than one vector-genome junction per fragment [3, 22]. This last strategy decreases the noise due to artifacts, such as chimeras generated during the library preparation and sequencing but makes the analysis unable to detect rearranged AAV genome integration (see Section 13.2.3.3).

13.2.3.5 Clonality Analysis

Due to the stochastic nature of the integration process, a single integration site marks univocally a single clone. The number of IS present at a particular time point in a given sample is thus a representation of the clonal configuration at that moment.

If some IS are found more frequently than others, then it can be inferred that those IS may be a sign of clonal expansion of the cell harboring that integration. The assessment that addresses the reconstruction and comparison of the clonality of a set of samples is composed of two procedures: (1) the rebuilding of the clonality of any single IS detected within a sample (IS clonality) and (2) the calculation of the clonality of the sample (clonal diversity).

The first approach associates with each IS a number representing the total number of times each clone is present in the sample. The straightforward derivation of IS clonality is by assessing the number of times a particular IS is detected in the given sample [3, 18, 89]. This approach, however, may be unable to distinguish between reads from different clones or PCR amplification. Other counting strategies can reduce this issue by exploiting experimental evidence and adding a few assumptions [19, 20, 27, 39]. For example, libraries that originate from random DNA shearing (e.g. by sonication) are less prone to contain identical fragments from the same clones, thus reducing the count bias introduced by the PCR [63]. On the other hand, this approach requires adjustment when the sequencing depth increases to a level where the sonicated fragments are saturated [63].

The clonal diversity is calculated from the IS clonality employing methods used in ecology for measuring biodiversity. Its main aim is to compare the clonal repertoire in the same individual and, as an example, identify the surge of a clone, intercepting potential adverse events caused by insertional mutagenesis as soon as possible. The simplest and frequently used measure of diversity is clonal abundance [27, 40, 78], which determines the number of times a clone is present in the sample compared to the total number of clones. When the clonal abundance of a specific IS increases over a defined threshold, a red flag is raised, triggering more detailed investigations. With this regard, several diversity indexes are employed, such as the Shannon index [90], the polyclonal monoclonal diversity (PMD) [91], and the shape-constrained splines (SCS) method [92] in GTx for tracking and comparing clonal diversity across samples and studies.

13.2.3.6 Genotoxic Integrations

The emergence of a clone that expands during longitudinal sampling triggers a major concern for an insertional mutagenesis event [93]. It is a common practice to assign each IS to the closest gene within a specific range [1, 13, 15, 18, 20, 27]. In this way, the analysis can focus only on the integrations close to a subset of cancer-associated genes. This may seem a straightforward process, however, selecting a comprehensive and meaningful set of detrimental genes is challenging. Aside from gamma-retroviral studies that established a specific (limited) group of genes as a source of potential insertional mutagenesis [94, 95], for more comprehensive analysis different, not overlapping collections of cancer genes, are curated by multiple databases [96–98]. An additional layer of complexity in the field stems from the fact that not all the cancer genes in one organism are found in a different organism. Moreover, the accuracy of annotation between different species genomes (cross annotation) depends on the computational strategy used and impacts the study outcome.

13.3 Assessing the Biologic Relevance of AAV Integration Profile

The Rian locus, primarily miR341, have shown to be rAAV vector integration site that is responsible for HCC development in neonatal mice [1, 2, 10, 20, 30]. This locus encodes for many regulatory noncoding RNAs that are expressed highly in neonatal mice, compared with adult mice. AAV vector integration in Rian locus in neonatal mice leads to dysregulation of the flanking genes and regulatory elements. In line with no HCC identified in adult mice, Rian integrations were not found in adult mice in many studies [13, 15]. Large animals and human genomes lack the Rian locus, especially an ortholog of mir341. A homolog of the delta-like homolog 1-deiodinase type 3 (DLK1-DIO3) region where the Rian locus is in murine genome exists on human chromosomal 14. However, no evidence indicates that the genome integration has occurred in the DLK1-DIO3 region in AAV treated large animals and humans.

While most data from adult mice, nonrodents (dogs and NHP), and humans have suggested that the integration of AAV vectors is a rare event, often in the random sites across the genome with no hotspots or clusters, preferred regions of integration in rodents, large animals and human have been identified to be chromosomal breakage sites, DNA palindromic regions, active genes, GC-rich regions, and CpG islands [13, 99], presumably due to highly active genes or regions where chromosome is unwound by transcriptional machinery and DNA strand is easily accessible for breakage and random integration to occur.

The biological consequence of AAV integration in a genome depends on the sites of the integration and the genes affected. The integration event can be clinically silent and the cells with the integration may stay unchanged in terms of biological functions if the integration does not change the protein expression, structure, and/or functions. However, the integration may lead to a genotoxic event if insertions and deletions subsequently change the expression of genes related to cell cycle or cell proliferation, or interfere with the chromosome stability. It is worth to mention that multiple genetic mutations are needed to sustain the clonal expansion and transform to a full malignancy, especially in the cellular program that favors self-renewal replication over proliferation in connection with differentiation or senescence. The process may take years to develop, but no animal model is fully translatable and there are no other species that can be reasonably studied as long as humans.

It is known that hepatocytes can become polyploid during normal liver homeostasis, which could have been scored in the current assays as low-level clonal expansion. In addition, clonal expansion could happen in the normal appearing hepatocytes during the aging process or in presence of chronic infection of HBV. Aging is also considered as the strongest risk factor for neoplastic disease in humans. The nature of clonal expansion, if detected in preclinical animal models or patients under AAV treatment after long time follow-up, needs to be carefully examined to determine the relevance to AAV genome integration and to differentiate from aging. AAV integration and clonal expansion in the absence of tumor formation have been observed in hemophilia dogs after long-term follow-up of AAV treatment [27]. It is currently unknown if the clonal expansions detected in the dogs were pre-malignant and could result in malignancies if the dogs had lived longer. However, hemophilic dogs treated with AAV GTx and followed for more than 10 years did not identify hepatic neoplasia, suggesting the clonal expansion observed in these dogs may have been age-related [26, 32–34].

Assessing the risk of an integration profile is based not only on the location of the integration within the genome, but more importantly the functional consequence of the integration events. If there are no apparent changes in cell behavior (e.g. increased cell proliferation) or alteration of gene expression that may lead to subsequent oncogenic transformation the integration event may be considered non adverse. To proactively understand potential tumorigenesis risk caused by AAV genome integration, liver tissues from AAV vector-treated preclinical animals may be profiled for genome integration and clonal expansion. The genes with concerning integration events may be further evaluated for expression changes and downstream functional alteration. Human genetics and pharmacology database may be utilized to evaluate the long-term adverse effects and carcinogenesis potential of the loci harboring vector integration [100, 101]. If clonal expansion and tumors are identified after AAV treatment, the relevance of genome integration to the tumor formation needs to be assessed at molecular level. Genetic materials should be obtained from the cancer tissue biopsy and adjacent healthy tissue and examined for the presence of AAV vector DNA and insertional events. The molecular events, including vector copy numbers, genome integration profile, and the resulting gene expression changes in the tumor tissue may be compared with those in the adjacent normal tissue. If high copy numbers of AAV vector DNA and insertional events are detected in the tumor, the cellular genes near the IS may be examined further for the dysregulation that may lead to the functional changes related to cell cycle regulation, cellular proliferation, genome stability, and oncogenesis. These molecular and cellular signals together with evidence of clonal expansion and pathological changes will be evaluated to identify the associations among insertional events, clonal expansion, and tumor development. In some situations, the direct causality of integration events on tumor development will not be discernable, but the weight of evidence (WOE) combining cellular and molecular, physiological, and pathological signals, as well as patients predisposed genetic background, preexisting medical conditions related to cancer risk and cotreatments will be holistically evaluated to identify the possible causes.

13.4 Conclusion and Future Direction

The data on AAV integration and functional consequence when viewed in totality across species and the years of clinical experience with AAV therapeutics indicates the human risk of tumorigenesis following AAV gene therapy is likely to be low. The literature reviewed in this chapter on long-term follow-up of AAV gene therapy in nonrodents and humans indicates that AAV integration poses minimal risk for hepatocellular tumor and consequently the hepatic tumors observed in mice treated with AAV vectors are unlikely to translate to oncogenic risk in man. The most compelling data set indicating that AAV gene therapy is unlikely to pose human oncogenic risk is the experience gained from patients treated with AVXS-101/Zolgensma. Zolgensma is administered at a high dose of 2E14 vg/kg. This vector uses the strong chicken-beta-actin promoter (CBA) and is administered to infants less than two years of age, both of which are attributes that mouse studies [2] indicate have the highest risk of hepatocellular tumor development. There is no indication in the publicly available literature that patients treated with Zolgensma are prone to developing neoplasia, which is consistent with more than two decades of human clinical trial experience and long-term follow-up with AAV, indicating that there is no neoplastic risk in humans. One caveat to this conclusion is that Zolgensma is a self-complementary AAV vector and where HCC has been observed in mice, the vector was a single-stranded AAV vector. Differences in integration profile and tumorigenesis in mice related to single vs. self-complementary AAV vectors is an area for further investigation.

An import aspect of understanding the biologic consequences of AAV integration is the methods used for assessing integration. These methods are evolving to more comprehensively assess the integration of different components of the DNA delivered by the vector. To date, the most frequently used methods have relied on PCR-based methods (e.g. LM-PCR) that focus on identifying the integration site nearby a small portion of the vector DNA (e.g. ITR). While PCR methods may be more sensitive, new methods are evolving that are designed to capture the diversity of DNA that is delivered in the vector capsid using homologous probes to capture the vector DNA followed by NGS to identify the associated genomic DNA.

The expanding use of AAV GTx and the interest in the scientific and regulatory community on insertional mutagenic risk suggest that additional work needs to be done to understand the human relevance of the HCC findings in mice. Additional studies in species other than mice need to be done to assess the relevance of HCC observed in mice. To date, there have been no long-term studies in rats assessing the neoplastic potential of AAV. An extensive review of the experience gained from decades of conducting two species (mice and rats) carcinogenic studies to assess human carcinogenic risk of chemicals has found that when mice are the only species to develop liver tumors the observation is not considered relevant for human safety assessment [102–104].

In order to develop a better understanding of potential human risk associated with AAV genomic integration, there needs to be additional investigation into: (1) understanding species differences in neoplastic outcome following AAV treatment, (2) a more comprehensive understanding of the forms of AAV-delivered DNA that integrates into the host cell genome, and (3) an assessment of altered cell function associated with AAV integration. Ultimately, it will be data from long-term follow-up of humans treated with AAV gene therapy that will provide definitive data on the risk of AAV-associated HCC observed in mice.

References

1 Donsante, A., Miller, D.G., Li, Y. et al. (2007). AAV vector integration sites in mouse hepatocellular carcinoma. *Science* 317 (5837): 477.

2 Chandler, R.J., LaFave, M.C., Varshney, G.K. et al. (2015). Vector design influences hepatic genotoxicity after adeno-associated virus gene therapy. *J. Clin. Invest.* 125 (2): 870–880.

3 Gil-Farina, I., Fronza, R., Kaeppel, C. et al. (2016). Recombinant AAV integration is not associated with hepatic genotoxicity in nonhuman primates and patients. *Mol. Ther.* 24 (6): 1100–1105.

4 Cellular, Tissue, and Gene Therapies Advisory Committee September 2-3, 2021 Meeting Announcement, (2021).

5 ASGCT. Integration/Insertion Considerations for AAV-based Gene Therapy Vectors: ASGCT Recommendations based on the Society's AAV Integration Roundtable event, (2021). https://asgct.org/global/documents/advocacy/2021-fda-liaison-meeting/final-aav-integration-slides.aspx (accessed 9 January 2023).

6 Nakai, H., Yant, S.R., Storm, T.A. et al. (2001). Extrachromosomal recombinant adeno-associated virus vector genomes are primarily responsible for stable liver transduction in vivo. *J. Virol.* 75 (15): 6969–6976.

7 Donsante, A., Vogler, C., Muzyczka, N. et al. (2001). Observed incidence of tumorigenesis in long-term rodent studies of rAAV vectors. *Gene Ther.* 8 (17): 1343–1346.

8 Bell, P., Wang, L., Lebherz, C. et al. (2005). No evidence for tumorigenesis of AAV vectors in a large-scale study in mice. *Mol. Ther.* 12 (2): 299–306.

9 Bell, P., Moscioni, A.D., McCarter, R.J. et al. (2006). Analysis of tumors arising in male B6C3F1 mice with and without AAV vector delivery to liver. *Mol. Ther.* 14 (1): 34–44.

10 Zhong, L., Malani, N., Li, M. et al. (2013). Recombinant adeno-associated virus integration sites in murine liver after ornithine transcarbamylase gene correction. *Hum. Gene Ther.* 24 (5): 520–525.

11 Reiss, J. and Hahnewald, R. (2011). Molybdenum cofactor deficiency: mutations in GPHN, MOCS1, and MOCS2. *Hum. Mutat.* 32 (1): 10–18.

12 Embury, J.E., Frost, S., Charron, C.E. et al. (2008). Hepatitis virus protein X-phenylalanine hydroxylase fusion proteins identified in PKU mice treated with AAV-WPRE vectors. *Gene Ther. Mol. Bio.* 12: 69–76.

13 Li, H., Malani, N., Hamilton, S.R. et al. (2011). Assessing the potential for AAV vector genotoxicity in a murine model. *Blood* 117 (12): 3311–3319.

14 Wang, P.R., Xu, M., Toffanin, S. et al. (2012). Induction of hepatocellular carcinoma by in vivo gene targeting. *PNAS* 109 (28): 11264–11269.

15 Rosas, L.E., Grieves, J.L., Zaraspe, K. et al. (2012). Patterns of scAAV vector insertion associated with oncogenic events in a mouse model for genotoxicity. *Mol. Ther.* 20 (11): 2098–2110.

16 Kao, C.Y., Yang, S.J., Tao, M.H. et al. (2013). Incorporation of the factor IX Padua mutation into FIX-Triple improves clotting activity in vitro and in vivo. *Thromb. Haemost.* 110 (2): 244–256.

17 Walia, J.S., Altaleb, N., Bello, A. et al. (2015). Long-term correction of Sandhoff disease following intravenous delivery of rAAV9 to mouse neonates. *Mol. Ther.* 23 (3): 414–422.

18 Kang, H.R., Gjorgjieva, M., Smith, S.N. et al. (2019). Pathogenesis of hepatic tumors following gene therapy in murine and canine models of glycogen storage disease. *Mol. Ther. Methods Clin. Dev.* 15: 383–391.

19 Dalwadi, D.A., Torrens, L., Abril-Fornaguera, J. et al. (2021). Liver injury increases the incidence of HCC following AAV gene therapy in mice. *Mol. Ther.* 29 (2): 680–690.

20 Li, Y., Miller, C.A., Shea, L.K. et al. (2021). Enhanced efficacy and increased long-term toxicity of CNS-directed, AAV-based combination therapy for Krabbe disease. *Mol. Ther.* 29 (2): 691–701.

21 Sabatino, D.E., Bushman, F.D., Chandler, R.J. et al. (2022). Evaluating the state of the science for adeno-associated virus integration: An integrated perspective. *Mol. Ther.* 30 (8): 2646–2663.

22 Nowrouzi, A., Penaud-Budloo, M., Kaeppel, C. et al. (2012). Integration frequency and intermolecular recombination of rAAV vectors in non-human primate skeletal muscle and liver. *Mol. Ther.* 20 (6): 1177–1186.

23 Mattar, C.N.Z., Gil-Farina, I., Rosales, C. et al. (2017). In utero transfer of adeno-associated viral vectors produces long-term factor IX levels in a cynomolgus macaque model. *Mol. Ther.* 25 (8): 1843–1853.

24 Spronck, L., de Haan, M., Heijink, L. et al. (2020). Assessment of vector integration of AAV5-hFIX in mice and non-human primates indicates No association with tumorigenic risk. *Res. Pract. Thromb. Haemost.* 4.

25 Sullivan, L., Zahn, M., Gil Farina, I. et al. (2021). Rare genomic integrations of AAV5-hFVIII-SQ occur without evidence of clonal activation or gene-specific targeting. *Mol. Ther.* 29: 425.

26 Niemeyer, G.P., Herzog, R.W., Mount, J. et al. (2009). Long-term correction of inhibitor-prone hemophilia B dogs treated with liver-directed AAV2-mediated factor IX gene therapy. *Blood* 113 (4): 797–806.

27 Nguyen, G.N., Everett, J.K., Kafle, S. et al. (2021). A long-term study of AAV gene therapy in dogs with hemophilia A identifies clonal expansions of transduced liver cells. *Nat. Biotechnol.* 39 (1): 47–55.

28 Batty, P., Sihn, C.R., Ishida, J. et al. (2020). Long-term vector genome outcomes and immunogenicity of AAV FVIII gene transfer in the hemophilia A dog model [abstract]. *Res. Pract. Thromb. Haemost.* 4.

29 Batty, P., Fong, S., Franco, M. et al. (2020). Frequency, location and nature of AAV vector insertions after long-term follow up of FVIII transgene delivery in a Hemophilia A dog model [abstract]. *Res. Pract. Thromb. Haemost.* 4: 82.

30 Daly, T.M., Ohlemiller, K.K., Roberts, M.S. et al. (2001). Prevention of systemic clinical disease in MPS VII mice following AAV-mediated neonatal gene transfer. *Gene Ther.* 8 (17): 1291–1298.

31 Nathwani, A.C., Rosales, C., McIntosh, J. et al. (2011). Long-term safety and efficacy following systemic administration of a self-complementary AAV vector encoding human FIX pseudotyped with serotype 5 and 8 capsid proteins. *Mol. Ther.* 19 (5): 876–885.

32 Nichols, T.C., Whitford, M.H., Arruda, V.R. et al. (2015). Translational data from adeno-associated virus-mediated gene therapy of hemophilia B in dogs. *Hum. Gene Ther. Clin. Dev.* 26 (1): 5–14.

33 Arruda, V.R., Schuettrumpf, J., Herzog, R.W. et al. (2004). Safety and efficacy of factor IX gene transfer to skeletal muscle in murine and canine hemophilia B models by adeno-associated viral vector serotype 1. *Blood* 103 (1): 85–92.

34 Sabatino, D.E., Armstrong, E., Edmonson, S. et al. (2004). Novel hemophilia B mouse models exhibiting a range of mutations in the Factor IX gene. *Blood* 104 (9): 2767–2774.

35 Srivastava, A. (2016). Adeno-associated virus: the naturally occurring virus versus the recombinant vector. *Hum. Gene Ther.* 27 (1): 1–6.

36 Kuzmin, D.A., Shutova, M.V., Johnston, N.R. et al. (2021). The clinical landscape for AAV gene therapies. *Nat. Rev. Drug Discovery* 20 (3): 173–174.

37 Wellman, J.A., Mingozzi, F., Ozelo, M.C. et al. (2012). 69. Results from the long-term follow-up of severe hemophilia B subjects previously enrolled in a clinical study of AAV2-FIX gene transfer to the liver. *Mol. Ther.* 20: S28–S29.

38 George, L.A., Ragni, M.V., Rasko, J.E.J. et al. (2020). Long-term follow-up of the first in human intravascular delivery of AAV for gene transfer: AAV2-hFIX16 for severe hemophilia B. *Mol. Ther.* 28 (9): 2073–2082.

39 La Bella, T., Imbeaud, S., Peneau, C. et al. (2020). Adeno-associated virus in the liver: natural history and consequences in tumour development. *Gut* 69 (4): 737–747.

40 Nault, J.C., Datta, S., Imbeaud, S. et al. (2015). Recurrent AAV2-related insertional mutagenesis in human hepatocellular carcinomas. *Nat. Genet.* 47 (10): 1187–1193.

41 Qin, W., Xu, G., Tai, P.W.L. et al. (2021). Large-scale molecular epidemiological analysis of AAV in a cancer patient population. *Oncogene.* 40 (17): 3060–3071.

42 Tatsuno, K., Midorikawa, Y., Takayama, T. et al. (2019). Impact of AAV2 and hepatitis B virus integration into genome on development of hepatocellular carcinoma in patients with prior hepatitis B virus infection. *Clin. Cancer Res.* 25 (20): 6217–6227.

43 Guidance for Industry (2006). Gene therapy clinical trials – observing subjects for delayed adverse events. Docket No. 2005D-0310.

44 Lewis, T. (2021). The quest to overcome gene therapy's failures. *Nature.*

45 Assaf, B.T. and Whiteley, L.O. (2018). Considerations for preclinical safety assessment of adeno-associated virus gene therapy products. *Toxicol. Pathol.* 46 (8): 1020–1027.

46 US Food and Drug Administration (FDA) Guideline on the quality, non-clinical and clinical aspects of gene therapy medicinal products. (2018).

47 US Food and Drug Administration (FDA) Draft Guideline on quality, non-clinical and clinical requirements for investigational advanced therapy medicinal products in clinical trials. (2019).

48 Guidance for Industry (2020). Long term follow-up after administration of human gene therapy products.

49 EPAR (2012). Summary for the Public: Glybera (Alipogene Tiparvovec).

50 European Medicines Agency New gene therapy for rare inherited disorder causing vision loss recommended for approval [Press release]. 21 September 2018, 2018.

51 European Medicines Agency New gene therapy to treat spinal muscular atrophy [Press release]. 27/03/2020, 2020.

52 Food and Drug Administration (FDA) (2017). FDA approves novel gene therapy to treat patients with a rare form of inherited vision loss. [press release]. December 18, 2017.

53 Food and Drug Administration (FDA) (2019). FDA approves innovative gene therapy to treat pediatric patients with spinal muscular atrophy, a rare disease and leading genetic cause of infant mortality [Press release].

54 Food and Drug Administration (FDA) (2021). Cellular, Tissue, and Gene Therapies Advisory Committee (CTGTAC) Meeting #70 Toxicity Risks of Adeno-associated Virus (AAV) Vectors for Gene Therapy (GT).

55 Tai, P.W.L., Xie, J., Fong, K. et al. (2018). Adeno-associated virus genome population sequencing achieves full vector genome resolution and reveals human-vector chimeras. *Mol. Ther. Methods Clin. Dev.* 9: 130–141.

56 Zhang, J., Yu, X., Herzog, R.W. et al. (2021). Flies in the ointment: AAV vector preparations and tumor risk. *Mol. Ther.* 29 (9): 2637–2639.

57 Guerin, K., Rego, M., Bourges, D. et al. (2020). A novel next-generation sequencing and analysis platform to assess the identity of recombinant adeno-associated viral preparations from viral DNA extracts. *Hum. Gene Ther.* 31 (11-12): 664–678.

58 Hutt, J.A., Assaf, B.T., Bolon, B. et al. (2022). Scientific and regulatory policy committee points to consider: nonclinical research and development of in vivo gene therapy products, emphasizing adeno-associated virus vectors. *Toxicol. Pathol.* 50 (1): 118–146.

59 Kustikova, O.S., Modlich, U., and Fehse, B. (2009). Retroviral insertion site analysis in dominant haematopoietic clones. *Methods Mol. Biol.* 506: 373–390.

60 Schmidt, M., Schwarzwaelder, K., Bartholomae, C. et al. (2007). High-resolution insertion-site analysis by linear amplification-mediated PCR (LAM-PCR). *Nat. Methods* 4 (12): 1051–1057.

61 Gabriel, R., Eckenberg, R., Paruzynski, A. et al. (2009). Comprehensive genomic access to vector integration in clinical gene therapy. *Nat. Med.* 15 (12): 1431–1436.

62 Harkey, M.A., Kaul, R., Jacobs, M.A. et al. (2007). Multiarm high-throughput integration site detection: limitations of LAM-PCR technology and optimization for clonal analysis. *Stem Cells Dev.* 16 (3): 381–392.

63 Berry, C.C., Gillet, N.A., Melamed, A. et al. (2012). Estimating abundances of retroviral insertion sites from DNA fragment length data. *Bioinformatics* 28 (6): 755–762.

64 Brugman, M.H., Suerth, J.D., Rothe, M. et al. (2013). Evaluating a ligation-mediated PCR and pyrosequencing method for the detection of clonal contribution in polyclonal retrovirally transduced samples. *Hum. Gene Ther. Methods* 24 (2): 68–79.

65 Biasco, L., Rothe, M., Buning, H. et al. (2018). Analyzing the genotoxicity of retroviral vectors in hematopoietic cell gene therapy. *Mol. Ther. Methods Clin. Dev.* 8: 21–30.

66 Firouzi, S., Lopez, Y., Suzuki, Y. et al. (2014). Development and validation of a new high-throughput method to investigate the clonality of HTLV-1-infected cells based on provirus integration sites. *Genome Med.* 6 (6): 46.

67 Pouzolles, M., Machado, A., Guilbaud, M. et al. (2020). Intrathymic adeno-associated virus gene transfer rapidly restores thymic function and long-term persistence of gene-corrected T cells. *J. Allergy Clin. Immunol.* 145 (2): 679–97 e5.

68 Senis, E., Mosteiro, L., Wilkening, S. et al. (2018). AAVvector-mediated in vivo reprogramming into pluripotency. *Nat. Commun.* 9 (1): 2651.

69 Cameron, D.L., Jacobs, N., Roepman, P. et al. (2021). VIRUSBreakend: viral integration recognition using single breakends. *Bioinformatics* 37 (19): 3115–3119.

70 Huser, D., Weger, S., and Heilbronn, R. (2002). Kinetics and frequency of adeno-associated virus site-specific integration into human chromosome 19 monitored by quantitative real-time PCR. *J. Virol.* 76 (15): 7554–7559.

71 Altschul, S.F., Gish, W., Miller, W. et al. (1990). Basic local alignment search tool. *J. Mol. Biol.* 215 (3): 403–410.

72 Kent, W.J. (2002). BLAT--the BLAST-like alignment tool. *Genome Res.* 12 (4): 656–664.

73 Wang, Q., Jia, P., and Zhao, Z. (2013). VirusFinder: software for efficient and accurate detection of viruses and their integration sites in host genomes through next generation sequencing data. *PLoS One* 8 (5): e64465.

74 Ho, D.W., Sze, K.M., and Ng, I.O. (2015). Virus-Clip: a fast and memory-efficient viral integration site detection tool at single-base resolution with annotation capability. *Oncotarget* 6 (25): 20959–20963.

75 Afzal, S., Wilkening, S., von Kalle, C. et al. (2017). GENE-IS: time-efficient and accurate analysis of viral integration events in large-scale gene therapy data. *Mol. Ther. Nucleic Acids* 6: 133–139.

76 Huser, D., Gogol-Doring, A., Lutter, T. et al. (2010). Integration preferences of wildtype AAV-2 for consensus rep-binding sites at numerous loci in the human genome. *PLoS Pathog.* 6 (7): e1000985.

77 Gauttier, V., Pichard, V., Aubert, D. et al. (2013). No tumour-initiating risk associated with scAAV transduction in newborn rat liver. *Gene Ther.* 20 (7): 779–784.

78 Kaeppel, C., Beattie, S.G., Fronza, R. et al. (2013). A largely random AAV integration profile after LPLD gene therapy. *Nat. Med.* 19 (7): 889–891.

79 Lecomte, E., Tournaire, B., Cogne, B. et al. (2015). Advanced characterization of DNA molecules in rAAV vector preparations by single-stranded virus next-generation sequencing. *Mol. Ther. Nucleic Acids* 4: e260.

80 Tai, W., Li, J., Corey, E. et al. (2018). A ribonucleoprotein octamer for targeted siRNA delivery. *Nat. Biomed. Eng.* 2 (5): 326–337.

81 Kessler, P.D., Podsakoff, G.M., Chen, X. et al. (1996). Gene delivery to skeletal muscle results in sustained expression and systemic delivery of a therapeutic protein. *PNAS* 93 (24): 14082–14087.

82 Afione, S.A., Conrad, C.K., Kearns, W.G. et al. (1996). In vivo model of adeno-associated virus vector persistence and rescue. *J. Virol.* 70 (5): 3235–3241.

83 Xiao, X., Li, J., and Samulski, R.J. (1996). Efficient long-term gene transfer into muscle tissue of immunocompetent mice by adeno-associated virus vector. *J. Virol.* 70 (11): 8098–8108.

84 Schultz, B.R. and Chamberlain, J.S. (2008). Recombinant adeno-associated virus transduction and integration. *Mol. Ther.* 16 (7): 1189–1199.

85 Huser, D., Gogol-Doring, A., Chen, W. et al. (2014). Adeno-associated virus type 2 wild-type and vector-mediated genomic integration profiles of human diploid fibroblasts analyzed by third-generation PacBio DNA sequencing. *J. Virol.* 88 (19): 11253–11263.

86 Calabria, A., Beretta, S., Merelli, I. et al. (2020). gamma-TRIS: a graph-algorithm for comprehensive identification of vector genomic insertion sites. *Bioinformatics* 36 (5): 1622–1624.

87 Slotkin, R.K. (2018). The case for not masking away repetitive DNA. *Mobile DNA* 9: 15.

88 Peccoud, J., Lequime, S., Moltini-Conclois, I. et al. (2018). A survey of virus recombination uncovers canonical features of artificial chimeras generated during deep sequencing library preparation. *G3 (Bethesda)* 8 (4): 1129–1138.

89 LaFave, M.C., Varshney, G.K., Gildea, D.E. et al. (2014). MLV integration site selection is driven by strong enhancers and active promoters. *Nucleic Acids Res.* 42 (7): 4257–4269.

90 Aiuti, A., Biasco, L., Scaramuzza, S. et al. (2013). Lentiviral hematopoietic stem cell gene therapy in patients with Wiskott-Aldrich syndrome. *Science.* 341 (6148): 1233151.

91 Afzal, S., Gil-Farina, I., Gabriel, R. et al. (2019). Systematic comparative study of computational methods for T-cell receptor sequencing data analysis. *Briefings Bioinf.* 20 (1): 222–234.

92 Del Core, L., Cesana, D., Gallina, P. et al. (2022). Normalization of clonal diversity in gene therapy studies using shape constrained splines. *Sci. Rep.* 12 (1): 3836.

93 Braun, C.J., Boztug, K., Paruzynski, A. et al. (2014). Gene therapy for Wiskott-Aldrich syndrome--long-term efficacy and genotoxicity. *Sci. Transl. Med.* 6 (227): 227ra33.

94 Hacein-Bey-Abina, S., Von Kalle, C., Schmidt, M. et al. (2003). LMO2-associated clonal T cell proliferation in two patients after gene therapy for SCID-X1. *Science* 302 (5644): 415–419.

95 Ott, M.G., Schmidt, M., Schwarzwaelder, K. et al. (2006). Correction of X-linked chronic granulomatous disease by gene therapy, augmented by insertional activation of MDS1-EVI1, PRDM16 or SETBP1. *Nat. Med.* 12 (4): 401–409.

96 Huret, J.L., Minor, S.L., Dorkeld, F. et al. (2000). Atlas of genetics and cytogenetics in oncology and haematology, an interactive database. *Nucleic Acids Res.* 28 (1): 349–351.

97 Akagi, T. (2004). Oncogenic transformation of human cells: shortcomings of rodent model systems. *Trends Mol. Med.* 10 (11): 542–548.

98 Tate, J.G., Bamford, S., Jubb, H.C. et al. (2019). COSMIC: the catalogue of somatic mutations in cancer. *Nucleic Acids Res.* 47 (D1): D941–D947.

99 Nakai, H., Montini, E., Fuess, S. et al. (2003). AAV serotype 2 vectors preferentially integrate into active genes in mice. *Nat. Genet.* 34 (3): 297–302.

100 Deaton, A.M., Fan, F., Zhang, W. et al. (2019). Rationalizing secondary pharmacology screening using human genetic and pharmacological evidence. *Toxicol. Sci.* 167 (2): 593–603.

101 Liu, X., Zhang, Y., Ward, L.D. et al. (2021). A proteomic platform to identify off-target proteins associated with therapeutic modalities that induce protein degradation or gene silencing. *Sci. Rep.* 11 (1): 15856.

102 Alden, C.L., Smith, P.F., Piper, C.E. et al. (1996). A critical appraisal of the value of the mouse cancer bioassay in safety assessment. *Toxicol. Pathol.* 24 (6): 722–725.

103 Osimitz, T.G., Droege, W., Boobis, A.R. et al. (2013). Evaluation of the utility of the lifetime mouse bioassay in the identification of cancer hazards for humans. *Food Chem. Toxicol.* 60: 550–562.

104 Goodman, J.I. (2018). Goodbye to the bioassay. *Toxicol. Res. (Camb).* 7 (4): 558–564.

14

Detection and Quantification of Genome Editing Events in Preclinical and Clinical Studies

Marina Falaleeva[1*], *Shengdar Tsai*[2], *Kathleen Meyer*[1], *and Yanmei Lu*[1]

[1]*Nonclinical Department, Sangamo Therapeutics, Richmond, CA, USA*
[2]*Department of Hematology, St Jude Children's Research Hospital, Memphis, TN, USA*

14.1 Introduction

Genome editing technologies, such as engineered nucleases, base editors, or prime editors, can make permanent genomic modifications to patient cells. These modifications can result in the disruption of the sequence of a disease-associated/mutated gene, correction of a mutated gene, or insertion of a corrective gene into a precise genomic location. While these technologies remain relatively new and defining efficacy, durability, and safety in the clinic is ongoing, genome editing holds enormous potential for treating both inherited and acquired disorders. To evaluate efficacy and long-term safety, both on- and off-target editing should be evaluated and monitored during preclinical and clinical development.

This chapter focuses on the different types of engineered nucleases used in clinical studies and their mechanisms of action leading to induction of double-strand breaks (DSBs). A brief overview of US Food and Drug Administration (FDA) and European Medicines Agency (EMA) regulatory guidance for assessing on- and off-target nuclease activity is provided, as well as a summary of initial clinical studies employing engineered nucleases for potential therapeutic benefit. The workflow to evaluate the efficiency of editing at the intended sites as well as at potential off-target sites is described. A detailed overview of methodologies is provided to evaluate the activity of engineered nucleases by quantifying short indels.

Drug Development for Gene Therapy: Translational Biomarkers, Bioanalysis, and Companion Diagnostics, First Edition. Edited by Yanmei Lu and Boris Gorovits.

Methods for detecting large genomic rearrangements will also be described. A special focus will be given to bioanalytical characterization of next-generation sequencing (NGS) assays for detection of short indels. The methods described in this chapter are applicable to both *ex vivo* and *in vivo* genomic editing strategies. The methodologies used to characterize gene correction, transgene, and vector integration as well as use of base editors, primer editors, and epigenic regulators are described elsewhere [1, 2].

14.1.1 Genome Editing Modalities and Molecular Outcomes

There are four main types of engineered nucleases currently available in a drug developers' toolbox – meganucleases [3, 4], zinc finger nucleases (ZFNs) [5], transcription activator like effector nucleases (TALENs) [6], and clustered regularly interspaced short palindromic repeats (CRISPR)-associated nucleases (Cas) [7, 8]. Meganucleases, ZFNs, and TALENS use DNA-binding protein domains to recognize specific DNA sequences while CRISPR-Cas nuclease employs guide RNA molecules (gRNA) to target specific DNA sequences through Watson-Crick base pairing [9]. Regardless of the mechanism of DNA recognition, these technologies are engineered to create a DSB at a specific site within the genome. Nuclease-induced DSBs then trigger one of two main cellular repair mechanisms – homolog-directed repair (HDR) and non-homologous end joining (NHEJ). HDR is a high-fidelity DNA repair mechanism that relies upon a donor DNA template containing sequences homologous to the cleaved ends of the DSB. When supplied together with an engineered nuclease, the donor DNA template can be used to correct disease-causing mutations, to insert a therapeutic gene into a genomic safe harbor locus or into its endogenous locus [10]. NHEJ is error-prone repair mechanism that can lead to insertion or deletion of a small number of nucleotides (Figure 14.1) or insertion of larger, nonspecific fragments of the donor DNA template without involvement of the homology arms [10]. Generation of indels can be leveraged for therapeutic applications by introducing mutations at a specific site to disrupt the DNA sequence of a target gene, which can then result in desired clinical outcomes. For example, editing and subsequent disruption of the BCL11A enhancer gene by engineered nucleases has shown clinical benefit in β-thalassemia and sickle cell anemia (see Section 14.1.2). The frequency of indels at the intended site thus can serve as biomarkers to measure the efficiency of gene editing and as surrogate potency for the desired biological outcome.

Apart from editing at the intended genomic site, unintended nuclease-induced DSBs can result in associated genotoxic events such as off-target indels, inversions, or translocations. Small insertions and deletions can result in frameshift mutations resulting in lack of production of specific protein or production of a truncated non-functional protein. Translocations can occur when the same cell

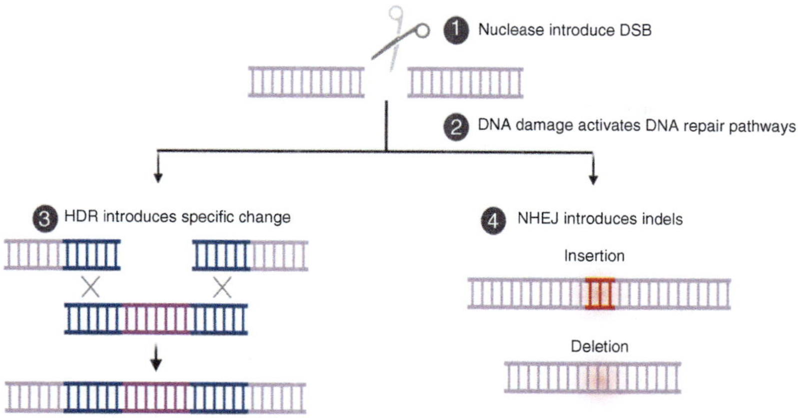

Figure 14.1 Molecular outcomes of gene editing events by engineered nucleases. Engineered nucleases introduce a double-stranded break (DSB) in genomic DNA (step 1). This DNA damage activates intrinsic DNA repair pathways with a cell (step 2). In the presence of a donor template, homology-directed repair (HDR) can result in integration of donor transgene into the genome through homologous recombination, at mostly G2 and S phases of the cell cycle (step 3). Non-homologous end joining (NHEJ), error-prone repair, is active throughout the cell cycle and does not require a repair template. This repair mechanism introduces small nucleotide insertions and deletions (indels) at the DSB site (step 4). Source: Created with BioRender.com.

contains more than one DSB e.g. one at the intended and one at an off-target site or between two off-target sites. Chromosomal rearrangements were also observed when only one DSB was detected at the intended site [11, 12].

In addition to off-target editing due to non-specific recognition of DNA sequences, other potential genotoxic events have been described during editing. This includes nonspecific single-stranded DNA cleavage by CRISPR-Cas12a [13] and large deletions at the site of intended editing [11, 14]. Lastly, all genome editing approaches implementing synthetic DNA repair templates are susceptible to random integration at DSBs independent of nuclease activity [15].

The health risks related to off-target editing in the clinic are not well understood, particularly when off-target editing occurs at a very low frequency or in intergenic and/or intronic genomic regions.

Due to the potential risk of genotoxic events posed by DSB introduction, the genome editing field is exploring other means of editing genes or modulating gene activity. Indeed, in addition to their endonuclease activity, zinc finger proteins (ZFPs) and CRISPR-Cas9 can be engineered to create other therapeutic modalities. ZFPs and catalytically inactive Cas9 can be fused to transcription factors that modulate mRNA transcription [16, 17] without inducing DSBs. For example, ZFP-transcriptional factors have been used for allele-selective transcriptional repression of mutant HTT for the treatment of Huntington's disease.

Further engineering of CRISPR-Cas proteins has created a diversity of editing modalities: Cas9 protein was altered to form a nickase that introduces a single-stranded break in DNA [7], base editors that can induce transition (and certain transversion) mutations [18, 19, 20, 21], and prime editors that use Cas9 nickase fused to a reverse transcriptase and prime editing guide RNA (pegRNA) to mediate targeted small insertions, deletions as well as substitutions [22].

14.1.2 Clinical Trials Using Genome Editing Technologies

Clinical therapeutic applications of genome editing comprise *ex vivo* and *in vivo* gene modifications. There are presently over 50 clinical studies utilizing *ex vivo* and *in vivo* genome editing strategies listed in clinical trials.gov. For *ex vivo* editing approach, hematopoietic cells (e.g. autologous hematopoietic stem and progenitor cells (HSPC), autologous CAR-T cells) are collected from the patients, modified and expanded *ex vivo,* and then reinfused into patients. The efficiency and specificity of genomic editing can be assessed in the drug product and samples collected from patients post infusion. For *in vivo* editing approach, genome editing components are delivered systemically or locally to patients via lipid nanoparticles or recombinant adeno-associated virus (rAAV), then expressed in target cells such as hepatocytes (Figure 14.2), followed by editing of the genomic target. Here tissue biopsies are collected, if possible, for assessing editing efficiency.

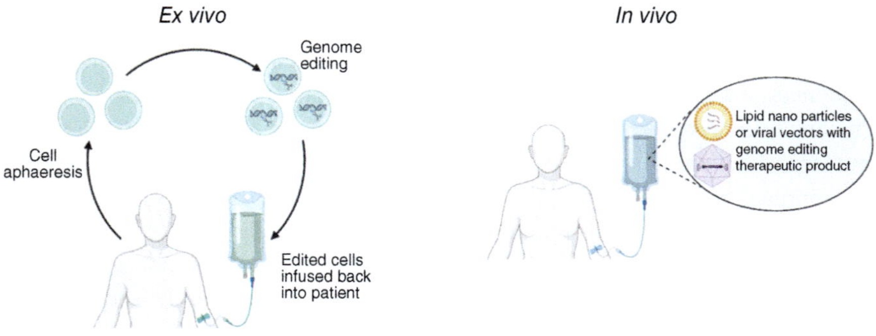

Figure 14.2 *Ex vivo* and *in vivo* genome editing for clinical applications. Left: Ex vivo genome editing. Cells are isolated from a patient, edited, activated/expended, and infused back into patient. Right: In vivo genome editing. Engineered nucleases are delivered by viral or nonviral approaches to the patient systemically. Source: Created with BioRender.com.

ZFNs pioneered the genome editing field [5] with the first *ex vivo* and *in vivo* clinical studies. In initial preclinical efforts, autologous T cells were modified *ex vivo* to disrupt the C–C motif chemokine receptor 5 (*CCR5*) gene, with the aim to induce resistance to HIV infection [23] (NCT03617198) as disruption of CCR5 restricts the ability of HIV to enter CD4+ T cells through the cell surface expressed CCR5 receptor. This was followed by several gene-edited cell therapy clinical studies with HIV patients (add references). Both ZFN and CRISPR-Cas9 technologies were used to disrupt the BCL11A erythroid-specific enhancer in autologous HSPCs to reactivate fetal hemoglobin (HbF) expression. This disruption consisted of NHEJ derived indels resulting from the nuclease-derived targeted DSB [24]. The expression of HbF was expected to ameliorate the symptoms of both beta thalassemia and sickle cell anemia. In these studies, cells were collected from individual patients, edited *ex vivo* and then reinfused into patients after myeloablation (ClinicalTrials.gov Identifier: NCT03655678; NCT03745287; NCT03653247) (Figure 14.2). Further examples of such *ex vivo* editing include CRISPR-Cas9, TALENs and meganucleases to engineer autologous and allogenic chimeric antigen receptor T lymphocytes (CAR-T) for antitumor immunity (ClinicalTrials.gov Identifier: NCT02735083; NCT02808442; NCT02746952; NCT03081715; NCT02793856; NCT04244656; NCT04035434; NCT04142619; NCT03190278; NCT04150497; NCT04649112).

The first-in-human *in vivo* editing studies utilized ZFNs and corrective transgene components packaged in recombinant AAVs and delivered intravenously to permanently modify patient hepatocytes. Expression of the ZFNs was driven by a liver-specific promoter. The ZFNs targeted the albumin intron 1 locus in hepatocytes, a safe harbor site, and following induction of a DSB resulted in the insertion of a corrective transgene at the albumin locus (Figure 14.2). This strategy was applied for the treatment of mucopolysaccharidosis type I (MPS I; Clinical Trials.gov Identifier NCT02702115), mucopolysaccharidosis type II (MPS II; ClinicalTrials.gov Identifier NCT03041324), and hemophilia B (ClinicalTrials.gov Identifier NCT02695160). In the MPS II clinical study, one patient showed transient plasma transgene protein at therapeutic levels, but expression was diminished due to a suspected immune response, indicated by elevated levels of alanine transaminase and aspartate transferase. Protein expression did not reach therapeutic levels in other patients in the studies [25]. In another study, CRISPR-Cas9 mRNA and single guide RNA were targeted to the liver using apolipoprotein E-modified lipid nanoparticles to address transthyretin amyloidosis through permanent disruption of the transthyretin gene (TTR) [26]. The study showed a 96% reduction of TTR in patient serum. If this treatment is proven to be durable, it is expected to improve disease symptoms.

14.2 Regulatory Guidance on Engineered Nuclease On- and Off-target Assessment

As interest is exponentially increasing for the development of new therapies derived from engineered nucleases, harmonized regulatory guidance is needed to ensure the appropriate characterization of safety profiles and risk assessments of these genomic medicines. The FDA issued a draft guidance in 2022 to provide recommendations for assessment of safety and quality of products incorporating genome editing in human somatic cells to support Investigational New Drug (IND) applications [27]. The aim of this guidance is to assist in translation of gene editing products from laboratory bench to clinical studies by providing recommendations for assessing the safety and quality as well as addressing the potential risks for these products. Some of the risks associated with genome editing include off-target editing, unintended consequences as well as the unknown long-term effects of on- and off-target editing.

Preclinical studies are recommended to identify and characterize the risk of genome editing at on- and off-target loci, including identification of off-target editing activity, including type, frequency, and location of all off-target editing events. Although no specific methods for assessment of off-target editing are noted, instead multiple orthogonal methods are recommended for identification of potential off-target sites, including an unbiased genome-wide analysis. Potential off-target sites should be verified using methods with adequate sensitivity to detect low-frequency events. The acceptable sensitivity for detecting low-frequency events was not specified by the agency and would likely depend on specific genomic location and risk posed by the editing. The analytical methodologies used for off-target evaluation need to be well described in regulatory submissions, including bioanalytical parameters such as sensitivity, specificity, accuracy, precision, and description of the reference materials. An assessment of genomic integrity is also advised, including evaluation of potential chromosome rearrangements, large insertions and deletions, integration of exogenous DNA, and potential oncogenicity or insertional mutagenesis. Evaluation of the biological consequences associated with on- and off-target editing is also necessary, as feasible. In addition, characterization is needed for the kinetic profile of genome editing components expression and editing activity. For clinical studies, consideration should be given for adequate monitoring of any off-target editing and adequate assessment of the outcomes of unintended consequences of on- and off-target editing.

The 2020 European Medicines Agency (EMA) guidance also emphasizes the importance of characterizing on- and off-target editing in the genetically modified cells [28]. Since genome editing is a rapidly evolving field, EMA recommends using current scientific knowledge for selecting a strategy for evaluating on- and off-target activity. Additionally, on-target genome editing should be characterized to establish that the target site is correctly edited and no unintended changes have occurred. To evaluate off-target activity, in silico prediction and at least one sensitive and well-characterized experimental assay should be used in the cell type that will be used therapeutically, and an unbiased genome-wide evaluation of off-target activity *in vitro* is also necessary. The chosen assay strategy should be justified, and the sensitivity of the methods should be indicated. In addition, the possibility of large genomic rearrangements needs to be evaluated based on the actual profile of on- and off-target edits. The risk associated with off-target activity and large genomic rearrangements should be assessed in the therapeutic cell type. EMA and FDA guidance documents do not recommend specific methods or acceptable levels of assay sensitivity, most likely due to the evolving nature of the methods to assess off-target activity.

With multiple clinical trials ongoing and forthcoming, sensitive and specific methods are needed to assess genome editing outcomes and potential risks to patients. Best practices for indel assessment are being discussed in the genome editing, bioanalytical, and safety assessment field, yet the number of publications and white papers remain limited [29], while other genetic outcomes such as long rearrangements are yet to be discussed in the bioanalytical literature.

The National Institute of Standards (NIST)-led Genome Editing Consortium was recently organized [30], and encompasses collaboration between NIST, industry experts, academia, and other government agencies. One of the aims of this consortium is to establish tools for understanding reproducibility, performance, and comparability of the assays used for detecting genome editing outcomes. This work is currently ongoing, and its results are highly anticipated by the field.

This review discusses the current and evolving methods for these assessments of on- and off-target activity and genomic integrity. Some of these methods are more focused on *in vitro* preclinical assessments and others for assessing editing in *in vivo* preclinical studies as well as in clinical studies.

14.3 Strategies and Methodologies to Evaluate On-target and Off-target Activities

14.3.1 Strategies to Evaluate Off-target Sites in Preclinical and Clinical Studies

An approach to select the therapeutic genome editing lead candidate with the least potential off-target sites can be divided into several phases. First, in the discovery phase, a broad range of methods are utilized in combination to identify the genome-wide activity of engineered nucleases and nominate a list of candidate off-target sites. Next, in the validation phase, targeted high-throughput sequencing or other comparable approaches can be used to measure editing frequencies in relevant cell types or samples from preclinical *in vivo* studies. Finally, the validated sites can then be quantitatively measured using well-characterized or qualified bioanalytical assays in patient samples from clinical studies (Figure 14.3).

Broadly, methods for discovering the unintended genome-wide off-target activity of genome editors can be divided into *in silico*, cellular, and biochemical categories. *In silico* nomination methods are easy and inexpensive, but their ability to accurately rank off-target activity remains unreliable [31]. Cellular methods are the most direct but have limitations in terms of sensitivity. Biochemical methods are the most sensitive but may nominate sites that are not modified in cells at frequencies above assay limits of detection and may lack influences of chromatin on genomic DNA structure and accessibility. To identify off-target sites as comprehensively as possible, multiple orthogonal methods in multiple donors are recommended [27].

Here, we describe cellular and biochemical methods for discovering the genome-wide off-target activity of editors with an emphasis on those that have been more frequently used to characterize therapeutic genome editing candidates.

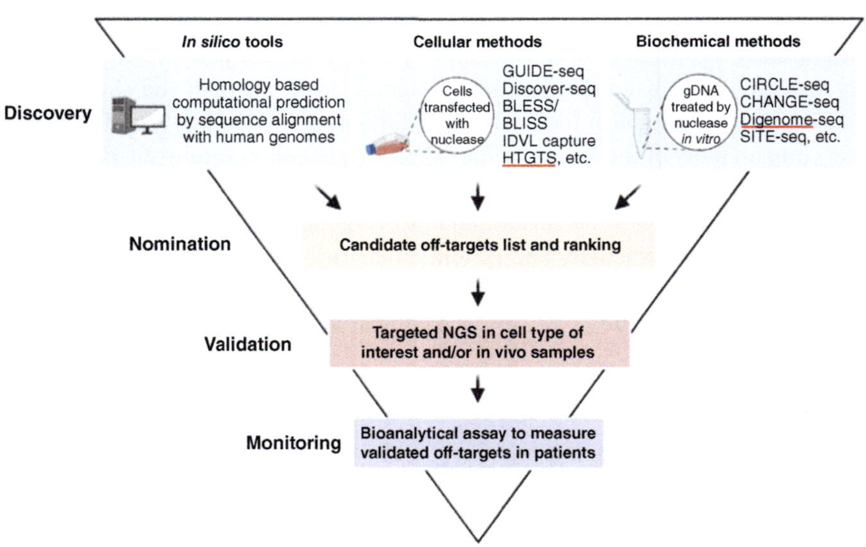

Figure 14.3 Workflows for evaluating off-target activities during genome editor drug development. Characterization of off-target activities is divided into different phases. Discovery can include *in silico* prediction, cellular and biochemical methods that result in candidate off-target list followed by validation of candidate off-target sites using targeted NGS in the cell type of interest. The validated sites are then quantitatively measured during the clinical studies. Source: Created with BioRender.com.

14.3.2 Techniques to Identify Genome Wide Off-target Sites

Cellular methods to define the genome-wide off-target activity of genome editing nucleases have become broadly adopted to characterize the specificity of editors for therapeutic treatment of diseases such as inherited childhood blindness and sickle cell disease.

There are now a number of cellular methods to define genome-wide off-target activity of genome editing nucleases that include: GUIDE-seq [32–34], Discover-seq [35], Discover-seq+ [36], BLESS/BLISS [37, 38], IDLV capture [39], and HTGTS [40].

Of these cellular methods, DNA end-capture methods like GUIDE-seq have become the most widely used. GUIDE-seq is based on the principle of efficient integration of end-protected short DNA tags into the sites of nuclease-induced DNA DSBs followed by tag-specific amplification and high-throughput sequencing of flanking genomic DNA (gDNA). Its advantages are that it is fairly sensitive, with the capability of detecting off-target sites with mutation rates as low as 0.1%, although it is not as sensitive as some biochemical assays (Figure 14.4A).

It works well in some therapeutically relevant cell types, such as T cells [33, 41] and fibroblasts. Some limitations are that there may be dsDNA-related toxicity or variable integration rates in some cell types, such as human hematopoietic stem cells or iPS cells. Quantitative tag integration proportional to indel mutation frequencies also implies that linear increases in number of input genomes and sequencing would be required to scale GUIDE-seq to higher sensitivity. GUIDE-seq and related end-capture-based methods have been used to analyze the specificity of ZFNs and TALENS as well [42, 43].

A number of biochemical methods to define the genome-wide off-target activity of genome editors have also been developed. The advantages of biochemical methods are that they have the potential to be more sensitive and scalable to many target sites.

Biochemical methods for defining the genome-wide activity of therapeutic genome editors include CIRCLE-seq [44], CHANGE-seq [45], Digenome-seq [46], and SITE-seq [47]. Digenome-seq was one of the first biochemical methods to be developed and is based on the principle of whole-genome sequencing of nuclease-modified gDNA and bioinformatic of reads that have a signature of editing, such as uniform start positions. An advantage is that it is simple to practice and PCR-free; limitations are that it requires large amounts of sequencing and it may be challenging to distinguish background reads that line up by chance from true signal (Figure 14.4B).

CIRCLE-seq, CHANGE-seq, and SITE-seq are all methods for selectively sequencing nuclease-modified gDNA. CIRCLE-seq and CHANGE-seq achieve this by generation of libraries of highly purified, circularized gDNA followed by treatment with Cas9 ribonucleoprotein complex. Only gDNA circles that have been linearized will have ends that are available for adapter ligation and sequencing. CHANGE-seq is a streamlined high-throughput improvement on the original CIRCLE-seq method that utilizes Tn5 tagmentation to eliminate a large number of molecular biology steps. In the SITE-seq protocol, sequencing adapters are sequentially added to high-molecular-weight gDNA that has been treated with Cas9 digestion and fragmentation, respectively. Biotinylated adapter containing DNA is pulled down via streptavidin affinity purification (Figure 14.4C).

In direct comparisons of cellular GUIDE-seq and biochemical CHANGE-seq methods, we found that sites detected exclusively by CHANGE-seq could be confirmed in cells and concluded that CHANGE-seq is more sensitive, a likely general property of biochemical methods where superphysiologic ratios of enzyme to gDNA substrates can be used to maximize detection sensitivity.

Currently, using a combination of cellular and biochemical methods can provide the best mix of direct detection and sensitivity. In the future, optimizing methods like GUIDE-seq for more cell types will be important to support the increasing range of cells that are being edited for clinical applications.

14.3.3 Targeted Approaches to Measure Short Insertions and Deletions

Insertions and deletions are the footprints of nuclease activity – editing efficiency and specificity can be assessed by analyzing indels at the on-target and off-target genomic locations. Editing efficiency is usually expressed as the percentage of DNA alleles in the study sample that possess indels, i.e. are edited. Since NHEJ mechanism of repair is often imprecise, it results in a polyclonal cell mixture with a variety of different indels. Detection of indel sequences in conjunction with quantification of each clone's frequency is a safety evaluation of clonal amplification.

There are several molecular biology techniques that were either repurposed or specifically developed to detect the short indels introduced by engineered nucleases. All these methods rely on enrichment of the genomic locus containing the edited site. A majority of indels introduced by engineered nucleases are under 50 bp [48, 49]. Most commonly the enrichment is performed by PCR using primers flanking the edited site. Once the edited site is enriched it can be analyzed using four main groups of approaches – Next-Generation Sequencing (NGS), Endonuclease Mismatch Cleavage Assays (EMC), digital droplet PCR (ddPCR), Sanger sequencing combined with sequence trace decomposition and Indel Detection by Amplicon Analysis (IDAA) (Table 14.1). This section will provide an overview of each technique, the context of use, and technique's advantages and disadvantages.

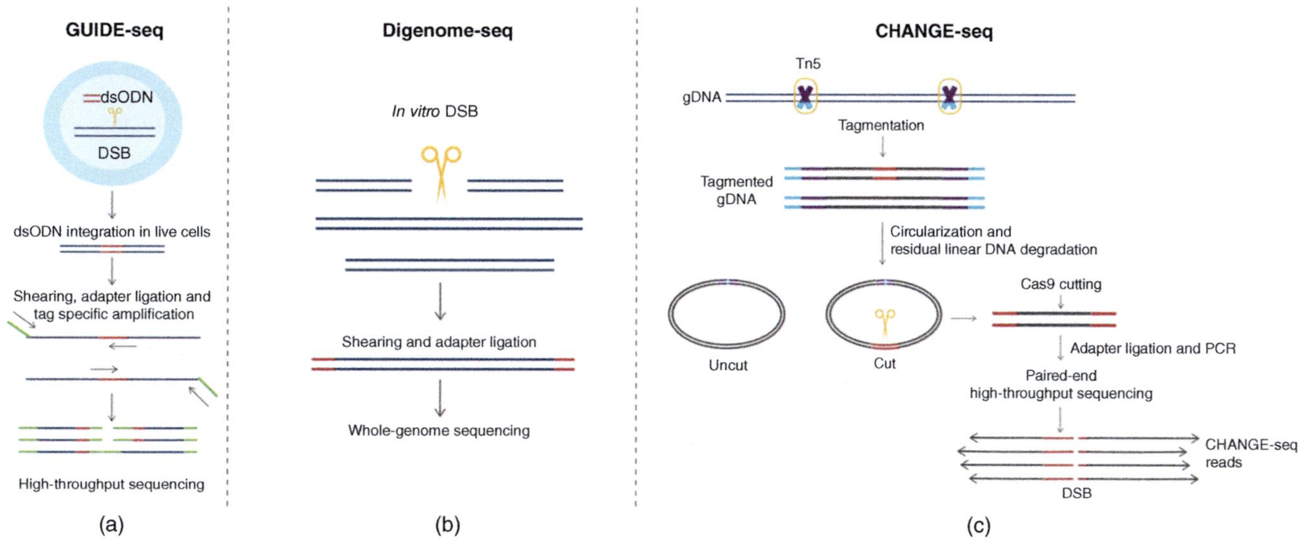

Figure 14.4 Techniques to identify genome-wide off-target sites. (a) GUIDE-seq is based on the principle of efficient integration of short, end-protected oligodeoxynucleotide (dsODN) tags into the sites of nuclease-induced DSBs. Genomic DNA shearing, adapter ligation, and tag-specific amplification are performed to map cellular DSBs. GUIDE-seq reads are proportional to indel mutation frequencies in cells. (b) Genomic DNA is treated with Cas9 ribonucleoprotein complexes, sheared, ligated to adapters for whole-genome sequencing. After sequencing, sites with uniform ends that are likely created by Cas9 in vitro cleavage are identified with a bioinformatic algorithm. (c) CHANGE-seq is based on the principle of selective sequencing of nuclease-modified genomic DNA. Genomic DNA is tagmented with Tn5 to add adapters for circularization. Genomic DNA is circularized by intramolecular ligation and excess linear DNA removed by exonuclease treatment. Genomic DNA circles are treated with Cas9 and only circles that have been cut will have ends available for adapter ligation, PCR, and high-throughput sequencing. CHANGE-seq reads map outward from the breakpoint location.

Table 14.1 Targeted approaches to measure short insertions and deletions.

Approach group	Readout	Ability to detect all types of short indels	Sensitivity	Ability to analyze multiple sites in one reaction	Turnaround time	Sample complexity	Context of use
NGS	Exact DNA sequence	Likely all types of short indels	≈0.001–0.2%	Hundreds of edited sites	3 days–1 week	High and low complexity samples	In vitro, animal and clinical samples
ddPCR	Modification level without sequence information	Unknown	≈0.2–0.5%	Several sites	1 day	High and low complexity samples	In vitro, animal and clinical samples
EMC	Modification level without sequence information	Limited in detecting single base indels	≈0.5%	One site	1 day	High complexity samples	In vitro samples with ≤10% indels
TIDE and ICE	Identifies indel types with limited sequence information	Limited in detecting some indels	≈2.5%	One site	3 days–1 week	High and low complexity samples	In vitro samples
IDAA	Limited sequence information	Limited in detecting single base indels	≈0.1–1%	One site	1 day	High and low complexity samples	In vitro samples

Targeted Next-generation Sequencing for Indel Detection

NGS is currently accepted in the field as the gold standard for high-throughput detection of genome editing events and is likely the most accurate analysis platform. It is the only method that allows for simultaneous quantification of both on- and off-target edits in multiple samples. Since its first use in detecting *CCR5* modification introduced by ZFNs using Roche 454 [50] and Illumina sequencing [51], it is now widely adopted for all other genome editing platforms.

Whole genome sequencing is rarely used to assess genome editing due to limited sensitivity and high cost. More commonly a targeted sequencing approach is employed where the edited sites are first enriched and then sequenced, followed by a bioinformatics analysis that determines the exact sequences and frequencies of the indels. Targeting the specific editing sites allows a higher depth of sequencing, resulting in a higher sensitivity and confidence of adjudicating indels. In addition, the number of individual samples to be pooled into one sequencing run can be increased, resulting in lower sequencing costs per sample and shorter turn-around time [52, 53].

Sequencing is commonly performed using sequencing-by-synthesis on MiSeq or NextSeq instruments from Illumina though the use of other technologies such as Oxford Nanopore and PacBio sequencing is increasing [54, 55]. The use of PacBio sequencing is limited due to specialized sample preparation and high artifactual rate of long-range PCR (PCR chimeras and amplicon length-dependent biases).

This section will focus on targeted NGS performed using sequencing-by-synthesis from Illumina for products enriched by two commonly used techniques–amplicon sequencing and hybrid capture.

Amplicon Sequencing

Amplicon sequencing relies on two sequential PCR reactions (Figure 14.5A). The first PCR reaction is a locus-specific PCR where the edited site and flanking sequences (approximately 75 bp on each side) are amplified with primers containing overhang sequences. An aliquot of the first PCR reaction is then taken into a second PCR reaction where overhang sequences are recognized by the second primer pair, which adds sample specific barcodes and the P5 and P7 universal Illumina sequencing adaptors. The resulting PCR products are then analyzed for purity using gel electrophoresis, pooled before purification and normalization and sequenced. The sample specific barcodes are 6–8 nucleotide sequences that are unique for each PCR reaction. Their presence allows for the pooling of multiple samples into the same sequencing run. Post sequencing, the barcodes allow the assignment of sequencing reads to individual samples bioinformatically. This enables amplicon sequencing to analyze multiple samples in a single run.

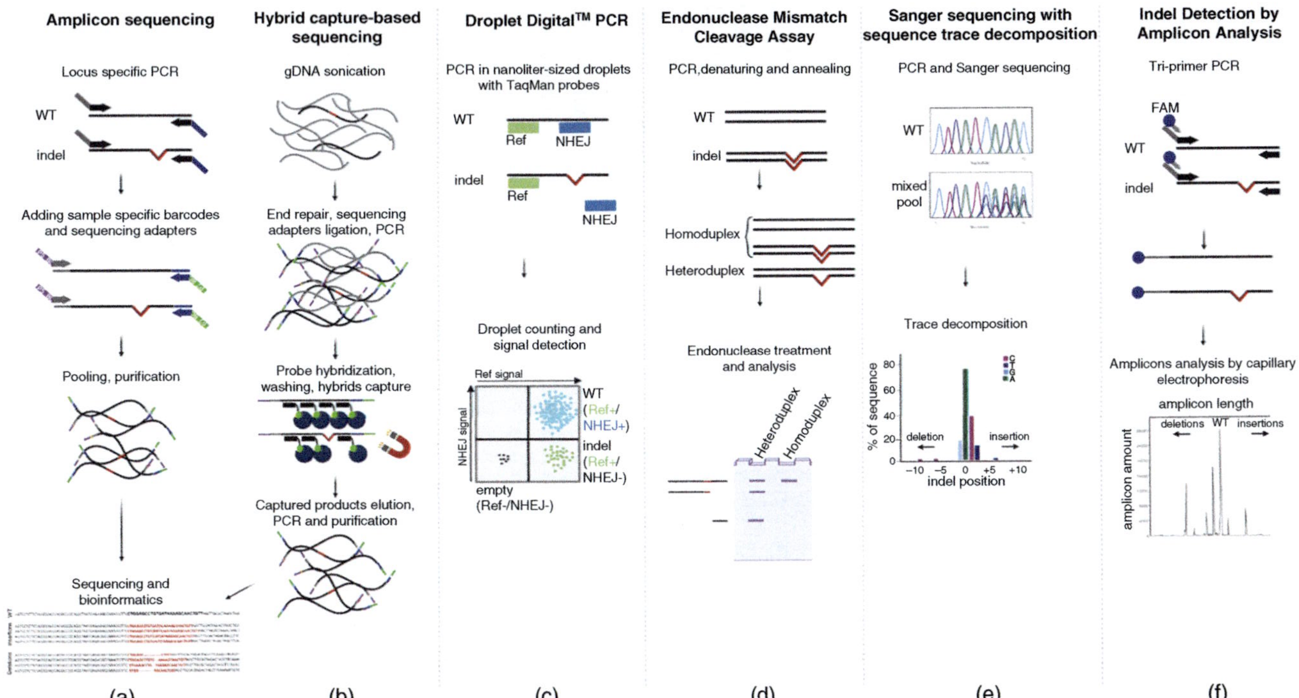

Figure 14.5 Targeted approaches to measure short insertions and deletions. (a) Amplicon sequencing; (b) Hybrid capture-based sequencing; (c) Droplet Digital PCR (one of various approaches for indel detection is illustrated); (d) Endonuclease Mismatch Cleavage Assays; (e) Sanger sequencing combined with sequence trace decomposition (TIDE workflow is illustrated); (f) Indel Detection by Amplicon Analysis (IDAA). Note, all methods require extracted gDNA which is not illustrated here. Source: Created with BioRender.com.

When performing multiplexed amplicon sequencing, the design of the primers requires special attention. Primers should be highly specific to the queried sites without cross-reacting with each other to minimize primer-dimer formation and amplification at sites other than target site. In addition, to ensure equal sequencing coverage across all amplicons, the ratio between different primer pairs needs to be optimized to prevent preferential amplification of some amplicons. There are several technologies that are addressing these concerns by using highly developed primer design algorithms. rhAmpSeq combines primer design algorithm with increased PCR specificity by using RNase H-dependent PCR [56, 57]. Though this technology is marketed for CRISPR-based editing, rhAmpSeq can be readily applied to detect indels introduced by other nucleases if alternative sequencing analysis software is used. This technology enables the examination of hundreds of edited sites in the same sample. In rhAmpSeq each primer contains an RNA base at or near the 3′-end of the primer and lacks the free 3′-OH group, thereby needing "activation" to be extended by DNA polymerase. Once perfect complementarity between primer and its target DNA is achieved, the primer:target heteroduplex is cleaved at the 5′-side of the RNA base by thermally stable RNase H2 supplied as a part of PCR master mix. The resulting 3′-OH group then allows DNA polymerase to carry out the primer extension. Due to the requirement for primers to be "activated" by first binding to the target sequence, primer-dimer formation and false amplification of similar sequences are greatly reduced.

The advantages of amplicon sequencing techniques are relative ease of design and implementation. In addition, the library preparation protocol is simple, quick, and not labor-intensive. Perhaps the main disadvantage of amplicon sequencing is the difficulty of integrating unique molecular identifies (UMIs) into the workflow. UMIs are molecular barcodes that comprise of short sequences used to uniquely tag each molecule in a sample prior to PCR amplification. UMIs can be used to remove sequencing reads that arise from PCR duplicates as PCR duplicates will have the same UMI. This allows for accurate allele frequency detection as PCR duplicates can falsely overrepresent different allele frequencies if more PCR copies were generated for one variant compared to other variants present in a sample.

A recent publication showed that UMI tagging can be performed using the primer extension reaction with one UMI primer prior to PCR. The extension product is then purified using AMPure XP beads and PCR-amplified using the universal primer and the gene-specific reverse primer [54]. This application has yet to be broadly used.

Hybrid Capture-based Sequencing

A second enrichment technique is referred to as hybrid capture [52] (Figure 14.5B). Briefly, gDNA is sheared into 120–200 bp DNA fragments using sonication. DNA fragments are then end-repaired and ligated with sequencing adaptors that include sample specific barcodes and P5 and P7 universal Illumina sequencing adaptors. The ligation products are then subjected to PCR to generate sufficient material for the capture reaction. The library of amplified genomic fragments is hybridized with biotinylated RNA or DNA probes tiling across the candidate edited site and flanking sequences. The hybridization products are then purified using streptavidin beads, amplified, purified, and sequenced. This hybrid capture approach is compatible with UMI tagging, which can address artifacts arising from PCR and sequencing errors. The disadvantages of hybrid capture approaches are requirements for substantial assay optimization, including gDNA sheering conditions and PCR conditions. In addition, DNA capture is less efficient for GC- and AT-rich sequences. Since off-target sites are often located in intronic and intergenic regions the efficient probe development to accommodate all edited sites can be challenging.

NGS Sensitivity for Indel Detection and Quantification

In general, the sensitivity of indel detection using NGS depends on sequencing depth (how many sequencing reads correspond to each analyzed sample) and background signal (modifications measured in non-edited sample).

There is an inverse correlation between level of editing and sequencing depth requirements – samples with a low editing frequency require a higher number of sequencing reads for accurate detection. So far, no systematic study has been published to investigate the exact relationship between modification frequency and number of sequencings reads that are needed for accurate quantification. A coverage of >1000 paired reads per edited site was shown to allow for detection of indels down to 0.5% with a 40% coefficient of variation (CV) [57]. In another publication, the coverage of >5000 paired reads per edited site was suggested to detect indels down to 0.2%, but the precision of detection was not specified [58].

NGS background is defined as detected modifications in unedited samples. Such background can arise due to errors introduced during PCR amplification and sequencing. The background noise reported for Illumina sequencing is 0.1–0.3% [59, 60]. Background noise was also shown to increase when poor quality gDNA was used [61]. Investigation of background modification levels across 273 CRISPR edited sites showed that the background varies from 0% to 1.0%, with 98% of sites having background indels ranging from 0% to 0.4% [57]. One of the ways to decrease the background bioinformatically is to shorten the sequence "window" where indels are quantified. Several published software packages stress the importance of quantifying indels in a specific sequence "window" where modification is expected to occur when high sensitivity of detection is desired. Such a "window" is specific to each type of engineered nuclease and needs to be experimentally established. When the optimal "window" was applied, a 60%

decrease in background (false-positive indel signal) was seen while retaining >98% indels in treated samples [57]. The optimal sequence "window" can be established by careful analysis of oligo incorporation using unbiased genome-wide sequencing methods e.g. GUIDE-seq. The exact genomic position of incorporated oligo can be used to establish the correct sequence "window" to analyze and quantify indels.

The sensitivity of indel quantification can be thought to depend on the particular NGS approach (e.g. amplicon sequencing vs. hybrid capture). Head-to-head comparison of hybrid capture and multiplexed amplicon sequencing (rhAmpSeq) for edited sites introduced by CRISPR-Cas9 showed high data similarity; the rhAmpSeq analysis was more sensitive likely due to the higher sequencing depth used [58]. Recently, Kurgan et al. investigated the sensitivity of the rhAmpSeq assay [57]. They prepared a mixture of synthetic DNA containing insertions, deletions, and SNPs to simulate the editing of the *HPRT1* genomic locus by CRISPR-Cas9 nuclease. This mixture was then serially diluted with wild-type synthetic DNA to create samples with 0–100% "modifications," then analyzed using the rhAmpSeq protocol with a sequencing depth of >40,000 reads per sample. The authors found that at 1% modification and lower, there was a ≥20% deviation in detected modification levels compared to that expected. The accuracy of detection was improved by applying a background correction. This allowed the detected modifications down to 0.1%, within 10% of expected values. Whether such high sensitivity can be achieved in real biological samples for all investigated sites requires additional investigation. Indeed, the authors emphasized the need for improving library construction chemistry such as incorporating UMIs and coming up with sophisticated background correction techniques to improve the accuracy and precision of indel calling.

The assay LOD is often defined as the indel frequency that can be detected as a statistically significant difference between edited and unedited samples. In two publications, the edited sites were defined as confirmed when the difference in indel frequencies between treated and untreated samples was higher than 0.16% [62] and 0.20% [58]. In the scenario where untreated samples show no background indels, the highest possible theoretical sensitivity would be 0.16% and 0.20% indels, respectively.

Miller et al. demonstrated high indel sensitivity (≈0.001–0.025% indels with 50% CV) at 100 AAVS1 off-target sites assessed as independent PCR reactions and sequenced using optimized conditions on NextSeq Illumina [63]. High sensitivity was achieved through increasing sequencing depth and decreasing background signal to 0.01% indels by implementing multiple improvements. Sequencing depth was increased so that each input DNA allele was sequenced at least ten times (≥200,000 reads per replicate per target site), and multiple technical replicates of edited samples and up to 24 replicates of unedited background samples were run. The occurrence of sequencing artifacts was decreased by reducing the cluster density on the flow cell during the sequencing reaction and precisely defining the sequence "window" where editing is expected to occur. This method is amenable for *in vitro* studies where multiple biological replicates of treated and untreated samples can be run.

Bioanalytical Characterization of NGS for Indel Detection and Quantification

The FDA guidance for bioanalytical method validation [64] underlines the importance of analytical method characterization supporting regulatory submissions. This includes bioanalytical parameters, such as sensitivity, specificity, accuracy, precision, and description of the reference materials. There is currently no specific FDA guidance for analytical method characterization for off-target detection. This section will focus on our case study of indel quantification in ZFN-edited hematopoietic stem and progenitor cells using NGS, specifically, how to generate material for making quality controls (QCs) . Reference standards and QCs are important reagents for quantification of most detection methodologies. QCs are commonly used to develop and characterize the assay and to monitor run-to-run variability when drug product or post-infusion patient samples are analyzed. There are several approaches for QC creation that have been explored in the field. One approach is to use engineered cell lines by introducing edits at the same genomic locations as expected in study samples (cell therapy drug product as well as animal and human samples post-infusion). Such cell lines can be generated by transfecting immortalized cells with studied engineered nucleases to cause a high level of indels, followed by the propagation and banking of these cell lines. The level of indels can be determined by NGS and, ideally, by an orthogonal methodology e.g. ddPCR. QC samples containing different levels of indels then can be generated by mixing a cell line with a known frequency of editing together with WT cells at different ratios [25] or isolated gDNA.

An alternative approach relies on preparing a mixture of synthetic DNA containing the anticipated edits with wild-type synthetic DNA to create samples with various modification levels [57]. The synthetic DNA mixtures can then be spiked into gDNA isolated from WT animal or human tissues to include background matrix of the study samples.

The synthetic DNA approach and edited cells approach both have their respective advantages and limitations. Engineered cell lines can be cumbersome to generate and maintain, which substantially extend assay development timelines, especially if multiple editing sites are explored (on- and off-target editing in the same assay). Multiple cell lines with individual edits can be pooled or the same cell line can be sequentially edited. However, the clear advantage of engineered control cell lines is that they better represent the complexity of study samples and editing outcomes. They can be easily propagated, and more QC materials can be made as needed using the same stock material. Though synthetic DNA can be implemented faster than engineered cells, this approach suffers from its artificial nature. In cell or tissue study samples, indel sequences at a known editing site typically are heterogenous in nucleotide sequence and length. Usually, only a few indel sequences are mimicked by the synthetic DNA, while others are not represented. In addition, synthetic DNA is usually short and therefore an easier target for PCR amplification when compared to the more complex gDNA matrix present in study samples. Some DNA sequences might be difficult to synthesize due to their sequence composition (homopolymers, stem-loop structures, etc.) which may result in a low yield and high level of sequence errors in synthetic DNA. This can be a common issue for off-target sites that are often found in non-coding regions with sequence compositions that may be difficult to synthesize. When synthetic DNA is ordered, it should be carefully evaluated how much material is required to support testing for the entire clinical trial to avoid batch effects.

Therefore, the engineered control cell line approach may be better suited when establishing method feasibility and accuracy if timeline permits. A synthetic DNA approach can be implemented at the sample testing stage with the purpose of assessing run-to-run variability.

In summary, targeted NGS sequencing provides the frequency and exact sequence of indels. It can be used for high-throughput quantification of both on- and off-target edits with high sensitivity in one biological sample. There are several targeted NGS technologies available: a particular approach should be chosen based on required sensitivity, number of analyzed sites, and the amount of sample available for analysis. To support clinical sample monitoring, NGS assays can be analytically characterized to establish assay sensitivity, variability, and acceptance criteria.

14.3.3.1 Droplet Digital™ PCR

Droplet Digital™ PCR (ddPCR™) is a technique that allows for absolute quantitation of target DNA in complex biological matrices. The PCR reaction is fractionated into 20,000 nanoliter-size droplets. As a result of partitioning, target DNA molecules get distributed across the droplets so that each droplet contains zero, one, or a few copies of DNA templates. PCR amplification is then carried out to the plateau phase and the droplets are counted as positive and negative reactions by a droplet reader. The quantity of target DNA can be calculated using the fraction of positive droplets and Poisson statistics [65, 66]. This methodology was successfully adopted to assess NHEJ-derived indels and implemented in various research studies (GEF-dPCR – gene-editing frequencies digital PCR or DSB-ddPCR – DSB ddPCR) [67–69] (Figure 14.5B). These techniques utilize two different TaqMan probes carrying two different fluorophores (FAM and HEX) targeting the same PCR product and simultaneously detecting wild-type and edited alleles.

One TaqMan probe is designed to quantify the total number of alleles present in the sample (NHEJ insensitive reference probe, indicated as "Ref" on Figure 14.5C) and another TaqMan probe is designed to quantify the number of edited alleles (NHEJ sensitive probe, indicated as "NHEJ" on Figure 14.5C). Due to the imprecise nature of NHEJ, edited sites contain indels of diverse sequence and therefore, it is not feasible to design a probe that can efficiently bind all possible sequences at the edited site. Instead, the "drop out" probe strategy is employed where the NHEJ sensitive probe is specific to the wild-type version of the modification site and only produces fluorescent signal if DNA is not modified. Therefore, the number of edited alleles is derived from the lack of signal from the NHEJ sensitive probe. The data are presented as a two-dimensional plot where the wild-type DNA signal is derived from NHEJ insensitive/NHEJ sensitive double positive droplets and the edited DNA signal is derived from the droplets with a decrease or loss of the signal from the NHEJ sensitive probe and positive signal from the NHEJ insensitive probe.

This ddPCR method was shown to be highly sensitive with a limit of detection (LOD) of 0.2–0.5% indels [69, 70]. The detention of indel frequency was also accurate when compared with NGS and flow cytometry with reported differences in as low as 4% [68]. To achieve such high analytical performance, the assay needs to be carefully optimized. This includes fine tuning of PCR conditions and determining probe sequences that provide the highest possible signal/noise ratio. Ideally, the edited site should be sequenced so that the NHEJ sensitive probe is designed accordingly [71]. Miyaoka et al. extended this assay to simultaneously quantify both HDR-mediated point mutation correction and NHEJ-mediated insertions and deletions [72, 73].

Due to the limit of the droplet number employed by the ddPCR platform (total 20,000 droplets), the gDNA amount used in the experiments should be carefully titrated to provide the appropriate resolution between NHEJ-insensitive and NHEJ-sensitive probes. High amounts of gDNA are not recommended as it may result in decreased assay sensitivity [70].

In summary, ddPCR can be a useful tool for quantification of NHEJ-derived indels as well as HDR-mediated point mutations and is characterized to have high sensitivity and a quick turnaround time (one day). It can be a cost-effective option to monitor genome editing when the precise sequence information of the edits is not required.

14.3.3.2 Endonuclease Mismatch Cleavage Assays

Enzymatic detection of DNA mutations has been possible since the discovery of proteins, such as CEL-I nuclease, which can cleave ssDNA present in mismatched regions of dsDNA [74]. Since its development by Sangamo Therapeutics in the mid-to-late 2000s [75], EMC has become one of the most commonly used methods to detect indels introduced by engineered nucleases. EMC uses mismatch-detection nucleases that selectively cleave DNA heteroduplexes but leave DNA homoduplexes intact [76] (Figure 14.5D). Firstly, gDNA is isolated from the edited sample and the genomic locus flanking the edited site is amplified using PCR. The PCR product is then denatured and reannealed resulting in DNA heteroduplexes containing mismatches where one DNA strand is wild-type and another DNA strand contains an indel. This mixture is then treated with a nuclease that recognizes and cleaves the dsDNA at the site of mismatch. Lastly, the products of this reaction are resolved using methods such as gel electrophoresis. The approximate indel frequency is then estimated by quantifying the cleaved and intact DNA as represented by bands of different mobility (Figure 14.5D). Optimal conditions for heteroduplex formulation between indel-bearing and wild-type DNA strands occurs when indel levels are relatively low (around 10%) and wild-type amplicons are in excess. Therefore, a simple formula can be applied to quantify percent modification (percent modification = fraction of cleaved bands/2). Once modification levels are higher, the formula needs to be corrected to account for the fact that indel-bearing strands do not exclusively anneal with wild-type strands [77]. There are several commercially available endonucleases that can be used for the EMC assay – Surveyor™ nuclease [78], T7 endonuclease I [76], CEL-I endonuclease [79], T4 endonuclease VII [80], and endonuclease V [81]. The endonuclease should be carefully selected to suit the application as each enzyme has it is own properties that can impact the sensitivity, efficiency, and specificity of the EMC assay. The method used to separate the products of the cutting reaction can also impact the assay's performance. For example, up to 3% editing can be detected using Surveyor™ nuclease and WAVE HSD HPLC [78] while up to 0.5% editing can be detected using T7 endonuclease I and gel electrophoresis in polyacrylamide gel [82].

EMC has several limitations: false-positive signals can be detected if the analyzed genomic locus contains a single nucleotide polymorphism (SNP) that results in heteroduplex formation and enzymatic cleavage in unedited sample. Therefore, each investigated locus is tested for the absence of cutting using gDNA isolated from the naive sample. The EMC assay is also known to underestimate the editing efficiency due to the insensitivity of endonucleases to cleave heteroduplexes containing single base indel events – common editing outcomes for CRISPR/Cas9-based genome editing [83]. In addition, this assay is unsuitable for detecting high modification rates as mentioned above and low complexity samples (clonal or close to clonal cell lines) because mutated amplicons are able to reanneal and be resistant to endonuclease [83].

In summary, the EMC assay is simple, cost-effective, and has a fast turnaround time (one day). However, it has several limitations that have restricted its use to academic settings or the early stages of drug development.

14.3.3.3 Sanger Sequencing Combined with Sequence Trace Decomposition

There are several methods for indel quantification that rely on low-throughput sequencing. The targeted region together with the flanking DNA sequence can be amplified using PCR followed by Sanger sequencing. The PCR product can be either sequenced directly or first subcloned and then sequenced. Subcloning includes PCR product ligation with a plasmid, bacterial culture transformation, and spreading over the agar plates. Agar plates with the colonies can be sent to a commercial sequencing provider that returns Sanger sequence traces that are usually analyzed manually. Subcloning is usually performed to increase the quality of Sanger sequence traces as PCR products may contain components inhibiting the Sanger sequencing reaction. This procedure is straightforward but labor-intensive and comparatively lengthy. It is not recommended for samples with a low level of indels or high complexity samples with various indels at the edited site because many colonies will need to be sequenced to gain an accurate picture of editing frequency. Therefore, the Sanger sequencing approach is usually selected for clonal indel sequencing.

Notwithstanding these limitations, Sanger sequence traces can be used to quantify CRISPR-introduced indels in high-complexity samples using the methods of Tracking of Indels by Decomposition (TIDE) or Inference of CRISPR Edits (ICE) analyses [84, 85] (Figure 14.5E). In these assays, the region of interest is amplified by PCR from both edited and wild-type samples. PCR products are then purified, quantified, and Sanger sequenced. High quality Sanger sequencing data are essential to ensure successful computational analysis, therefore PCR products are usually carefully purified from the components of the PCR reaction and primer dimers. The sequencing from both ends of the amplicon is recommended to increase confidence in results. The Sanger sequences traces and gRNA sequence are uploaded to the software, which then compares sequencing traces corresponding to the untreated sample (wild-type sequences) with the mixture of sequencing traces obtained from the edited sample. The software then reports a bar graph showing indel size and frequency, including frequencies of individual indels without providing the exact sequence of identified indels. A head-to-head comparison between ICE and NGS showed a good agreement where reported indel frequencies were largely similar across a variety of edited sequences [84, 86]. Similar results were reported for comparing TIDE and NGS where TIDE identified the same main indel types (where the indel frequency was ≥5%) and showed a negative bias of about 10–20% [83]. Both TIDE and ICE are reproducible, reporting nearly identical results in replicate experiments with a LOD of about 2.5% [86]. Some of the reported disadvantages of TIDE and ICE were poorer performance in detecting longer indels and decreased sensitivity for detecting indels in high-complexity samples. In addition, TIDE was shown to diverge from NGS data substantially when performed on individual clones containing both insertions and deletions in the same clone [83]. TIDE and ICE software are designed specifically for CRISPR and do not allow analysis for indels introduced by other genome editors.

The main limitation of TIDE and ICE is that they do not provide the exact sequence of identified indels. Due to the simplicity and free accessibility of the software, TIDE and ICE methodologies were widely adopted in the academic CRISPR field and are currently used for high-throughput screening; however, its quantitative limitations have prevented it from being adopted in therapeutic genome editing efforts nearing the clinic.

14.3.3.4 Indel Detection by Amplicon Analysis (IDAA)

Indel Detection by Amplicon Analysis (IDAA) takes advantage of the ability of capillary electrophoresis to resolve small size differences between PCR amplicons [87] (Figure 14.5F). The optimized PCR labeling workflow was able to achieve resolution down to a single base-pair [88]. In IDAA, the region of interest containing the edited site is PCR amplified using three primers - two of which are standard gene-specific primers and the third is a 5′ fluorescently-labeled universal primer complementary to an overhang on the forward primer. PCR reaction is carried out using touchdown PCR conditions with optimized primers ratio to yield uniformly labeled PCR amplicons with high purity. The labeled amplicons are then resolved using standard capillary electrophoresis and peaks are called using standard capillary electrophoresis software (for detailed protocol please see Ref. [89]). IDAA showed superior performance compared to EMC – IDAA is more sensitive, provides information about indel size, able to detect indels in homozygous clones. IDAA showed similar results when compared to NGS with a reported sensitivity of 0.1–1% [89]. Similar to TIDE, IDAA miscalled indels in multiple clones containing both insertions and deletions [83]. In addition, IDAA was not always able to call −1 and +1 nt peaks from the wild-type peak, but it was possible to do by manual inspections of the peaks [83]. This technique can be a cost-effective alternative to NGS when the exact sequence of the indels is not required.

In summary, NGS-based methods became gold standard techniques for indel detection. NGS provides the frequency and exact sequence of indels with high sensitivity. Exact indel sequences may be required when evaluating the formation of clonality in cell therapy products. Due to the high cost of NGS and long turnaround time other methods can be advantageous. These methods can be applied when exact information about indel sequence is not required.

14.3.4 Technologies to Measure Large Genomic Rearrangements

In addition to unintended editing resulting in indels at off-target sites, larger genomic rearrangements such as large chromosomal deletions, inversions, and translocations have been detected at on- and off-target sites [12] regardless of engineered nucleases type [11, 90]. Translocations can occur when two or more DSBs are present on the same or two different chromosomes. In this case, one DSB is introduced intentionally by an engineered nuclease acting at a therapeutic site while another DSB is ectopic and introduced due to off-target activities. Translocations were also detected with only one DSB at the on-target location when the target region contained substantial sequence homology to another region located elsewhere in the genome [11, 12]. Large chromosomal deletions, inversions, and chromothripsis were also detected at the intended editing sites [11, 12, 14, 91]. Several studies showed that such mutations at the intended editing site may lead to detrimental effects in mitotically active cells [11, 14].

While to date large genomic rearrangements have only been observed in preclinical *in vitro* experiments, the potential impact of such events during therapeutic editing compels us to understand their genesis in greater detail and develop technologies for their monitoring (Table 14.2). There are several techniques used for genotoxicity assessment of edited cell therapy products. Historically, karyotyping analysis using G-banded chromosome analysis has been used to evaluate

Table 14.2 Technologies to measure large genomic rearrangements.

Assay group	Technologies	Sensitivity	Advantage	Disadvantage	Note
Cytogenetic	Karyotype Microscopy (G-banded chromosome analysis), FISH	Low	GLP methods available, cost-effective, well established	Low Sensitivity	Gold standard method
Genome-wide molecular assays	Targeted NGS	High	Provides exact sequence at translocation site	Complexity, require knowledge of one of the translocation partners, likely difficult to outsource	Useful at the discovery stage
Targeted molecular assays	qPCR, ddPCR	High	Relatively easy to set up, cost-effective	Need to know the exact sequence at rearrangement site	Confirmation and quantification of rearrangements discovered by genome-wide molecular assays

potential structural and numerical chromosome aberrations following genome editing. Newer techniques incorporate fluorescence *in situ* hybridization (FISH) to identify and evaluate chromosomal rearrangements. Additionally, a soft agar transformation assay using human fibroblasts has been used to assess potential tumorigenicity risk of edited cells [92]. Karyotyping provides a high-level view of chromosome aberrations, it has a fast turnaround time and is cost-effective. Though karyotyping is still widely used in industry, it has low sensitivity due to limited resolution and the low number of events that can be screened in one sample.

Recently, NGS-based techniques have been developed to identify large rearrangements with sensitivity and specificity superior to karyotyping assays [93]. These techniques include anchored multiplex PCR sequencing (AMP-seq) [93], linear amplification mediated high-throughput genomic translocations sequencing (LAM-HTGTS) [40], uni-directional targeted sequencing (UDiTaS) [94], and chromosomal aberrations analysis by single targeted linker-mediated PCR sequencing (CAST-seq) [40]. These methods are designed to identify chromosomal aberrations involving on- and off-target sites with prior knowledge of at least one fusion partner.

AMP-seq was designed to assess single nucleotide variants, insertions, deletions, copy number changes, and translocations in clinical molecular diagnostic assays. It was then adopted by the genome editing community to analyze large rearrangements in samples treated with engineered nucleases. This method can detect fusion transcripts from RNA input or translocation using gDNA. Double-stranded cDNA or fragmented gDNA are end-repaired, adenylated, and ligated with a half-functional adapter that contains sample-specific barcode. This step results in all DNA having the same 5′- and 3′-ends. Purified products are then PCR amplified with a target-specific primer and a primer complementary to a portion of the universal ligated adapter. This enables enrichment of target sequences where only one part of the sequence is known. The following nested PCR reaction accomplishes target enrichment and the addition of sequences fully functional for Ion Torrent or Illumina sequencing. Authors note that RNA-based detection provides potentially easier unique alignment, deeper coverage, and information about expressed fusion transcript sequence. However, RNA-based detection will miss transcriptionally unactive translocation detection. AMP-seq is amenable to fresh and formalin-fixed paraffin-embedded (FFPE) material.

LAM-HTGTS [40] is based on linear amplification PCR (LAM-PCR), which is used for characterizing unknown DNA adjacent to known DNA. In LAM-PCR primer is designed to bind known DNA sequence and produce multiple copies of known DNA linked to unknown DNA sequence in unidirectional manner. gDNA is first fragmented, then a biotinylated primer is designed to bind known sequences in a translocation event and generate multiple copies of the translocation junction. The 5′-biotinylated copy is then captured on streptavidin magnetic beads, ligated with bridge adapters followed by two rounds of nested PCR with adaptor and locus-specific primers that add sample-specific barcodes and sequences necessary for Illumina sequencing. After the first PCR step, a restriction enzyme is used to reduce the background from unrearranged target alleles. Final libraries are sequenced using Illumina's MiSeq platform. A bioinformatic pipeline then identifies and maps translocation junctions. This method is designed to detect off-target editing and unbalanced chromosomal aberrations and measure their relative frequencies. Yin et al. modified LAM-HTGTS by decreasing number of linear amplification cycles from 80 to 1, introducing ligation with 14-base pair random UMIs labeling each target copy, deeper sequencing with HiSeq, and improving bioinformatic analysis. This method was named primer-extension-mediated sequencing (PEM-seq) [95, 96]. The introduction of molecular barcodes allowed to distinguish PCR duplicates from the original templates and together with improved bioinformatics analysis increased the assay sensitivity. PEM-seq can be applied for measuring short indels. The high amount of gDNA (20–100 μg) required for LAM-HTGTS or PEM-seq makes them more suitable when testing material can be generated in large quantities.

UDiTaS [94] was developed for simultaneous measurement of small indels and larger structural rearrangements, such as large deletions, inversions, and translocations. This approach employs "tagmentation" where a custom-designed Tn5 transposon is used to simultaneously shear gDNA and add universal priming sites to each fragmented DNA. After two rounds of PCR that enriches the target region and adds sample-specific and Illumina barcodes, libraries are sequences on Illumina instruments. Using engineered cell lines and plasmid constructs with known indels and large rearrangements, the authors showed that this assay was able to measure down to 0.1% indel, large deletion and inversion events (≈1.1 kilobase), and had 0.01% LOD for balanced, acentric, and dicentric fusions between homologous or unrelated chromosomes. Head-to-head comparison of UDiTaS and AMP-seq was performed using plasmid surrogates mimicking a wild-type locus, a large deletion, and inversion spiked into mouse gDNA (mouse matrix). The data showed high agreement between these two methods, both being linear between 2200 and 714,000 genome equivalents. UDiTaS requires 50 ng of gDNA to carry out the reaction and authors noted that assay sensitivity can be increased by increasing gDNA input and sequencing depth. The impact of increased gDNA input on "tagmentation" efficiency was not discussed but needs to be considered when developing this assay. Authors underlined the importance of carefully assessing this assay for each locus using plasmid DNA or engineered cell lines due to the potential sequence bias in tagmentation efficiency.

For CAST-seq [11], gDNA is fragmented followed by linker ligation, in a similar fashion to AMP-seq, followed by three PCR reactions. The first PCR reaction is performed with a primer that binds to the on-target sequence, a primer that recognizes ligated linker sequence and "decoy primers" that bind the on-target sequence to prevent amplification from target alleles that are not involved in rearrangements. The first PCR reaction will only result in a product if the binding sites of the decoy primers were lost due to a large rearrangement such as a translocation or large deletions. The second and third PCR reactions introduce adapters and barcode sequences necessary for NGS. The inclusion of "decoy primers" reduced background and increased the sensitivity of the assay by five-fold. The number of translocation events in the sample is calculated by bioinformatic analysis that takes into account that nuclease-induced DSB can be processed differently before translocation, each translocation having a distinct fusion point and linker ligation point. CAST-seq requires 500 ng of gDNA and is able to detect large genomic rearrangements at a frequency of 0.006%.

Amit et al. [97] recently showed that translocations between on- and off-target sites can be discovered from rhAmpSeq sequencing data using CRISPECTR software (please find detailed description of rhAmpSeq in "Amplicon sequencing" section). Because multiplexed PCR reaction contains primers targeting on- and off-target sites, it will generate fusion PCR products originating from translocated alleles. Careful bioinformatic analysis can then report the translocation frequencies. More investigation needs to be done to understand the applicability of this approach to real biological samples. Special attention needs to be paid to unedited controls as PCR may generate fusion PCR products even without translocated DNA alleles.

Long-range PCR with dual UMI approach [98] was applied for the detection and quantification of large insertions, deletions, and local chromosomal rearrangements. Five- to 6-kilobase genomic regions around edited site were amplified and simultaneously tagged at 5′- and 3′-ends with terminal UMIs using PCR reaction consisting of two cycles (UMI-tagging PCR), followed by two consecutive PCR reactions to amplify UMI-tagged molecules and add sample-specific barcodes for multiplexed sequencing. The resulting PCR product was subjected to SMRT-seq library preparation and sequencing on PacBio sequencer. The assay was benchmarked using mixture of plasmids containing wild-type sequence at 80% frequency and eight plasmids containing artificial long deletion ranging from 921 to 4416 bp at 20% frequency for all variants. The results showed high agreement between expected and experimental results, validating the utility of long-range sequencing for large deletions discovery and quantification. More in-depth investigation is required to show the accuracy and precision of this assay for detecting indels, large insertions, and deletions of different length in the same sample.

All the techniques described above can be used for structural characterization and the discovery of large chromosomal rearrangements. Secondary techniques

Table 14.3 Comparison of molecular assays for genome-wide assessment of genomic rearrangements.

Methods	Input	Reduction of background	Sensitivity	Readout
AMP-seq	Total RNA	None	Unknown	Chromosomal rearrangements
LAM-HTGTS	20–100 μg gDNA	Restriction enzyme	Unknown	Off-target editing and chromosomal rearrangements
PEM-seq	20–50 μg gDNA	None	Unknown	Off-target editing and chromosomal rearrangements
UDiTaS	50 ng gDNA	None	0.01–0.1%	Off-target editing and chromosomal rearrangements
CAST-seq	500 ng gDNA	Decoy primer	0.006%	Chromosomal rearrangements

such as qPCR or ddPCR can then be used to quantify the frequency of structural variants once their genomic sequences are known [11].

Table 14.3 provides overview of molecular assays for genome-wide assessment of genomic rearrangements. There is currently no consensus in the genome editing field on what technology should be used to monitor for large rearrangements. In addition, a specific study with benchmark samples is needed to assess and directly compare the currently available technologies (Figure 14.6).

14.3.5 Discussion

On- and off-target editing assessment is an essential part of pharmacology and safety evaluation during preclinical and clinical development of genome editing products. Because the target genomic site is known, the intended on-target editing activity is typically measured by targeted sequencing analysis. The most widely used approach is target enrichment by PCR amplification (or hybrid capture) followed by NGS. For off-target detection, early engagement with the health authorities is recommended to plan a series of studies to assess the safety and toxicity of untended editing activities. These studies include combining orthogonal *in silico* and *in vitro* experimental analyses to perform genome-wide off-target site identification to determine a ranking list of potential off-target sites, performing targeted sequencing to validate the sites in *in vitro* and/or *in vivo* preclinical studies, and quantifying the editing events of validate sites in clinical trial samples to monitor the off-target activities. Depending on the nuclease editing technology platform, the mechanism of action of the drug and the patient population, the appropriate unbiased and targeted methodologies with adequate assay sensitivities can be

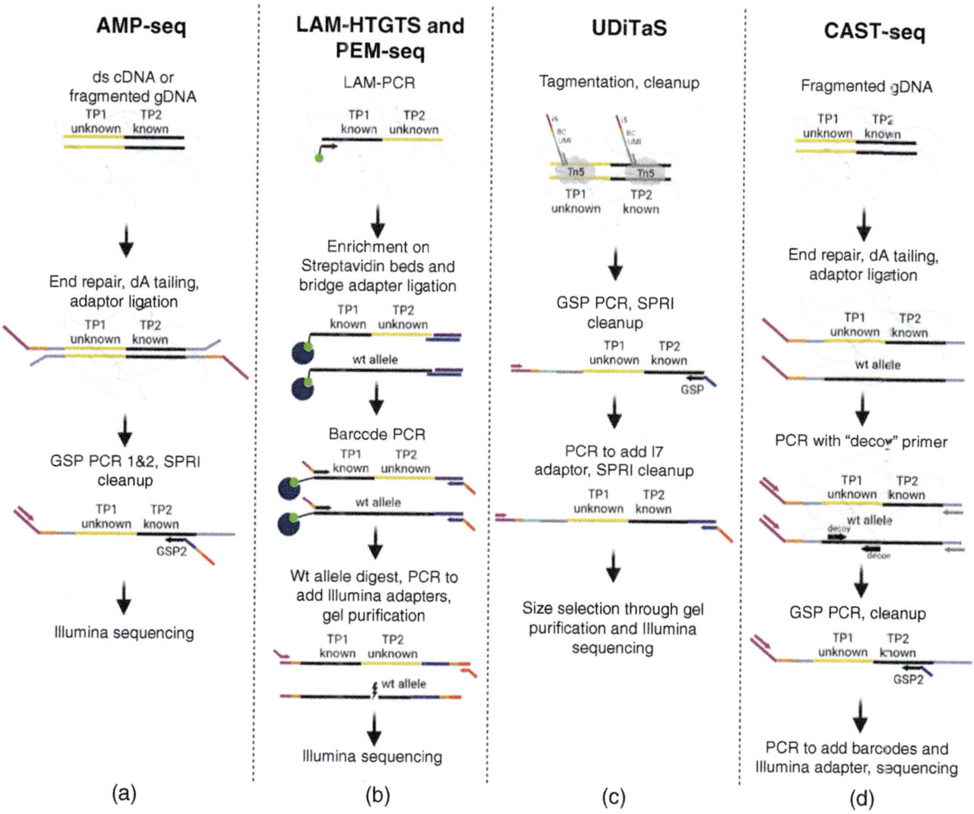

Figure 14.6 Technologies to measure large genomic rearrangements. (a) AMP-seq. Workflow compatible with Illumina sequencing indicated. Two PCR step where universal forward primer and gene-specific primer (GSP) are used shown as one step where GSP2 is indicated. (b) LAM-HTGTS workflow is shown. PEM-seq is different from LAM-HTGTS in number of linear amplification cycles, introduction of 14-base pair random UMIs labeling each target copy, sequencing depth, and bioinformatic analysis. (c) UDiTaS. An example of method usage for translocation detection is shown. (d) CAST-seq. AMP-seq, LAM-HTGTS and CAST-seq start with DNA fragmentation using sonication which is not shown. TP – translocation partner, GSP – gene-specific primer, SPRI – Solid-phase reversible immobilization, LAM-PCR – linear-amplification–mediated PCR, pale gray indicates background gDNA. Source: Created with BioRender.com.

successfully designed to support regulatory filing and clinical development [92, 99]. For example, unbiased methods such as GUIDE-Seq and related methodologies utilizing double-stranded oligonucleotide capture have broad applications, among other methods, to identify potential off-target sites of CRISPR, ZFN, and TALEN genome editing platforms. In addition to off-target profiling, large genomic rearrangements such as chromosomal translocation between on- and off-target sites also need to be evaluated as part of genotoxicity risk assessment. The traditional cytogenetic tests (karyotyping and FISH) can be conducted in compliance with the agency guidelines and are still the gold standard to provide information on structural and numeric chromosome aberrations. Recently, developed NGS-based techniques offer better sensitivity and granularity of DNA sequences at translocation sites. These molecular methods are expected to gain broader use to complement cytogenetic testing.

14.4 Concluding Remarks

Genome editing technologies are evolving rapidly with multiple studies being carried out in the clinic. Recent advances in detecting on- and off-target editing help to characterize the efficacy and safety of genome editing tools. Safety assessment includes orthogonal evaluation of off-target activities encompassing short indels, large deletions and insertions as well as chromosome translocations. We will continue to learn about the mechanisms and factors affecting the occurrence of these off-target effects and how best to detect these molecular events. Recent regulatory guidance from US FDA and the EMA provides scientists and drug developers insight into assessing safety, quality, and potential risks for therapies utilizing engineered nucleases. These engineered nucleases offer great promise for treating acquired and genetic diseases, thus underline the importance of these evolving molecular methods as part of safety and risk assessment for patients.

References

1 Hendel, A., Fine, E.J., Bao, G., and Porteus, M.H. (2015). Quantifying on- and off-target genome editing. *Trends Biotechnol.* 33 (2): 132–140.

2 Saeed, S., Khan, S.U., Khan, W.U. et al. (2023). Genome editing technology: a new frontier for the treatment and prevention of cardiovascular diseases. *Curr. Probl. Cardiol.* 48 (7): 101692.

3 Silva, G., Poirot, L., Galetto, R. et al. (2011). Meganucleases and other tools for targeted genome engineering: perspectives and challenges for gene therapy. *Curr. Gene Ther.* 11 (1): 11–27.

4 Epinat, J.-C., Arnould, S., Chames, P. et al. (2003). A novel engineered meganuclease induces homologous recombination in yeast and mammalian cells. *Nucleic Acids Res. Suppl.* 31 (11): 2952–2962.

5 Urnov, F.D., Miller, J.C., Lee, Y.-L. et al. (2005). Highly efficient endogenous human gene correction using designed zinc-finger nucleases. *Nature* 435 (7042): 646–651.

6 Bogdanove, A.J. and Voytas, D.F. (2011). TAL effectors: customizable proteins for DNA targeting. *Science.* 333 (6051): 1843–1846.

7 Jinek, M., Chylinski, K., Fonfara, I. et al. (2012). A programmable dual-RNA-guided DNA endonuclease in adaptive bacterial immunity. *Science* 337 (6096): 816–821.

8 Sapranauskas, R., Gasiunas, G., Fremaux, C. et al. (2011). The Streptococcus thermophilus CRISPR/Cas system provides immunity in Escherichia coli. *Nucleic Acids Res. Suppl.* 39 (21): 9275–9282.

9 Gaj T, Sirk SJ, Shui S-L, Liu J. Genome-editing technologies: principles and applications. Cold Spring Harb Perspect Biol. 2016;8(12).

10 Laoharawee, K., DeKelver, R.C., Podetz-Pedersen, K.M. et al. (2018). Dose-dependent prevention of metabolic and neurologic disease in murine MPS II by ZFN-mediated in vivo genome editing. *Mol. Ther.* 26 (4): 1127–1136.

11 Turchiano, G., Andrieux, G., Klermund, J. et al. (2021). Quantitative evaluation of chromosomal rearrangements in gene-edited human stem cells by CAST-Seq. *Cell Stem Cell* 28 (6): 1136–1147.

12 Boutin, J., Cappellen, D., Rosier, J. et al. (2022). ON-target adverse events of CRISPR-Cas9 nuclease: more chaotic than expected. *CRISPR J.* 5 (1): 19–30.

13 Chen, J.S., Ma, E., Harrington, L.B. et al. (2018). CRISPR-Cas12a target binding unleashes indiscriminate single-stranded DNase activity. *Science.* 360 (6387): 436–439.

14 Kosicki, M., Tomberg, K., and Bradley, A. (2018). Repair of double-strand breaks induced by CRISPR-Cas9 leads to large deletions and complex rearrangements. *Nat. Biotechnol.* 36 (8): 765–771.

15 Merrihew, R.V., Marburger, K., Pennington, S.L. et al. (1996). High-frequency illegitimate integration of transfected DNA at preintegrated target sites in a mammalian genome. *Mol. Cell. Biol.* 16 (1): 10–18.

16 Gilbert, L.A., Larson, M.H., Morsut, L. et al. (2013). CRISPR-mediated modular RNA-guided regulation of transcription in eukaryotes. *Cell* 154 (2): 442–451.

17 Zeitler, B., Froelich, S., Marlen, K. et al. (2019). Allele-selective transcriptional repression of mutant HTT for the treatment of Huntington's disease. *Nat. Med.* 25 (7): 1131–1142.

18 Kurt, I.C., Zhou, R., Iyer, S. et al. (2020). CRISPR C-to-G base editors for inducing targeted DNA transversions in human cells. *Nat. Biotechnol.* 39 (1): 41–46.

19 Gaudelli, N.M., Komor, A.C., Rees, H.A. et al. (2017). Programmable base editing of A•T to G•C in genomic DNA without DNA cleavage. *Nature* 551 (7681): 464–471.

20 Komor, A.C., Zhao, K.T., Packer, M.S. et al. (2017). Improved base excision repair inhibition and bacteriophage Mu Gam protein yields C:G-to-T:A base editors with higher efficiency and product purity. *Sci. Adv.* 3 (8): eaao4774.

21 Komor, A.C., Kim, Y.B., Packer, M.S. et al. (2016). Programmable editing of a target base in genomic DNA without double-stranded DNA cleavage. *Nature* 533 (7603): 420–424.

22 Anzalone, A.V., Randolph, P.B., Davis, J.R. et al. (2019). Search-and-replace genome editing without double-strand breaks or donor DNA. *Nature* 576 (7785): 149–157.

23 Tebas, P., Stein, D., Tang, W.W. et al. (2014). Gene editing of CCR5 in autologous CD4 T cells of persons infected with HIV. *N. Engl. J. Med. Overseas Ed.* 370 (10): 901–910.

24 Psatha, N., Reik, A., Phelps, S. et al. (2018). Disruption of the BCL11A erythroid enhancer reactivates fetal hemoglobin in erythroid cells of patients with β-thalassemia major. *Mol. Ther. Methods Clin. Dev.* 10: 313–326.

25 Harmatz, P., Prada, C.E., Burton, B.K. et al. (2022). First-in-human in vivo genome editing via AAV-zinc-finger nucleases for mucopolysaccharidosis I/II and hemophilia B. *Mol. Ther.* 30 (12): 3587–3600.

26 Gillmore, J.D., Gane, E., Taubel, J. et al. (2021). CRISPR-Cas9 In vivo gene editing for transthyretin amyloidosis. *N. Engl. J. Med. Overseas Ed.* 385 (6): 493–502.

27 Human Gene Therapy Products Incorporating Human Genome Editing; Draft Guidance for Industry [Internet]. 2023 [cited 2023Apr18]. Available from: https://www.fda.gov/media/156894/download

28 Guideline on quality, non-clinical and clinical aspects of medicinal products containing genetically modified cells [Internet]. 2023 [cited 2023Apr18]. Available from: https://www.ema.europa.eu/en/documents/scientific-guideline/guideline-quality-non-clinical-clinical-aspects-medicinal-products-containing-genetically-modified_en-0.pdf

29 Corsaro, B., Yang, T.-Y., Murphy, R. et al. (2021). 2020 white paper on recent issues in bioanalysis: vaccine assay validation, qPCR assay validation, QC for CAR-T flow cytometry, NAb assay harmonization and ELISpot validation (Part 3 – Recommendations on immunogenicity assay strategies, NAb Assays, biosimilars and FDA/EMA immunogenicity guidance/guideline, gene & cell therapy and vaccine assays). *Bioanalysis* 13 (6): 415–463.

30 Plant, A.L., Camp, C., Elliott, J.T. et al. (2022). The role of the national institute of standards in measurement assurance for cell therapies. *Cell Ther.* 609–625.

31 Bao, X.R., Pan, Y., Lee, C.M. et al. (2021). Tools for experimental and computational analyses of off-target editing by programmable nucleases. *Nat. Protoc.* 16 (1): 10–26.

32 Malinin, N.L., Lee, G., Lazzarotto, C.R. et al. (2021). Defining genome-wide CRISPR-Cas genome-editing nuclease activity with GUIDE-seq. *Nat. Protoc.* 16 (12): 5592–5615.

33 Tsai, S.Q., Zheng, Z., Nguyen, N.T. et al. (2014). GUIDE-seq enables genome-wide profiling of off-target cleavage by CRISPR-Cas nucleases. *Nat. Biotechnol.* 33 (2): 187–197.

34 Tsai, S.Q., Topkar, V.V., Joung, J.K., and Aryee, M.J. (2016). Open-source guideseq software for analysis of GUIDE-seq data. *Nat. Biotechnol.* 34 (5): 483.

35 Wienert, B., Wyman, S.K., Richardson, C.D. et al. (2019). Unbiased detection of CRISPR off-targets in vivo using DISCOVER-Seq. *Science.* 364 (6437): 286–289.

36 Zou, R.S., Liu, Y., Gaido, O.E.R. et al. (2023). Improving the sensitivity of in vivo CRISPR off-target detection with DISCOVER-Seq. *Nat. Methods* .

37 Crosetto, N., Mitra, A., Silva, M.J. et al. (2013). Nucleotide-resolution DNA double-strand break mapping by next-generation sequencing. *Nat. Methods* 10 (4): 361–365.

38 Yan, W.X., Mirzazadeh, R., Garnerone, S. et al. (2017). BLISS is a versatile and quantitative method for genome-wide profiling of DNA double-strand breaks. *Nat. Commun.* 8: 15058.

39 Gabriel, R., Lombardo, A., Arens, A. et al. (2011). An unbiased genome-wide analysis of zinc-finger nuclease specificity. *Nat. Biotechnol.* 29 (9): 816–823.

40 Frock, R.L., Hu, J., Meyers, R.M. et al. (2014). Genome-wide detection of DNA double-stranded breaks induced by engineered nucleases. *Nat. Biotechnol.* 33 (2): 179–186.

41 Maeder, M.L., Stefanidakis, M., Wilson, C.J. et al. (2019). Development of a gene-editing approach to restore vision loss in Leber congenital amaurosis type 10. *Nat. Med.* 25 (2): 229–233.

42 Paschon, D.E., Lussier, S., Wangzor, T. et al. (2019). Diversifying the structure of zinc finger nucleases for high-precision genome editing. *Nat. Commun.* 10 (1): 1133.

43 Cui, Z., Liu, H., Zhang, H. et al. (2021). The comparison of ZFNs, TALENs, and SpCas9 by GUIDE-seq in HPV-targeted gene therapy. *Mol. Ther. Nucleic Acids* 26: 1466–1478.

44 Tsai, S.Q., Nguyen, N.T., Malagon-Lopez, J. et al. (2017). CIRCLE-seq: a highly sensitive in vitro screen for genome-wide CRISPR-Cas9 nuclease off-targets. *Nat. Methods* 14 (6): 607–614.

45 Lazzarotto, C.R., Malinin, N.L., Li, Y. et al. (2020). CHANGE-seq reveals genetic and epigenetic effects on CRISPR-Cas9 genome-wide activity. *Nat. Biotechnol.* 38 (11): 1317–1327.

46 Kim, D., Bae, S., Park, J. et al. (2015). Digenome-seq: genome-wide profiling of CRISPR-Cas9 off-target effects in human cells. *Nat. Methods* 12 (3): 237–243. 1 p following 243.

47 Cameron, P., Fuller, C.K., Donohoue, P.D. et al. (2017). Mapping the genomic landscape of CRISPR-Cas9 cleavage. *Nat. Methods* 14 (6): 600–606.

48 Koike-Yusa, H., Li, Y., Tan, E.-P. et al. (2013). Genome-wide recessive genetic screening in mammalian cells with a lentiviral CRISPR-guide RNA library. *Nat. Biotechnol.* 32 (3): 267–273.

49 Mv, O., Capurso, D., Carter, M.M. et al. (2016). DNA repair profiling reveals nonrandom outcomes at Cas9-mediated breaks. *Mol. Cell* 63 (4): 633–646.

50 Perez, E.E., Wang, J., Miller, J.C. et al. (2008). Establishment of HIV-1 resistance in CD4+ T cells by genome editing using zinc-finger nucleases. *Nat. Biotechnol.* 26 (7): 808–816.

51 Wang, J., Friedman, G., Doyon, Y. et al. (2012). Targeted gene addition to a predetermined site in the human genome using a ZFN-based nicking enzyme. *Genome Res.* 22 (7): 1316–1326.

52 Mamanova, L., Coffey, A.J., Scott, C.E. et al. (2010). Target-enrichment strategies for next-generation sequencing. *Nat. Methods* 7 (2): 111–118.

53 Kozarewa, I., Armisen, J., Gardner, A.F. et al. (2015). Overview of target enrichment strategies. *Curr. Protoc. Mol. Biol.* 112: 7.

54 Bi, C., Wang, L., Yuan, B. et al. (2020). Long-read inspanidual-molecule sequencing reveals CRISPR-induced genetic heterogeneity in human ESCs. *Genome Biol.* 21 (1): 213.

55 Park, S.H., Cao, M., Pan, Y. et al. (2022). Comprehensive analysis and accurate quantification of unintended large gene modifications induced by CRISPR-Cas9 gene editing. *Sci. Adv.* 8 (42): eabo7676.

56 Dobosy, J.R., Rose, S.D., Beltz, K.R. et al. (2011). RNase H-dependent PCR (rhPCR): improved specificity and single nucleotide polymorphism detection using blocked cleavable primers. *BMC Biotech.* 11: 80.

57 Kurgan, G., Turk, R., Li, H. et al. (2021). CRISPAltRations: a validated cloud-based approach for interrogation of double-strand break repair mediated by CRISPR genome editing. *Mol. Ther. Methods Clin. Dev.* 21: 478–491.

58 Chaudhari, H.G., Penterman, J., Whitton, H.J. et al. (2020). Evaluation of homology-independent CRISPR-Cas9 Off-target assessment methods. *CRISPR J.* 3 (6): 440–453.

59 Pfeiffer, F., Gröber, C., Blank, M. et al. (2018). Systematic evaluation of error rates and causes in short samples in next-generation sequencing. *Sci. Rep.* 8 (1): 10950.

60 Fox, E.J., Reid-Bayliss, K.S., Emond, M.J., and Loeb, L.A. (2014). Accuracy of next generation sequencing platforms. *Next Gener. Seq. Appl.* 1.

61 McElhoe, J.A. and Holland, M.M. (2020). Characterization of background noise in MiSeq MPS data when sequencing human mitochondrial DNA from various sample sources and library preparation methods. *Mitochondrion* 52: 40–55.

62 Jo, S., Das, S., Williams, A. et al. (2022). Endowing universal CAR T-cell with immune-evasive properties using TALEN-gene editing. *Nat. Commun.* 13 (1): 3453.

63 Miller, J.C., Patil, D.P., Xia, D.F. et al. (2019). Enhancing gene editing specificity by attenuating DNA cleavage kinetics. *Nat. Biotechnol.* 37 (8): 945–952.

64 Bioanalytical Method Validation – FDA Guidance for Industry [Internet]. 2023 [cited 2023Apr18]. Available from: https://www.fda.gov/media/70858/download.

65 Vogelstein, B. and Kinzler, K.W. (1999). Digital PCR. *PNAS* 96 (16): 9236–9241.

66 Hindson, B.J., Ness, K.D., Masquelier, D.A. et al. (2011). High-throughput droplet digital PCR system for absolute quantitation of DNA copy number. *Anal. Chem.* 83 (22): 8604–8610.

67 Rose, J.C., Stephany, J.J., Valente, W.J. et al. (2017). Rapidly inducible Cas9 and DSB-ddPCR to probe editing kinetics. *Nat. Methods* 14 (9): 891–896.

68 Mock, U., Machowicz, R., Hauber, I. et al. (2015). mRNA transfection of a novel TAL effector nuclease (TALEN) facilitates efficient knockout of HIV co-receptor CCR5. *Nucleic Acids Res. Suppl.* 43 (11): 5560–5571.

69 Miyaoka, Y., Chan, A.H., Judge, L.M. et al. (2014). Isolation of single-base genome-edited human iPS cells without antibiotic selection. *Nat. Methods* 11 (3): 291–293.

70 Mock, U., Hauber, I., and Fehse, B. (2016). Digital PCR to assess gene-editing frequencies (GEF-dPCR) mediated by designer nucleases. *Nat. Protoc.* 11 (3): 598–615.

71 Bennett, E.P., Petersen, B.L., Johansen, I.E. et al. (2020). INDEL detection, the 'Achilles heel' of precise genome editing: a survey of methods for accurate profiling of gene editing induced indels. *Nucleic Acids Res. Suppl.* 48 (21): 11958–11981.

72 Miyaoka, Y., Berman, J.R., Cooper, S.B. et al. (2016). Systematic quantification of HDR and NHEJ reveals effects of locus, nuclease, and cell type on genome-editing. *Sci. Rep.* 6: 23549.

73 Miyaoka, Y., Mayerl, S.J., Chan, A.H., and Conklin, B.R. (2018). Detection and quantification of HDR and NHEJ induced by genome editing at endogenous gene loci using droplet digital PCR. *Methods Mol. Biol.* 1768: 349–362.

74 Till, B.J., Burtner, C., Comai, L., and Henikoff, S. (2004). Mismatch cleavage by single-strand specific nucleases. *Nucleic Acids Res. Suppl.* 32 (8): 2632–2641.

75 Guschin, D.Y., Waite, A.J., Katibah, G.E. et al. (2010). A rapid and general assay for monitoring endogenous gene modification. *Methods Mol. Biol.* 649: 247–256.

76 Mashal, R.D., Koontz, J., and Sklar, J. (1995). Detection of mutations by cleavage of DNA heteroduplexes with bacteriophage resolvases. *Nat. Genet.* 9 (2): 177–183.

77 Miller, J.C., Holmes, M.C., Wang, J. et al. (2007). An improved zinc-finger nuclease architecture for highly specific genome editing. *Nat. Biotechnol.* 25 (7): 778–785.

78 Qiu, P., Shandilya, H., D'Alessio, J.M. et al. (2004). Mutation detection using surveyor nuclease. *BioTechniques* 36 (4): 702–707.

79 Oleykowski, C.A., Mullins, C.R.B., Godwin, A.K., and Yeung, A.T. (1998). Mutation detection using a novel plant endonuclease. *Nucleic Acids Res. Suppl.* 26 (20): 4597–4602.

80 Youil, R., Kemper, B.W., and Cotton, R.G. (1995). Screening for mutations by enzyme mismatch cleavage with T4 endonuclease VII. *PNAS* 92 (1): 87–91.

81 Gao, H., Huang, J., Barany, F., and Cao, W. (2006). Switching base preferences of mismatch cleavage in endonuclease V: an improved method for scanning point mutations. *Nucleic Acids Res. Suppl.* 35 (1): e2.

82 Zhu, X., Xu, Y., Yu, S. et al. (2014). An efficient genotyping method for genome-modified animals and human cells generated with CRISPR/Cas9 system. *Sci. Rep.* 4: 6420.

83 Sentmanat, M.F., Peters, S.T., Florian, C.P. et al. (2018). A survey of validation strategies for CRISPR-Cas9 editing. *Sci. Rep.* 8 (1): 888.

84 Conant, D., Hsiau, T., Rossi, N. et al. (2022). Inference of CRISPR edits from sanger trace data. *CRISPR J.* 5 (1): 123–130.

85 Brinkman, E.K., Chen, T., Amendola, M., and Steensel, B. (2014). Easy quantitative assessment of genome editing by sequence trace decomposition. *Nucleic Acids Res. Suppl.* 42 (22): e168.

86 Brinkman, E.K., Kousholt, A.N., Harmsen, T. et al. (2018). Easy quantification of template-directed CRISPR/Cas9 editing. *Nucleic Acids Res. Suppl.* 46 (10): e58.

87 Schuelke, M. (2000). An economic method for the fluorescent labeling of PCR fragments. *Nat. Biotechnol.* 18 (2): 233–234.

88 Yang, Z., Steentoft, C., Hauge, C. et al. (2015). Fast and sensitive detection of indels induced by precise gene targeting. *Nucleic Acids Res. Suppl.* 43 (9): e59.

89 Lonowski, L.A., Narimatsu, Y., Riaz, A. et al. (2017). Genome editing using FACS enrichment of nuclease-expressing cells and indel detection by amplicon analysis. *Nat. Protoc.* 12 (3): 581–603.

90 Brunet, E., Simsek, D., Tomishima, M. et al. (2009). Chromosomal translocations induced at specified loci in human stem cells. *PNAS* 106 (26): 10620–10625.

91 Leibowitz, M.L., Papathanasiou, S., Doerfler, P.A. et al. (2021). Chromothripsis as an on-target consequence of CRISPR-Cas9 genome editing. *Nat. Genet.* 53 (6): 895–905.

92 Ando, D. and Meyer, K. (2017). Gene editing: regulatory and translation to clinic. *Hematol. Oncol. Clin. North Am.* 31 (5): 797–808.

93 Zheng, Z., Liebers, M., Zhelyazkova, B. et al. (2014). Anchored multiplex PCR for targeted next-generation sequencing. *Nat. Med.* 20 (12): 1479–1484.

94 Giannoukos, G., Ciulla, D.M., Marco, E. et al. (2018). UDiTaS™, a genome editing detection method for indels and genome rearrangements. *BMC Genomics* 19 (1): 212.

95 Yin, J., Liu, M., Liu, Y. et al. (2019). Optimizing genome editing strategy by primer-extension-mediated sequencing. *Cell Discovery* 5: 18.

96 Liu, Y., Yin, J., Gan, T. et al. (2022). PEM-seq comprehensively quantifies DNA repair outcomes during gene-editing and DSB repair. *STAR Protoc.* 3 (1): 101088.

97 Amit, I., Iancu, O., Levy-Jurgenson, A. et al. (2021). CRISPECTOR provides accurate estimation of genome editing translocation and off-target activity from comparative NGS data. *Nat. Commun.* 12 (1): 3042.

98 Karst, S.M., Ziels, R.M., Kirkegaard, R.H. et al. (2021). High-accuracy long-read amplicon sequences using unique molecular identifiers with Nanopore or PacBio sequencing. *Nat. Methods* 18 (2): 165–169.

99 Cavagnaro, J.A. and Cosenza, M.E. (2021). Translational medicine. *Transl. Med.*

Section IV

Companion Diagnostic Development for Gene Therapy

15

Introduction to Companion Diagnostics for Gene Therapy

Paul Bartel[1] *and Jennifer Granger*[2]

[1]*Companion Diagnostics, Myriad Genetics, Inc., Salt Lake City, UT, USA*
[2]*PharmaDx, ARUP Laboratories, Salt Lake City, UT, USA*

15.1 Introduction to Companion Diagnostics

A companion diagnostic (CDx) is a medical device that is required by regulatory agencies to determine patient eligibility for a drug or biological product [1]. Most often, a CDx is used to identify patients who are likely to benefit from treatment with a particular therapeutic. CDx products are in vitro diagnostics (IVD), which the Food and Drug Administration (FDA) defines as the reagents, instruments, and systems intended for use in diagnosis of disease or other conditions. They may be used in a manner to cure, mitigate, treat, or prevent disease. In the United States, CDx are typically categorized as Class III in vitro diagnostics (CDx are Class C in the EU), which is the highest risk category. As such, a CDx will require premarket approval (PMA) prior to full commercial marketing.

The concept of a CDx was introduced through the approval by the US Food and Drug Administration (FDA) of the HER2 assay for the drug trastuzumab in 1998 [2]. However, it was not until later in 2014 when the FDA issued "Guidance for Industry: In Vitro Companion Diagnostic Devices" [3] that the FDA more formally recognized the CDx. This guidance document was intended to assist pharmaceutical companies with an early assessment of the need for a CDx, i.e. during the drug development process, with the intention of CDx and therapeutic co-development offering more rapid access to treatments for patients.

Drug Development for Gene Therapy: Translational Biomarkers, Bioanalysis, and Companion Diagnostics, First Edition. Edited by Yanmei Lu and Boris Gorovits.

Other regulatory agencies have also recognized the need for companion diagnostics including the European Medicines Agency (EMA), which introduced the In Vitro Diagnostic Devices Regulation (IVDR) that included a new classification system for CDx in 2017 [4], and the Pharmaceuticals and Medical Devices Agency (PMDA) of Japan, which published a "Notification on Approval Application for *In Vitro* Companion Diagnostics and Corresponding Therapeutic Products" [5] in 2013 defining a CDx as an in vitro diagnostic reagent or device essential for the safe and effective use of corresponding therapeutic product.

The following sections describe the role of the CDx in gene therapy, strategies for CDx development, and considerations for their commercialization. In addition, Chapters 16 and 17 provide additional details on the validation and regulatory considerations for CDx development, respectively.

15.2 Role in Gene Therapy

CDx have been used to ensure the best chance of success through a personalized medicine approach to choosing the appropriate drug for a given target population [6]. To date, over 50 companion diagnostic devices have been approved by the US FDA, the majority of which are intended for indications within the oncology space [7]. In rare instances, the FDA has also approved cancer therapeutics with so-called complementary diagnostics, which are defined as diagnostics that may predict a more favorable benefit ratio but which are deemed non-essential due to the overall drug benefit (Progression Free Survival [PFS], Overall Survival [OS]) within the indication [6].

The requirement for CDx tests has recently begun to expand beyond oncology as gene therapy (GTx) emerges as a promising new approach in personalized medicine with the potential to cure inherited disorders with a single treatment. GTx functions at the molecular level by introducing therapeutic genes into target cells using a delivery vehicle, or vector. Vectors can be non-viral, such as those found in DNA and mRNA strategies, or viral, which are typically recombinant viral vectors that are unable to replicate in the host cell [8].

One of the most promising classes of viral vectors currently being developed for GTx is the adeno-associated virus (AAV) [9]. However, since AAVs are endemic to the human population, the determination of pre-existing immunity against an AAV GTx is essential for patient safety and drug efficacy. In current GTx clinical trials, early bioanalytical methods are largely being utilized to identify patients with higher levels of preexisting immunity against a specific vector that could neutralize the corresponding drug product and reduce the level of efficacy. Seroprevalence studies (as discussed in Chapter 6: Bioanalytical Methods to Detect Pre-existing and Post-administration Humoral Immune Responses Against

AAV Capsid Proteins) have demonstrated both age-related increases in total and neutralizing antibody titers against various AAV capsids. Extensive preclinical data, as well as emerging clinical experience, illustrate that higher preexisting antibody levels in the blood may be able to neutralize a significant proportion of a given AAV dose. AAV clinical experience to date has established that these products are well tolerated and harbor acceptable safety margins, yet health authorities and industry are engaged in ongoing discussions/negotiations to maintain an acceptable benefit-risk scenario for intended target populations and therefore may require a CDx. Further recommendations regarding the need for a CDx for the development and use of a GTx product are outlined in an FDA guidance [10]. However, within the AAV GTx space, CDx remain in their infancy with only two CDx approved; one in Japan and the other in US and EU.

15.3 Overall Strategy

The preferred, and more expedient, approach is co-development of the CDx and the therapeutic drug product. This was the approach used for HercepTest™, the CDx developed alongside trastuzumab for the treatment of breast cancer. Through early collaborations, the performance of the device can be assessed both analytically and clinically to ensure that eligible patients will receive the appropriate therapy [11]. One of the ways that this can be achieved is through use of the analytically validated version of the assay in clinical trials testing to screen patients for enrollment into the trial, which facilitates early assessment of device safety and efficacy as it pertains to the therapeutic. Additionally, co-development can lead to contemporaneous approval of the CDx so that the diagnostic can be made available for use when the drug therapy is approved.

However, there may be cases where it is not possible to conduct contemporaneous development of both the in vitro diagnostic and the drug due to the novelty of the test analyte, emerging safety or efficacy issues in development of the therapeutic or in the case where an existing device has already been developed by a manufacturer. In these instances, approval of the therapy may be delayed until marketing authorization of the CDx is received.

15.4 Development Process

There are several gating stages that define the development pathway of an in vitro diagnostic that is validated for use in clinical trials for the drug product and ultimately submitted for regulatory approval as a CDx as illustrated in Figure 15.1. Throughout each of the phases of device development, the project team, generally

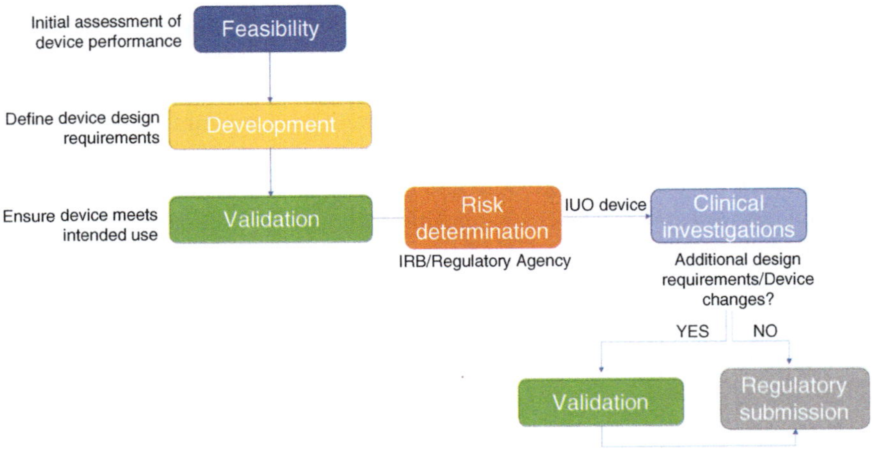

Figure 15.1 Process diagram for development of a companion diagnostic.

comprised of members from the IVD manufacturer and drug developer, should meet to discuss program requirements and results.

It is important to note that IVDs intended for use as a CDx are generally considered to be Class III medical devices, and as such, must be developed under the regulatory requirements of 21 CFR 820 [12] in the United States. In the European Union (EU), CDx are categorized as Class C and must be developed to comply with the IVDR. Compliance with 21 CFR and IVDR requires organizations to maintain an effective Quality Management System (QMS) that addresses applicable regulatory requirements. Conformance to ISO 13485 [13] is often used to help comply with these regulatory requirements. ISO 13485 is a voluntary global standard that is an approach accepted by regulators to assure that a company is meeting QMS expectations for medical devices. Although both 21 CFR 820 and ISO 13485 [13] pertain to the Quality Management System (QMS for medical device manufacturing, there are some differences between these regulations. However, in 2022, the US FDA recently published a proposed rule to harmonize 21 CFR 820 to the ISO 13485 QMS standard [14].

In addition to the aforementioned QMS regulations, another important prerequisite is the accreditation of the laboratory that is developing, validating, and performing the CDx. In the United States, laboratories should hold accreditation through the College of American Pathologists (CAP) or the Clinical Laboratory Improvement Amendments (CLIA), which is regulated through the Centers for Medicare and Medicaid Services (CMS). These organizations ensure that laboratories meet quality standards for medical testing, which is performed under the supervision of a medical director.

During CDx development, it is common to perform feasibility studies designed to assess analytical performance and identify potential concerns at an early stage. Additionally, these studies can guide the development team toward design requirements that are required to meet the needs of both the device and the therapeutic.

After initial feasibility, a more formal design development process can begin. This stage includes detailed planning and documentation of the outputs from the device design requirements. Where possible, studies intended to support device requirements should be designed using published standards such as those from the Clinical and Laboratory Standards Institute (CLSI) or direct guidance received from a regulatory agency. Procedures for the use of the device in the clinical setting and the manufacturing process should also be optimized during this stage of development.

Once the design phase is complete, transfer of the device into production can occur. Production of the device generally follows two configurations: (1) a single-site model, where commercial testing would be performed at one location, or (2) a distributed kit format, where the test is able to be performed at any qualified laboratory. It is at this point that numerous validation activities are performed to ensure that the device meets its intended use and, subsequently, is ready for use in the clinical setting. Validation studies should encompass all inputs from the device design requirements and may include software, analytical specifications, and process validations. Device validation should be rigorous and can take several months to complete depending upon the design requirements. For additional detail on the analytical, as well as clinical, validation of CDx, please refer to Chapter 16.

At the conclusion of validation, the device is designated as "Investigational Use Only" (IUO) and can be considered validated for use in clinical studies of the targeted therapy. However, use of the IUO device in clinical studies is predicated by assessment of risk to the patient. Risk determination for the device is captured through several pathways. As the process followed for the drug, an Institutional Review Board (IRB), academic or commercial, is required to weigh benefit and risk prior to use of the IVD in clinical testing. An additional consideration of risk can be achieved through submission of an application to the regulatory agency requesting their assessment of risk for use of the device in a clinical investigation. In the United States, a Significant Risk Determination (SRD) can be made through a submission to the FDA [15]. In this submission, the device manufacturer will provide the agency with information about the clinical study design as well as the device and its intended use. Should the FDA conclude that a device poses a Significant Risk (SR), an Investigational Device Exemption (IDE) is required prior to initiation of the clinical study. The IDE application contains additional detail and data supporting the development and validation of the device for its intended use and must demonstrate device compliance with IDE regulations (21 CFR 812).

Risk assessment is also recognized for clinical trials using a medical device in the EU, and a Performance Evaluation Application (PEA) must be submitted to competent authorities at the country level prior to use of the device in the clinical study.

During clinical investigations, the IUO device is used to screen patients for enrollment into the study and is the point at which data will be generated for the performance, efficacy, and safety of the device. At this stage, it may become evident that additional device requirements or modifications are needed from either the analytical or manufacturing perspective. All changes are subject to additional verification and validation studies and, if significant, will require notification to regulatory agencies if the device is being used under an IDE or PEA.

Near the conclusion of clinical investigations, the collected data, especially that from pivotal studies, will be subjected to rigorous statistical analysis as a function of several categorical variables (e.g. demographics, prophylactic treatments) and continuous variables (e.g. age, diagnostic test results). This assessment of clinical data, along with a summary of safety and efficacy of the drug, will be provided to regulatory agencies in a marketing application.

After completion of device development, the project team will prepare a marketing application for submission to regulatory agencies for approval of the device as a companion diagnostic. Although not a strict requirement, the preference is for concurrent submission for both the drug and device, therefore project teams from both the device and drug sides often work closely to ensure accurate information is captured in both applications.

Device marketing applications (e.g. PMA, 510(k), and the Humanitarian Device Exemption (HDE) (*US*), Technical File (EU)) are a complete summary of activities performed to ensure that a device meets the requirements of its intended use in a safe and effective manner. These submissions will include a technical section for non-clinical studies, such as those performed during device validations. Clinical investigations are also presented and include study protocols, safety and effectiveness data, and adverse events for both the device and the drug. Additional information about the QMS, software, and device labeling will also be included.

After submission, regulatory agencies will review all materials and provide feedback or make additional requests for information. The duration of this process can vary depending on the complexity of the device and can be tied to drug approval timelines.

15.5 Considerations for Commercialization

Early planning for the post-market setting helps ensure smooth commercial launch of the CDx and drug therapy. Once again, this process requires dedicated collaboration between device and therapy manufacturers and discussion should begin prior to submission of a marketing application.

One of the first, and perhaps most important, considerations for a successful commercialization strategy is logistics. The workflow of specimen collection, transport, testing, and results may seem straightforward, but can also meet unforeseen obstacles. Teams may find it useful to engage healthcare systems prior to launch to request feedback about their processes and determine the need for education and support programs. Furthermore, any marketing collateral used in support of the CDx and therapeutic should be co-developed to ensure consistency in the information provided to healthcare providers and patients.

Depending upon the testing location for the CDx, additional complications may arise with global programs, which may require more nuanced strategies. Teams may encounter country-specific import and export requirements that may require additional support and can add to shipment transit times. Data privacy regulations and reimbursement requirements can also differ depending on the location of the patient.

15.6 Conclusion

The requirement for a CDx may be imposed by regulatory bodies for the marketing approval of a GTx product. The development of a CDx can be complex and typically involves extensive interactions between the diagnostic company, the drug developer, and regulatory bodies to fashion a successful approach. The following chapters go into additional detail about the validation of CDx products (see Chapter 16) and regulatory considerations for CDx (see Chapter 17).

References

1 US Food and Drug Administration, 2018. Companion Diagnostics. https://www.fda.gov/medical-devices/in-vitro-diagnostics/companion-diagnostics (accessed 17 April 2023).
2 HERCEPTIN (trastuzumab) (1998). http://www.accessdata.fda.gov/drugsatfda_docs/label/2010/103792s5250lbl.pdf (accessed 17 April 2023).
3 US Food and Drug Administration (2014). In vitro companion diagnostic devices, guidance for industry and food and drug administration staff.
4 Regulation (EU) (2017). 2017/746 of the European Parliament and of the Council, Official Journal of the European Union, 60.
5 Japan Pharmaceuticals and Medical Devices Agency (2013). Notification on approval application for in vitro companion diagnostics and corresponding therapeutic products.

6 Scheerens, H., Malong, A., Bassett, K. et al. (2017). Current status of companion and complementary diagnostics: strategic considerations for development and launch. *Clin. Transl. Sci.* 10 (2): 84–92.
7 US Food and Drug Administration (2022). List of cleared or approved companion diagnostic devices (in vitro and imaging tools), https://www.fda.gov/medical-devices/in-vitro-diagnostics/list-cleared-or-approved-companion-diagnostic-devices-in-vitro-and-imaging-tools (accessed 17 April 2023).
8 Sayed, N., Allawadhi, P., Khurana, A. et al. (2022). Gene therapy: comprehensive overview and therapeutic applications. *Life Sci.* 294.
9 Wang, D., Tai, P.W.L., and Gao, G. (2019). Adeno-associated virus vector as a platform for gene therapy delivery. *Nat. Rev. Drug Discovery* 18: 358–378.
10 US Food and Drug Administration (2020). Human gene therapy for rare disease.
11 US Food and Drug Administration (2016). Principles for codevelopment of an in vitro companion diagnostic device with a therapeutic product.
12 US Food and Drug Administration (2022). Code of Federal Regulations Title 21, Subchapter H, Medical Devices.
13 International Standards Organization (2016). ISO 13485:2016, Medical devices – quality management systems – requirements for regulatory purposes.
14 US Federal Register, Vol. 87, No. 36, February 23, 2022, Proposed Rules, Medical Devices; Quality System Regulation Amendments.
15 US Food and Drug Administration (2006). Information sheet guidance for IRBs, clinical investigators, and sponsors, significant risk and nonsignificant risk medical device studies.

16

Validation for Gene Therapy Companion Diagnostics

Karen L. Richards and Kennon Daniels

Precision for Medicine, Bethesda Metro Center, Bethesda, MD, USA

16.1 Introduction

Gene therapy has sparked great interest among researchers, healthcare providers, and patients alike because it offers the possibility of new cures, particularly for rare diseases with a genetic basis. However, the field is in its nascent stages, and the ideal methods and solutions for unlocking the full potential of gene therapies (GTx) are still being developed. In recent years, the requirement for development of tests to detect antibodies against the vectors used to deliver the human vector-based GTx product to patients to appropriately select patients for eligibility to receive the GTx and to support market authorization of the GTx has been introduced by Food and Drug Administration (FDA). This includes a requirement to make such tests available as a companion diagnostic (CDx) requiring market authorization at the same time as the GTx approval. In this chapter, we explore general principles for validating a CDx to optimize the likelihood of preclinical and clinical trial development success. We describe regulatory guidelines and explain how CDx sponsors can ensure a scientifically valid diagnostic development plan to support contemporaneous premarket approval (PMA) by FDA.

16.1.1 Overview of FDA Oversight for the Use of Assays in Gene Therapy Clinical Trials and the Path to Commercialization with Corresponding Level of Validation

Considering the hurdles to successful GTx production and increasing efforts to raise production, the US Food and Drug Administration (FDA) has provided

Drug Development for Gene Therapy: Translational Biomarkers, Bioanalysis, and Companion Diagnostics, First Edition. Edited by Yanmei Lu and Boris Gorovits.

Table 16.1 FDA guidance documents for industry that provide recommendations for vector testing, preclinical development, clinical trial design, and FDA approval of gene therapies.

Guidance document	Published date
Final Guidance for Industry: Preclinical Assessment of Investigational Cellular and Gene Therapy Products	2013
Guidance for Industry: Expedited Programs for Regenerative Medicine Therapies for Serious Conditions	2019
Final Guidance for Industry: Human Gene Therapy for Retinal Disorders	2020
Final Guidance for Industry: Human Gene Therapy for Rare Diseases	2020
Final Guidance for Industry: Human Gene Therapy for Hemophilia	2020
Final Guidance for Industry: Long-term Follow-up After Administration of Human Gene Therapy Products	2020
Final Guidance for Industry: Chemistry, Manufacturing, and Control (CMC) Information for Human Gene Therapy Investigational New Drug Applications	2020
Final Guidance for Industry: Testing of Retroviral Vector-Based Human Gene Therapy Products for Replication-Competent Retrovirus During Product Manufacture and Patient Follow-up	2020
Final Guidance for Industry: Interpreting Sameness of Gene Therapy Products Under the Orphan Drug Regulations	2021
Draft Guidance for Industry: Studying Multiple Versions of a Cellular or Gene Therapy in an Early-Phase Clinical Trial	2021

Source: Adapted from FDA [1].

guidance for industry regarding cellular and GTx products. Table 16.1 highlights some of the guidance documents (drafts and final guidances) related to the development and application of GTx published by the FDA.

In several of these guidance documents, the FDA refers to the development of a CDx to detect antibodies against the human vector-based GTx product to appropriately select patients for its clinical trial and to support market authorization. However, there are currently no standards for manufacturers developing a CDx product to select eligible patients for GTx trials. Therefore, CDx manufacturers must design, develop, and validate these CDx assays to meet other published standards (discussed in detail below). Another important resource in CDx development is the Q-Submission program [2], which provides a mechanism for interactive feedback with the FDA on the important analytical verification and clinical validation studies required for CDx approval through the premarket approval

(PMA) process. By working closely with the FDA during a CDx's development and validation phases, the likelihood of a contemporaneous approval of that CDx and its corresponding GTx increases – a critical factor for the success of patients who will benefit from the GTx.

16.1.2 Summary of Validation Requirements for Gene Therapy Companion Diagnostics (GTx CDx)

Like any other in vitro diagnostic (IVD) regulated by the FDA, GTx CDx must follow analytical and clinical performance validation requirements. As mentioned previously, the CDx claim for an assay requires PMA for commercialization since assays with a CDx claim are of the highest risk type. Hence, these assays must follow FDA requirements per Clinical and Laboratory Standards Institute (CLSI) Guidelines. However, because the studies required to validate a CDx under CLSI guidelines are very thorough and time-consuming, sponsors often choose to first validate their product as Clinical Trial Assay (CTA) and meet FDA requirements for submission as an Investigational Device Exemption (IDE), which requires less rigorous validation than a PMA submission for a CDx. The laboratory conducting these validations will need to assess performance of the CTA in accordance with the Clinical Laboratory Improvement Amendments (CLIA) program prior to CLSI validation studies.

IVD manufacturers typically take a two-phase approach to validate the CDx: First, they conduct an initial analytical validity performance assessment of a CTA to ensure accurate and reliable test results where the initial analytical sensitivity results can be used to collect safety and efficacy data. This is followed by clinical data evaluation to select the clinical cutoff and to validate CLSI standards for final IVD CDx configuration and use in the pivotal trial to allow patient selection or stratification.

16.1.3 Role of CDx in Therapeutic Development and Unique Challenges to Validating GTx CDx

Eligibility criteria for patients undergoing GTx can be evaluated based on expected risks and potential benefits determined from preclinical studies. As a result, inclusion of patients with varying severities of disease should be considered carefully. Healthy volunteers should be excluded from most GTx trials. Early-phase GTx trials may sometimes only enroll patients who do not have any other acceptable treatment options. Additionally, patients who may have characteristics that influence the safety or efficacy of the therapy may also be excluded from trials, as these can affect results [3].

When used with GTx, a CDx can help inform treatment decisions. Thus, identifying the appropriate CDx has been proposed in multiple guidelines relevant to gene therapy [4–6]. CDxs are often IVD devices that provide information essential for safe and effective use of a corresponding drug or biologic.

A CDx can help identify patients who are likely to benefit from therapy or those likely to experience treatment-related adverse events. These tools may facilitate the monitoring of treatment response, enabling healthcare providers to adjust therapy and achieve improved safety or effectiveness [7]. A few examples of approved CDxs include polymerase chain reaction kits to detect mutations in patients and immunohistochemistry or enzyme-linked immunosorbent assays to detect protein expression related to disease or treatment. Current regulatory guidance recommends the development and use of CDx assays to assess not only GTx safety but efficacy as well. For these uses, CDxs can be split into two categories:

- **Tests used to confirm genetic disorders:** For diseases caused by a genetic defect, genetic testing should be performed. In the absence of a reliable, readily available means of obtaining the necessary genetic diagnosis, a CDx may be needed and should be considered early in development of the GTx [4].
- **Tests to evaluate preexisting antibodies:** To ensure the therapeutic potential of a GTx product, sponsors should consider developing CDxs to detect total antibodies (TAbs) and/or neutralizing antibodies (NAbs) in patient serum or plasma. If CDxs are needed to appropriately select patients for clinical trials and, ultimately, for treatment, then submission of the marketing application for the CDx and the biologics license application for the GTx should be coordinated to support contemporaneous marketing authorizations [4].

16.1.4 Key Considerations for Developing GTx CDx

Ideally, CDx development should occur in parallel with drug development [8]. As with development of a CDx for any other type of drug, development of GTx CDx should begin with a clear definition of the assay's use, what it measures, and the risks and benefits associated with it. In addition, it is important to define which patient population(s) would benefit from use of the assay in conjunction with therapy [8]. The investigational device exemption (IDE) for CDxs used in clinical studies is based on level of risk in that the IDE regulation distinguishes between nonsignificant and significant device risks [9]:

- IDE Exempt: CDx has no direct effect on treatment.
- Nonsignificant-risk (NSR) Abbreviated IDE: A wrong result from the CDx does not constitute a safety risk.
- Significant-risk (SR) IDE: A wrong result from the CDx constitutes a safety risk.

Sponsors are advised to carefully consider the risks associated with using a CDx in any clinical study and the potential impact of CDx results on decisions to treat patients with investigational GTx products. It is particularly important to enter phase 2 clinical trials with a proposed clinical cutoff based on results from prior analytical and clinical studies. Then, based on how the proposed clinical cutoff performs in phase 2 studies, the cutoff may then be validated in phase 3 for registrational studies. If the cutoff used in the phase 2 study is set too conservatively, it may be difficult to demonstrate nonsignificant risk and defend use of a higher cutoff in the phase 3 studies to support registration. While not impossible to demonstrate, a change in the cutoff could require supportive evidence to minimize the risk to subjects, such as a small trial prior to the larger registrational trial. Therefore, the data to support an appropriate clinical cutoff in phase 2 should be carefully considered.

16.2 Development of CTAs for Use in GTx Clinical Trials

There are several factors to consider for a CTA (i.e. a potential CDx) that will be used in a GTx clinical trial. The CTA's development timing and corresponding quality system requirements must be considered with the long-term view for whether the CTA will become a CDx for market entry and commercialization. Further, how the CTA is used to make decisions for subjects enrolled in the GTx clinical trial as well as the GTx trial phase are important considerations for the CTA's validation strategy.

For example, the GTx being used in first-in-human trials where an analytical cutoff for the CTA is based on neutralizing antibody (NAb) titers at the limit of detection, but a clinical cutoff for commercial application is not yet known? In this example, assay validation requirements may be limited to analytical sensitivity and precision based on how the CTA may initially be used in the GTx clinical trial.

Another example could be when GTx are being used in a rare diseases but pivotal clinical trials, where the clinical cutoff may be determined through previous studies, and the CTA is being used to enroll subjects. In this example, analytical validation of the assay would need to be more robust, and other factors described below must be evaluated.

16.2.1 Stratification vs. Selection

An important consideration for any clinical trial where a CTA is to be used, and for which a CDx may eventually be made available to the market, is whether the assay is used for stratification vs. selection. Stratification means that the CTA is used to assign subjects to different arms of the GTx clinical trial, but results from the CTA itself are not used to determine the treatment that the subject receives [8].

In contrast, selection means that results from the CTA itself are used to determine the treatment of the subject receives during the GTx clinical trial and is often part of the inclusion/exclusion criteria for subject eligibility.

Understanding how the CTA/CDx will be used in the GTx trial (stratification vs. selection) can influence not only the eventual regulatory pathway for market entry of the CDx, but also the risk determination (and therefore assay validation requirements) for use of the CTA in the GTx trial.

16.2.2 Regulatory Risk Determination: Significant or Nonsignificant?

Determining the risk associated with CTA use in a GTx clinical trial is a critical step for preparing Institutional Review Boards (IRBs) and potentially the FDA to review a clinical trial protocol with an associated CTA. In the section above, one critical assessment is whether the CTA will be used to either stratify or select subjects in a clinical trial to receive the investigational GTx.

To understand and make a risk determination, it is important to first understand how the FDA views risks associated with medical devices (including IVDs as CTAs) [10]. The FDA classifies medical device studies into three types: significant risk (SR), nonsignificant risk (NSR), and exempt, all in accordance with the 21 CFR 812 regulation.

A significant risk device under 21 CFR 812.3(m) means an investigational device or CTA that:

1) Is intended as an implant and presents a potential for serious risk to the health, safety, or welfare of a subject;
2) Is purported or represented to be for use supporting or sustaining human life and presents a potential for serious risk to the health, safety, or welfare of a subject;
3) Is for a use of substantial importance in diagnosing, curing, mitigating, or treating disease, or otherwise preventing impairment of human health and presents a potential for serious risk to the health, safety, or welfare of a subject; or
4) Otherwise presents a potential for serious risk to the health, safety, or welfare of a subject.

Definitions 3 and 4 are the most relevant to CTAs used in investigational GTx trials. The concern about CTA use is that incorrect test results can present a significant risk when they lead to mismanagement of a subject's care. In the context of GTx, an incorrect result could mean either that an eligible subject would not receive the investigational GTx or that an otherwise ineligible subject receives the investigational GTx. Incorrect results could also lead to incorrect selection/non-selection of subjects into the GTx clinical trial, which might impact the data supporting the safety and efficacy of the CTA and GTx. In most instances, the risks associated with the GTx will outweigh the risks of incorrect CTA results. Still, it is important that this risk is minimized through proper validation that is documented in a risk/benefit determination document.

When the CTA is intended to stratify subjects into different arms of a GTx study, results from the CTA itself are not being used to determine treatment of the subject. Therefore, such studies are normally determined to be NSR, and while IRB approval is still required, FDA approval of a CTA as an NSR would likely not be required. In contrast, when the CTA is intended for used to select subjects eligible to receive the investigational GTx, inaccurate results could put subjects at potential risk. Therefore, the CTA is generally considered an SR device subject to both IRB and FDA approval for use in the trial through submission of an IDE application. Supportive evidence for an IDE application will include sufficient data demonstrating that the CTA provides accurate results and answers to the following questions [11], included in the application:

1) Will use of the results from an investigational IVD (e.g. a CTA) lead to some study subjects foregoing or delaying treatment that is known to be effective?
2) Will use of the results from an investigational IVD expose study subjects to safety risks (e.g. adverse events from the investigational therapeutic product) that exceed the risks encountered with the control arm therapy or standard of care?
3) Is it likely, based on existing knowledge about the relationship between the biomarker and the investigational therapeutic product, that incorrect results from the investigational IVD would present a potential for serious risk to study subjects?
4) Does use of the investigational IVD require invasive sampling that is not part of the standard of care?

A nonsignificant risk device study under the IDE regulation 21 CFR 812 is one that does not meet the above definition of an SR device study and is subject to the abbreviated IDE requirements under §812.2 (b). These include labeling per the IDE regulation ((§812.5) and must bear the statement "CAUTION – Investigational Device. Limited by Federal (or United States) law to investigational use."; IRB Approval; Informed Consent per 21 CFR 50; monitoring to protect the human subjects and assure compliance with approved protocols; maintaining specific records and reports as required by the IDE regulations (§812.2(b)(1)(v)) and prohibitions related to the promotion, marketing, and misrepresentation of the investigational device (§812.7).

An exempt study may be a study evaluating consumer preference testing or use of an already FDA-cleared device in a study as long as there is no collection of safety or effectiveness data [12].

When it may be unclear whether the CTA is SR or NSR, study sponsors may request a study risk determination under the FDA's Q-Submission Program [2]. Study sponsors can then submit supportive documentation to the FDA to determine whether use of their CTA in the GTx trial results in an SR or NSR study. Under the current guidance document, the FDA has 90 days to respond to the sponsor with a determination. One benefit of this approach is that if the study sponsor receives an NSR designation from the FDA, this documentation can be used when submitting study protocols to IRBs for review and approval. This can prevent delays from IRBs if they see the FDA has deemed the study NSR.

16.2.3 CTA Design Considerations

When considering the design of a CTA for use in an investigational GTx clinical trial, there are factors to consider beyond the science to support use of the CTA. For example, if the CTA will become a CDx, the design of the CTA should be modeled in accordance with 21 CFR 820, and elements of design control requirements should be in place. Depending on where the CTA is in the development cycle, documentation must support the requirements set forth by the CTA and include the various development, verification, and validation requirements for the CTA to be submitted to FDA as a CDx.

Further, if the CTA will become a CDx, the design of the CTA to be used in the investigational GTx clinical trial should be under a "design freeze." This generally means the following:

- The critical reagents and suppliers have been identified;
- The test method and data analysis specifications have been developed and documented; and
- The key equipment used to execute the test method has been identified, installed, and qualified for use during the clinical trial regardless of phase.

For any studies in which the investigational GTx and CTA have been or will be used, it is also critical that subjects are consented for use of leftover samples in various ongoing research and that results from testing subject samples using the CTA can be used in a regulatory submission. This added consent can often be overlooked by sponsors. The availability of leftover samples from the clinical trial allows for any bridging studies to be conducted between the CTA and final CDx, should changes to the design be necessary. However, at the time that the CTA is being used in a registrational GTx clinical trial, sponsors should strive to be using a CTA that will not undergo a design change prior to submission as a CDx.

16.2.4 CTA Validation Requirements

Currently, there are no established rules or guidance on how much testing is required for a CTA to be used in an investigational GTx clinical trial. To use an assay as a CTA, the IVD manufacturer will need to validate the assay per the CLIA program regulations (42 CFR 493.1253). Data derived from the CLIA validation should be used to support the IDE application, including evaluating accuracy, precision, analytical sensitivity and specificity, reportable range, reference interval, and any additional performance metric to ensure accurate test performance. Even with these data, CTA validation must still meet rigorous standards for safe and effective use of the CTA to stratify or selection patients into the investigational GTx clinical trial; however, the FDA may require fewer studies with smaller sample sizes and replicates. A key consideration in determining the level of validation required for the CTA will depend on how the CTA is used in the investigational GTx clinical trial and the risks associated with erroneous results as assessed above in Section 16.2.2. Additionally, the same Q-submission process described in Section 16.2.2 can be leveraged for submission of a "pre-IDE" meeting with the FDA to outline how the CTA will be validated for use in the investigational GTx clinical trial.

16.3 Best Practices for Sample Banking and Consent of Subjects

From a strategy perspective, it is strongly recommended to bank clinical samples from all subjects tested not only in the pivotal trial but also across all phases of the GTx clinical trials, including patients excluded in the trials. The samples are likely from a rare patient population and not easily accessible from standard biobanks. However, in order to do so, GTx sponsors must include the proper informed consent that will allow testing of the patient samples for enrollment in the current trial as well as downstream testing for analytical validation of the final CDx or for a bridging study where testing of all pivotal study samples (positive and negative or above and below the clinical cutoff) will need to be conducted with the final CDx.

16.3.1 Validation Strategies for CDxs for Commercial Use

As described above for the CTA, there are several factors to consider when validating a CDx for commercial use. The CDx's development timing and the impact of any design changes between the CTA and CDx must be assessed. For example, if there are differences between the CTA and CDx that raise the level of a design change, all prior analytical and clinical data generated on the CTA must be assessed for impact on the CDx. This impact assessment includes review of the data generated with the CTA against the design changes to the CDx to determine whether additional studies are required.

Other factors to consider are whether the CDx and CTA are manufactured by the same company vs. undergoing a design transfer and manufacturing scale-up that would require additional validation studies to be performed.

16.4 Design Considerations

The design of a CDx should be considered early on in its development program. As discussed above, limiting the differences between the CTA and CDx will make it easier for sponsors to validate the CDx for commercial use. As stated in Section 16.2.3 for the CTA, the CDx must also conform to the requirements stated in 21 CFR 820, and elements of design control requirements must be in place. If the designs of the CDx and CTA are the same, then the design control documentation used during the development of the CTA may be leveraged for the CDx. However, if the CDx is significantly different from the CTA, the design control documentation to support the CDx must comprehensively address the design of the CDx from feasibility through validation. In this case, the CTA may provide supportive information for documenting feasibility of the CDx.

16.4.1 Single-site vs. Distributable Kit

The typical product configuration for a CDx is an IVD kit that can be sold to and run in any CLIA laboratory in the United States and potentially abroad. However, GTx CDx that evaluate NAbs for each patient requires working cell banks to run the assays. This component of CDx does not align well with a distributable kit model due to potential variability from cell bank to cell bank. Therefore, we recommend a single-site testing model for commercialization in which all components and quality system requirements for performing these assays are tested at a single laboratory in a controlled environment. Therefore, the sponsor would submit a PMA for the assay run out of a single site for use as the GTx CDx.

16.4.2 Validation Requirements

To submit a CDx for approval by the FDA, the CDx must be validated in accordance with specific standards and best practices known to be acceptable by the

FDA in the absence of a published standard. The CLSI publishes specific standards for how diagnostic tests should be validated and which typically specify the number of samples, lots, replicates, etc. to be tested and methods for analysis of test results. It is the obligation of the sponsor to set acceptance criteria for validation studies and document the requirements of the validation as part of the analytical validation study protocol and report.

Table 16.2 below provides the general list of required studies that must be performed to support validation of a CDx for submission to the FDA. Depending on whether the CDx is qualitative, semi-quantitative, or quantitative, some studies may

Table 16.2 Typical studies required for validation and FDA submission of a CDx.

Study	CLSI guideline	CLSI standard name
Reference Range Study	EP28-Ac3	Defining, Establishing, and Verifying Reference Intervals in the Clinical Laboratory; Approved Guideline—Third Edition.
Analytical Sensitivity (LoB/D/Q)	EP17-A2	Evaluation of Detection Capability for Clinical Laboratory Measurement Procedures; Approved Guideline—Second Edition.
Accuracy	EP12-A2	User Protocol for Evaluation of Qualitative Test Performance.
Precision	EP05-A3 EP12.A2	Evaluation of Precision of Quantitative Measurement Procedures; Approved Guideline—Third Edition. User Protocol for Evaluation of Qualitative Test Performance.
Endogenous & Exogenous Interferences	EP07-A3 EP37Ed1E	Interference Testing in Clinical Chemistry, 3rd Edition. Supplemental Tables for Interference Testing in Clinical Chemistry, 1st Edition.
Cross Reactivity	EP07-A3 EP37Ed1E	Interference Testing in Clinical Chemistry, 3rd Edition. Supplemental Tables for Interference Testing in Clinical Chemistry, 1st Edition
Carryover/ Cross Contamination	N/A	Evaluation of carry-over and cross-contamination between wells and its effect on assay results. Specifically, the study should evaluate the potential for carry-over between wells with high levels of nAbs to wells with low nAbs by alternating high- and low-level samples in a checkerboard pattern or columns in a plate.

(Continued)

Table 16.2 (Continued)

Study	CLSI guideline	CLSI standard name
Sample Stability	N/A	Evaluation of all possible sample storage conditions, including long-term frozen stability, short-term stability (room temperature and/or refrigerated), freeze/thaw cycles, and transport stability.
Reagent Shelf-Life Stability	EP25-A	Evaluation of Stability of In Vitro Diagnostic Reagents.
Reagent Open Vial Stability	EP25-A	Evaluation of Stability of In Vitro Diagnostic Reagents.
Tube Type Equivalency Study	EP14-A2	Evaluation of Matrix Effects.
Linearity	EP06Ed2	Evaluation of Linearity of Quantitative Measurement Procedures.
High-Dose Hook Effect	N/A	Determine if the assay is susceptible to a high-dose hook effect and at what concentrations of AAV inhibitory antibodies the high-dose hook effect occurs.

or may not apply. But as a general rule, for a CDx with a specific clinical cutoff for which results may be expressed as "positive/negative" or $>/\leq$ a given level, the FDA expects that samples used in validation studies are within $\pm$ 20% of the clinical cutoff. The cutoff must be determined prior to conducting these studies, which is informed by the GTx partner based on GTx clinical trials prior to the pivotal study. And for a CDx for which there is no specific clinical cutoff but for which results will be reported in a quantitative manner, the FDA expects that samples used in validation studies include multiple samples across the measuring range of the CDx.

A strategy for CDx validation considers that there are no anticipated changes between the CTA and CDx. In this case, during the CTA validation phase, sponsors may want to conduct some of the studies listed below to leverage results from the CTA validation toward submission of the CDx, saving time and effort at the end of the development cycle.

16.5 Bridging Studies

A bridging study is only required when the final, locked CDx was not utilized for patient inclusion/exclusion in the pivotal clinical study with the GTx. A bridging study evaluates efficacy of the therapeutic product in subjects selected by the CDx by assessing both concordance and discordance between the final CDx and the CTA used in the pivotal clinical study while using the same specimens from subjects who were tested for the investigational GTx in the pivotal clinical study. Statistical analysis of performance for the final CDx needs to consider any potential impact of missing samples not available for the bridging study. The ability of the CDx to predict the efficacy of the therapeutic product can be supported indirectly by high analytical concordance with the CTA on a large number of representative samples, including samples from subjects excluded from the investigational GTx clinical trial because they were below or above the threshold established by the CTA. Thus, the FDA's assessment of the clinical validity of the CDx relies on extrapolating the clinical performance characteristics of the CTA to the clinical performance characteristics of the CDx [8].

Bridging studies can prove challenging if the sponsor has not retained samples from the investigational GTx clinical trial, including samples from patients who were in and excluded from the trial, or if retained samples have not been demonstrated to have long-term sample stability under the specified storage conditions. This is a critical point in that lack of sample stability data will preclude use of those samples in the downstream bridging study, which is important for GTx for rare diseases where there may be a limited number of patients who even qualify for clinical study enrollment.

As discussed above, the ideal scenario for the sponsor is when the CTA used in the investigational GTx registrational or pivotal clinical trial is the same as the eventual CDx to be submitted to the FDA and for commercial launch. However, it is not always possible for the CTA to be the same as the CDx. In this event, the performance criteria in the bridging study will need to consider any differences between the CTA and CDx without undermining the ability of the study to demonstrate clinical validity of the CDx and show that results obtained with the CTA are equivalent to results obtained with the CDx.

The ideal bridging study is one in which all samples tested with the CTA are retested with the CDx, valid test results are obtained, and those results are used to assess comparative performance conducted under an approved protocol with prespecified statistical analyses. If only a subset of samples is available for retesting, the sponsor should ensure that the characteristics of the subset adequately reflect the characteristics that affect test performance and that the characteristics of the subjects that may affect therapeutic product efficacy (e.g. patient demographics stage of disease, stratification factors) are proportionally preserved in the retest sample set when compared to the samples in the original set. For the bridging study, the CDx must test both positive and negative samples or samples from patients that were included and excluded from the pivotal trial for the bridging study.

16.6 Commensurate Regulatory Review and Approval of GTx CDx

It is the goal of the GTx and CDx sponsors, along with the regulatory authorities, to have commensurate approval of the CDx PMA with the GTx Biologics License Application (BLA). This requires close collaboration between the GTx and CDx sponsors early in the development of the GTx to ensure that an appropriate CTA to CDx product development plan and strategy can be developed and executed, including the regulatory submissions to the FDA for both the GTx and CDx. For the CDx, the pre-submission process will be critical to gain valuable guidance and input from the FDA prior to conducing the assay validation studies per CLSI guidelines and preparing the FDA for the upcoming modular PMA submission. Timing and coordination across both FDAs Centers for Biologics Evaluation & Research (CBER) and Centers for Device & Radiological Health (CDRH) is critical to ensure contemporaneous approval.

However, it is within the FDA's discretion to determine whether the GTx and CDx must be approved at the same time. Because GTx sponsors may come late to the CDx sponsor, the CDx sponsor may be unable to complete the required design control documentation and analytical validation per the CLSI guidelines to support parallel submissions to FDA. Therefore, frequent communications with the FDA on study progress and submission timing are essential for helping ensure potential GTx approval with the promise of a CDx.

16.7 Concluding Sections

16.7.1 Summary of Validation Considerations for CTAs/CDx in GTx Clinical Trials

The early steps of development of a CTA for a GTx are critical in ensuring long-term success in commercialization through contemporaneous approval with the GTx or as a post-market commitment for the GTx. The GTx and CDx sponsors need to determine the type of assay as well as how the assay will be used in the final patient journey for treatment in clinical trials. This will inform the risk determination and positioning with the FDA's IDE requirements. The assay should be analytically validated to CLIA as early as possible to ensure the assay is accurate and reproducible. Further, the quality system should be put into place prior to CLIA validation and continued throughout the product development as this is critical for design control. All patients from first-in-human through pivotal studies should be consented such that the samples are consented to and available for testing prospectively and also in downstream analytical validation studies or even a bridging study if the final CDx is not used for testing in the pivotal study. The stability of the banked samples needs to be shown over time and therefore, short-term and long-term sample stability studies should be started during the CLIA validation studies. This will support the sample handling during the trials and storage of the samples over time.

16.7.2 Summary of Validation Considerations for CTAs/CDx to Enable GTx Marketing

Following CLIA validation of the CTA, the next step is to conduct a performance validation per CLSI guidelines of the final version of the CDx. This is a comprehensive set of studies meant to evaluate all components of assay performance. Some of these studies will require clinical samples, and contrived sample panels cannot be used, which again speaks to the importance of banking specimens. It will also be critical to know the clinical cutoff at which to validate the assay's performance.

The goal of proper planning in assay development should be that the final version of the CDx is locked prior to the pivotal GTx study and that all CLSI validation is conducted prior to using the CDx in that study. Therefore, a bridging study for inclusion in the PMA would not be required. Further, the CDx sponsor's design control documentation should address the design of the CDx from feasibility through validation. Timing for reviews and requirements should be coordinated across both CDRH and CBER to ensure contemporaneous approval of both submissions (PMA and BLA).

References

1 U.S. Food and Drug Administration (FDA) (2021). Cellular and gene therapy guidances; Dec [cited 2022 Aug 24]. https://www.fda.gov/vaccines-blood-biologics/biologics-guidances/cellular-gene-therapy-guidances (accessed 28 September 2023).

2 U.S. Food and Drug Administration (FDA) (2021). Requests for feedback and meetings for medical device submissions: the Q Submission Program; guidance for industry and Food and Drug Administration staff; Jan [cited 2022 Aug 24]. https://www.fda.gov/media/114034/download (accessed 28 September 2023).

3 U.S. Food and Drug Administration (2015). Considerations for the design of early-phase clinical trials of cellular and gene therapy products; guidance for industry. U.S. Food and Drug Administration. Center for Biologics Evaluation

and Research; June [cited 2019 Mar 11]. https://www.fda.gov/media/106369/download (accessed 28 September 2023).

4 U.S. Food and Drug Administration (2020). Human gene therapy for rare diseases; guidance for industry. Center for Biologics Evaluation and Research; Jan [cited 2019 Mar 11]. https://www.fda.gov/media/113807/download (accessed 28 September 2023).

5 U.S. Food and Drug Administration (2020). Human gene therapy for hemophilia; guidance for industry. Center for Biologics Evaluation and Research; Jan [cited 2019 Mar 11]. https://www.fda.gov/media/113799/download (accessed 28 September 2023).

6 U.S. Food and Drug Administration (2018). Human gene therapy for retinal disorders; draft guidance for industry. Center for Biologics Evaluation and Research; Jul [cited 2019 Mar 11]. https://www.fda.gov/media/113814/download (accessed 28 September 2023).

7 U.S. Food and Drug Administration (FDA) (2018). In vitro companion diagnostic devices. https://www.fda.gov/regulatory-information/search-fda-guidance-documents/in-vitro-companion-diagnostic-devices (accessed 27 September 2018).

8 U.S. Food and Drug Administration 2016 Principles for codevelopment of an in vitro companion diagnostic device with a therapeutic product; draft guidance for industry and Food and Drug Administration staff. Center for Devices and Radiological Health; Jul [cited 2022 Aug 24]. https://www.fda.gov/media/99030/download (accessed 28 September 2023).

9 U.S. Food and Drug Administration (2020). IDE approval process; Nov [cited 2022 Aug 24]. https://www.fda.gov/medical-devices/investigational-device-exemption-ide/ide-approval-process (accessed 28 September 2023).

10 U.S. Food and Drug Administration (2006). Significant risk and nonsignificant risk medical device studies. https://www.fda.gov/regulatory-information/search-fda-guidance-documents/significant-risk-and-nonsignificant-risk-medical-device-studies (accessed 06 September 2018).

11 U.S. Food and Drug Administration (2017). Investigational IVDs in therapeutic product clinical investigations. Center for Devices and Radiological Health; Dec [cited 2022 Aug 24]. https://www.fda.gov/media/109464/download (accessed 28 September 2023).

12 U.S. Food and Drug Administration (2006). Information sheet guidance for sponsors, clinical investigators, and IRBs frequently asked questions statement of investigator. https://www.fda.gov/regulatory-information/search-fda-guidance-documents/information-sheet-guidance-sponsors-clinical-investigators-and-irbs-frequently-asked-questions (accessed 01 Feburary 2022).

17

Regulatory Considerations for Gene Therapy Companion Diagnostics

Mica Elizalde[1] *and Paul Bartel*[2]

[1] *Regulatory Digital Health, Merck Sharp & Dohme LLC, Rahway, NJ, USA*
[2] *Myriad Genetics, Companion Diagnostics, Salt Lake City, UT, USA*

17.1 Introduction

Because of the critical role that companion diagnostics (CDx) play in the selection of a patient's therapeutic treatment, the development, approval, and maintenance of these diagnostics is highly regulated in a number of markets. In this chapter, the requirements for regulatory approval in the United States and (European Union) EU will be covered in some detail, while other regulated markets (e.g. United Kingdom, Japan, China) are noted but not covered in detail. The agencies responsible for regulating CDx in several key markets are presented in Table 17.1.

This chapter also includes discussion of topics such as global regulatory strategies, alternative CDx development strategies, commercialization, and CDx for rare diseases. For overall review of the role of CDx in gene therapy and strategies for CDx development and considerations for commercialization, see Chapter 15: Introduction to Companion Diagnostics for Gene Therapy.

17.2 US FDA

In the United States, the Center for Devices and Radiological Health (CDRH), a branch of the Food and Drug Administration (FDA), is responsible for marketing

Drug Development for Gene Therapy: Translational Biomarkers, Bioanalysis, and Companion Diagnostics, First Edition. Edited by Yanmei Lu and Boris Gorovits.

Table 17.1 Regulation of companion diagnostics in key markets.

Market	Regulatory agency
United States of America (US)	Food and Drug Administration (FDA)
European Union (EU)	Individual Member State National Competent Authority
Canada	Health Canada (HC)
United Kingdom (UK)	Medicines and Healthcare products Regulatory Agency (MHRA)
Japan	Pharmaceuticals and Medical Device Agency (PMDA)
People's Republic of China (China)	National Medical Products Administration (NMPA)

authorization of CDx, as well as administering oversight of manufacturing, performance and safety of such devices. The FDA has issued guidance documents including "Guidance for Industry: In Vitro Companion Diagnostic Devices" and "Principles for Codevelopment of an In Vitro Companion Diagnostic Device with a Therapeutic Product" to assist companies in the development of CDx [1, 2].

17.2.1 Clinical Trials for Investigational Device Exemption

The Investigational Device Exemption (IDE) is the regulatory pathway for clinical trials conducted in the United States that utilize a CDx. The IDE pathway defines three types of device studies: exempt, nonsignificant Risk (NSR), or significant risk (SR) [3]. It is the responsibility of the CDx manufacturer to determine the study risk type that is applicable to their device. When making this assessment, the manufacturer has the option to consult with the FDA by submitting a study risk determination pre-submission. The study risk determination process is a request from the manufacturer to the FDA to determine if the clinical study would be a SR, NSR, or exempt. The requirements of each IDE regulation risk status are outlined below.

Exempt device studies do not have to follow the requirements listed in 21 CFR 812, with the exception of disqualification of clinical investigators from the study (21 CFR 812.119) [4]. For a clinical trial to be considered an exempt device study, the device must meet the requirements of 21 CFR 812.2(c):

- A device that was in commercial distribution prior to May 28, 1976 and is used within its intended use.
- A device that has received FDA clearance or approval and is used within its intended use.

- A diagnostic device that,
  - Is noninvasive or does not require invasive sampling.
  - Does not introduce energy into a subject.
  - Is not used as a diagnostic procedure or is used with confirmation by another, medically established diagnostic product or procedure.
  - Is undergoing consumer preference testing, testing of a modification, or testing of a combination of two or more devices in commercial distribution.
- A custom device that is not being used to determine safety or effectiveness

If the manufacturer determines that use of their device in the study meets the requirements of a NSR device study, the manufacturer and trial sponsor are required to follow the abbreviated requirements for 21 CFR 812.2(b). This obligates them to the following:

- Label the device as investigational use only.
- Obtain and maintain institutional review board (IRB) approval of the investigation.
- Obtain informed consent for each patient in accordance with 21 CFR 50 and 21 CFR 56.
- Conduct monitoring of the device investigation.
- Maintain the records of the clinical trial, including adverse events records and reports.
- Maintain and submit progress and final reports.
- Do not promote or commercialize the investigational device.

For a CDx that is being used in a therapeutic clinical trial, most of the abbreviated requirements are already being completed by or on behalf of the drug sponsor. What is worth noting is that these activities are typically conducted with the therapeutic in mind and not necessarily with thought of the device. One consideration is that the informed consent document should also speak to the use of the investigational device in the study, including any risks to the patient from the use of the device. In addition, monitoring of clinical trial sites that are conducted by the drug sponsor does not always include the clinical trial sites for the device.

If the manufacturer determines the device meets the definition of a SR study, or the IRB does not agree with the manufacturer's assessment of NSR, the full requirements of 21 CFR 812 must be followed. This includes the additional requirement of submission of an IDE application to the FDA. An IDE application includes elements such as:

- Device description
- Analytical validation studies
- Information regarding previous clinical studies
- Description of the manufacturing process
- Risk summary of the device

- Labeling material
- Investigator and clinical site information

Once the IDE application is submitted, the FDA has 30 days to approve, conditionally approval, or reject the application. If approval is granted, the manufacturer can initiate the clinical trial once IRB approval is received. If the FDA grants a conditional approval, the manufacturer can still initiate the clinical trial once IRB approval is received; however, they will have to fulfill the additional requests received in the conditional approval letter. These requests could be additional validation studies, updates to labeling, or responses to information requests. If rejected, the manufacturer cannot initiate the clinical trial and the manufacturer must remediate any concerns identified by the FDA and resubmit the application once resolved.

Interventional clinical trials for AAV-mediated gene therapies that utilize a CDx will typically be considered either NSR or SR. Some questions that should be assessed would be:

- Is the sample collection process invasive? Some testing only requires blood collection, while others may require a fresh biopsy to be collected within certain time limits.
- What is the risk to the patient in the event of a false positive or false negative? For patients that would be enrolled in the study based on the potentially false result, an overview and understanding of the risks associated with AAV-mediated gene therapy will be necessary to assess the patient risk. This includes treatment of emergent adverse events (TEAE) and immune-related adverse events (irAE). For patients who would not be enrolled in the study based on the potentially false result, an understanding of the existing treatment options that would be available to them would be necessary.
- What are the risks that enrolled patients will be exposed to? Similar to the discussion points in the above bullet on false-positive or false-negative results, an understanding of the risks that enrolled patients will be exposed to during the trial compared to standard of care treatment options must be considered in understanding the full risk profile of the CDx during the trial.
- How will the health care professional (HCP) monitor the patient during the clinical trial and will the patient continue to receive the same standard of care that they did prior to receiving the investigational treatment? HCPs that may continue to offer other treatment options to their patient during the trial may help to offset the device study risk status. Conversely, patients who must cease all other treatment or care options before, during, or after treatment with the AAV-mediated gene therapy may pose an increased device study risk status.

For more information on evaluating risks for a CDx in a clinical trial or understanding the differences between NSR or SR, refer to FDA draft guidance *Investigational IVDs Used in Clinical Investigations of Therapeutic Products* [5].

17.2.2 US FDA Marketing Authorization Pathways

CDx seeking marketing authorization in the US has two primary regulatory pathways that can be followed based on the classification of their device: Premarket Notification 510(k) or Premarket Approval (PMA). For programs that are seeking approval for a rare disease indication, a Humanitarian Device Exemption (HDE) authorization is an additional option that can be used and is explained below.

17.2.2.1 510(k) process

The 510(k) process is the marketing authorization pathway for non-exempt Class I and Class II *in vitro* diagnostics devices [6]. This process requires the manufacturer to demonstrate to the FDA that the device is as safe and effective as a predicate device. This is referred to as demonstrating substantial equivalence. The FDA defines a predicate device as a legally marketed device that is:

- A device that was placed on the market prior to May 28, 1976, referred to as a pre-amendment device.
- A Class I or Class II device that has already demonstrated and shown substantial equivalence through the 510(k) process or was approved and reclassified as Class I or Class II through the *de novo* process.

Substantial equivalence is demonstrated by the manufacturer by showing that their device has the same intended use as the predicate device and has the same technological characteristics as the predicate. If the device does not have the same technological characteristics as the predicate, then the device must be as safe and as effective as the predicate device. The contents of a 510(k) application include items such as:

- Device description
- Comparison of the subject device to the predicate device
- Performance specification and testing results
- Labeling
- Shelf life of the device
- Performance testing and results

At the end of the 510(k) process, the FDA will determine if the device is substantially equivalent (SE) to the predicate device. If SE is demonstrated, the device will receive clearance and will be able to be commercially marketed once the clearance has been issued. This will also allow the device to be used as a predicate device for future submissions. FDA typically provides its SE determination within 90 days; however, this timeline can be extended if the FDA determines they need additional information and place the review on hold. Premarket inspection of a device manufacturing facilities is not typically required for a 510(k) review and SE determination, though the manufacturer will be subject to FDA quality system inspections after receiving 510(k) clearance.

17.2.2.2 PMA Process

For Class III devices, a manufacturer must submit a PMA application to the FDA [7]. A PMA goes through a formal review and approval process with the FDA, similar to a New Drug Application (NDA) or Biologics Licensing Agreement (BLA). The PMA process is the most robust device review process at the FDA. The manufacturer must demonstrate that the device is both safe and effective for the patient population based on its intended use. Due to the large amount of information included in a PMA, most information submitted in a PMA falls into one of the five sections. FDA has also issued a number of PMA guidance documents including "Premarket Approval Application Modular Review" [8] and "Acceptance and Filing Reviews for Premarket Approval Applications (PMAs)" [9].

- Administrative information supporting the application, such as device description, summary of the contents, marketing history, and financial declarations.
- Quality system information showing the device meets the requirements of the quality system regulation, 21 CFR 820.
- Nonclinical studies, both a summary of what was completed and the complete test protocols, plans, and reports.
- Clinical study overview that demonstrates the studies were completed in accordance with the relevant IDE requirements, conclusions of safety and effectiveness, and complete line data from the studies.
- Software information should also be included if the device is or contains software components that are required for the device to operate as intended.

At the end of the PMA process, if approved, the FDA will issue either an approval letter or an approvable letter. An approval letter is issued based on draft labeling and the device is approved on the condition that the manufacturer submits final labeling before marketing the device. An approvable letter means that the device substantially meets the requirements for approval and FDA believes they can approve the device if specific information is supplied by the manufacturer to FDA or specific conditions are agreed to by the manufacturer. FDA typically provides its approval determination within 180 days from submission; however, this timeline can be extended if the FDA determines they need additional information and place the review on hold.

17.2.2.3 HDE Process

An HDE application can be used for devices that are intended for rare disease populations, a rare disease being defined by the Orphan Drug Act of 1984 as "a disease or condition that affects fewer than 200,000 people in the United States" [10]. An HDE application is most similar to a PMA application, with a few exceptions, most notable is that an HDE is exempt from the requirements of demonstrating effectiveness [11]. Because of this, devices approved by HDE are required to contain a statement in their labeling stating that the effectiveness of the device has not been demonstrated. Two other important considerations when determining an HDE authorization are: their eligibility for profit and the continual monitoring of the device's Annual Distribution Number (ADN).

Devices authorized under the HDE pathway cannot be sold for profit, or for an amount that exceeds the cost of research, development, production, and distribution. If the device is priced above $250, a report by an independent certified public accountant or an attestation by a responsible individual of the organization verifying the cost is required to be submitted to the FDA. There are two exceptions to this rule that are outlined under section 520(m)(6)(A)(i) of the FD&C Act:

- The device is intended for the treatment or diagnosis of a disease or condition that occurs in pediatric patients or in a pediatric subpopulation, and such device is labeled for use in pediatric patients or in a pediatric subpopulation in which the disease or condition occurs; or
- The device is intended for the treatment or diagnosis of a disease or condition that does not occur in pediatric patients or that occurs in pediatric patients in such numbers that the development of the device for such patients is impossible, highly impracticable, or unsafe.

As only devices that are intended for rare disease populations are eligible for the HDE pathway, the total number of devices sold must stay below the FDA-determined ADN for the device. The ADN is typically 8,000 devices which assumes one device is needed to treat, diagnose, and/or cure one patient. However, if more than one device is needed, then the ADN will be increased accordingly. The HDE holder is required to monitor the number of devices sold per year and report this to the FDA in their periodic report, also known as an annual report. If the number of devices sold exceeds the established ADN, the HDE holder is required to immediately notify FDA. If at any point after obtaining an HDE authorization the disease status changes and is no longer considered a rare disease, the HDE holder will need to submit and receive a PMA authorization in order to keep their device commercially available.

17.2.2.4 Differences Between 510(k) and PMA

Some differences in review between a 510(k) and PMA are the level of detail and information that is expected to be included in the application. While both pathways have an expectation for nonclinical performance data, the PMA submission is required to include the test protocols and full testing reports that include testing results, discussion, and a summary of any testing deviations that occurred. A 510(k) submission may not be required to have conducted a clinical study, while a PMA application has the requirement for clinical data with the expectation of a statistical analysis to be completed and the full line data for the study to be supplied.

A PMA will also include a premarket inspection of the submitter's facility to verify the device was developed in accordance with the quality system regulation, 21 CFR 820. This inspection is scheduled after the review of the quality system portion of the application is completed and all outstanding questions have been resolved.

17.2.3 US FDA Pre-submission Feedback

During the CDx development process, the manufacturer has the option to seek feedback from the FDA on their device by submitting a Pre-Submission Request [12]. A Pre-Submission, referred to as a Pre-Sub, is a formal written feedback request from the CDx manufacturer to the FDA to solicit feedback. During the written request, the manufacturer can request to schedule a 1-hour meeting with the agency, to take place after the receipt of the formal feedback. All interactions during the Pre-Sub, requests, briefing materials, feedback, and meeting minutes, are documented and retained to guide for future interactions, submissions, or applications.

The Pre-Sub process is entirely voluntary, but it is highly recommended that it be utilized. Through a Pre-Sub, the manufacturer can reach agreement with the agency on topics such as analytical validation, clinical validation, requirements for an IDE or PMA application, classification, or labeling. For a Pre-Sub, the manufacturer will submit a cover letter and briefing book to the agency. The cover letter will contain the manufacturer's contact information, method of feedback (written feedback only or written feedback followed by a meeting), dates and times of the meeting (if requested), purpose of the Pre-Sub, device description, or intended use. The briefing book will contain the relevant information that is necessary to obtain useful and thorough feedback from the agency. A first Pre-Sub request for a device may contain a detailed summary of the device, prospective patient population, and intended disease area, in addition to relevant information pertaining to the questions being asked by the reviewers. Conversely, Pre-Sub requests that are following up on previous interactions may contain the device's intended use statement with a summary of any pertinent changes before going into information specific to the questions and/or topics being addressed. For more information regarding Pre-Subs, including example questions, refer to FDA guidance Requests for feedback and meetings for medical device submissions: The Q-Submission Program [12].

17.3 European Union

17.3.1 European Union Clinical Trials

In 2017, the EU released Regulation (EU) 2017/746 *in vitro* diagnostic regulation (IVDR) [13]. The IVDR sets out requirements for *in vitro* diagnostics in the EU. This includes in vitro diagnostics that are performed ex-EU for which results are provided to patients that are within the EU, which is referred to as distance sales. Chapter VI of the IVDR covers the requirements for clinical evidence, performance evaluation, and performance studies. Under Article 58(1) it states that a performance study application and approval is required when the study involves:

- Surgically invasive sampling only for the purpose of the performance study
- Is an interventional performance study where the test results may influence patient management decisions or be used to guide treatment.
- The study involves additional invasive procedures or risks.

In addition, Article 58(2) specifies that a CDx performance study using leftover samples will need to also generate a performance study application and follow the notification process, for which approval does not need to be received but the application is still required to be submitted.

The application for the performance study must include all the elements outlined in Annex XIV includes:

- Applicant information, including authorized representative and notified body.
- Device information and overview of the device procedure.
- Identification of all countries the study will be performed in, both EU Member States and non-EU member states.
- Clinical performance study plan or clinical protocol.
- Analytical performance of the device.
- Clinical data, both from previously conducted studies as well as literature sources.
- Benefit-risk analysis and known risks.
- Evidence of meeting the relevant requirements in the general safety and performance requirements (GSPR) outlined in Annex I, including a signed statement by the manufacturer stating this requirement has been met.
- Ethics committee approval.
- Proof of insurance to cover subjects in the case of injury.
- Informed consent documents.
- Summary of data protection and privacy.

Under the IVDR, the application is to be submitted using the electronic system set up by the European Commission. This electronic system is referred to as EUDAMED. At the time of writing this chapter, the performance study application module of Eudamed was not launched and available for use. In lieu of using EUDAMED, MDCG 2022-12, *Guidance on harmonized administrative practices and alternative technical solutions until EUDAMED is fully functional (for Regulation (EU) 2017/746 on in vitro diagnostic medical devices)*, stated that the performance study applications should follow each member state's national procedure. This means that a performance study application must be submitted to each member state in which the clinical trial is conducted.

The review process for a performance study application under IVDR can range from 10 days up to 80 days. Once the application is submitted, the reviewing competent authority has 10 days, with an optional five-day extension, to notify the sponsor (i.e. the submitter of the application) that the application is within scope of IVDR and is complete. If the competent authority does not think the application is incomplete or that the application does not fall into scope of the IVDR, the sponsor has 10 days to respond. After responding, the competent authority has an additional 5 days to review the updated response and/or application with an optional five-day extension. This first review is considered the validation of the application. Once the application has been validated, the competent authority has 45 days to review and notify the sponsor of the approval status of the application, with an optional 20-day extension for the purpose of consulting with experts. For CDx, studies using leftover samples are able to initiate the clinical trial after the application validation date assuming that they have received ethics committee approval. For CDx clinical trials not using leftover samples, the competent authority approval and ethics committee approval must both be received prior to initiating the clinical trial.

In October 2021, the European Commission amended the transitional provisions of IVDR to provide additional time for some diagnostic classes to conform to the new IVDR requirements [14]. While this did not specifically amend performance study application requirements, it did allow for CDx with a valid CE mark to continue to be used until 26 May 2026. This would allow a CDx that was CE marked and used within its intended use to be used in a clinical trial without a performance study application.

With the newness of IVDR, both therapeutic and device manufacturers are uncertain of how the requirements should be interpreted. One area of ambiguity is around who should be the submitter of the performance study application. A majority of the elements required for a performance study application would be known by the device manufacturer; however, with the scope of IVDR being broadened to include any *in vitro* diagnostic that may influence patient management decisions and IVDR not having an option to submit an application based on risk (i.e. FDA NSR studies being exempt from an IDE application), not all *in vitro* diagnostic manufacturers have the capabilities to submit a performance study application. This is leading to some pharmaceutical companies taking on the role of submitting the performance study application for the *in vitro* diagnostic.

17.3.2 European Union Marketing Authorization Pathways

Under IVDR, the regulatory pathway or conformity route chosen depends on the classification of the device. There are four conformity pathways that can be used either as standalone or in combination with one another. CDx are classified as a

Class C device under IVDR and therefore have two options to obtain a CE mark, a quality management system (QMS) and technical documentation according to Annex IX or a type examination according to Annex X combined with a production quality assurance according to Annex XI.

QMS and technical documentation (Annex IX): This conformity route involves a full assessment of the manufacturer's QMS and a review of the device's technical documentation by a notified body. This is the route most often selected by manufacturers as it provides opportunities for changes to the device without invalidating the CE mark and is similar to the type of review done by other health authorities for marketing approval. There is no timeline provided in the regulations specifying how long the notified body has from receipt of the application to issuance of the CE Mark. The review timeline will likely vary based on the readiness of the applicant and quality of the technical documentation.

Type examination (Annex X), including technical documentation review, combined with production quality assurance (Annex XI): In this conformity route, the manufacturer will supply a representative device to the notified body for examination along with a review of its technical documentation. The notified body will also conduct an examination of the production and testing capabilities of the manufacturer. Upon successful completion of both activities, a notified body will issue two CE certificates to the manufacturer, one for each annex. This conformity pathway is ideal for devices that are locked with no changes expected. A change in the device may invalidate the CE mark and require a new type examination submission. There is no timeline provided in the regulations specifying how long the notified body has from receipt of the application to issuance of the CE Mark. The review timeline will likely vary based on the readiness of the applicant and quality of the technical documentation.

For CDx, the notified body is also expected to consult with the competent authority reviewing the therapeutic or with EMA in the event the centralized procedure is being used for approval. The consultation with the competent authority or EMA is to take place after the notified body has reviewed the application and determines that it conforms with the IVDR. Once submitted, the competent authority or EMA has 60 days to review the CDx and provide their scientific opinion to the notified body. This review also has the option to be extended for an additional 60 days. If after approval the CDx manufacturer makes changes to the device, they must notify the notified body of the changes. The notified body then determines if the device needs to be submitted for review and, if so, they must consult with the competent authority or EMA on the changes. If consulted, the competent authority or EMA has 30 days to review and provide their scientific opinion of the change to the notified body.

The technical documentation generated for a device under IVDR consists of two key documents: the general safety and performance requirements (GSPR) outlined in Annex I and the summary of technical documentation (STED) outlined in Annex II.

The GSPR is a list of all requirements an *in vitro* diagnostic device must meet in order to bear the CE mark. The GSPR is broken into three chapters:

1) General Requirements: Items addressing risk management and general device design development.
2) Requirements Regarding Performance, Design, and Manufacture Requirements: specific performance characteristics requirements for non-clinical, clinical, sterility, environment, software, energy, and self or near-patient testing
3) Requirements Regarding Information Supplied with the Device: The required elements of device labeling and instructions for use.

When the GSPR is completed by the manufacturer, this is most often seen as a table with each requirement from Annex I presented. The manufacturer then documents if the requirement was applicable or not applicable, documenting a reason or justification if deemed not applicable. For applicable requirements, the manufacturer must then document the method or standard used to fulfill the requirement and provide a reference to the objective evidence that the requirement was met. An example would be in relation to requirements of risk management, the manufacturer would be expected to include citation to ISO 14971 and provide a reference to the device's risk management file.

The STED is a narrative document drawn up by the manufacturer to provide an overall summary of the device. The STED is expected to present the device description, summary of nonclinical performance testing, clinical summaries and data, overview of manufacturing methods and qualification testing, and an overview of the software development and system. This document is expected to include references to the device history files that support each of the elements above. Both the STED and GSPR are expected to be reviewed and updated throughout the life of the device. The updated files may also need to be supplied to the notified body and competent authority if requested.

17.4 Other Regulated Markets

In addition to the United States and EU, a number of other markets require regulatory approval for CDx products, notable among these are the United Kingdom, Canada, Japan, and China.

- MHRA regulates CDx used in the United Kingdom. This includes requirements for clinical trial applications and seeking a UKCA Mark for commercial approval. The medical device regulation in the United Kingdom is in transition as after

Brexit, MHRA announced that they were intending to introduce new regulation for governing medical devices, including CDx, which were separate from those established by the EU Commission (i.e. MDR and IVDR). At the time this chapter was written, the MHRA had released transitional provisions to allow for the continued acceptance of the EU CE Mark, under both the previous directive (98/79/EC) and the current regulation (IVDR 2017/746). MHRA is aiming for core aspects of the future medical device regulations to apply in July 2025 [15].

- Health Canada has a robust *in vitro* diagnostic device regulation where they have classification rules around CDx and guidance documents speaking to their requirements [16].
- In Japan, regulatory oversight of CDx products is provided by the Pharmaceuticals and Medical Devices Agency (PMDA). Similar to other markets, the preference is for co-development of the therapeutic with the associated CDx. PMDA provides several guidance documents for CDx, including "Notification on approval application for in vitro CDx and corresponding therapeutic products" [17].
- CDx products in China are regulated by the National Medical Products Administration (NMDA) and there is limited guidance in this region for CDx development [18]. China does not typically allow for shipment of samples outside of China for the purposes of testing; therefore, it is recommended to have an in-country presence or partner to coordinate with.

17.4.1 Global Regulatory Strategy

A global regulatory strategy (GRS) document looks at the device for what is currently being developed and thinks wholistically about the device and its marketing future. The document should be generated at the beginning of the device development and updated regularly throughout its lifecycle. By creating the GRS early in the device development, it ensures all parties are including country-specific requirements that may be needed at a future date and could potentially cause rework or delays. For example, some countries require in-country clinical trials to be completed before authorization of the device. Knowing these requirements early in the development process will minimize any delays or additional clinical trials that may be needed.

The GRS document can also keep track of changes the device or development strategy experiences over the development process. A CDx can take anywhere from 2 to 10 or more years from initiation of development to receipt of marketing approval. A central document that tracks the key discussion items and decisions made concerning the strategy can be extremely valuable in the later years of development. The GRS should also record requests, decisions, or disagreements with regulatory agencies that need to be addressed at the time of submission for marketing authorization. For example, more complex CDx programs may have three or more pre-submissions that take place with FDA prior to submission of a PMA and the rationale behind the development team's responses should be recorded. The FDA review team may provide the manufacturer with requests for the PMA in their first pre-submission, which will need to be addressed during the PMA submission.

17.5 Development Strategy with the Therapeutic

CDx can be developed as a codeveloped product or a follow-on device. In co-development, both products are being developing simultaneously and the CDx is used in the pivotal clinical trial and sometimes in earlier clinical studies with the objective of a simultaneous submission and approval of the drug and the device. When the CDx is developed as a follow-on device, the gene therapy may already be approved or in the review process with the regulatory agency while the CDx either has not been developed or is still in development. Conversely, a CDx may already be approved or commercially available and the gene therapy is not yet approved or not cross-referenced in the intended use for the device. For AAV-mediated gene therapy products, it is strongly recommended by FDA in multiple gene therapy-specific guidance documents that a CDx be co-developed with the therapy [19–21].

When codeveloping a CDx and gene therapy, it is advisable for the gene therapy manufacturer to identify the key requirements of the CDx early in the clinical development (Figure 17.1). In an ideal situation, the CDx requirements are being assessed during the Phase 1 clinical trial when the first human study can be used to assess the true benefit of the CDx to the safety and efficacy of the patient population and define key performance requirements of the device, such as assay sensitivity. If the Phase 1 study shows that a CDx is necessary for patient safety or benefit, the Phase 2 study can be used to further develop the CDx assay and use it in the clinical trial for patient management decision making, typically seen as a test used for inclusion/exclusion screening. The CDx and gene therapy manufacturer can even begin having pre-submission or scientific advice meetings with the regulatory authorities on the further development needs of the CDx for the program.

During drug development, it is understood that the ideal situation does not always occur. In some cases, the Phase 2 study results may be sufficient to receive accelerated approval. The CDx may be used during the pivotal study but need to go through substantial changes to be ready for a marketing authorization submission. In other situations, the commercial diagnostic partner may not be the same diagnostic company used during the clinical trial. In this case, the commercial diagnostic partner may be developing their CDx as the pivotal study is being

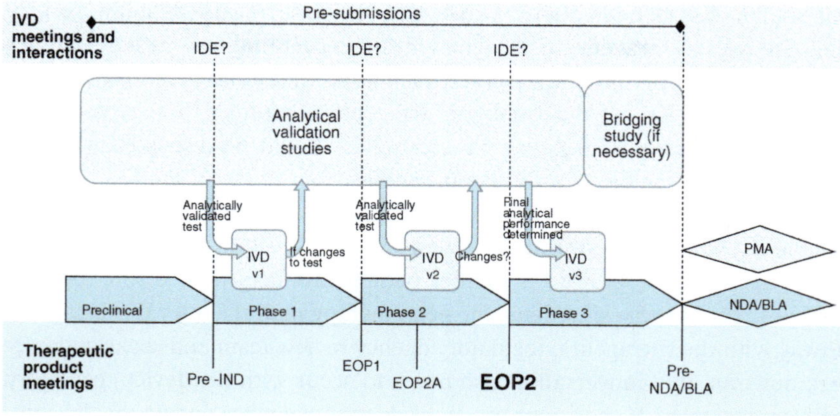

Figure 17.1 Codevelopment of a therapeutic and companion diagnostic from FDA Guidance Principles for Codevelopment of an In Vitro Diagnostic Device with a Therapeutic Product [2]. Source: https://www.fda.gov/media/99030/download .Public domain.

conducted or after the study has already been completed. In both situations, the CDx manufacturer will be required to conduct a bridging study.

In a bridging study situation, the residual samples (i.e. leftover samples) from the pivotal study are retained and evaluated by the commercial CDx assay. The results obtained from the commercial CDx (CDx IVD) are compared against the results obtained by the assay used in the clinical trial. This is done to clinically validate the commercial CDx assay. If a bridging study is necessary, it is recommended that the CDx and gene therapy manufacturers meet with the FDA and EMA to discuss and agree on the design of the bridging study and target percent agreement values.

17.5.1 Considerations for Rare Disease Indications

When developing a CDx for a rare disease, there are constraints that need to be considered during development that may not be present for a common or prominent disease indication.

One of the first considerations would be sample availability for use in the nonclinical validation of the CDx. As rare diseases will have a smaller patient population, this can cause limitations on the samples available to conduct suitable verification and validation studies for the CDx. The CDx manufacturer will be forced to use surrogate samples for some of their nonclinical performance studies. These can be contrived samples, samples of alternate sources or even samples from similar disease indications. However, even if surrogate samples are procured, the regulatory agency reviewers will still expect that true patient samples be used in the key performance studies, such as accuracy and precision. The use of surrogate samples and the number of true patient samples that are necessary for conducting nonclinical performance studies should be discussed with the FDA or EMA early in the CDx development process.

Another consideration would be the limited amount of clinical data that may be available to support the clinical validation of the CDx. Rare diseases may often receive accelerated approval on smaller patient numbers than a full approval would require. In these situations, the gene therapy manufacturer discusses this pathway with the therapeutic regulatory agency review team and seeks endorsement; however, this conversation also needs to occur with the device regulatory agency review team. If these conversations do not take place during the development timeline, the CDx approval may jeopardize the approval of the gene therapy.

17.6 Partner Relationship

17.6.1 Importance of the Partner Relationship

It may be obvious but is worth stating that the relationship between the CDx manufacturer and the gene therapy manufacturer is a different relationship than that typically associated with vendors or contract research organizations (CROs). In the case of development of a CDx the device and gene therapy are two sides of the same coin where both will be necessary for the benefit of the HCPs and patients. Also, in this collaboration, both parties are required to share information that is not typically provided outside of the company, such as development and approval timelines, adverse events and safety information, clinical raw data, and feedback received from regulatory agencies. But, for the partnership to be successful, and therefore the gene therapy and CDx to be successful, the two teams need to work together in a partnership.

The selection of the CDx partner by the gene therapy manufacturer is a critical process that should not be rushed and be very deliberate. Some questions that the gene therapy manufacturer should ask themselves before selecting the partner would be:

Will the CDx be codeveloped or developed as a follow-on device?

Will the same assay be used from Phase 1 through commercial launch or will the commercial CDx be brought later in the development timelines?

- Does the CDx partner have experience with the biomarker or analyte in question? Do they have the technical and/or clinical expertise needed to support the assay and clinical development program?

- Do they have experience with the technology?
- Do they have experience with the disease indication?
- Does the CDx partner already have a similar device that is either commercially available or available to be used as a design prototype?
- Does the CDx need to be a distributable kit or is a single site assay more applicable for the commercial needs?
- Can the CDx partner support the commercial volume of the gene therapy?
- Can the CDx partner support a global regulatory strategy? Do they have a presence in the necessary countries or relationship with the key stakeholders that may be needed (i.e. in country authorized representatives or delegated marketing authorization holders)?

These questions should be assessed wholistically and on a case-by-case basis. The needs of the gene therapy manufacturer may be different for each product, indication, or country. It is also suggested that the gene therapy company employ staff or consultants who have expertise in CDx development. The regulations expectations for a gene therapy are very different than those of a CDx. Having the in-house expertise can help facilitate the development process to ensure that both sides have all that is needed to achieve the product goals.

17.7 Commercial and Post-Approval Considerations

A commercialization strategy for the CDx should be prepared well before marketing approval is obtained in order to ensure rapid deployment of the test. The strategy must consider matters, such as reimbursement, logistics, marketing, and regulatory approvals, in additional markets. Post-approval the CDx may evolve as reagents and instruments are updated or changed, or laboratory facilities are moved and these changes typically require some type of regulatory action. And as a regulated product, CDx may be subject to various reporting obligations to the relevant authorities.

17.7.1 Future Proofing the Companion Diagnostic

A common scenario is for an already approved CDx to be used in clinical development for a different drug, indication, or patient population than its original approval. For example, the FoundationOne CDx (Foundation Medicine, Inc.), cobas EGFR Mutation Test v2 (Roche Molecular Systems, Inc.), and BRACAnalysis CDx (Myriad Genetic Laboratories, Inc.) have claims for multiple drugs and indications as shown on the FDA's website, List of Cleared or Approved CDxDevices (In Vitro and Imaging Tools) [22]. The use of an already approved test during clinical development will typically incur lower risk than development of a new test.

When using an already approved CDx in a new gene therapy clinical development program, the CDx should still be assessed to determine if it is suitable for the new indication. This includes assessing if the previously validated cut-off is suitable for the new disease or patient population. Another consideration would be if the new indication would use a different sample type or sample volume (i.e. pediatric applications).

17.7.2 Modifications of the Companion Diagnostic

In the years following marketing, modification to the CDx may be required because of changes in reagent or instrumentation availability. Other potential changes that may impact a CDx include software updates and moving the site of a manufacturing or testing facility. These changes to the device typically require additional validation and reporting or submissions to regulatory agencies.

17.8 Final Word

This chapter provides only a brief overview of the regulatory considerations for CDx products. Additional details can be obtained from guidance documents published by regulatory agencies or through consultation with regulatory agencies and industry experts. The views and opinions expressed in this chapter are those of the authors personally and are not necessarily the views and opinions of their respective employers.

References

1 Food and Drug Administration (FDA) (2014). In vitro companion diagnostic devices guidance for industry and food and drug administration staff. August 2014.
2 Food and Drug Administration (FDA) (2016). Principles for codevelopment of an in vitro companion diagnostic device with a therapeutic product. July 2016.
3 Food and Drug Administration (FDA) (2006). Information sheet guidance for IRBs, clinical investigators, and sponsors significant risk and nonsignificant risk medical device studies. January 2006.
4 Investigational Device Exemption (United States), 21 CFR Part 812.
5 Food and Drug Administration (FDA) (2017). Investigational IVDs used in clinical investigations of therapeutic products draft guidance for industry, FDA staff, sponsors and IRBs. December 2017.

6 Food and Drug Administration (FDA) (2022). Premarket Notification 510(k). https://www.fda.gov/medical-devices/premarket-submissions-selecting-and-preparing-correct-submission/premarket-notification-510k (accessed 17 October 2023).

7 Food and Drug Administration (FDA) (2016). Premarket approval (PMA). https://www.fda.gov/medical-devices/premarket-submissions-selecting-and-preparing-correct-submission/premarket-approval-pma (accessed 17 October 2023).

8 Food and Drug Administration (FDA) (2003). Guidance for industry and FDA staff premarket approval application modular review. November 2003.

9 Food and Drug Administration (FDA) (2019). Acceptance and filing reviews for premarket approval applications (PMAs) guidance for industry and food and drug administration staff. December 2019.

10 Food and Drug Administration (FDA) (2022). Humanitarian device exemption. October 3, 2022. https://www.fda.gov/medical-devices/premarket-submissions-selecting-and-preparing-correct-submission/humanitarian-device-exemption (accessed 17 October 2023).

11 Food and Drug Administration (FDA) (2019). Humanitarian device exemption (HDE) program guidance for industry and food and drug administration staff. September 6, 2019.

12 Food and Drug Administration (FDA) (2023). Requests for feedback and meetings for medical device submissions: the Q-submission program. June 2023.

13 Regulation (EU) 2017/746 of the European Parliament and of the Council of 5 April 2017 on in vitro diagnostic medical devices and repealing Directive 98/79/EC and Commission Decision 2010/227/EU2017/746.

14 European Union Commission (2021). Regulation (EU) 2017/746 of the European Parliament and of the Council of 5 April 2017 on in vitro diagnostic medical devices and repealing Directive 98/79/EC and Commission Decision 2010/227/EU. October 2021.

15 Medicines & Healthcare products Regulatory Agency (2023). Standard implementation of the future regulations. https://www.gov.uk/government/publications/implementation-of-the-future-regulation-of-medical-devices-and-extension-of-standstill-period/implementation-of-the-future-regulations (accessed 27 July 2023).

16 Government of Canada (2016). Guidance document: guidance for the risk-based classification system for in vitro diagnostic devices (IVDDs). https://www.canada.ca/en/health-canada/services/drugs-health-products/medical-devices/application-information/guidance-documents/guidance-document-guidance-risk-based-classification-system-vitro.html (accessed 17 October 2023).

17 Pharmaceutical and Food Safety Bureau (2013). Ministry of Health, Labour and Welfare. Notification on approval application for in vitro companion diagnostics and corresponding therapeutic products. July 2013.

18 National Medical Products Administration (NMPA) (2023). http://english.nmpa.gov.cn/ (accessed 17 October 2023).

19 Food and Drug Administration (FDA) (2020). Human gene therapy for rare diseases guidance for industry. January 2020.

20 Food and Drug Administration (FDA) (2020). Human gene therapy for hemophilia guidance for industry. January 2020.

21 Food and Drug Administration (FDA) (2015). Considerations for the design of early-phase clinical trials of cellular and gene therapy products guidance for industry. June 2015.

22 Food and Drug Administration (FDA) (2023). List of cleared or approved companion diagnostic devices (in vitro and imaging tools). May 23, 2023. https://www.fda.gov/medical-devices/in-vitro-diagnostics/list-cleared-or-approved-companion-diagnostic-devices-in-vitro-and-imaging-tools (accessed 6 October 2023).

Section V

Regulatory Perspectives on Gene Therapy

18

Current Regulatory Landscape for Gene Therapy Product Development and the Role of Biomarkers

Laura I. Salazar-Fontana PhD[1] *and Mike Havert PhD*[2]

[1]*LAIZ Regulatory Science Consulting, Lausanne, Switzerland*
[2]*Gene Therapy Partners, LLC, Arlington, Virginia, USA*

18.1 Introduction

Scientists have long ago linked genes to biological traits and features. Hershey and Chase demonstrated in 1952 that DNA was the basis for inheritance and the early experiments that followed provided the basic understanding of how DNA is passed from parent to progeny and used as a set of instructions for a cell [1]. Visionary scientists hypothesized that genetic modification through the introduction of exogenous DNA could be the basis for disease treatments [2]. By providing new genetic coding sequences, one could reprogram cells to perform a new function or restore a missing function. Instead of providing a protein, a chemical drug, or a metabolite, these scientists were dreaming of ways to alter genes. It was in July of 2012 that the European Medicines Agency (EMA) granted approval for the first gene therapy product, Glybera®, an adeno-associated viral vector engineered to deliver a corrected form of the enzyme lipoprotein lipase to the muscle tissue of patients suffering from a severe deficiency in this enzyme. Five years later, three approvals by the U.S. Food and Drug Administration (FDA) marked a breakout in which two *ex vivo* gene-modified products (Kymriah®, Yescarta®) and one direct *in vivo* gene transfer product (Luxturna®) were approved. Collectively termed gene therapy (GT), these products established a new pharmaceutical classification and the formal recognition of genetic modification of human cells as a new treatment paradigm.

Drug Development for Gene Therapy: Translational Biomarkers, Bioanalysis, and Companion Diagnostics, First Edition. Edited by Yanmei Lu and Boris Gorovits.

Yet, making changes to the genome of living human cells does not come without risk. The path to these GT approvals was long and winding. The development of gene delivery vectors, such as replication-defective retrovirus and adenovirus, coupled with encouraging results in preclinical disease models, led to early initiation of several National Institutes of Health (NIH) funded clinical trials in the 1990s. Unfortunately, these early trials exposed serious treatment-related toxicities, such as inflammatory responses to *in vivo* administered viral vectors and malignancies caused by retroviral-mediated insertional activation of proto-oncogenes [3, 4]. These setbacks fueled more basic research in virology, immunology, cell biology, and model development, which ultimately led to successful clinical translation using engineered lentiviral (LVV) and adeno-associated viral (AAV) vectors. In the time since 2012, we have had 19 gene therapy approvals worldwide, including genetically modified cell therapy products, and 28 GT are currently under regulatory evaluation [5]. These approved products have had to meet a high standard for demonstrating that clinical benefit clearly outweighted the potential risks associated with genetic modification and the potential long-term consequences of this new pharmaceutical modality.

This chapter provides an overview of the biomarkers that have been used from the start of gene therapy trials in support of regulatory filings and how they have evolved over time. The future for GT remains promising as we are better able to design and select safer and more effective product candidates based on the knowledge gained from multiple biomarkers. The value of evaluating new, particularly those related to immune toxicities, is further discussed.

18.2 What is Gene Therapy?

Gene therapy products (GTP) are a diverse group of biotherapeutics that are generally developed to treat conditions for which there are limited or no effective treatments [6].

The FDA has regulated what it calls human gene therapy as biologic products since 1993 [7]. FDA guidance describes gene therapy as products seeking to modify or manipulate the expression of a gene or to alter the biological properties of living cells for therapeutic use [8, 9, 10, 11]. Through the transcription or the translation of transferred genetic material, or by specifically altering host (human) genetic sequences. Some examples of gene therapy products include nucleic acids (e.g. plasmids, *in vitro* transcribed ribonucleic acid (RNA)), genetically modified microorganisms (e.g. viruses, bacteria, fungi), engineered site-specific nucleases used for human genome editing (e.g. CRISPR/Cas9), and *ex vivo* genetically modified human cells. Whereas EMA defines gene therapy medicinal products (GTMP) as a biological medicinal product that consists of a recombinant nucleic acid that once administered to human beings can regulate, repair, replace, add, or delete a genetic sequence; thus, GTMP mediate their therapeutic, prophylactic, or diagnostic effect through the recombinant nucleic acid sequence they contain, or through the product of genetic expression of this sequence [12]. Although there is no universal definition of GTP, the World Health Organization (WHO) has recently published a document aimed to achieve worldwide regulatory consensus on this definition [13]. For now, both developers and regulators distinguish GTP from therapeutic proteins (so-called "biotech" products) by categorizing them as "advanced" therapy products, and regulatory agencies anticipate the same level of quality, safety, and efficacy evidence to attain a positive benefit:risk assessment to support the approval and commercialization of a new GTP/GTMP [6].

18.3 Biomarkers Defined

Biomarkers are objective and quantifiable characteristics of biological processes [14]. The first definition was formally proposed by the National Institutes of Health (NIH) biomarkers working group back in 1998 and amended by the International Program on Chemical Safety, led by the WHO in coordination with the United Nations (UN), in 2001. The NIH working group defined biomarkers as a "characteristic that is objectively measured and evaluated as an indicator of normal biological processes, pathogenic processes, or pharmacological responses to a therapeutic intervention" [15]. Whereas, the WHO definition added to the NIH definition the utility that these measurements could have to predict the outcome and/or incidence of a disease by taking into consideration the effects of treatment, interventions, and environmental factors, such as pollutants, in the biological processes, and classified them into biomarkers of exposure, effect, and susceptibility [16].

But it was not until 2004 with the announcement of FDA's critical path initiative, that biomarkers could be accepted as part of the preliminary proof of safety and effectiveness of new therapeutic products with the goal of accelerating product development. The aim of this initiative was to enable the combination of new scientific and predictive methods to ease the transition between the laboratory concept to clinical development culminating in product commercialization [17].

Biomarkers can provide researchers and regulators with interim evidence concerning the safety and efficacy of a given treatment while more definitive clinical data are collected. For example, surrogate endpoints can be accepted as a preliminary proof of efficacy if they have been well-characterized for any given biological process. Yet for a biomarker to be considered a surrogate endpoint, there must be solid scientific evidence (e.g. epidemiological, therapeutic, and/or pathophysiological) demostrating that the chosen biomarker can also consistently and accurately predict a clinical outcome, either a benefit or a harm. In this case, the biomarker can be proposed as a surrogate endpoint to the regulatory agencies to support product development, but acceptance will be granted on a case-by-case basis, and very rarely to replace a clinical endpoint.

The use of biomarkers in the development of gene therapy products is further discussed in the following sections.

18.4 Early Gene Therapy Biomarkers

The first human GT clinical protocol approved by the NIH involved the genetic marking of *ex vivo* expanded lymphocytes. Researchers at the National Cancer Institute (NCI) used a murine gamma retrovirus (Maloney murine leukemia retrovirus, produced using PA317/LNL6-c8) encoding a bacterial neomycin resistance gene (NeoR) to tag tumor-infiltrating lymphocytes (TILs) extracted from a metastatic melanoma tumor sample [18]. As the first openly deliberated gene transfer into humans, a number of practical, safety and ethical considerations were pondered before NIH approved and initiated this study. These considerations actually paved the road to future studies by and define three key biological response signals that have ever since used as relevant biomarkers for GTP.

The first of the biological responses was and still is the detection of replication-competent virus. Because, GTP has traditionally relied on modified viral vectors (for gene transfer and genetic modification of human cells they may revert to virulence and possibly become pathogenic when administered to humans. As such, it is important to understand whether these GTP contain self-replicating viruses that might harm patients or others exposed to treatments, such as caregivers or family contacts, specially if the chosen viral vector is an engineered retrovirus. The presence of replication-competent virus may be difficult to detect in the final drug product as is the persistence of vector sequences with the potential for mobilization and/or activation through recombination due to environmental triggers. Therefore, replication-competent virus testing has been used as a quality attribute and as required patient safety biomarker for all retroviral/lentiviral vector GT investigational studies. Currently, screening for replication-competent retrovirus (RCR) or replication-competent lentivirus (RCL) infection in study subjects is performed by either serologic detection of retroviral specific antibodies or analysis of patient peripheral blood mononuclear cells (PBMC) by PCR for retroviral specific DNA sequences. Positive screen tests are followed by a direct coculture assay to obtain and characterize the infectious viral isolate [19]. Recommendations for sample collection include a pretreatment time, followed by collection after three (3), six (6), and twelve (12) months post administration, and yearly for up to fifteen (15) years (FDA) or until data indicate that there is no longer risk to be followed with a minimum of 5 years for viral vectors with no risk for integration, latency or reactivation potential (EMA) [20]. All commercial products approved so far have assessed long-term impacts on secondary malignancies and the potential for RCR/RCL generation in 15/20-year patient registry studies that consisted of 500 and 2500 patients [21]. The Center for International Blood and Marrow Transplant Research (CBMTR) has managed registry studies in for the United States and Japan and similarly the European Bone Marrow Transplant (EBMT) has outlined plans for registry studies in the European Union (EU) [22]. To date, substantial amount of data on the safety of retroviral vectors in clinical applications has been collected. To date, no RCR/RCL or delayed adverse event related to replication-competent virus has been reported in subjects who have received retroviral vector-based treatments [23]. Current guidance by FDA suggests that sponsors may discontinue RCL/RCR testing at some point after some initial data has been collected with some limited exemptions.

A second biomarker used in this first NIH-approved study was aimed to track the exogenous nucleic acid contained in the genetically modified lymphocytes. Both Southern blot and PCR for the bacterial NeoR transgene were used as overlapping semi-quantitative assays to distinguish exogenous nucleic acid (the bacterial neomycin resistance gene) inserted into marked lymphocytes from sequences present in the human genome. The investigators found that the genetically modified TILs were able to survive at the tumor site and in circulation for months, although the results were somehow limited in their ability to compare and quantitatively assess TIL numbers.

Nowadays, improvements to the early PCR assays allow more quantitative assessments. Real-time PCR, or quantitative PCR (qPCR), is now a well-established technology that measures the accumulation of DNA products amplified during a PCR reaction. Also, digital PCR (dPCR) offers further improvements in the sensitivity and precision of this approach. These quantitative PCR methodologies have become the regulatory standard for today's biodistribution and shedding studies.

Evaluating the biodistribution of a gene therapy vector is also an important first step in the design and understanding of a GT to support its potential therapeutic effect and safety risk. Biodistribution studies are performed to determine the dissemination and the GT persistence in both target and non-target tissues upon direct *in vivo* administration. Two FDA guidance documents outline the need for biodistribution studies and there is a current effort underway to draft an ICH document to harmonize approaches [8, 24, 25]. FDA guidance suggests that a biodistribution study be done in combination with an animal toxicology study because the location and persistence of the GT vector can help with the interpretation of toxicology assessments. In addition, the extent of long-term follow-up (LTFU) in patients may be guided by product biodistribution studies in animals. It also recommends developers to sample a panel of tissues, at a minimum (blood, injection site(s), gonads, brain, liver, kidneys, lung, heart, and spleen) and assess vector copy number with a sensitivity of not less than 50 copies per microgram of cellular DNA (or about 10e5 cells). General recommendation is that five (5) animals/sex/timepoint and for collections to include tissues prior to peak, steady-state (i.e. plateau), and declining (if feasible) [8, 25]. Similar considerations for the concurrent evaluation of biodistribution and toxicity in pre-clinical studies are contemplated by EMA with a clear recommendation to adequately justify the choice of endpoints and biomarkers predictive of toxicity in animals [26].

Many GT developers now go beyond a minimal biodistribution assessment and screen different product candidate to select those with optimal cell/tissue targeting in an animal model. In these situations, developers will likely assess not only vector uptake but also gene expression using a variety of tissue imaging, immunohistochemistry, and *in situ* hybridization techniques to ensure that not only is the vector delivered to the correct tissue but also that a target efficacy threshold for transgene (TG) expression can be reached.

Non-clinical biodistribution studies may also inform the design of clinical shedding studies. Clinical shedding studies may be used to understand potential environmental release of a gene-modified product and potential transmission to others within the environment. Many times, the most practical means to determine whether a GT is shed is by PCR.

As the application of detection methods, such as quantitative PCR, has evolved since the first GT trials to include biodistribution and shedding, so has the ability to detect the location of a TG and its integration sites within a host genome. Southern blot analysis and restriction fragment length polymorphism have been used in early gene mapping studies. Early retroviral insertion site analysis began with the isolation of genomic DNA, digestion by sequence-specific restriction endonucleases, and linear PCR amplification using a retroviral-specific primer (LAM-PCR). The measurement of the clonality of insertion sites is a current FDA recommendation to assess oligoclonality and blood dysplasia, which could be a potential early indication for cancer.

A third important biomarker was the detection of a biological activity of the genetically-marked cells. Although in the TIL study, demonstrating that the modified T cells had anti-tumor activity or retained Neo resistance could not be assessed, it was hypothesized that Neo resistance could be used as a selection marker for more potent tumor-directed lymphocytes in the future. Demonstration of *in vivo* functional activity for the genetically modified cells would have to wait for the second NIH-approved human gene transfer study. Shortly after the 1988 TIL approval, a second protocol was approved by the NIH that involved the gene correction of peripheral blood cells for the treatment of severe combined immune deficiency (SCID) caused by the adenosine deaminase (ADA) deficiency. As before with the gene marked TIL, this approach used retroviral-mediated gene transfer and careful consideration was given to the risks associated with a genetically modified virus and the generation of RCR. As with the TIL study, biomarkers for RCR indicated no RCR was generated in study subjects and the biodistribution and persistence of genetically modified cells was assessed for up to 10 years in humans. Interestingly, a biomarker of functional activity of the gene-modified cells was assessed by measuring ADA activity present in peripheral blood samples [27]. This second study reported results for all three biomarkers and it has set the standard for many years to come. It was not until the mid-2010s that this set of biomarkers was substantially reevaluated and expanded with the introduction of high-dose systemic AAV gene therapy and the expansion of LVV gene-modified cell products.

18.5 Current Expectations for Gene Therapy Biomarkers

Gene therapy has evolved significantly since the early 1990's, and *in vivo* gene delivery has moved from a concept to reality. Initial experiments with *in vivo* gene therapy suggested that the immunotoxicity of the delivery system was a significant concern and limited the effectiveness of the gene therapy [28]. Currently, AAV vectors are widely used for *in vivo* GT because they are non-pathogenic in humans, are replication-defective, mostly non-integrative (episomal localization), and can transduce a large variety of tissues depending on the selected serotype [29, 30]. Five AAV-mediated gene therapies are available to patients as of 2023, these include Luxturna®, Zolgensma®, Roctavaria®, Hemgenix® and Elevidys®. Even with this promising set of approvals, treatment-emergent serious adverse events (TESAEs) have fuel interest in understanding and developing immune toxicity related biomarkers (discussed below under "Immune-toxicity biomarkers").

The field has continued to make progress with *ex vivo* gene-modified cells as well. It was apparent from early clinical trials that genetically modified cells could be detected for long periods of time (up to 10 years), and in fact tracking the expansion and persistence of gene-marked anti-CD19-CAR T cells has correlated with treatment responses and appears to be the best predictor of CAR T efficacy [31]. However, gene-modified cell products also had the potential to trigger delayed adverse events, including cancer and autoimmunity. The field has moved to "safer" approaches that cause less genotoxicity. Even still, for all gene therapy products approved by FDA that are designed to modify the human genome (Kymriah, Yescarta, Tecartus, Breyanzi, Abecma, Carviktyi, Zenteglo, Skysona), registry studies must be conducted with lengthy LTFU to monitor patients for delayed adverse events and secondary malignancy. With this LTFU requirement, recent findings suggest that even with these safer approaches, clonal expansion, myelodysplastic syndrome (MDS), and in some cases cancers are reported at some frequency [32]. Therefore, another developing area of biomarker research is insertion site analysis (ISA) and measures of cellular oligoclonality which maybe precursors to cancer.

18.6 Safety Biomarkers for Gene Therapy Products

18.6.1 Immune Toxicities to *in vivo* gene therapy

Serious treatment-emergent adverse events (TEAEs), including fatal hepatoxicities, dorsal root ganglia (DRG) toxicity, thrombotic microangiopathies (TMA), myocarditis, and cytokine release syndrome, have been reported from *in vivo* gene therapy clinical trials using AAV vectors, which in the worst circumstances, have resulted in the death of study subjects [33, 34].

Certain pre-clinical and clinical manifestations of these TEAEs are closely connected to immune toxicities derived from the activation of innate (complement and Toll-Like Receptor (TLR) systems) and adaptive immune responses (Antigen-presenting, B and T cells) elicited against the delivery vector (viral capsid), the expression plasmid (ssDNA, ssRNA, bacterial RNA), and/or the expressed TG protein [35]. The activation of the classical complement pathway has been associated with TMA cases in GT treatment of spinal muscular dystrophy (SMA) and Duchenne muscular dystrophy (DMD) [33]. Complement activation can occur by either direct binding of the C1q protein to the viral capsid or upon activation of Fc receptors to pre-existing anti-AAV antibodies-viral capsid complexes. Adaptive immune responses to AAV capsids are evidenced by the signifcant prevalence of pre-existing anti-AAV antibody (AVA) titers, treatment-boosted AVA titers, and detection of AAV-specific CD8+ T-cell c responses upon in vivo GTP administration [35, 36, 37]. Also CRS, can occur as a consequence of robust activation of the Toll-Like Receptor family (TLR) of receptors. At the molecular level, the presence of pathogen-associated molecular patterns (PAMPs) exhibited in the viral capsid and in the expression plasmids containing the TG can be recognized by members of the TLR family and initiate MyD88-mediated expression and secretion of pro-inflammatory cytokines such as type I interferons (IFN), IL6, and TNF, [35]. Newly developed AAV products are now engineered to reduce the abundance of PAMPs with the aim to eliminate activation of TLR-mediated pro-inflammatory responses. As an example, expression plasmids containing reduced or no CpG motifs result in abrogation of TLR9 activation and minimal liver damage when systemically administered into mice [38, 39]. Although is still not clear what is minimal threshold of un-methylated motifs needed to abrogate TLR9-mediated responses, product candidates are now engineered to reduce the content of CpG islands in the oepn reading frames (ORF) of the expression vector [40].

A turning point for *in vivo* gene therapy came in 2014, when researchers reported sustained but low-level expression of Factor IX (FIX) in Hemophilia B patients [41]. Earlier studies of using systemic AAV vector-mediated gene therapy for Hemophilia B resulted in an increase in liver transaminases timed with a loss of FIX expression, and concomitant AAV capsid-specific cellular immune response [42]. This was the first study to suggest that immune suppression may be an effective mitigation strategy to avoid immune responses and prevent killing of transduced cells. This hypothesis was later put to the test in a second clinical trial, again in Hemophilia B, where further observations were made related to increased alanine transaminase (ALT), loss of TG expression, and detection of T cell response [43]. A tapering course of glucocorticoid (prednisolone) showed a decline in ALT levels and stabilization of protein expression. Interestingly, the obseved AAV capsid-specific cellular responses were no longer detectable at later time points [43]. This data suggested that the capsid-specific immune cellular response was likely directed against the transduced target tissue (e.g. liver) inducing cellular cytotoxicity and release of liver transaminase, and these two clinical biomarkers were relevant for elucidating the mechanisms behind the loss of FIX expression.

Anti-TG cellular responses have also been documented in the recent analysis of myocarditis in multiple Duchenne Muscular Dystrophy (DMD) trials [44]. Several DMD patients receiving a single intravenous dose of an AAV vector carrying a gene to correct the expression of mini-dystropin developed symptoms of myocarditis around a month following dosing [45]. These patients experienced significantly increased muscle weakness and myocarditis, that improved upon various rounds of immunosuppresive treatments. A combined effort between researchers and developers, demonstrated that those DMD patients carrying exon deletions affecting the N-terminal region of the dystrophin gene were the ones suffering from loss of efficacy and myocarditis. The most plausible explanation is that the immune system of these patients lacks tolerance to the corrected version of the mini-dystrophin protein introduced by the GT. Hence, the TG is recognized as a neo-antigen by the immune system which, in turn, mounted an anti-TG-specific cellular response (TG-specific CD8+ T cells) responsible for the observed tissue damage. It is not clear why all patients with an apparent at-risk genotype did not develop symptoms, but biomarker assays to detect transgene reactive T-cell responses may be relevant to monitor in patients with null mutations or large deletions of the protein the GT intends to repair.

Although activation of cellular immune response to AAV vectors (both capsid and transgene) can trigger cellular toxicity, it does not appear to be the only mechanism by which AAV gene therapies mediate cellular damage. Selecting a high dose AAV (10e13 to 10e14 vector genomes (vg) per kilogram (kg) of body weight) has been associated with an increased number in renal and hepatic toxicities in clinical studies [46]. One mechanism contributing to cytotoxicity is endoplasmic reticulum (ER) stress response and unfolded protein response, where cells are overwhelmed and cannot handle the translational demands imposed by gene therapy treatment. Hordeaux and colleagues demonstrated that reducing transgene expression in DRG neurons by including a specific microRNA target sequence prevented neuronal degeneration and axonopathy following CNS administration of the vector [47]. The authors, as well as several other published studies, reported that modulation of inflammatory responses through coadministration of immunosuppressives was ineffective in reducing the severity and incidence of the AAV-mediated pathology suggesting a toxic effect mediated by the TG level of expression eliminating the potential for confounding immune responses [48, 49, 50]. For neurotoxicity, serum neurofilament light chain has been identified as a promising biomarker for neuronal damage in nonclinical toxicology studies, and this may help clinical research programs design mitigation measures for potential neurotoxocity in clinical studies [51].

As mentioned at the beginning of this section, cytotoxicity and tissue damage may also be caused secondary to arteriole and capillary endothelial pathology and microvascular thrombosis caused by thrombotic microangiopathy (TMA) [52]. Although the exact mechanisms for TMA are not fully understood, complement activation is thought to be involved in this process [53]. This is why activation of the classical complement pathway may be used as exploratory safety biomarker in early studies seeking to understand the contribution of pre-existing AVA to potential immune toxicities.

Current regulatory expectations are founded in the pre-clinical and clinical findings described earlier. FDA and EMA acknowledge that because immune responses can pose serious safety risks, measuring cellular and humoral immune responses to both the vector and the TG-encoded protein, or to the gene-editing components, are acceptable approaches to address clinical safety outcomes [8, 24, 26, 54]. Other monitoring strategies may include periodic clinical, laboratory testing (i.e. pro-inflammatory cytokines), and imaging assessments [55]. Some of the principles enunciated for the evaluation of immunogenicity of therapeutic proteins can be applied to predict, evaluate, and mitigate the immunogenicity risks of GTP. For example, the immunoassays developed and validated to monitor AVA and anti-TG antibody responses can be validated according to recommendations described in the 2019 FDA Guidance on Immunogenicity Testing of Therapeutic Protein Products – Developing and Validating Assays for Anti-Drug Antibody Detection [56]. Similarly, using immunogenicity risk assessment principles can help define a fit-for-purpose bioanalytical strategy and implementation of therapeutic interventions to reduce robust immune responses. The co-development of *in vitro* diagnostics (IVD) has been discussed in another chapter of this book. Yet, the utility of using IVD to determine pre-existing AVA titers as exclusion criteria for patient enrolment and/or treatment remains unclear. Reducing the incidence of immune-related adverse events in single dose treatments may still be mitigated through rational drug design and concurrent implementation of therapeutic interventions specially for those patients for which gene therapy is the only available course of action [57].

As per the duration of LTFU studies, health authorities recommend that a risk-based approach is followed, with a clear identification of delayed adverse events risk factors [8]. For instance, for gene therapy, besides the validation of anti-AVA and anti-TG antibody assays, it may be pertinent to apply a risk-based approach to identify patient and product-related immunogenicity factors and define the final bioanalytical strategy [58]. As evidenced by the myocarditis events observed in the DMD population treated with an *in vivo* GTP, the evaluation of adaptive immune responses to the TG product may be advisable for a patient population with null mutations where central tolerance to the native protein may be absent. Similar to what has been implemented for rare disease patients receiving enzyme-replacement therapy (ERT) products [59], a pre-dose immune-suppressive conditioning regimen (i.e. glucocorticoids) followed by a post-dosing treatment with a combination of immuno-suppressants (i.e. rituximab, rapamycin), and provisions for early administration of IL-6 and complement system inhibitors (i.e. tocilizumab, eculizumab), can considerably reduce unwanted immune responses and improve the efficacy of the treatment.

New-generation AAV GT candidates are been developed using gene editing and codon optimization technologies to select candidates with reduced immunogenicity risk, improved gene expression levels, and lower off-target toxicity [40, 60]. This new generation of products is likely to redefine regulatory recommendations but for the moment and until further clinical data are collected, requirements for the validation of safety biomarkers should be discussed in advance with the pertinent health authority.

18.6.2 Immune Toxicities to Ex Vivo GT

Ex vivo gene therapy products were generally able to overcome the immuntoxicity of *in vivo* gene delivery mostly because lentiviral vectors LV have been used *ex vivo* to transduce human cells *and* therefore avoid oncogenicity concerns. Autologous *ex vivo* gene therapy products are by definition "self." The components used for genetic modification (e.g. viral vector) can be washed away during manufacturing and prior to final product infusion. In addition, many *ex vivo* modified cell therapy treatments include a lymphodepleting (or ablating) conditioning regimen that is administered to the patient to "make space" and allow engraftment of the transferred cells, considerably reducing the magnitude of otherwise robust immune responses [61, 62, 63].

Nonetheless, early studies documented immunogenicity to autologous *ex vivo* genetically modified cells and postulated that this immune response negatively impacted efficacy [27]. It is plausible that residual virus or manufacturing components and transgenes (non-self, truncated, or misfolded proteins) trigger unwanted immune reactions that may limit a gene therapy application. Xenogeneic responses to chimeric antigen receptor (CAR) T cells have been documented in a number of studies. In a CAR T-cell product for colorectal cancer, the development of an anti-CAR antibody response coincided with rapid clearance of subsequent CAR T-cell infusions [64]. For another CAR T-cell product directed against CD19, early CAR clearance was described to be related with relapses [65]. Anti-CAR antibodies against repeatedly dosed, transiently transduced CAR T cells directed against mesothelin were shown to lead to anaphylaxis in at least one case [66]. Immune responses to CAR T have been linked to the presence of non-human sequences present in the CAR construct and the cases where immunological responses to CAR T products have been reported employed a murine derived single chain variable fragment (scFv) [67, 68]. New CAR T constructs are fully humanized molecules and carry a reduced risk for inducing anti-CAR T antibodies. The development of "off-the-shelf" CAR T therapy products is bound to replace the use of autologous cells and the consequent risk of graft-versus-host disease risk disease leading to poor engraftment of the product. This risk should be explore in this new modality of CAR T products.

18.6.3 Long-Term Risks

Long-term risks associated with genetic modification have always been a consideration in the development of gene therapy products. This has led to the development of safer LVV and AAV vectors. But even with these vectors, genetic and immune toxicities remain of concern and need to be further elucidated [32]. Although the described toxicities have only precluded development of some GTP, additional data should be collected to examine if biomarkers for tracking genetically modified cells, their clonal predominance, immune toxicities and possible treatment options will be of greater importance in the future.

18.7 Concluding Remarks

The future for GT remains promising as we are better able to design and select safer and more effective treatments based on the evaluation of key biological processes or biomarkers. As the field gains more experience, additional biomarker development may shed light into the development of new GT products with improved safety and efficacy. These new biomarkers may include those linked to immune activation and toxicity and will eventually help improve the safety and efficacy of new GT treatments.

References

1 Hershey, A.D. and Chase, M. (1952). Independent functions of viral protein and nucleic acid in growth of bacteriophage. *J. Gen. Physiol.* 36 (1): 39–56.

2 Merril, C.R., Geier, M.R., and Petricciani, J.C. (1971). Bacterial virus gene expression in human cells. *Nature* 233 (5319): 398–400.

3 Hacein-Bey-Abina, S., Von Kalle, C., Schmidt, M. et al. (2003). LMO2-associated clonal T cell proliferation in two patients after gene therapy for SCID-X1. *Science.* 302 (5644): 415–419.

4 Raper, S.E., Chirmule, N., Lee, F.S. et al. (2003). Fatal systemic inflammatory response syndrome in a ornithine transcarbamylase deficient patient following adenoviral gene transfer. *Mol. Genet. Metab.* 80 (1-2): 148–158.

5 PharmaIntelligence A. (2022). Gene, cell, & RNA therapy landscape Q1 quarterly data report. https://asgct.org/global/documents/asgct-pharma-intelligence-q1-2022-report.aspx (accessed 25 September 2023).

6 Salazar-Fontana, L.I. (2022). A regulatory risk-based approach to ATMP/CGT development: integrating scientific challenges with current regulatory expectations. *Front. Med. (Lausanne).* 9: 855100.

7 Application of current, statutory authorities to human somatic cell therapy products and gene therapy products; Notice Federal Register1993 [53248-51]. https://fda.report/media/76647/Application-of-Current-Statuatory-Authorities-to-Human-Somatic-Cell-Therapy-Products-and-Gene-Therapy-Products.pdf (accessed 25 September 2023).

8 FDA (2020). Guidance for industry long term follow up after administration of human gene therapy products. https://www.fda.gov/regulatory-information/search-fda-guidance-documents/long-term-follow-after-administration-human-gene-therapy-products (accessed 25 September 2023).

9 FDA (2020). Guidance for industry human gene therapy for rare diseases. https://www.fda.gov/regulatory-information/search-fda-guidance-documents/human-gene-therapy-rare-diseases (accessed 25 September 2023).

10 FDA (2021). Draft guidance for industry human gene therapy for neurodegenerative diseases https://www.fda.gov/regulatory-information/search-fda-guidance-documents/human-gene-therapy-neurodegenerative-diseases (accessed 25 September 2023).

11 FDA (2022). Draft guidance for industry human gene therapy products incorporatig human genome editing https://www.fda.gov/regulatory-information/search-fda-guidance-documents/human-gene-therapy-products-incorporating-human-genome-editing (accessed 25 September 2023).

12 Directive 2001/83/EC of the European Parliament and of the Council of 6 November 2001 on the Community code relating to medicinal products for human use. https://eur-lex.europa.eu/legal-content/en/ALL/?uri=CELEX%3A02001L0083-20190726 (accessed 25 September 2023).

13 WHO (2021). World Health Organization considerations on regulatory convergence of cell and gene therapy products https://cdn.who.int/media/docs/default-source/biologicals/ecbs/who-public-consultation_cgtp-white-paper_16_dec_2021.pdf?sfvrsn=18f6c549_5 (accessed 25 September 2023).

14 Strimbu, K. and Tavel, J.A. (2010). What are biomarkers? *Curr. Opin. HIV AIDS.* 5 (6): 463–466.

15 Biomarkers Definitions Working Group (2001). Biomarkers and surrogate endpoints: preferred definitions and conceptual framework. *Clin. Pharmacol. Ther.* 69 (3): 89–95.

16 Safety WIPoC (2001). Biomarkers in risk assessment: validity and validation. https://inchem.org/documents/ehc/ehc/ehc222.htm#1.0 (accessed 25 September 2023).

17 FDA (2004). FDA's critical path initiative http://wayback.archive-it.org/7993/20180125035414/https:/www.fda.gov/ScienceResearch/SpecialTopics/CriticalPathInitiative/ucm076689.htm (accessed 25 September 2023).

18 Rosenberg, S.A., Aebersold, P., Cornetta, K. et al. (1990). Gene transfer into humans–immunotherapy of patients with advanced melanoma, using tumor-infiltrating lymphocytes modified by retroviral gene transduction. *N. Engl. J. Med.* 323 (9): 570–578.

19 FDA (2020). Guidance for industry testing of retroviral vector-based human gene therapy products for replication competent retrovirus during product manufacture and patient follow-up. https://www.fda.gov/media/113790/download (accessed 25 September 2023).

20 EMA (2009). Guideline on follow-up of patients administered with gene therapy medicinal products. https://www.ema.europa.eu/en/documents/scientific-guideline/guideline-follow-patients-administered-gene-therapy-medicinal-products_en.pdf (accessed 25 September 2023).

21 Auletta, K.J., Chen, M., and Shaw, B.E. (2021). Current use and outcome of hematopoietic stem cell transplantation: CIBMTR US summary slides. https://cibmtr.org/CIBMTR/Resources/Summary-Slides-Reports.

22 EBMT. CAR-T data collection initiative. https://www.ebmt.org/registry/car-t-data-collection-initiative (accessed 25 September 2023).

23 Cornetta, K., Duffy, L., Turtle, C.J. et al. (2018). Absence of replication-competent lentivirus in the clinic: analysis of infused T cell products. *Mol. Ther.* 26 (1): 280–288.

24 FDA (2013). Guidance for industry preclinical assessment of investigational cellular and gene therapy products. https://www.fda.gov/regulatory-information/search-fda-guidance-documents/preclinical-assessment-investigational-cellular-and-gene-therapy-products (accessed 25 September 2023).

25 ICH (2021). ICH guideline S12 on nonclinical biodistribution considerations for gene therapy products – draft. https://www.ema.europa.eu/en/documents/regulatory-procedural-guideline/ich-guideline-s12-nonclinical-biodistribution-considerations-gene-therapy-products-step-2b_en.pdf (accessed 25 September 2023).

26 EMA (2018). Guideline on the quality, non-clinical and clinical aspects of gene therapy medicinal products. https://www.ema.europa.eu/en/documents/scientific-guideline/guideline-quality-non-clinical-clinical-aspects-gene-therapy-medicinal-products_en.pdf (accessed 25 September 2023).

27 Muul, L.M., Tuschong, L.M., Soenen, S.L. et al. (2003). Persistence and expression of the adenosine deaminase gene for 12 years and immune reaction to gene transfer components: long-term results of the first clinical gene therapy trial. *Blood* 101 (7): 2563–2569.

28 Crystal, R.G. (2020). My pathway to gene therapy. *Hum. Gene Ther.* 31 (5-6): 273–282.

29 Carter, B.J. (2004). Adeno-associated virus and the development of adeno-associated virus vectors: a historical perspective. *Mol. Ther.* 10 (6): 981–989.

30 Srivastava, A. (2016). In vivo tissue-tropism of adeno-associated viral vectors. *Curr. Opin. Virol.* 21: 75–80.

31 Guedan, S., Luu, M., Ammar, D. et al. (2022). Time 2EVOLVE: predicting efficacy of engineered T-cells – how far is the bench from the bedside? *J. ImmunoTher. Cancer* 10 (5): e003487.

32 FDA (2022). Cellular, tissue, and gene therapies advisory committee June 9-10, 2022 briefing document. https://www.fda.gov/media/159009/download (accessed 25 September 2023).

33 Ertl, H.C.J. (2022). Immunogenicity and toxicity of AAV gene therapy. *Front. Immunol.* 13.

34 Verdera, H.C., Kuranda, K., and Mingozzi, F. (2020). AAV vector immunogenicity in humans: a long journey to successful gene transfer. *Mol. Ther.* 28 (3): 723–746.

35 Shirley, J.L., de Jong, Y.P., Terhorst, C., and Herzog, R.W. (2020). Immune responses to viral gene therapy vectors. *Mol. Ther.* 28 (3): 709–722.

36 Calcedo, R., Morizono, H., Wang, L. et al. (2011). Adeno-associated virus antibody profiles in newborns, children, and adolescents. *Clin. Vaccine Immunol.* 18 (9): 1586–1588.

37 Calcedo, R., Vandenberghe, L.H., Gao, G. et al. (2009). Worldwide epidemiology of neutralizing antibodies to adeno-associated viruses. *J. Infect. Dis.* 199 (3): 381–390.

38 Faust, S.M., Bell, P., Cutler, B.J. et al. (2013). CpG-depleted adeno-associated virus vectors evade immune detection. *J. Clin. Invest.* 123 (7): 2994–3001.

39 Konkle, B.A., Walsh, C.E., Escobar, M.A. et al. (2021). BAX 335 hemophilia B gene therapy clinical trial results: potential impact of CpG sequences on gene expression. *Blood* 137 (6): 763–774.

40 Martino, A.T., Suzuki, M., Markusic, D.M. et al. (2011). The genome of self-complementary adeno-associated viral vectors increases Toll-like receptor 9-dependent innate immune responses in the liver. *Blood* 117 (24): 6459–6468.

41 Nathwani, A.C., Reiss, U.M., Tuddenham, E.G. et al. (2014). Long-term safety and efficacy of factor IX gene therapy in hemophilia B. *N. Engl. J. Med.* 371 (21): 1994–2004.

42 Manno, C.S., Pierce, G.F., Arruda, V.R. et al. (2006). Successful transduction of liver in hemophilia by AAV-Factor IX and limitations imposed by the host immune response. *Nat. Med.* 12 (3): 342–347.

43 Nathwani, A.C., Tuddenham, E.G.D., Rangarajan, S. et al. (2011). Adenovirus-associated virus vector–mediated gene transfer in hemophilia B. *N. Engl. J. Med.* 365 (25): 2357–2365.

44 Mendell, J.R., Campbell, K., Rodino-Klapac, L. et al. (2010). Dystrophin immunity in duchenne's muscular dystrophy. *N. Engl. J. Med.* 363 (15): 1429–1437.

45 Carsten, G., Bonnemann, B.A.B., Braun, S. et al. (2022). A collaborative analysis by clinical trial sponsors and academic experts of anti-transgene SAEs in studies of gene therapy for DMD - 2022 ASGCT Annual Meeting Abstracts. *Mol. Ther.* 30 (4): 1–592.

46 Flotte TRE-i-C (2020). Revisiting the "new" inflammatory toxicities of adeno-associated virus vectors. *Hum. Gene Ther.* 31 (7-8): 398–399.

47 Hordeaux, J., Buza, E.L., Dyer, C. et al. (2020). Adeno-associated virus-induced dorsal root ganglion pathology. *Hum. Gene Ther.* 31 (15-16): 808–818.

48 Hordeaux, J., Hinderer, C., Goode, T. et al. (2018). Toxicology study of intra-cisterna magna adeno-associated virus 9 expressing iduronate-2-sulfatase in rhesus macaques. *Mol. Ther. Methods Clin. Dev.* 10: 68–78.

49 Hordeaux, J., Wang, Q., Katz, N. et al. (2018). The neurotropic properties of AAV-PHP.B are limited to C57BL/6J mice. *Mol. Ther.* 26 (3): 664–668.

50 Rosenberg, J.B., Chen, A., De, B.P. et al. (2021). Safety of direct intraparenchymal AAVrh.10-mediated central nervous system gene therapy for metachromatic leukodystrophy. *Hum. Gene Ther.* 32 (11-12): 563–580.

51 Fader, K.A., Pardo, I.D., Kovi, R.C. et al. (2022). Circulating neurofilament light chain as a promising biomarker of AAV-induced dorsal root ganglia toxicity in nonclinical toxicology species. *Mol. Ther. Methods Clin. Dev.* 25: 264–277.

52 Wilson J. (2021). Adeno-associated virus-related toxicities in nonhuman primates. https://www.fda.gov/media/151950/download (accessed 25 September 2023).

53 Chand, D.H., Zaidman, C., Arya, K. et al. (2021). Thrombotic microangiopathy following onasemnogene abeparvovec for spinal muscular atrophy: a case series. *J. Pediatr.* 231: 265–268.

54 FDA (2015). Guidance for industry considerations for the design of early-phase clinical trials of cellular and gene therapy products.

55 FDA (2021). Briefing Document Cellular, Tissue, and Gene Therapies Advisory Committee (CTGTAC) Meeting #70 Toxicity Risks of Adeno-associated Virus (AAV) Vectors for Gene Therapy (GT). https://www.fda.gov/media/151599/download (accessed 25 September 2023).

56 FDA (2019). Guidance for Industry Immunogenicity Testing of Therapeutic Protein Products — Developing and Validating Assays for Anti-Drug Antibody Detection.

57 Zolgensma Package Insert https://www.novartis.com/us-en/sites/novartis_us/files/zolgensma.pdf (accessed 25 September 2023).

58 FDA (2014). Guidance for Industry Immunogenicity Assessment for Therapeutic Protein Products. https://www.fda.gov/media/85017/download (accessed 25 September 2023).

59 Kishnani, P.S., Dickson, P.I., Muldowney, L. et al. (2016). Immune response to enzyme replacement therapies in lysosomal storage diseases and the role of immune tolerance induction. *Mol. Genet. Metab.* 117 (2): 66–83.

60 Wang, D., Tai, P.W.L., and Gao, G. (2019). Adeno-associated virus vector as a platform for gene therapy delivery. *Nat. Rev. Drug Discovery* 18 (5): 358–378.

61 Gattinoni, L., Finkelstein, S.E., Klebanoff, C.A. et al. (2005). Removal of homeostatic cytokine sinks by lymphodepletion enhances the efficacy of adoptively transferred tumor-specific CD8+ T cells. *J. Exp. Med.* 202 (7): 907–912.

62 Jensen, M.C., Popplewell, L., Cooper, L.J. et al. (2010). Antitransgene rejection responses contribute to attenuated persistence of adoptively transferred CD20/CD19-specific chimeric antigen receptor redirected T cells in humans. *Biol. Blood Marrow Transplant.* 16 (9): 1245–1256.

63 Muranski, P., Boni, A., Wrzesinski, C. et al. (2006). Increased intensity lymphodepletion and adoptive immunotherapy–how far can we go? *Nat. Clin. Pract. Oncol.* 3 (12): 668–681.

64 Hege, K.M., Bergsland, E.K., Fisher, G.A. et al. (2017). Safety, tumor trafficking and immunogenicity of chimeric antigen receptor (CAR)-T cells specific for TAG-72 in colorectal cancer. *J. ImmunoTher. Cancer* 5: 22.

65 Ruella, M. and Maus, M.V. (2016). Catch me if you can: leukemia escape after CD19-directed T cell immunotherapies. *Comput. Struct. Biotechnol. J.* 14: 357–362.

66 Maus, M.V., Haas, A.R., Beatty, G.L. et al. (2013). T cells expressing chimeric antigen receptors can cause anaphylaxis in humans. *Cancer Immunol. Res.* 1 (1): 26–31.

67 Khan, A.N., Chowdhury, A., Karulkar, A. et al. (2022). Immunogenicity of CAR-T cell therapeutics: evidence, mechanism and mitigation. *Front. Immunol.* 13.

68 Wagner, D.L., Fritsche, E., Pulsipher, M.A. et al. (2021). Immunogenicity of CAR T cells in cancer therapy. *Nat. Rev. Clin. Oncol.* 18 (6): 379–393.

Index

Drug Development for Gene Therapy: Translational Biomarkers, Bioanalysis, and Companion Diagnostics, First Edition. Edited by Yanmei Lu and Boris Gorovits.

C

q

r

S